RNA Technologies

More information about this series at http://www.springer.com/series/8619

Nikolaus Rajewsky • Stefan Jurga • Jan Barciszewski
Editors

Systems Biology

Springer

Editors
Nikolaus Rajewsky
Max Delbrück Center for Molecular Medicine
Berlin Institute for Medical Systems Biology
Berlin-Buch,
Berlin, Germany

Stefan Jurga
Nanobiomedical Center
Adam Mickiewicz University
Poznań, Poland

Jan Barciszewski
Institute of Bioorganic Chemistry
Polish Academy of Sciences
Poznań, Poland

ISSN 2197-9731 ISSN 2197-9758 (electronic)
RNA Technologies
ISBN 978-3-030-06559-1 ISBN 978-3-319-92967-5 (eBook)
https://doi.org/10.1007/978-3-319-92967-5

Printed on acid-free paper

This Springer imprint is published by the registered company Springer Nature Switzerland AG
The registered company address is: Gewerbestrasse 11, 6330 Cham, Switzerland

Preface

Systems Biology

The nineteenth and the twentieth century—a time during which our knowledge about how organisms function on a cellular and molecular level started to explode—witnessed the emergence of many new branches of biology such as cell biology, developmental biology, evolutionary biology, biochemistry, genetics, epigenetics, and molecular biology. Each of them focuses on a different aspect of the mechanisms and principles governing living organisms.

Systems biology brings the findings of these disciplines together with the aim to develop a holistic rather than reductionist understanding of cells, organisms, and ecosystems. Its goal is to understand the networks of individual biological components and to decipher how these networks and regulatory circuits interact to form living systems. A deep understanding of biological systems is achieved by gaining insight into their structure, dynamics, and control mechanisms. Systems biology represents a highly integrative and interdisciplinary approach. In addition to biology and medicine, it heavily relies on computer sciences and mathematics while also involving chemistry and physics.

The concept of systems biology emerged during the early twentieth century, when the notion became more and more accepted that biological systems follow physical and mechanical laws, elegantly outlined in D'Arcy Thompson's work "On Growth and Form," 1917. Other theories and discoveries contributed to the refinement of this concept during the course of the twentieth century. Notable examples include Conrad Waddington, who characterized networks of genes, cells, and tissues as decision-making dynamical systems; Ludwig von Bertallanffy with his "Outline of General Systems Theory" in 1950; Alan Lloyd Hodgkin and Andrew Fielding Huxley, who in 1952 spearheaded mathematical modeling of biological systems by describing how an action potential moves along the axon of a neuronal cell; Jacques Jacob and Francois Monod, who, when conducting their famous research on gene regulatory elements in the 1960s, concluded that mechanisms of gene regulation could form a variety of networks endowed with any desired degree of stability; as well as Eric Davidson and Roy Britten, who in 1969 pioneered the concept of gene regulatory networks. The term systems biology is attributed to systems theorist Mihajlo Mesarovic. He coined it in 1966 when hosting the international symposium "Systems Theory and Biology" at the Case Institute of Technology in Cleveland, OH. With the *Institute for Systems Biology* in Seattle and the *Systems Biology Institute* in Tokyo, the first systems biology institutes were founded in the year 2000, and many others followed.

The rise of systems biology as a key biological discipline in the new millennium was fueled by the preceding and concurrent development of high-throughput technologies such as genomics, transcriptomics, proteomics, metabolomics, and epigenomics. Omics technologies required novel specialized devices and experimental workflows as well as accompanying computational tools and mathematical models. The latter, which are needed to integrate the wealth of the generated data, were made possible thanks to the simultaneous vast expansion of computing power. Vice versa, systems biology continues to be a driving force behind the constant development and improvement both of experimental techniques and equipment to extract large amounts of qualitative and quantitative information from complex biological samples and of the bioinformatic pipelines necessary to obtain meaningful biological insights. An example of a more recent technological advancement in systems biology is represented by the development of single-cell omics technologies over the last decade, which now permit us to study the molecular make-up and dynamics of tissues and entire organisms at single-cell resolution.

A current challenge in systems biology is the integration of different regulatory levels such as genetic, epigenetic, and posttranscriptional gene regulation and the comprehension of the interplay between these levels. The long-term goal is the deduction of predictive models that enable us to foresee how cells and organisms change over time and in response to external stimuli or perturbations. Machine learning and artificial intelligence are going to be essential in the development of such multidimensional models that take spatial and temporal information into account. Being able to predict the fate of cells, tissues, organs, and organisms would be extremely powerful, since it would not only provide us with a fundamental understanding of how life works on a molecular and cellular level, but would also be a huge step forward for personalized medicine. It would allow us to foresee the course of human diseases and to choose the most effective therapies for each patient.

This book illustrates how systems biology is instrumental in advancing our knowledge about the principles of cellular and tissue organization. Themes covered include regulation of gene expression by genome structure, RNA-binding proteins, RNA–RNA interactions, noncoding RNAs, transcriptomics, epigenomics, metabolomics, posttranscriptional gene regulation, systems biology in health and disease, experimental and computational tools for systems biology research, computational methods for multidimensional data analysis, and integration as well as the deduction of predictive models.

The chapters will provide the reader with examples of how important scientific questions are addressed in systems biology and of bioinformatic tools designed to reach valuable conclusions from the abundance of the generated information.

Berlin, Germany Nikolaus Rajewsky
Berlin, Germany Verena Maier
Poznań, Poland Stefan Jurga
Poznań, Poland Jan Barciszewski

Contents

Systems Biology of Genome Structure and Dynamics

Zahra Fahmi, Sven A. Sewitz, and Karen Lipkow

Contents

Abstract Our view of the packed genome inside a nucleus has evolved greatly over the past decade. Aided by novel experimental and bioinformatic analysis techniques and detailed computational models, fundamental insights into the structure and dynamics of chromosomes have been gained. This has revealed that genome organisation has an essential role in controlling genome function during normal growth, cellular differentiation, and stress response, showing that, overall, 3D reorganisation is tightly linked to changes in gene expression. Chromatin, which is composed of DNA and a large number of different chromatin-associated proteins and RNAs, is often chemically modified, in patterns that affect gene expression. It has become clear that this highly interconnected system requires computational simulations to gain an understanding of the underlying system-wide mechanisms.

In this review, we describe different modelling approaches that are used to investigate both the linear and spatial arrangement of chromatin. We illustrate how dynamic computer simulations are used to study the mechanisms that control and maintain genome architecture and drive changes in this structure. We focus on

Z. Fahmi · S. A. Sewitz
Nuclear Dynamics Programme, Babraham Institute, Cambridge, UK

K. Lipkow (✉)
Nuclear Dynamics Programme, Babraham Institute, Cambridge, UK

Cambridge Systems Biology Centre, University of Cambridge, Cambridge, UK
e-mail: karen.lipkow@babraham.ac.uk

N. Rajewsky et al. (eds.), *Systems Biology*, RNA Technologies,
https://doi.org/10.1007/978-3-319-92967-5_1

models of the dynamics of epigenetic modifications, of protein–DNA interactions, and the polymer dynamics of chromosomes. These approaches provide reliable frameworks to integrate additional biological data; enable accurate, genome-wide predictions; and allow the discovery of new mechanisms.

Keywords Chromatin organisation · Computational model · Histone modification · Facilitated diffusion · Polymer · Chromatin loop · Self-organisation

1 Background

Intensive studies over the past decades have revealed multiple levels of organisation in eukaryotic genomes. The DNA wraps around eight histone proteins to make a nucleosome, the fundamental subunit of the chromatin fibre (van Holde 1989; Ramakrishnan 1997; Sewitz et al. 2017b). In mammals, the chromatin then folds to build higher genomic structures of different scales such as sub-megabase topologically associated domains (TADs), megabase A and B compartments, and chromosomal territories (Bonev and Cavalli 2016; Sewitz et al. 2017b). The nucleus is a highly crowded environment with efficiently packed and organised chromatin and hundreds to thousands of protein species, engaged in various types of interactions, such as protein–protein, DNA–protein, chromatin–chromatin, and chromatin–lamina interactions. It is now known that these interactions play an important role in controlling the organised structure and regulating the transcriptional activity of the genome (Gómez-Díaz and Corces 2014; Long et al. 2016; Flavahan et al. 2016) and that the structure changes upon differentiation and internal and external conditions (Guidi et al. 2015; Javierre et al. 2016; Sewitz et al. 2017a; Lazar-Stefanita et al. 2017). However, a comprehensive view of the mechanisms that drive organisation and dynamics of this highly complex system remains elusive.

Many research projects have investigated the linear arrangement of DNA, identifying the local regulatory elements that modulate transcription, such as transcription factor binding sites and their consensus sequences (Levine and Tjian 2003), enhancers (Long et al. 2016), histone modifications (Smolle and Workman 2013), and sites of DNA methylation (Schübeler 2015). Activator and repressor proteins recruit enzymes, such as histone acetyltransferase or histone deacetylase, that modify histones. Histone modifications control gene expression by altering the local chromatin structure and inhibiting or attracting DNA-binding factors (Dindot and Cohen 2013). In addition, DNA methylation can repress transcription through blocking the binding of transcription factors or mediating the binding of repressors (Jaenisch and Bird 2003).

It has more recently become possible to quantitatively investigate 3D genome architecture using live-cell microscopy, and chromosome conformation capture techniques, such as 3C, 4C, 5C, Hi-C, and Capture Hi-C (Schmitt et al. 2016b). This has greatly enhanced our understanding of gene regulatory mechanisms, by showing how the three-dimensional organisation of the genome influences gene regulation (Babu et al. 2008; Cavalli and Misteli 2013; Zuin et al. 2014; Lupiáñez et al. 2015; Dixon et al. 2016; Schmitt et al. 2016a). Many genes occupy preferred non-random positions within the nucleus: in mammals, gene-poor or transcriptionally inactive regions are located close to the nuclear envelope in most cell types, whereas gene-rich or transcriptionally active regions prefer to localise at the borders of chromosome territories, away from the nuclear periphery (Foster and Bridger 2005; Nagano et al. 2013). Manipulating the position of genes can also affect their activity; for human and mouse cells, it has been shown that relocating genes from their normal position to regions close to the nuclear periphery results in gene silencing (Reddy et al. 2008; Finlan et al. 2008). The single-celled eukaryote *S. cerevisiae* displays a mosaic arrangement of heterochromatin and euchromatin at the nuclear periphery, with active genes located close to nuclear pores (Casolari et al. 2004) and inactive genes associated with other parts of the nuclear periphery and the nuclear centre (Zimmer and Fabre 2011).

This organisation is achieved within a highly dynamic nucleoplasm (Misteli 2001; Vazquez et al. 2001; Lanctôt et al. 2007). For example, in mammalian cells, GFP-tagged proteins were measured to diffuse with diffusion coefficients of 0.24–0.53 $\mu m^2\ s^{-1}$, taking 24–54 s to travel 5 μm, a distance almost equal to the radius of the nucleus (Phair and Misteli 2000). Tagged chromosomal loci in living *S. cerevisiae* cells move more than 0.5 μm, equivalent to half of nuclear radius, within a few seconds (Heun et al. 2001). There is now evidence that the dynamics of the heterogeneous chromatin fibre contributes to thermodynamically driven 3D self-organisation (Sewitz et al. 2017a).

Investigation of chromatin organisation in space and time by novel experimental techniques has unravelled some of the key features of this intricate system of how genome structure relates to the function of the genome. To further study the dynamics of chromosome structures, particularly aspects that are not amenable to experimental analysis, scientists have adopted modelling approaches. Models provide the most direct way to explore mechanisms, as all components, interactions, reactions, and forces are defined, and any observed behaviour must be a consequence of these. During recent years, a wide range of models of the full or partial genome have been developed to analyse the interplay of genome structure and function. In this review, we categorise these models into three major groups: models of epigenetic modification dynamics, protein–DNA models, and polymer-based models.

2 Models of Epigenetic Modification Dynamics

Histone proteins can be covalently modified on several residues after translation (Allfrey et al. 1964), which leads to the recruitment of transcriptional regulatory proteins and structural proteins over a local chromatin region. For example, the combined deacetylation and methylation of the lysine at position 9 of histone H3 (H3K9) is required to create a binding site for the Swi6/HP1 silencing factor

(Nakayama et al. 2001; Shankaranarayana et al. 2003). Binding of silencing factors facilitates the modification of histones on adjacent nucleosomes, and sequential rounds of epigenetic modification and protein binding lead to the spreading of heterochromatin over a chromatin region (Grewal and Moazed 2003). Specialised boundary elements inhibit the heterochromatin extension and therefore separate silent and active chromatin domains (West et al. 2002; Labrador and Corces 2002).

To understand the mechanisms behind the epigenetic memory of monostable domains, predictive models have investigated the behaviour of H3K9 methylation domains (Hathaway et al. 2012; Hodges and Crabtree 2012; Müller-Ott et al. 2014; Erdel and Greene 2016). Simulations at single-nucleosome resolution showed that confined and heritable steady states of histone marks can be achieved by modelling linear propagation of histone modifications from nucleation sites to adjacent nucleosomes. Turnover of modified nucleosomes could also happen simultaneously (Hathaway et al. 2012; Hodges and Crabtree 2012). In contrast, another model assumed loop-driven spreading of histone marks with sparse nucleation sites. By adjusting parameters such as modification rates, the model was shown to be robust against replication (Erdel and Greene 2016), and the response towards transient perturbations was in line with experimental data (Müller-Ott et al. 2014).

Genomic regions of high epigenetic dynamics are bistable states, characterised by the presence of both activating and repressive histone marks (Bernstein et al. 2006). They have been observed for confined chromatin domains in various cell types (Rohlf et al. 2012; Tee et al. 2014). To study the features and dynamics of these states, several computational models have been developed (Dodd et al. 2007; Sedighi and Sengupta 2007; David-Rus et al. 2009; Micheelsen et al. 2010; Mukhopadhyay et al. 2010; Angel et al. 2011; Dodd and Sneppen 2011; Berry et al. 2017). In these models, a region of chromatin is represented as a sequence of nucleosomes. At every time step, each nucleosome has a state or a rate of histone modification based on its histone marks, with rules that govern state transitions or changes in rates. These models have shown that nonlinear positive feedback loops are required for robust and heritable bistable epigenetic states. Positive feedback loops arise when modifications of one nucleosome stimulate the modifications of other nucleosomes. The required nonlinearity can be achieved in different ways: (1) via the cooperativity of two or more nucleosomes with the same histone marks, which recruit histone modifiers on other nucleosomes (Dodd et al. 2007; Sedighi and Sengupta 2007; David-Rus et al. 2009; Micheelsen et al. 2010; Mukhopadhyay et al. 2010; Angel et al. 2011; Dodd and Sneppen 2011); (2) through two-step feedback loops, where the switch of histone modification states of nucleosomes occurs via an intermediate state, i.e. the state first changes to the intermediate state and then to the favoured state (Dodd et al. 2007; Angel et al. 2011; Berry et al. 2017); (3) through the local transcription rate, which can be affected by silencing, in turn leading to a change in the local modification rate (Sedighi and Sengupta 2007); and (4) through interactions with non-neighbour nucleosomes (Dodd et al. 2007). Another mathematical model with a 1D array of nucleosomes has been formulated to study the dynamics of histone modification in bivalent domains, where active and repressive histone marks coexist on nucleosomes (Ku et al. 2013). These domains are important elements in stem cells, and according to the model's prediction, their formation process is generally slow. The model also suggested that a coordinated set of parameters, such as recruitment and exchange rates of marks, leads to established and maintained bivalent domains over several cell cycles.

3 Protein–DNA Models

Transcription factors (TF) affect the transcriptional activity of specific genes through binding to specific DNA sequences (Ptashne and Gann 2002). It has been proposed that these proteins search for their target sequences through facilitated diffusion (Berg et al. 1981, 1982; Berg and von Hippel 1985), i.e. alternating rounds of 3D diffusion in the solution, sliding along the DNA, short-range excursions called hopping, and intersegmental transfer between DNA segments. The characteristics of this search mechanism have been widely studied, and computational models of different scales have brought new insights into its dynamics. All models discussed in this section have focused on facilitated diffusion of TFs.

At the most detailed, atomistic level, molecular dynamics (MD) simulations have been used to explain how, e.g. the *lac* repressor protein (LacI) moves along DNA (Marklund et al. 2013) and how it identifies its target site (Furini et al. 2013). LacI is modelled to take a helical path to probe the DNA, with its DNA-binding interface being insensitive to modest bends in DNA conformation. The hydrogen bonds formed between the DNA and the LacI interface are dynamic and flexible, allowing fast sliding of the protein (Marklund et al. 2013). This was found to enable the protein to probe the DNA quickly and reach the proximity of the target site. Once the specific DNA sequence is bound, it becomes significantly slower, resulting in the formation of a stable protein–DNA structure and a drop in enthalpy (Iwahara and Levy 2013; Furini et al. 2013). Another fine-grained MD simulation has proposed that binding of the CSL (CBF1/Suppressor of Hairless/LAG-1) protein to the DNA can transmit a signal through the protein structure according to the bound sequences. This influences the inter-domain dynamics of the protein and consequently its functional activities (Torella et al. 2014).

The effects of DNA conformation on the dynamics of TF proteins probing the DNA were explored via coarse-grained MD simulations, where proteins interact with the DNA via electrostatic interactions (Bhattacherjee and Levy 2014a, b). The geometry of DNA was tuned by two factors, curvature and the degree of helical twisting. Highly curved or highly twisted DNA was seen to lead to a decrease in sliding frequencies and an increase in hopping events (Bhattacherjee and Levy 2014a). In addition, introducing curvatures in the DNA conformation was found to increase the frequency of jumping events of a multidomain protein between distant DNA sites. However, curvature does not necessarily result in faster search kinetics as sliding happens less often (Bhattacherjee and Levy 2014b). Hence, an optimal DNA conformation can lead to a balanced number of searching events and maximal probing of DNA.

To investigate the role of nonspecific DNA–protein interactions during the search for specific target sites, Monte Carlo simulations were adopted (Das and Kolomeisky 2010; Tabaka et al. 2014; Mahmutovic et al. 2015). It was argued that the binding of the LacI repressor to nonspecific DNA is controlled by either activation or steric effects instead of being limited by diffusion (Tabaka et al. 2014; Mahmutovic et al. 2015). Furthermore, it was shown that for efficient and fast probing of DNA, moderate ranges of nonspecific binding energies and protein concentrations are required (Das and Kolomeisky 2010). The necessity for moderate DNA–protein binding strength has been indicated for proteins with different subdiffusive motions using simulations based on Brownian dynamics (Liu et al. 2017).

Large-scale computer simulations have been performed to study the search kinetics of transcription factors both in prokaryotic and eukaryotic cells. Software called GRiP (Gene Regulation in Prokaryotes) (Zabet and Adryan 2012a) provides a simulation framework for analysing the stochastic target search process of TF proteins. In GRiP the DNA is modelled as a string of base pairs, and TFs are highly diffusing components that interact with DNA sequences or with each other. This framework has been utilised to build a detailed model of facilitated diffusion, where TF orientation on the DNA, cooperativity of TFs, and crowding were incorporated (Zabet and Adryan 2012b). A similar model was adopted to dissect the effects of biologically relevant levels of mobile and immobile crowding on TF performance in a bacterial cell (Zabet and Adryan 2013): immobile crowding fixed on the DNA raises the occupancy of target sites significantly, whereas both mobile and immobile crowding have negligible impacts on the mean search time. Another model of the bacterial genome has taken two types of crowding molecules into account (Brackley et al. 2013). Proteins which bind to and move along DNA (1D crowding) do not change the search time significantly, even at very high densities. However, crowding molecules diffusing freely in 3D space increase the frequency of 1D sliding of TFs along DNA, while they enhance the robustness of the search time against any change in protein–DNA affinity.

A different approach based on the Gillespie stochastic simulation algorithm has been developed to analyse the influence of macromolecular crowding on gene expression in stem cells (Golkaram et al. 2017). The crowding was assumed to be correlated with the local chromatin density, which was calculated using Hi-C data. Diffusive TFs and RNA polymerases were only moving in the proximity of promoters, as crowding would not allow them to diffuse to other regions between rebindings. The model predicted that an increase in chromatin density during development leads to a rise in transcriptional bursting and subsequently heterogeneous expression of genes in a cell population.

Our lab has developed a computational model of TF motions in eukaryotes (Schmidt et al. 2014; Sewitz and Lipkow 2016) using the particle-based simulator Smoldyn (Andrews et al. 2010). This model has considered different types of movements for TFs: 3D diffusion, sliding, hopping, and intersegmental transfer. Among others, it showed the importance of intersegmental transfer, and it provided an explanation for the size of nucleosome-free regions on the DNA, which improve the process of TFs binding to their targets. Similar to a prokaryotic model (Tabaka et al. 2014), inclusion of 1D diffusion reduced the time to find the target sites by one and two orders of magnitude.

Finally, the complexity of gene regulation in higher eukaryotes has motivated the study of evolutionary dynamics of the TF repertoire and their binding preferences. A stochastic model based on duplication and mutation of genes suggested that more complex organisms with higher number of genes have higher levels of redundancy of TF binding (Rosanova et al. 2017).

4 Polymer-Based Models

The dynamic nature of the chromatin fibre lends itself to simulating chromatin as an extended, highly mobile polymer. Several studies have extended concepts developed in physics and applied them to the analysis of chromatin (Tark-Dame et al. 2011; Koslover and Spakowitz 2014; Shukron and Holcman 2017). This has led to an understanding of genome-wide data of chromosome folding and their interactions with each other and with other nuclear elements. In all models presented here, the chromatin fibre is a diffusing and self-avoiding chain of beads arranged in 3D space.

4.1 Models Based on Chromatin Loops

Chromatin loops have been observed in both eukaryotes and prokaryotes (Hofmann and Heermann 2015), and their vital regulatory impact has been demonstrated. A number of these models have suggested that chromatin loops are formed mainly by interactions between specific protein complexes like condensin (Cheng et al. 2015) or CTCF (Tark-Dame et al. 2014). These models have successfully reproduced the experimentally observed genome compaction. In addition, the importance of balance between short-range and long-range loops for controlling the changes in chromosomes structure has been revealed (Tark-Dame et al. 2014). It has furthermore been indicated that the dynamic bridges between condensin complexes bring about the intrachromosomal interactions during both interphase and mitosis in budding yeast (Cheng et al. 2015).

Other models have explored the general effects of protein interactions on chromatin structure. A heteropolymer model incorporated proteins implicitly, by mapping different epigenetic states onto the beads. Specific interactions between beads of the same state were differentiated from nonspecific interactions between any pair of beads (Jost et al. 2014). The model predicted that inter-TAD interactions are highly dynamic, which was in line with Hi-C results. It also predicted the fast formation of TADs, followed by a slow and long process of compaction (Jost et al. 2014). The lattice version of this model (Olarte-Plata et al. 2016), and another heteropolymer model (Ulianov et al. 2016) with active or inactive epigenomic states for beads, confirmed stronger self-attraction for inactive domains (Ulianov et al. 2016; Olarte-Plata et al. 2016) and an increase in their compaction as the domain size grows (Olarte-Plata et al. 2016). Other models based their assignment on levels of gene activity, with highly active or less active states assigned according to their expression levels (Jerabek and Heermann 2012). Highly active chromatin sections had low interaction strength, while less active ones had higher interaction affinity. The average distances between genomic loci, the average volume ratio between highly active and less active regions, and the positioning of highly active loci close to the boundary of chromosome territories were all in line with experimental measurements. In another work the polymer model was informed by protein binding sites and histone modifications (Brackley et al. 2016) and produced a population of genome conformations, which predicted the 3D distances between selected genomic sites on the globin locus in mouse ES cells.

In addition, polymer models based on protein interactions and without relying on predetermined information for the state of chromatin beads were developed (Giorgetti et al. 2014; Tiana et al. 2016; Chiariello et al. 2016). Using iterative Monte Carlo simulations and comparisons to the measured contact frequencies, the parameters of the models were optimised, and ensembles of chromatin configurations were achieved (Giorgetti et al. 2014; Tiana et al. 2016; Chiariello et al. 2016). These models correctly estimated the contact frequencies of TADs (Giorgetti et al. 2014; Chiariello et al. 2016) and the mean 3D distances between labelled loci upon perturbations of specific sites (Giorgetti et al. 2014). Combined with live-cell measurements, it has been suggested that changes in TAD conformations happen fast enough (in a much shorter time frame than the cell cycle) to facilitate dynamic interactions between regulatory elements, such as enhancer–promoter interactions (Tiana et al. 2016). A homopolymer model (Doyle et al. 2014), which implemented chromatin loops in the proximity of enhancer and promoter elements, indicated that the loops can either facilitate or insulate the enhancer–promoter interactions significantly. It was shown that the regulatory effect of the loop was dependant on the relative positions of loop anchors. To minimise the reliance on specific biological data, a heteropolymer model was built based on hierarchical folding and statistical physics of disordered systems (Nazarov et al. 2015). This model has two types of monomers that can interact with each other. By tuning the 1D sequence of monomers and the temperature controlling the folding, the simulated contact maps achieved a resemblance to Hi-C data.

Besides the notion that direct interactions between bound proteins shape chromatin loops, another mechanism, called loop extrusion, has been proposed (Nasmyth 2001; Alipour and Marko 2012; Sanborn et al. 2015; Fudenberg et al. 2016). This model calls for the action of extruding machines, possibly condensin or cohesin complexes, to bind and move along the DNA in opposite directions (Nasmyth 2001; Alipour and Marko 2012; Sanborn et al. 2015; Fudenberg et al. 2016). This leads to the extrusion of DNA loops until domain boundaries, occupied by CTCF proteins, are reached (Sanborn et al. 2015; Fudenberg et al. 2016). This mechanism can account for the compaction and folding of mitotic chromosomes (Nasmyth 2001; Alipour and Marko 2012). Furthermore, in combination with polymer physics, the model reproduced the observed decay of contact probabilities with increasing genomic distance, leading to simulated contact maps consistent with Hi-C data. It also predicted the changes in contact frequencies and 3D distances between loci due to CTCF and cohesin perturbations (Sanborn et al. 2015; Fudenberg et al. 2016).

4.2 Models Based on Supercoiling

Different levels of unconstrained supercoiling have been observed for chromatin (Kouzine et al. 2013; Naughton et al. 2013), and it has been reported that transcription leads to supercoiling (Wu et al. 1988; Kouzine et al. 2008; Papantonis and Cook 2011). To explore the effects of supercoiling on genome organisation in both eukaryotic (Benedetti et al. 2014) and prokaryotic (Le et al. 2013) cells, detailed polymer models have been employed. In a eukaryotic model, borders of TADs were mapped to the chromatin fibre, and strong supercoiling was imposed to the intervening chromatin (Benedetti et al. 2014). This led to the formation of TADs and contact maps broadly consistent with 3C data. In a bacterial model, chromatin was simulated as a dense array of plectonemes that were attached to a back bone (Le et al. 2013). By inserting plectoneme-free regions in the model at the positions of highly expressed genes, the contact frequencies observed for chromosomal interaction domains were reproduced. Overall, supercoiling is essential for creating chromosomal interaction domains (Le et al. 2013) and topologically associated domains (Benedetti et al. 2014). Intriguingly, a recent model investigated the role of supercoiling introduced by the transcribing RNA polymerase (Racko et al. 2017): when both CTCF and cohesin were included in the simulation, cohesin rings were seen to accumulate at CTCF sites demarking TAD borders. These observations are also seen experimentally (Uusküla-Reimand et al. 2016). Under these conditions, supercoiled DNA loops were extruded, and the supercoiling was the driving force for extruding the DNA loops. This is interesting because until now it was unclear how the energetically expensive loop extrusion could be achieved. Now, RNA polymerase-generated supercoiling provides a credible and testable hypothesis.

4.3 Integrative Models and Self-Organisation

With significant amounts of genome-wide datasets becoming available, computational models of chromatin are becoming more sophisticated and feature-rich. Computational models have explored the role of this heterogeneity in self-organisation of the genome structure.

In budding yeast, highly expressed genes are less occupied by chromatin-associated proteins, whereas genes that show lower overall expression are bound more extensively (Sewitz et al. 2017a). Protein occupancy can affect the local physical properties of the chromatin segment by means of a range of parameters such as changes in mass, diameter, local viscosity (Jirgensons 1958; Oldfield and Dunker 2014), diffusion speed (Jerabek and Heermann 2012; Phillip and Schreiber 2013; Wollman et al. 2017), and electrical charge of chromatin. This has led to the development of heteropolymeric models which incorporate some of the underlying complexity and points towards protein occupancy being a causal factor in determining self-organisation of genome structure in yeast (Sewitz et al. 2017a).

A significant challenge in this area is to continue to develop physical models of heteropolymeric motion applicable to chromatin. In many instances, insights are mainly qualitative and require physical parameters that are known to be unphysiological. As an example, it was shown that two chromosomes that differed in temperature-driven mobility would separate via a process akin to phase separation (Loi et al. 2008). Chromatin segments that harboured more active genes were given a higher temperature. This model reproduced the experimentally observed chromosomal territories (Ganai et al. 2014), but only if a temperature difference of 20-fold was assumed. Using much longer chromosomal segments, similar phase separations could already be observed with much smaller differences in temperature, bringing the model in closer proximity to real-life biological systems (Smrek and Kremer 2017). Still, current models are not yet fully able to deal with the structural complexity that is the hallmark of chromatin.

5 Conclusion and Outlook

It is now evident that the study of chromatin structure is at a stage where computational models are not just an accessory but a required component of any thorough investigation. The advent of pervasive high-performance computing has made it possible to attempt whole genome simulations at moderate resolutions, or smaller genomes at higher resolutions. Two future strands of development are now visible. Firstly, an ever-increasing amount of relevant genomic data is making its way into computational simulations. This will lead to more complex models that incorporate genome-wide protein binding data, extended epigenetic data, and measures of local chromatin conformation. This will also push the theoretical descriptions in polymer physics, where we foresee that increased and intensive collaboration and exchange is necessary. This will be mutually beneficial, as both fields will fundamentally improve their understanding of an area of biological physics that underpins questions of gene regulation during development, in response to external changes, and, in cases of misregulation, disease. These efforts are just at the beginning and will require the combined expertise of computational scientists, physicists, and experimental biologists to fully unravel the complex dynamics that lead to chromatin self-organisation.

References

Alipour E, Marko JF (2012) Self-organization of domain structures by DNA-loop-extruding enzymes. Nucleic Acids Res 40:11202–11212

Allfrey VG, Faulkner R, Mirsky AE (1964) Acetylation and methylation of histones and their possible role in the regulation of RNA synthesis. Proc Natl Acad Sci U S A 51:786–794

Andrews SS, Addy NJ, Brent R, Arkin AP (2010) Detailed simulations of cell biology with Smoldyn 2.1. PLoS Comput Biol 6:e1000705

Angel A, Song J, Dean C, Howard M (2011) A polycomb-based switch underlying quantitative epigenetic memory. Nature 476:105–108

Babu MM, Janga SC, de Santiago I, Pombo A (2008) Eukaryotic gene regulation in three dimensions and its impact on genome evolution. Curr Opin Genet Dev 18:571–582

Benedetti F, Dorier J, Burnier Y, Stasiak A (2014) Models that include supercoiling of topological domains reproduce several known features of interphase chromosomes. Nucleic Acids Res 42:2848–2855

Berg OG, von Hippel PH (1985) Diffusion-controlled macromolecular interactions. Annu Rev Biophys Biophys Chem 14:131–160

Berg OG, Winter RB, von Hippel PH (1981) Diffusion-driven mechanisms of protein translocation on nucleic acids. 1. Models and theory. Biochemistry 20:6929–6948

Berg OG, Winter RB, von Hippel PH (1982) How do genome-regulatory proteins locate their DNA target sites? Trends Biochem Sci 7:52–55

Bernstein BE, Mikkelsen TS, Xie X et al (2006) A bivalent chromatin structure marks key developmental genes in embryonic stem cells. Cell 125:315–326

Berry S, Dean C, Howard M (2017) Slow chromatin dynamics allow polycomb target genes to filter fluctuations in transcription factor activity. Cell Syst 4:445–457.e8

Bhattacherjee A, Levy Y (2014a) Search by proteins for their DNA target site: 1. The effect of DNA conformation on protein sliding. Nucleic Acids Res 42:12404–12414

Bhattacherjee A, Levy Y (2014b) Search by proteins for their DNA target site: 2. The effect of DNA conformation on the dynamics of multidomain proteins. Nucleic Acids Res 42:12415–12424

Bonev B, Cavalli G (2016) Organization and function of the 3D genome. Nat Rev Genet 17:661–678

Brackley CA, Cates ME, Marenduzzo D (2013) Intracellular facilitated diffusion: searchers, crowders, and blockers. Phys Rev Lett 111:108101

Brackley CA, Brown JM, Waithe D et al (2016) Predicting the three-dimensional folding of cis-regulatory regions in mammalian genomes using bioinformatic data and polymer models. Genome Biol 17:59

Casolari JM, Brown CR, Komili S et al (2004) Genome-wide localization of the nuclear transport machinery couples transcriptional status and nuclear organization. Cell 117:427–439

Cavalli G, Misteli T (2013) Functional implications of genome topology. Nat Struct Mol Biol 20:290–299

Cheng TMK, Heeger S, Chaleil RAG et al (2015) A simple biophysical model emulates budding yeast chromosome condensation. eLIFE 4:e05565

Chiariello AM, Annunziatella C, Bianco S et al (2016) Polymer physics of chromosome large-scale 3D organisation. Sci Rep 6:29775

Das RK, Kolomeisky AB (2010) Facilitated search of proteins on DNA: correlations are important. Phys Chem Chem Phys 12:2999–3004

David-Rus D, Mukhopadhyay S, Lebowitz JL, Sengupta AM (2009) Inheritance of epigenetic chromatin silencing. J Theor Biol 258:112–120

Dindot SV, Cohen ND (2013) Epigenetic regulation of gene expression: emerging applications for horses. J Equine Vet Sci 33:288–294

Dixon JR, Gorkin DU, Ren B (2016) Chromatin domains: the unit of chromosome organization. Mol Cell 62:668–680

Dodd IB, Sneppen K (2011) Barriers and silencers: a theoretical toolkit for control and containment of nucleosome-based epigenetic states. J Mol Biol 414:624–637
Dodd IB, Micheelsen MA, Sneppen K, Thon G (2007) Theoretical analysis of epigenetic cell memory by nucleosome modification. Cell 129:813–822
Doyle B, Fudenberg G, Imakaev M, Mirny LA (2014) Chromatin loops as allosteric modulators of enhancer-promoter interactions. PLoS Comput Biol 10:e1003867
Erdel F, Greene EC (2016) Generalized nucleation and looping model for epigenetic memory of histone modifications. Proc Natl Acad Sci U S A 113:E4180–E4189
Finlan LE, Sproul D, Thomson I et al (2008) Recruitment to the nuclear periphery can alter expression of genes in human cells. PLoS Genet 4:e1000039
Flavahan WA, Drier Y, Liau BB et al (2016) Insulator dysfunction and oncogene activation in IDH mutant gliomas. Nature 529:110–114
Foster HA, Bridger JM (2005) The genome and the nucleus: a marriage made by evolution. Genome organisation and nuclear architecture. Chromosoma 114:212–229
Fudenberg G, ImakaevM LC et al (2016) Formation of chromosomal domains by loop extrusion. Cell Rep 15:2038–2049
Furini S, Barbini P, Domene C (2013) DNA-recognition process described by MD simulations of the lactose repressor protein on a specific and a non-specific DNA sequence. Nucleic Acids Res 41:3963–3972
Ganai N, Sengupta S, Menon GI (2014) Chromosome positioning from activity-based segregation. Nucleic Acids Res 42:4145–4159
Giorgetti L, Galupa R, Nora EP et al (2014) Predictive polymer modeling reveals coupled fluctuations in chromosome conformation and transcription. Cell 157:950–963
Golkaram M, Jang J, Hellander S et al (2017) The role of chromatin density in cell population heterogeneity during stem cell differentiation. Sci Rep 7:13307
Gómez-Díaz E, Corces VG (2014) Architectural proteins: regulators of 3D genome organization in cell fate. Trends Cell Biol 24:703–711
Grewal SIS, Moazed D (2003) Heterochromatin and epigenetic control of gene expression. Science 301:798–802
Guidi M, Ruault M, Marbouty M et al (2015) Spatial reorganization of telomeres in long-lived quiescent cells. Genome Biol 16:206
Hathaway NA, Bell O, Hodges C et al (2012) Dynamics and memory of heterochromatin in living cells. Cell 149:1447–1460
Heun P, Taddei A, Gasser SM (2001) From snapshots to moving pictures: new perspectives on nuclear organization. Trends Cell Biol 11:519–525
Hodges C, Crabtree GR (2012) Dynamics of inherently bounded histone modification domains. Proc Natl Acad Sci U S A 109:13296–13301
Hofmann A, Heermann DW (2015) The role of loops on the order of eukaryotes and prokaryotes. FEBS Lett 589:2958–2965
Iwahara J, Levy Y (2013) Speed-stability paradox in DNA-scanning by zinc-finger proteins. Transcription 4:58–61
Jaenisch R, Bird A (2003) Epigenetic regulation of gene expression: how the genome integrates intrinsic and environmental signals. Nat Genet 33(Suppl):245–254
Javierre BM, Burren OS, Wilder SP et al (2016) Lineage-specific genome architecture links enhancers and non-coding disease variants to target gene promoters. Cell 167:1369–1384
Jerabek H, Heermann DW (2012) Expression-dependent folding of interphase chromatin. PLoS One 7:e37525
Jirgensons B (1958) Optical rotation and viscosity of native and denatured proteins. X. Further studies on optical rotatory dispersion. Arch Biochem Biophys 74:57–69
Jost D, Carrivain P, Cavalli G, Vaillant C (2014) Modeling epigenome folding: formation and dynamics of topologically associated chromatin domains. Nucleic Acids Res 42:9553–9561
Koslover EF, Spakowitz AJ (2014) Multiscale dynamics of semiflexible polymers from a universal coarse-graining procedure. Phys Rev E Stat Nonlinear Soft Matter Phys 90:013304

Kouzine F, Sanford S, Elisha-Feil Z, Levens D (2008) The functional response of upstream DNA to dynamic supercoiling in vivo. Nat Struct Mol Biol 15:146–154

Kouzine F, Gupta A, Baranello L et al (2013) Transcription-dependent dynamic supercoiling is a short-range genomic force. Nat Struct Mol Biol 20:396–403

Ku WL, Girvan M, Yuan G-C et al (2013) Modeling the dynamics of bivalent histone modifications. PLoS One 8:e77944

Labrador M, Corces VG (2002) Setting the boundaries of chromatin domains and nuclear organization. Cell 111:151–154

Lanctôt C, Cheutin T, Cremer M et al (2007) Dynamic genome architecture in the nuclear space: regulation of gene expression in three dimensions. Nat Rev Genet 8:104–115

Lazar-Stefanita L, Scolari VF, Mercy G et al (2017) Cohesins and condensins orchestrate the 4D dynamics of yeast chromosomes during the cell cycle. EMBO J 36:2684–2697

Le TBK, Imakaev MV, Mirny LA, Laub MT (2013) High-resolution mapping of the spatial organization of a bacterial chromosome. Science 342:731–734

Levine M, Tjian R (2003) Transcription regulation and animal diversity. Nature 424:147–151

Liu L, Cherstvy AG, Metzler R (2017) Facilitated diffusion of transcription factor proteins with anomalous bulk diffusion. J Phys Chem B 121:1284–1289

Loi D, Mossa S, Cugliandolo LF (2008) Effective temperature of active matter. Phys Rev E Stat Nonlinear Soft Matter Phys 77:051111

Long HK, Prescott SL, Wysocka J (2016) Ever-changing landscapes: transcriptional enhancers in development and evolution. Cell 167:1170–1187

Lupiáñez DG, Kraft K, Heinrich V et al (2015) Disruptions of topological chromatin domains cause pathogenic rewiring of gene-enhancer interactions. Cell 161:1012–1025

Mahmutovic A, Berg OG, Elf J (2015) What matters for lac repressor search in vivo—sliding, hopping, intersegment transfer, crowding on DNA or recognition? Nucleic Acids Res 43:3454–3464

Marklund EG, Mahmutovic A, Berg OG et al (2013) Transcription-factor binding and sliding on DNA studied using micro- and macroscopic models. Proc Natl Acad Sci U S A 110:19796–19801

Micheelsen MA, Mitarai N, Sneppen K, Dodd IB (2010) Theory for the stability and regulation of epigenetic landscapes. Phys Biol 7:026010

Misteli T (2001) Protein dynamics: implications for nuclear architecture and gene expression. Science 291:843–847

Mukhopadhyay S, Nagaraj VH, Sengupta AM (2010) Locus dependence in epigenetic chromatin silencing. Biosystems 102:49–54

Müller-Ott K, Erdel F, Matveeva A et al (2014) Specificity, propagation, and memory of pericentric heterochromatin. Mol Syst Biol 10:746

Nagano T, Lubling Y, Stevens TJ et al (2013) Single-cell Hi-C reveals cell-to-cell variability in chromosome structure. Nature 502:59–64

Nakayama J, Rice JC, Strahl BD et al (2001) Role of histone H3 lysine 9 methylation in epigenetic control of heterochromatin assembly. Science 292:110–113

Nasmyth K (2001) Disseminating the genome: joining, resolving, and separating sister chromatids during mitosis and meiosis. Annu Rev Genet 35:673–745

Naughton C, Avlonitis N, Corless S et al (2013) Transcription forms and remodels supercoiling domains unfolding large-scale chromatin structures. Nat Struct Mol Biol 20:387–395

Nazarov LI, Tamm MV, Avetisov VA, Nechaev SK (2015) A statistical model of intra-chromosome contact maps. Soft Matter 11:1019–1025

Olarte-Plata JD, Haddad N, Vaillant C, Jost D (2016) The folding landscape of the epigenome. Phys Biol 13:026001

Oldfield CJ, Dunker AK (2014) Intrinsically disordered proteins and intrinsically disordered protein regions. Annu Rev Biochem 83:553–584

Papantonis A, Cook PR (2011) Fixing the model for transcription: the DNA moves, not the polymerase. Transcription 2:41–44

Phair RD, Misteli T (2000) High mobility of proteins in the mammalian cell nucleus. Nature 404:604–609
Phillip Y, Schreiber G (2013) Formation of protein complexes in crowded environments – from in vitro to in vivo. FEBS Lett 587:1046–1052
Ptashne M, Gann A (2002) Genes & signals. Cold Spring Harbor Laboratory Press, Cold Spring Harbor, NY
Racko D, Benedetti F, Dorier J, Stasiak A (2017) Transcription-induced supercoiling as the driving force of chromatin loop extrusion during formation of TADs in interphase chromosomes. Nucleic Acids Res 46:1648–1660
Ramakrishnan V (1997) Histone structure and the organization of the nucleosome. Annu Rev Biophys Biomol Struct 26:83–112
Reddy KL, Zullo JM, Bertolino E, Singh H (2008) Transcriptional repression mediated by repositioning of genes to the nuclear lamina. Nature 452:243–247
Rohlf T, Steiner L, Przybilla J et al (2012) Modeling the dynamic epigenome: from histone modifications towards self-organizing chromatin. Epigenomics 4:205–219
Rosanova A, Colliva A, Osella M, Caselle M (2017) Modelling the evolution of transcription factor binding preferences in complex eukaryotes. Sci Rep 7:7596
Sanborn AL, Rao SSP, Huang S-C et al (2015) Chromatin extrusion explains key features of loop and domain formation in wild-type and engineered genomes. Proc Natl Acad Sci U S A 112:E6456–E6465
Schmidt HG, Sewitz S, Andrews SS, Lipkow K (2014) An integrated model of transcription factor diffusion shows the importance of intersegmental transfer and quaternary protein structure for target site finding. PLoS One 9:e108575
Schmitt AD, Hu M, Jung I et al (2016a) A Compendium of chromatin contact maps reveals spatially active regions in the human genome. Cell Rep 17:2042–2059
Schmitt AD, Hu M, Ren B (2016b) Genome-wide mapping and analysis of chromosome architecture. Nat Rev Mol Cell Biol 17:743–755
Schübeler D (2015) Function and information content of DNA methylation. Nature 517:321–326
Sedighi M, Sengupta AM (2007) Epigenetic chromatin silencing: bistability and front propagation. Phys Biol 4:246–255
Sewitz S, Lipkow K (2016) Systems biology approaches for understanding genome architecture. Methods Mol Biol 1431:109–126
Sewitz SA, Fahmi Z, Aljebali L et al (2017a) Heterogeneous chromatin mobility derived from chromatin states is a determinant of genome organisation in *S. cerevisiae*. bioRxiv:106344. https://doi.org/10.1101/106344
Sewitz S, Fahmi Z, Lipkow K (2017b) Higher order assembly: folding the chromosome. Curr Opin Struct Biol 42:162–168
Shankaranarayana GD, Motamedi MR, Moazed D, Grewal SIS (2003) Sir2 regulates histone H3 lysine 9 methylation and heterochromatin assembly in fission yeast. Curr Biol 13:1240–1246
Shukron O, Holcman D (2017) Transient chromatin properties revealed by polymer models and stochastic simulations constructed from chromosomal capture data. PLoS Comput Biol 13:e1005469
Smolle M, Workman JL (2013) Transcription-associated histone modifications and cryptic transcription. Biochim Biophys Acta 1829:84–97
Smrek J, Kremer K (2017) Small activity differences drive phase separation in active-passive polymer mixtures. Phys Rev Lett 118:098002
Tabaka M, Kalwarczyk T, Hołyst R (2014) Quantitative influence of macromolecular crowding on gene regulation kinetics. Nucleic Acids Res 42:727–738
Tark-Dame M, van Driel R, Heermann DW (2011) Chromatin folding – from biology to polymer models and back. J Cell Sci 124:839–845
Tark-Dame M, Jerabek H, Manders EMM et al (2014) Depletion of the chromatin looping proteins CTCF and cohesin causes chromatin compaction: insight into chromatin folding by polymer modelling. PLoS Comput Biol 10:e1003877

Tee W-W, Shen SS, Oksuz O et al (2014) Erk1/2 activity promotes chromatin features and RNAPII phosphorylation at developmental promoters in mouse ESCs. Cell 156:678–690

Tiana G, Amitai A, Pollex T et al (2016) Structural fluctuations of the chromatin fiber within topologically associating domains. Biophys J 110:1234–1245

Torella R, Li J, Kinrade E et al (2014) A combination of computational and experimental approaches identifies DNA sequence constraints associated with target site binding specificity of the transcription factor CSL. Nucleic Acids Res 42:10550–10563

Ulianov SV, Khrameeva EE, Gavrilov AA et al (2016) Active chromatin and transcription play a key role in chromosome partitioning into topologically associating domains. Genome Res 26:70–84

Uusküla-Reimand L, Hou H, Samavarchi-Tehrani P et al (2016) Topoisomerase II beta interacts with cohesin and CTCF at topological domain borders. Genome Biol 17:182

van Holde KE (1989) Chromatin. Springer series in molecular and cell biology. Springer, New York

Vazquez J, Belmont AS, Sedat JW (2001) Multiple regimes of constrained chromosome motion are regulated in the interphase *Drosophila* nucleus. Curr Biol 11:1227–1239

West AG, Gaszner M, Felsenfeld G (2002) Insulators: many functions, many mechanisms. Genes Dev 16:271–288

Wollman AJ, Shashkova S, Hedlund EG et al (2017) Transcription factor clusters regulate genes in eukaryotic cells. eLIFE 6:e27451

Wu HY, Shyy SH, Wang JC, Liu LF (1988) Transcription generates positively and negatively supercoiled domains in the template. Cell 53:433–440

Zabet NR, Adryan B (2012a) GRiP: a computational tool to simulate transcription factor binding in prokaryotes. Bioinformatics 28:1287–1289

Zabet NR, Adryan B (2012b) A comprehensive computational model of facilitated diffusion in prokaryotes. Bioinformatics 28:1517–1524

Zabet NR, Adryan B (2013) The effects of transcription factor competition on gene regulation. Front Genet 4:197

Zimmer C, Fabre E (2011) Principles of chromosomal organization: lessons from yeast. J Cell Biol 192:723–733

Zuin J, Dixon JR, van der Reijden MIJA et al (2014) Cohesin and CTCF differentially affect chromatin architecture and gene expression in human cells. Proc Natl Acad Sci U S A 111:996–1001

A Systems Perspective of Complex Diseases: From Reductionism to Integration

Khushdeep Bandesh, Pawan K. Dhar, and Dwaipayan Bharadwaj

Contents

Abstract Complex systems exist across all levels of biological organization ranging from the simplest (subatomic realm) to most complex (individual organism to whole populations and beyond). This complex nature of both the common diseases

K. Bandesh
Academy of Scientific and Innovative Research, CSIR-Institute of Genomics and Integrative Biology, New Delhi, India
e-mail: khushdeep@igib.in

P. K. Dhar
Synthetic Biology Group, School of Biotechnology, Jawaharlal Nehru University, New Delhi, India
e-mail: pawandhar@mail.jnu.ac.in

D. Bharadwaj (✉)
Systems Genomics Laboratory, School of Biotechnology, Jawaharlal Nehru University, New Delhi, India
e-mail: db@jnu.ac.in

N. Rajewsky et al. (eds.), *Systems Biology*, RNA Technologies,
https://doi.org/10.1007/978-3-319-92967-5_2

and the human beings has kept researchers far from a holistic understanding of underlying biological processes. Over the past decade, there has been a rapid and vast accumulation of large scale high-throughput biological data at physiological, cellular, molecular, and submolecular levels. It includes genetic association studies of complex human diseases and traits, quantification of genome-wide RNA expression patterns, comprehensive profiling of cellular proteins and metabolites, gene regulatory information (DNA methylation, histone modifications, chromatin accessibility, evolutionary constraint, etc.), and characterization of networks of molecular interactions. The clinical utility of such enormous data demands interpretation and understanding at the biological level to reveal mechanistic insights of molecular etiology. An important element of this task is to complement the detailed pieces of biological information with new advanced methods of system integration and reconstruction. This requires conversion of actual biological systems into computational models to make reliable predictions of biological responses following targeted manipulation under untested conditions. The frequency at which signals are presently being discovered mandates a systematic and integrative "omics" approach to bridge the "genotype to phenotype" gap. The chapter highlights the fundamental ways to integrate high-quality biological data that await systemic interpretations.

Keywords Complex systems · Common diseases · High-throughput data · Computational models · Systematic interpretation

1 Introduction to Systems Biology

1.1 Establishing the Need

The classic reductionist approach in biological sciences, generally known by the terms like molecular biology and biochemistry, has led to generation of enormous "parts-data." The collection of data has been aided by the parallel development of sequencing, structural, and expression measurement technologies. From low-throughput data collection, the community has reached high-throughput data collection, storage and analytical technologies.

The enormous success of reductionist approach has helped to determine the composition of the system and individual correlation of parts with a given phenotype, in a large number of situations. However, it has also thrown up a major challenge, i.e., to understand collective behavior of thousands of parts working together to maintain the functioning and robustness of a cell and an organism. The big challenge is to construct a large virtual matrix where biological components interact virtually and help understand biological decisions various scales and granularity.

Back in 1944, Norbert Weiner foresaw the need for a new approach that focused on stitching individual parts to describe collective response and coined the term "Systems Biology." Though the idea was novel and path-building, the time was not yet ripe for launching a new approach, due to scarcity of data and computational resources.

The idea of systems approach again picked up in the mid-1960s and 1970s, when concepts like metabolic flux and control analysis gained traction. The aim was to study the flow of metabolites through a certain path/pathway and identify choke points that controlled the flux. A large body of literature during this time led to emergence of a new Biochemical Systems Theory.

The situation remained somewhat unchanged for the next few decades, till a new high-throughput technology of gene sequencing and expression measurement arrived. Biological sciences suddenly changed the stick shift and went into a higher gear of data gathering, management and analysis. The paradigm shift was greatly helped by parallel technological advancement in the computer industry. The storage got cheaper, processes got faster and algorithms were written to swim through oceans of data to find patterns.

The speed, scale and variety of data breathed life into Nobert Weiner's work of 1940s and "Systems Biology" as a formal discipline was launched. For many years the community debated on the concept, definition, scope and tools of the new systems approach. However, what emerged as a common thread was the acceptance that (a) collective behavior of biological parts was different than the sum-of-its-parts and (b) modeling in biology was essential to understand biological decisions, narrow down the range of experiments and generate hypothesis.

The biological community was beginning to sense the power of mathematics and computation that played a major role in the origin and evolution of engineering from physics. The need for modeling was also felt for the reasons that, on one hand, not enough experiments could be performed to collect all kinds of data in all kinds of contexts. On the other hand a lot of data in the published literature domain was inaccurate.

Here it may be relevant to introduce a few definitions.

1.2 What Is a Model?

A model is a representation of a system in a certain form that looks closest to the real life situation. The skeletal system of a model is made of components and their interactions. It is somewhat easy to define a static system in terms of components and interactions. However, the real challenge arrives when one moves from a static to a dynamic description, i.e., creating a movie out of snapshots arranged along a certain time series.

Modeling itself is an iterative process that goes on and on till experimental results match the modeling predictions. A model may be rigorous with mathematical representation or simply a sketch of nodes and arrows. It may depict a flow of information (as in metabolic pathways) or direction invariant (as in protein–protein interaction networks).

Furthermore, mathematical models may be deterministic (responses are predictable) or stochastic (responses are determined by probability distribution). Watching a model grow over an x-axis of time is called simulation. Adding mathematical muscles to a bare bone model is both an art and a science. One needs to be convinced of the flow of information in a certain way to adopt a certain modeling approach. Also, the choice of modeling method is governed by the kind of question one asks, the availability of data (qualitative to quantitative) and computational resources.

1.3 Steps in Building a Model

1. Make a parts list data from literature and annotate every part by including measurements, protocols, perturbations, constraints, and error bar. Here it is important to know if the data were independently confirmed.
2. Draw a parts-interaction map in the form of pathways. The map may represent translocation (ion channel), transformation (substrate–enzyme reaction), and binding events (transcription factor) in the form of nodes (molecules) and edges (interactions).
3. Use appropriate qualitative or quantitative methods to empower the power of conversation. Build conceptual, analytical models for simulation.

Apply perturbations at predefined points where phenotypic assays are possible and generate novel observations and hypothesis (Fig. 1).

1.4 Modeling Methods and Tools

Ever since the first conference of Systems Biology was held in Tokyo in 2000, a large number of tools have come up addressing various needs of the modeling community. Some of the most common resources and tools used are:

1. ***Pre Constructed Pathway Maps***
Kyoto Encyclopedia of genes and genomes http://www.genome.ad.jp/kegg/
BioCyc http://www.biocyc.org
BioCarta https://cgap.nci.nih.gov/Pathways/BioCarta_Pathways

2. ***Enzyme Databases***
BRENDA http://www.brenda-enzymes.info/
ExPASy https://www.expasy.org/

3. ***Tools for Constructing, Simulating, and Analyzing Pathways***
http://sbml.org/SBML_Software_Guide/SBML_Software_Summary

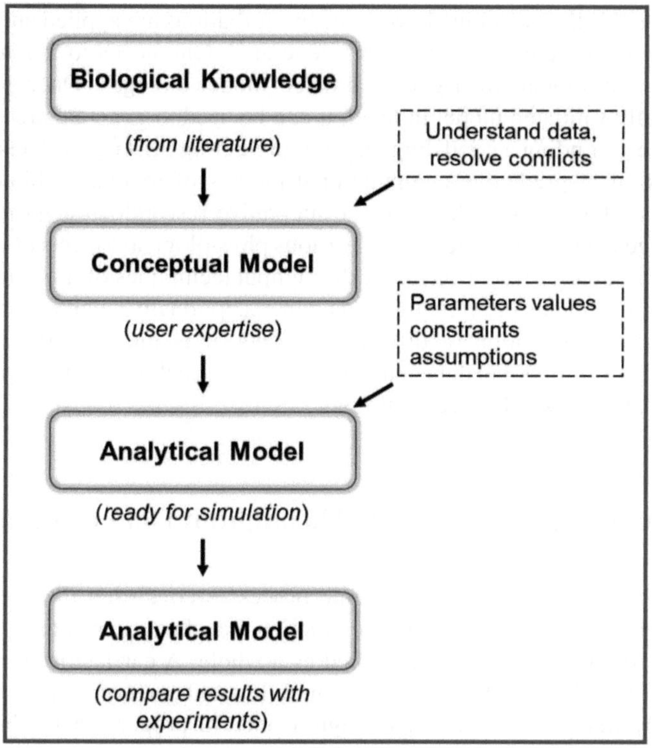

Fig. 1 A general strategy of building pathway models

In general, for modeling metabolic pathways, where large number of molecules interact (and data are frequently available) one uses ordinary differential equation based approach. For modeling gene expression events, where numbers are very less (transcription factor molecules) and fluctuations are high, the method of choice is stochastic. Some people also use ODE approach, as it comes with less computational cost. In situations where the large scale networks need to be modeled, rule based, fuzzy logic based, Boolean based and petri net approaches have been used with success. As the scale of the network increases in size the computational costs soar. To find the right balance, one may use a combination of qualitative (Boolean, rule based) and quantitative (ODE and stochastic) methods. Some of the issues that often emerge in quantitative modeling approaches are parameter estimation and optimization.

The need for a good parameter estimation method is felt more when the network data is incomplete, i.e., there is a space of unknown that needs to be considered and computed in the model. Several ODE and stochastic methods to estimate missing parameters are available. However, none of the methods can absolutely guarantee the accuracy of the output. One needs to feed in predicted data over and over again for optimization purposes. A fully parameterized and computationally optimized pathway model is then examined over time, perturbations are applied and the output is compared with the experimental data generated. One needs to iterate back and forth, tinker the model till a good validated model emerges. Once we have an experimentally validated model in hand, it can be used to generate predictions and hypothesis, e.g., finding a good drug target or predicting off-target effects and so on.

Formulation of predictive computational models of regulatory biological networks in complex diseases demands an integrative research strategy to articulate different large datasets collected across various physiological aspects of healthy and diseased individuals. Present-day high-throughput techniques of molecular biology facilitate large amounts of high quality data in exceptionally small time.

For effective integration of different datasets, it is intrinsically important to understand high-throughput data generation at various cellular and molecular levels that are valuable to sketch disease etiology.

2 Common Methods of High-Throughput Data Generation

Biological systems are complex dynamic processes consisting of several diverse entities in which each unit has a definite function that changes over time. A complex system can be easily simplified if studied as a whole. A car factory looks awfully complicated to a layman but for an automobile engineer every small process on the assembly line has a well-defined significance in proper functioning of a car. Likewise, in biology, an assimilation of diverse molecular signatures can effectively tackle the complexity of physiological systems in normal and perturbed conditions. In lieu of sufficient data from a single individual, a comprehensive picture of development of a disease needs procurement of many personal trajectories with some in healthy range and others diverting towards disease. Today, a plethora of large datasets encompassing several functional regulatory elements of complex diseases (DNA, RNA, proteins, regulators, metabolites, etc.) exist for various human populations. An introduction of systems concept in understanding complex disease biology is insightful to identify early signs or biomarkers for regular screening of healthy people for the disease (Fig. 2).

2.1 Genomic Data

Genomics aims to study the total DNA of a cell or organism. Human genome comprises 3.2 billion nucleotide base-pairs and nearly 19,836 protein coding genes (Harrow et al. 2012). Genes are basic units of inheritance across generations and hotspots of variations and mutations. Several aberrations are known to exist at DNA level—insertions, deletions, duplications, single nucleotide polymorphisms (SNPs), nucleotide repeats, copy number variations, etc. SNPs are the most common variations in DNA sequences among individuals. A SNP is a nucleotide variation at

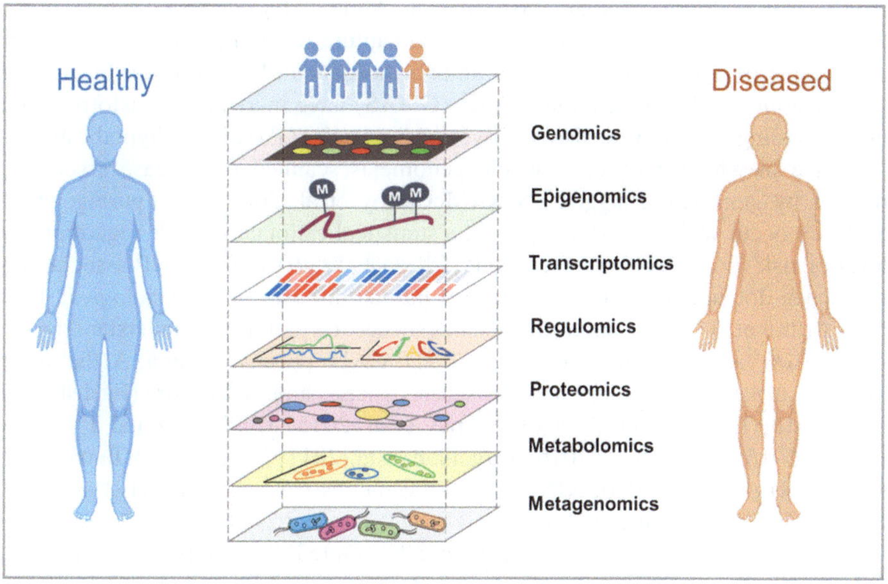

Fig. 2 An integrative approach for studying complex diseases

a single position in DNA that is altered in more than 1% of the population (NIH-Genetics Home Reference 2017a). For example, at SNP rs6520015 on chromosome 22, some individuals have cytosine (C) and others show thymine (T). The altered nucleotides are alleles of SNP. C and T are alleles of rs6520015. There are roughly ten million SNPs in human genome occurring once in every 300 bases (NIH-Genetics Home Reference 2017a).

Linkage analyses to identify causal variants in family based studies have been extremely fruitful for many single gene disorders, but failed for complex diseases (Altshuler et al. 2008). In consideration, a parallel approach for identification of genetic risk factors for complex diseases is to assess the correlation between the diseased status and frequency of alleles of specific genetic variant by comparing affected individuals with unaffected controls (Association study) (Cordell and Clayton 2005). Here, function of a gene important for the etiology of a disease may be affected by associated SNPs residing within the gene or in a nearby regulatory region. For instance, SNP rs599839 residing downstream of *SORT1* gene, has previously been reported to be associated with Coronary Artery Disease (CAD) and LDL-cholesterol levels (Wang et al. 2011). Presence of minor allele G results in increased levels of *SORT1* mRNA that further leads to increased uptake of LDL into cells. SNPs studies are preferably conducted in large human population sample sets in order to pin down fundamental genes or genomic regions of considered disease.

Previously, association of a gene with a particular disease was studied and analyzed individually (candidate gene approach). For instance, a research conducted in 1656 unrelated Indians tested three SNPs residing in a transcription activator gene-FOXA2 for association to type 2 diabetes (T2D) (Tabassum et al. 2008).

However, over the recent years, microarray technology has advanced substantially. DNA microarrays have been economical to capture difference in DNA sequence of millions of SNPs simultaneously among thousands of individuals. In view of this, Genome Wide Association Studies (GWAS) have incredibly transformed our understanding of complex diseases over past decade. GWAS is a hypothesis-free approach in which a person's whole genome is scanned for disease associated regions by genotyping tagged SNPs (McCarthy et al. 2008). By figuring which SNPs co-occur with disease symptoms, a statistical estimate is made regarding the level of risk associated with each SNP. Till date, 52,491 unique SNP associations have been documented for a multitude of human diseases and physiological traits (MacArthur et al. 2017). This approach has succeeded partially in understanding the genetics of various common diseases—T2D, CAD, obesity, asthma, Alzheimer's disease, stroke, inflammatory bowel disease, cancers, and many more (Fuchsberger et al. 2016; Tabassum et al. 2013; Nikpay et al. 2015; McPherson and Tybjaerg-Hansen 2016; Locke et al. 2016; Torgerson et al. 2011; Bertram and Tanzi 2009; Cauwenberghe et al. 2016; NINDS Stroke Genetics Network (SiGN) et al. 2016; Lange et al. 2017; Chang et al. 2014).

NHGRI-EBI GWAS catalog is a manually curated collection of all published GWAS for various diseases/traits conducted so far (MacArthur et al. 2017). Besides, a worldwide effort—International HapMap Project seeks to determine the frequency and common patterns of SNPs and other genetic variants in the genomes of populations of African, Asian and European ancestry by whole genome genotyping (The International HapMap Consortium 2003). Human populations are considerably different from one another in terms of anthropometric parameters, biochemical traits or resistance to a disease. There are disease SNPs that are specific to one population but non-polymorphic in another. Besides whole genome genotyping data, 1000 Genomes Project provides a comprehensive resource on human genetic variation by sequencing entire genomes of 2504 unrelated people from 26 different human populations (Sudmant et al. 2015). This integrated map of structural variants—insertions, deletions, duplications, inversions, and SNPs—is valuable for constructing personalized genomes.

2.2 *Epigenomic Data*

The epigenome of a cell constitutes a set of chemical modifications to the DNA and DNA associated proteins that govern gene expression without changing the DNA sequence (NIH-Genetics Home Reference 2017b). Human body has trillions of cells that perform specialized function in muscle, brain, eye, bones, gut, etc. Each of these cells carries basically the same DNA but drastically differ in terms of what set of genes are turned on/off in various cell types. A person's genome is a storehouse of instructions, whereas an epigenetic mark regulates how cells follow these instructions. Epigenetic signatures are inheritable but can get altered in response to environmental exposure or disease (NIH-Genetics Home Reference 2017b).

The most common and crucial epigenomic modification is methylation. A methyl group ($-CH_3$) is covalently added to the fifth carbon atom of cytosine ring of DNA by DNA methyltransferases. Addition of methyl groups lead to silencing of a gene and no protein is produced from that gene. Nearly 1.5% of human genomic DNA contains modified cytosines (5-methylcytosine) (Lister et al. 2009). Apart from transcription, DNA methylation regulates many cellular processes—chromatin structure, stability, genomic imprinting, embryonic development, etc. (Schübeler 2015). Altered DNA methylation have been implicated in metabolic diseases—obesity, T2D, atherosclerosis, non-alcoholic fatty liver disease, etc. (Zhao et al. 2012; Wahl et al. 2015; Sonne et al. 2017; Zaina et al. 2014; Ahrens et al. 2013; Giri et al. 2017). DNA methylation marks are generally captured by whole-genome bisulfite sequencing (Li and Tollefsbol 2011). Bisulfite treatment converts cytosines to uracils; however, methylated cytosines do not get converted. The method nicely isolates methylated cytosines in the genome.

Besides DNA methylation, posttranslational modifications of histone proteins also occur frequently throughout the genome. Histone proteins (H2A, H2B, H3, and H4) form the core of nucleosomes that represent the first level of chromatin organization. Histone modifications—methylation, phosphorylation, acetylation, ubiquitylation, and sumoylation—influence gene expression by altering chromatin structure or recruiting histone modifier (Bannister and Kouzarides 2011). H3 histone is the most modified histone that can dictate the type of chromatin (euchromatin or heterochromatin), pinpoint functional genomic elements (gene body, promoters, enhancers, or silencers) and whether these elements are in active or repressed state. Histone H3 is largely acetylated at lysines 9, 14, 18, 23, 27, and 56, methylated at arginine 2 and lysines 4, 9, 27, 36, and 79, and phosphorylated at serines 10, 28, and threonines 3, 11. Histone marks—H3K9ac, H3K27ac, H3K4me1, and H3K4me3 are active marks found at transcriptional enhancers and gene promoters. Tri-methylation of H3K36 is observed at the transcribed regions of a gene. H3K9me3 and H3K27me3 are repressive histone marks spotted respectively in heterochromatin and gene promoters. These marks are determined by chromatin immunoprecipitation and sequencing (ChIP-seq) method (Raha et al. 2010). It involves immunoprecipitation of formaldehyde cross-linked chromatin by specific histone antibodies followed by sequencing.

NIH Roadmap Epigenomics Project houses epigenome maps (DNA methylation and several histone modifications) of a large number of human tissues to facilitate the role of epigenomics in human diseases (Bernstein et al. 2010).

2.3 *Transcriptomic Data*

The transcriptome is the total mRNA content of a cell or an organism that reflects the genes expressed at any given moment. Unlike the genome of an organism, the transcriptome actively changes depending upon several cellular needs. There are nearly 55,406 full-length protein coding RNAs that get transcribed from merely 19,836 genes (Harrow et al. 2012). A gene may produce more than one variant of mRNA due to alternative splicing or RNA editing mechanisms. A comparison of transcriptomes of different cell types can help to understand how a cell functions normally and what changes in gene activity are introduced under diseased conditions. Like DNA microarrays, gene expression microarrays have been indispensable to generate high-throughput expression profiles of thousands of gene at the same time. In contrast to microarray based technology, sequence-based methods directly determine the cDNA (reversely transcribed mRNA) sequence (i.e., SAGE or CAGE). These are tag-based methods that generate sequence library to uniquely identify a transcript and its abundance (3′ short sequence tags used for SAGE; 5′ caps used as tags for CAGE) (Yamamoto et al. 2001; de Hoon and Hayashizaki 2008). Recently, the development of high-throughput next generation sequencing—RNA-seq—has enabled far higher coverage and greater resolution of the dynamicity of the transcriptome. RNA-seq involves deep sequencing of cDNA to capture a detailed and quantitative picture of gene expression and allele-specific expression to clearly differentiate physiological and pathological states (Wang et al. 2009).

Besides, protein-coding RNAs, eukaryotic genome codes for a plethora of gene regulatory non-protein coding RNAs—largely—micro RNAs (miRNAs), short interfering RNAs (siRNAs), piwi-interacting RNAs, and long noncoding RNAs (lncRNAs). These RNAs regulate gene expression at transcriptional and posttranscriptional levels. Among various classes of different small noncoding RNAs, miRNAs are the largest class. There are 1881 human miRNAs that have been reported till date (Harrow et al. 2012). MicroRNAs generally bind to a complementary sequence of a specific target mRNA to induce cleavage and degradation thereby blocking transcription. Next, lncRNAs form the largest class of noncoding RNAs. Nearly 27,908 lncRNAs have been documented till date (Harrow et al. 2012). LncRNAs are key players of genome regulation with immense regulatory potential ranging from transcription catalysis and remodeling to RNA mediated silencing of an entire chromosome (Goff and Rinn 2015). Dysregulation of miRNAs and lncRNAs has been implicated in a wide variety of diseases—diabetes, cardiovascular disease, asthma, Alzheimer's disease, kidney diseases, neurological disorders, and cancers (Mendell and Olson 2012; Kantharidis et al. 2011; Martinez-Nunez et al. 2014; Sun and Wong 2016; Akerman et al. 2017; Huarte 2015; Wapinski and Chang 2011; Chen and Zhou 2017).

Genotype-Tissue Expression Project (GTEx) provides large-scale gene expression data in 53 different human tissues and its relationship to genetic variations (The GTEx Consortium 2013). Besides, ENCODE project has yielded high-throughput microarray and RNA sequencing data for analyzing human gene expression (ENCODE 2017).

2.4 Regulomic Data

A regulome of the cell comprises DNA elements and proteins that regulate protein gene expression. Here, chromatin accessibility is an important aspect that dictates the activation and repression of genes. Transcriptionally active DNA represents open chromatin that is easily accessible to transcription factors, enzymes and regulatory proteins. In contrast, closed chromatin denotes densely packed inaccessible DNA. Open chromatin is directly analyzed by DNase-seq and ATAC-seq. In DNase-seq, chromatin is partially digested with DNase I endonuclease and size selection is used to enrich for highly sensitive chromatin fragments (Song and Crawford 2013). In ATAC-seq, instead of DNase I treatment, an engineered Tn5 transposase cleaves DNA and integrates prime sequences into cleaved genomic DNA (Buenrostro et al. 2015). Positioning of nucleosomes in genome can modify the availability of binding sites for transcriptional machinery, chromatin remodelers and other transcription factors. MNase-seq and FAIRE-seq are two widely used methods to determine nucleosome positioning. Micrococcal nuclease (MNase), an endo-exonuclease, progressively digests DNA until obstructed by a nucleosome (Cui and Zhao 2012). However in FAIRE-seq, chromatin is cross-linked by formaldehyde and the resulting sheared DNA is isolated by phenol–chloroform method (Giresi et al. 2007).

In addition, transcription of a gene is delimited by recruitment of specific transcription factor proteins at *cis*-regulatory elements—promoters and enhancers. Bound transcription factors (TFs) engage certain co-regulators that alter histone modifications. Transcription factor sites are globally determined by ChIP-seq against specific transcription factor.

Furthermore, interactions of transcription factors with specific DNA sequences within enhancer regions activate gene enhancers. Transcription factor binding site and enhancer region may lie physically distant in linear genome but spatially proximate in 3D cellular nucleus for interaction. Chromatin interactions in the genome are commonly mapped by ChIA-PET and 5C method. Chromatin interaction analysis with paired-end tag sequencing (ChIA-PET) generates RNA polymerase II and CTCF binding sites (Li et al. 2014). Correspondingly, chromosome conformation capture carbon copy (5C) maps 3D organization of chromatin domain through digestion using various restriction enzymes (Dostie et al. 2006).

Human RegulomeDB is a comprehensive browser that provides annotated and integrated information of the experimentally defined functional and biochemical regulatory elements of the human genome (Boyle et al. 2012). Data regarding chromatin accessibility, certain histone modifications, binding sites of various TFs, ChIA-PET, 5C data for different human cell-lines is publicly available as a part of ENCODE data (2017). Also, NIH Roadmap Epigenomics Project offers *DNase I* hypersensitivity data for a large number of human tissues (Bernstein et al. 2010). Transfac Matrix and Factor databases also provide free access to genomic binding sites of various transcription factors (Wingender et al. 1996).

2.5 Proteomic Data

The proteome comprises a set of all expressed proteins in a cell, tissue or organism. It is a dynamic flow of information from genes to protein pathways and networks within the cell and the organism. Proteomic investigation undoubtedly imparts better understanding of molecular basis of variability to disease susceptibility. However, it is complicated by proteome domain size (>100,000 proteins) and its dynamic concentration range varying from pg/mL (cytokines) to mg/mL (albumin). This problem results in a failure to detect less abundant proteins and inability to analyze protein complexes. A wide range of proteomic approaches exist starting from gel based 2D polyacrylamide gel electrophoresis (2D-PAGE) to modern high-throughput mass spectrometry and protein chips.

2D-PAGE is a primary tool in proteomics research that separates a complex mixture of thousands of proteins using two different properties (Issaq and Veenstra 2008). Proteins are separated due to difference in isoelectric point (pI) in first dimension and by relative molecular weight in second dimension. Taking into account, limited resolution of gel electrophoresis and inability to identify expressed protein, 2D-PAGE is coupled with mass spectrometry. Mass Spectrometry (MS) is high-throughput analytical technique that measures mass-to-charge (*m*/*z*) ratio of charged particles and allows proteins to be analyzed rapidly, accurately, high reproducibility and sensitivity at a relatively low cost (Baker 2010). Further advancements in MS technology have made it possible to directly identify a protein. Depending upon the pattern of peptide fragmentation and separation, various MS techniques have been developed (MS/MS—Tandem MS, LC/MS—Liquid Chromatography MS, GC/MS—Gas Chromatography MS) (Lee et al. 2012). Recently, a novel MS-based approach—iTRAQ—has allowed flexibility to multiplex eight different biological samples simultaneously in a single experiment. Isobaric Tag for Relative and Absolute Quantification (iTRAQ) labels primary amino groups in intact proteins and enables identification of differentially labeled proteins and accordingly their proteolytic peptides as single peaks in MS spectra while retaining important posttranslational modifications (Tweedie-Cullen and Livingstone-Zatchej 2008).

Besides identification and quantification of proteins, elucidation of underlying biological process in a disease requires the study of plausible molecular interactions of the dysregulated proteins. The most common method for detecting protein–protein interaction is Yeast Two Hybrid (Y2H) method (Serebriiskii 2010). Here, a protein is fused to a DNA binding domain and tested for interaction against a panel of coding sequence constructs fused to a transcription activated domain in living yeast cells. An activation of reporter gene records a positive protein–protein interaction. Y2H has been automated on high-throughput scale to generate large interactome maps in Drosophila, Arabidopsis, and Humans (Formstecher et al. 2005; Obrdlik et al. 2004; Stelzl et al. 2005; Rual et al. 2005). Apart from Y2H, MS coupled with biochemical methods of affinity purification have been a powerful tool to study interactomes at large scale level. In affinity purification MS (AP-MS), a protein is fused to an epitope-tag and is either immunoprecipitated by specific antibody or purified by affinity columns recognizing the tagged epitope (Dunham et al. 2012). In addition, protein microarrays have also contributed tremendously to proteomic research. Protein arrays are not only valuable for analyzing defined set of spotted proteins, but also essential to confirm a binary protein interaction (Sutandy et al. 2014). The arrays have been an asset to screen thousands of small molecule interactions with proteins to generate therapeutic drugs.

The Swiss-Prot section of UniProt knowledgebase houses manually annotated and track able information of human proteins (Bairocha and Apweiler 2000). Likewise, Protein Data Bank (PDB) provides information about 3D shapes of 42,523 proteins and complex assemblies (Berman et al. 2000). In addition, the Human Protein Atlas (HPA) database is an integrated knowledge resource that maps all proteins expressed in various cells, tissues and organs (Uhlén et al. 2005). The data has been distributed into three sections: Tissue Atlas—details distribution of proteins across all major human tissues and organs; Cell Atlas—subcellular localization of proteins in single cells; and Pathology Atlas—impact of proteins for survival of cancer patients. Another ongoing international project—Human Proteome Project (HPP) aims to map and characterize human proteome following systems approach (Legrain et al. 2011). The ProteomeXchange Consortium offers global coordinated submission of mass spectrometric proteomic data to existing proteomic repositories—PRIDE, Peptide Atlas, MassIVE, and jPost (Vizcaíno et al. 2014).

Furthermore, a comprehensive resource—STRING is database of all known and predicted, direct and indirect protein–protein interactions (Szklarczyk et al. 2015). Interactions in STRING are derived from various sources such as high-throughput lab experiments, genomic context predictions, co-expression data, automated data mining and other primary databases—BIND, DIP, HPRD, MINT, and INTACT.

2.6 Metabolomic Data

Metabolomics refers to the study of identification and quantification of small molecule metabolites and their interactions in a biological system (cell, tissue, organ, body fluid, or organism) under a given set of conditions. Unlike other "*omics*" measures, metabolome is the downstream product of gene transcription and therefore, closest to the studied biological phenotype.

A person's metabolome is extremely dynamic in nature and varies drastically with every moment in time, thus making sample profiling problematic and laborious. Also, small alterations in the transcriptome and proteome at a given time substantially amplify the changes in the metabolome. Presence of several different biological molecules, makes a metabolome physically and chemically more complex than other "*omes*."

Metabolomic profiling is often done through mass spectrometry and NMR spectroscopy. Unlike MS, nuclear magnetic resonance spectroscopy uses magnetic property of the atomic nucleus (spin) to determine its physical and chemical nature in presence of electromagnetic radiations (Marion 2013).

The Human Metabolome Database (HMDB) is most comprehensive resource of human small molecule metabolites (Wishart et al. 2007). There are nearly 114,100 metabolite entries including even highly abundant or relatively rare metabolites. In addition to extensive literature mining, the HMDB data is derived from hundreds of MS and NMR metabolomics analyses on urine, blood and cerebrospinal fluid samples. Similarly, Madison Metabolomics Consortium Database also provides MS and NMR data for metabolomics research (Cui et al. 2008). Besides, METLIN is a largest MS/MS metabolite database of various lipids, steroids, small peptides, carbohydrates, drug molecules, etc. (Smith et al. 2005).

2.7 Metagenomic Data

Microbes are the basic part of all life on earth. The conversion of key elements—carbon, nitrogen, and oxygen—into biologically accessible forms is largely directed by microbes. It has been estimated that nearly 90% of human cells are bacterial, fungal or else nonhuman (Turnbaugh et al. 2007). Microbes inhabit various regions of human body—buccal cavity, stomach, intestines, etc. A metagenome comprises a collective genome of microorganisms from an environmental sample that determine diversity and ecology of a particular environment. Metagenome drastically varies between organs, individuals, diseased and healthy states, dietary conditions, etc. Largely the metabolites of these metagenome interact with host metabolites in various levels in given time and space.

A global initiative NIH Human Microbiome Project catalogs human metagenome from different human body sites (Turnbaugh et al. 2007). Similarly, the Human Pan-Microbe Communities (HPMC) database is a manually curated, searchable, metagenomic resource that enables investigation of human gastrointestinal microbiota (Forster et al. 2016).

3 A Practical Example of Systems Biology Application

In recent times Type 2 Diabetes (T2D) is a global epidemic. As of year 2015, 415 million adult people were reported to diabetic worldwide leading to an expenditure of 12% of global health budget on diabetes (IDF 2015). Diabetes is among the foremost leading causes of death in most of the countries. It is characterized by hyperglycemia that results due to body's inability to produce and/or use insulin. A long-standing and uncontrolled diabetes advances to several microvascular (neuropathy, nephropathy, retinopathy) and/or macrovascular complications (coronary artery disease, stroke, myocardial infarction, atherosclerosis, fatty liver disease, etc.), thus contributing significantly to the disease burden.

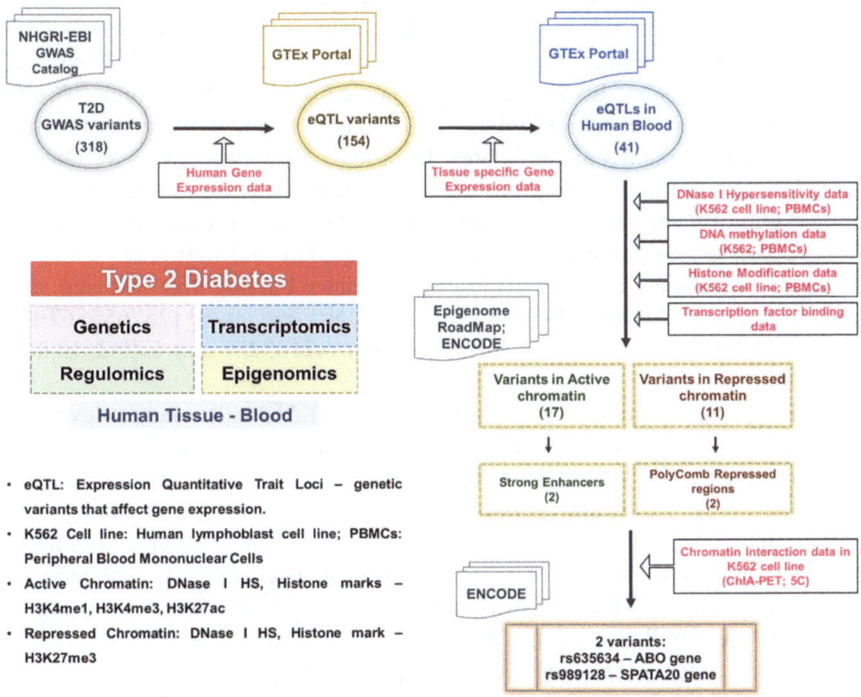

Fig. 3 A systems approach for selecting promising functional targets for type 2 diabetes

Over decades, ample amount of resources have been invested for T2D research, yet we are far from sketching out a comprehensive picture of the disease. Taking into account, the complex interplay of genetic and environmental factors in disease susceptibility, we envisage a systemic integration of publicly available high-throughput data from diverse biological levels to effectively pinpoint plausible candidates that could be easily tested for disease therapeutics.

At present, there are 318 reported T2D GWAS variants that were retrieved from NGHRI-EBI GWAS Catalog (Fig. 3).

We first tested the association of alleles of these genetic variants with gene expression levels (expression quantitative trait loci—eQTL). For this, single-tissue eQTL dataset of GTEx Analysis Release V7 Portal was probed. We grounded 154 T2D GWAS variants that were associated with variable gene expression in different human tissues. Keeping in mind that epigenomic and transcriptomic signatures drastically vary among tissues, we continued our analysis with human blood tissue only. So as a result of tissue specificity, we were left with 41 variants that served as eQTL for various genes in human blood tissue.

To study human blood tissue, publicly available epigenetic gene regulatory data is available for blood lymphocytes (peripheral blood mononuclear cells—PBMCs) and blood borne human cell line—K562 (human chronic myeloid leukemia cells).

Primarily, we checked the open or closed chromatin status by studying *DNase I* hypersensitivity marks at these 41 loci. DNA methylation marks and histone modifications (H_3K_{27}ac, H_3K_4me1, H_3K_4me3, H_3K_{36}me3, and H_3K_{27}me3) were also checked for PBMCs and K562 cells at Epigenome RoadMap Project and ENCODE. Additionally, we searched ENCODE transcriptional factor binding site dataset to identify putative functional candidates. In process we recovered 17 variants that resided in active chromatin regions and 11 variants located in repressed genome regions. Out of these variants, two variants (rs635634 and rs11257655) comprised strong enhancer regions and two variants (rs7163757 and rs989128) constituted Polycomb repressed chromatin. Transcription factor binding data revealed strong binding of GATA, FOXA1, EP300, CEBPC, JUND, etc. at rs11257655; CEBPB, GATA, EP300, FOS, STAT3, p_300, JUN D at rs7163757 and EZH2, EGR1 at rs989128.

Further physical interaction chromatin data for K562 cells was retrieved from ENCODE ChIA-PET and 5C datasets for these four variants. Finally, we obtained two variants—rs635634 and rs989128 which surely interact and regulate their respective eQTL associated genes—ABO and SPATA20 in 3D nucleus, thus biologically justifying the observed eQTL association. Hence, it can be summarized that the variant rs635634 resides within a robust enhancer element characterized by strong H_3K_4me1 and H_3K_{27}ac histone marks upstream of ABO gene to regulate its expression. Similarly, rs989128 is located in a CpG island downstream of SPATA20 gene that is repressed (H_3K_{27}me3) by EZH2 protein, a regulator of DNA methylation. Thus, a systematic integration of diverse biological datasets has aided in prioritizing two functional candidates from 318 associated T2D variants for successful therapeutic intervention.

This model was an example of systemic integration of trajectories from multiple publicly biological high-throughput datasets. Additionally, a holistic overview of T2D also requires multiple biological datasets for each studied personal trajectory. For instance, for a particular individual, a combination of its own genetic, epigenetic, transcriptomic, and metabolomic data would be highly fruitful to interpret disease biology.

In summary, our understanding of complex diseases is currently limited by lack of holistic overview of fundamental physiological processes. Systems biology serves as a roadway for extracting, integrating and interpreting valuable biological information from various large datasets to gain worthy insights into biology of complex diseases. Such a qualitative analysis can synergize with prior knowledge and predict what pathways/processes are disrupted—where and when to yield a specific biochemical phenotype which otherwise cannot be determined if individual datasets are studied. Subsequently, these findings can be used to assess suitability of various therapies to maintain or restore normal biological function.

4 Future Avenues

A total of 95 medicines were withdrawn from the US market (1960–1999) due to serious drug safety concerns (The Academy of Medical Sciences and The Royal Academy of Engineering 2007). Traditional methods of drug discovery are not helping.

By integrating the experimental data from parts-to-pathways, building models, and enabling targeted experiments, systems biology can help in reducing the drug discovery costs, drug repositioning; predicting on-target and off-target effects; shortening drug discovery life cycle; and finding new targets and effective drug combinations.

Despite a large body, the evolution of molecular pathogenesis in complex metabolic diseases like diabetes remains unknown. By building comprehensive and integrative in-silico models from epigenomic, transcriptomic, proteomic, metabolomic, and metagenomic data, an enhanced understanding of the disease etiology, intervention points, and drug–target interactions can be achieved.

References

Ahrens M, Ammerpohl O, von Schönfels W, Kolarova J, Bens S et al (2013) DNA methylation analysis in 6 nonalcoholic fatty liver disease suggests distinct disease-specific and remodeling signatures after bariatric surgery. Cell Metab 18:296–302

Akerman I, Tu Z, Beucher A, Rolando DMY, Sauty-Colace C et al (2017) Human pancreatic β cell lncRNAs control cell-specific regulatory networks. Cell Metab 25:400–411

Altshuler D, Daly MJ, Lander ES (2008) Genetic mapping in human diseases. Science 322(5903):881–888

Bairocha A, Apweiler R (2000) The SWISS-PROT protein sequence database and its supplement TrEMBL in 2000. Nucleic Acids Res 28:45–48

Baker M (2010) Mass spectrometry for biologists. Nat Methods 7:157–161

Bannister AJ, Kouzarides T (2011) Regulation of chromatin by histone modifications. Cell Res 21:381–395

Berman HM, Westbrook J, Feng Z, Gilliland G, Bhat TN et al (2000) The protein data bank. Nucleic Acids Res 28:235–242

Bernstein BE, Stamatoyannopoulos JA, Costello JF, Ren B, Milosavljevic A et al (2010) The NIH roadmap epigenomics mapping consortium. Nat Biotechnol 28:1045–1048

Bertram L, Tanzi RE (2009) Genome-wide association studies in Alzheimer's disease. Hum Mol Genet 18(R2):R137–R145

Boyle AP, Hong EL, Hariharan M, Cheng Y, Schaub MA et al (2012) Annotation of functional variation in personal genomes using RegulomeDB. Genome Res 22:1790–1797

Buenrostro J, Wu B, Chang H, Greenleaf W (2015) ATAC-seq: a method for assaying chromatin accessibility genome-wide. Curr Protoc Mol Biol 109:29.1–29.9

Cauwenberghe CV, Broeckhoven CV, Sleegers K (2016) The genetic landscape of Alzheimer disease: clinical implications and perspectives. Genet Med 18:421–430

Chang CQ, Yesupriya A, Rowell JL, Pimentel CB, Clyne M et al (2014) A systematic review of cancer GWAS and candidate gene meta-analyses reveals limited overlap but similar effect sizes. Eur J Hum Genet 22:402–408

Chen Y, Zhou J (2017) LncRNAs: macromolecules with big roles in neurobiology and neurological diseases. Metab Brain Dis 32:281–291
Cordell HJ, Clayton DJ (2005) Genetic association studies. Lancet 366(9491):1121–1131
Cui K, Zhao K (2012) Genome-wide approaches to determining nucleosome occupancy in metazoans using MNase-Seq. Methods Mol Biol 833:413–419
Cui Q, Lewis IA, Hegeman AD, Anderson ME, Li J et al (2008) Metabolite identification via the Madison metabolomics consortium database. Nat Biotechnol 26:162–164
de Hoon M, Hayashizaki Y (2008) Deep cap analysis gene expression (CAGE): genome-wide identification of promoters, quantification of their expression, and network inference. Biotechniques 44:627–632
Dostie J, Richmond TA, Arnaout RA, Selzer RR, Lee WL et al (2006) Chromosome conformation capture carbon copy (5C): a massively parallel solution for mapping interactions between genomic elements. Genome Res 16:1299–1309
Dunham WH, Mullin M, Gingras AC (2012) Affinity-purification coupled to mass spectrometry: basic principles and strategies. Proteomics 12:1576–1590
ENCODE: Encyclopedia of DNA Elements (2017) Available at https://www.encodeproject.org
Formstecher E, Aresta S, Collura V, Hamburger A, Meil A et al (2005) Protein interaction mapping: a *Drosophila* case study. Genome Res 15:376–384
Forster SC, Browne HP, Kumar N, Hunt M, Denise H et al (2016) HPMCD: the database of human microbial communities from metagenomic datasets and microbial reference genomes. Nucleic Acids Res 44(D1):D604–D609
Fuchsberger C, Flannick J, Teslovich TM, Mahajan A, Agarwala V et al (2016) The genetic architecture of type 2 diabetes. Nature 536:41–47
Giresi PG, Kim J, McDaniell RM, Iyer VR, Lieb JD (2007) FAIRE (Formaldehyde-Assisted Isolation of Regulatory Elements) isolates active regulatory elements from human chromatin. Genome Res 17:877–885
Giri AK, Bharadwaj S, Banerjee P, Chakraborty S, Parekatt V et al (2017) DNA methylation profiling reveals the presence of population-specific signatures correlating with phenotypic characteristics. Mol Genet Genomics 292:655–662
Goff LA, Rinn JL (2015) Linking RNA biology to lncRNAs. Genome Res 25:1456–1465
Harrow J, Frankish A, Gonzalez JM et al (2012) GENCODE: the reference human genome annotation for the ENCODE project. Genome Res 22:1760–1774
Huarte M (2015) The emerging role of lncRNAs in cancer. Nat Med 21:1253–1261
IDF (2015) Diabetes Atlas Edition: 7. Available at https://www.idf.org/e-library/welcome.html
Issaq H, Veenstra T (2008) Two-dimensional polyacrylamide gel electrophoresis (2D-PAGE): advances and perspectives. Biotechniques 44:697–700
Kantharidis P, Wang B, Carew RM, Lan HY (2011) Diabetes complications: the microRNA perspective. Diabetes 60:1832–1837
Lange KM, Moutsianas L, Lee JC, Lamb CA, Luo Y et al (2017) Genome-wide association study implicates immune activation of multiple integrin genes in inflammatory bowel disease. Nat Genet 49:256–261
Lee SY, Park NH, Jeong EK, Wi JW, Kim CJ et al (2012) Comparison of GC/MS and LC/MS methods for the analysis of propofol and its metabolites in urine. J Chromatogr B Analyt Technol Biomed Life Sci 900:1–10
Legrain P, Aebersold R, Archakov A, Bairoch A, Bala K et al (2011) The human proteome project: current state and future direction. Mol Cell Proteomics 10:M111.009993
Li Y, Tollefsbol TO (2011) DNA methylation detection: bisulfite genomic sequencing analysis. Methods Mol Biol 791:11–21
Li G, Cai L, Chang H, Hong P, Zhou Q et al (2014) Chromatin Interaction Analysis with Paired-End Tag (ChIA-PET) sequencing technology and application. BMC Genomics 15(Suppl 12):S11
Lister R, Pelizzola M, Dowen RH, Hawkins RD, Hon G et al (2009) Human DNA methylomes at base resolution show widespread epigenomic differences. Nature 462:315–322
Locke AE, Kahali B, Berndt SI, Justice AE, Pers TH et al (2016) Genetic studies of body mass index yield new insights for obesity biology. Nature 518:197–206

MacArthur J, Bowler E, Cerezo M, Gil L, Hall P et al (2017) The new NHGRI-EBI Catalog of published genome-wide association studies (GWAS Catalog). Nucleic Acids Res 45(D1):D896–D901

Marion D (2013) An introduction to biological NMR spectroscopy. Mol Cell Proteomics 12:3006–3025

Martinez-Nunez RT, Bondanese VP, Louafi F, Francisco-Garcia A, Rupani H et al (2014) A microRNA network dysregulated in asthma controls IL-6 production in bronchial epithelial cells. PLoS One 9:e111659

McCarthy MI, Abecasis GR, Cardon LR, Goldstein DB, Little J et al (2008) Genome-wide association studies for complex traits: consensus, uncertainty and challenges. Nat Rev Genet 9:356–369

McPherson R, Tybjaerg-Hansen A (2016) Genetics of coronary artery disease. Circ Res 118:564–578

Mendell JT, Olson EN (2012) MicroRNAs in stress signaling and human disease. Cell 148:1172–1187

NIH US National Library of Medicine (2017a) Genetics home reference. SNPs. Available at https://ghr.nlm.nih.gov/primer/genomicresearch/snp

NIH US National Library of Medicine (2017b) Genetics home reference. Epigenome. Available at https://ghr.nlm.nih.gov/primer/howgeneswork/epigenome

Nikpay M, Goel A, Won HH, Hall LM, Willenborg C et al (2015) A comprehensive 1000 Genomes–based genome-wide association meta-analysis of coronary artery disease. Nat Genet 47:1121–1130

NINDS Stroke Genetics Network (SiGN), International Stroke Genetics Consortium (ISGC) et al (2016) Loci associated with ischaemic stroke and its subtypes (SiGN): a genome-wide association study. Lancet Neurol 15:174–184

Obrdlik P, El-Bakkoury M, Hamacher T, Cappellaro C, Vilarino C et al (2004) K+ channel interactions detected by a genetic system optimized for systematic studies of membrane protein interactions. Proc Natl Acad Sci 101:12242–12247

Raha D, Hong M, Snyder M (2010) ChIP-Seq: a method for global identification of regulatory elements in the genome. Curr Protoc Mol Biol 21:Unit 21.19.1–21.19.14

Rual JF, Venkatesan K, Hao T, Hirozane-Kishikawa T, Dricot A et al (2005) Towards a proteome-scale map of the human protein-protein interaction network. Nature 437:1173–1178

Schübeler D (2015) Function and information content of DNA methylation. Nature 517:321–326

Serebriiskii I (2010) Yeast two-hybrid system for studying protein-protein interactions–stage 3: screen for interacting proteins. Cold Spring Harb Protoc 5:pdb.prot5431

Smith CA, O'Maille G, Want EJ, Qin C, Trauger SA et al (2005) METLIN: a metabolite mass spectral database. Ther Drug Monit 27:747–751

Song L, Crawford GE (2013) DNase-seq: a high-resolution technique for mapping active gene regulatory elements across the genome from mammalian cells. Cold Spring Harb Protoc 2:pdb.prot5384

Sonne SB, Yadav R, Yin G, Dalgaard MD, Myrmel LS et al (2017) Obesity is associated with depot-specific alterations in adipocyte DNA methylation and gene expression. Adipocyte 6:124–133

Stelzl U, Worm U, Lalowski M, Haenig C, Brembeck FH et al (2005) A human protein-protein interaction network: a resource for annotating the proteome. Cell 122:957–968

Sudmant PH, Rausch T, Gardner EJ, Handsaker RE, Abyzov A et al (2015) An integrated map of structural variation in 2504 human genomes. Nature 526:75–81

Sun X, Wong D (2016) Long non-coding RNA-mediated regulation of glucose homeostasis and diabetes. Am J Cardiovasc Dis 6:17–25

Sutandy FXR, Qian J, Chen CS, Zhu H (2014) Overview of protein microarrays. Curr Protoc Protein Sci 27:Unit 27.1

Szklarczyk D, Franceschini A, Wyder S, Forslund K, Heller D et al (2015) STRING v10: protein–protein interaction networks, integrated over the tree of life. Nucleic Acids Res 43(D1):D447–D452

Tabassum R, Chavali S, Dwivedi OP, Tandon N, Bharadwaj D (2008) Genetic variants of FOXA2: risk of type 2 diabetes and effect on metabolic traits in North Indians. J Hum Genet 53:957–965

Tabassum R, Chauhan G, Dwivedi OP, Mahajan A, Jaiswal A et al (2013) Genome-wide association study for type 2 diabetes in Indians identifies a new susceptibility locus at 2q21. Diabetes 62:977–986

The Academy of Medical Sciences and The Royal Academy of Engineering (2007) Systems Biology: a vision for engineering and medicine. Available at https://acmedsci.ac.uk/file-download/34677-1176712812.pdf

The GTEx Consortium (2013) The genotype-tissue expression (GTEx) project. Nat Genet 45:580–585

The International HapMap Consortium (2003) The international HapMap project. Nature 426:789–796

Torgerson DG, Ampleford EJ, Chiu GY, Gauderman JW, Gignoux CR et al (2011) Meta-analysis of genome-wide association studies of asthma in ethnically diverse North American populations. Nat Genet 43:887–892

Turnbaugh PJ, Ley RE, Hamady M, Fraser-Liggett CM, Knight R, Gordon JI (2007) The human microbiome project. Nature 449:804–810

Tweedie-Cullen RY, Livingstone-Zatchej M (2008) Quantitative analysis of protein expression using iTRAQ and mass spectrometry. Protocol Exchange. https://doi.org/10.1038/nprot.2008.89

Uhlén M, Björling E, Agaton C, Szigyarto CA, Amini B et al (2005) A human protein atlas for normal and cancer tissues based on antibody proteomics. Mol Cell Proteomics 4:1920–1932

Vizcaíno JA, Deutsch EW, Wang R, Csordas A, Reisinger F et al (2014) ProteomeXchange provides globally co-ordinated proteomics data submission and dissemination. Nat Biotechnol 32:223–226

Wahl S, Drong A, Lehne B, Loh M, Scott WR et al (2015) Epigenome-wide association study of body mass index, and the adverse outcomes of adiposity. Nature 541:81–86

Wang Z, Gerstein M, Snyder M (2009) RNA-Seq: a revolutionary tool for transcriptomics. Nat Rev Genet 10:57–63

Wang AZ, Li L, Zhang B, Shen GQ, Wang QK (2011) Association of SNP rs17465637 on chromosome 1q41 and rs599839 on 1p13.3 with myocardial infarction in an American Caucasian population. Ann Hum Genet 75:475–482

Wapinski O, Chang HY (2011) Long noncoding RNAs and human disease. Trends Cell Biol 21:354–361

Wingender E, Dietze P, Karas H, Knüppel R (1996) TRANSFAC: a database on transcription factors and their DNA binding sites. Nucleic Acids Res 24:238–241

Wishart DS, Tzur D, Knox C, Eisner R, Guo AC et al (2007) HMDB: the human metabolome database. Nucleic Acids Res 35(D1):D521–D526

Yamamoto M, Wakatsuki T, Hada A, Ryo A (2001) Use of serial analysis of gene expression (SAGE) technology. J Immunol Methods 250:45–66

Zaina S, Heyn H, Carmona FJ, Varol N, Sayols S et al (2014) DNA methylation map of human atherosclerosis. Cardiovasc Genet 7:692–700

Zhao J, Goldberg J, Bremner JD, Vaccarino V (2012) Global DNA methylation is associated with insulin resistance: a monozygotic twin study. Diabetes 61:542–546

Systems Biology of Bacterial Immune Systems: Regulation of Restriction-Modification and CRISPR-Cas Systems

Andjela Rodic, Bojana Blagojevic, and Marko Djordjevic

Contents

Abstract Restriction-modification (R-M) and CRISPR-Cas are bacterial immune systems which defend their prokaryotic hosts from invasive DNA. Understanding how these systems are regulated is necessary for both biotechnology applications, and for understanding how they modulate horizontal gene transfer (including acquisition of virulence factors). We here review results on modeling these systems which point to common general principles underlying their architecture and dynamical response, with particular emphasis on modeling methods. We show that the modeling predictions are in a good agreement with both in vitro measurements

---

A. Rodic
Institute of Physiology and Biochemistry, Faculty of Biology, University of Belgrade, Belgrade, Serbia

Multidisciplinary PhD program in Biophysics, University of Belgrade, Belgrade, Serbia

M. Djordjevic (✉)
Institute of Physiology and Biochemistry, Faculty of Biology, University of Belgrade, Belgrade, Serbia
e-mail: dmarko@bio.bg.ac.rs

B. Blagojevic
Institute of Physics, University of Belgrade, Belgrade, Serbia

N. Rajewsky et al. (eds.), *Systems Biology*, RNA Technologies,
https://doi.org/10.1007/978-3-319-92967-5_3

of promoter transcription activity and the first in vivo measurements of gene expression dynamics in R-M systems. Modeling induction of CRISPR-Cas systems is challenging, as signaling which leads to their activation is currently unknown. However, based on similarities between transcription regulation in CRISPR-Cas and some R-M systems, we argue that transcription regulation of much simpler (and better studied) R-M systems can be used as a proxy for CRISPR-Cas transcription regulation, allowing to in silico assess CRISPR-Cas dynamical properties. Based on the obtained results, we propose that mechanistically otherwise different bacterial immune systems, presumably due to a common function, share the same unifying principles governing their expression dynamics.

Keywords Thermodynamic modeling · Restriction-modification systems · CRISPR-Cas · Gene expression regulation · Regulatory dynamics

1 Introduction

Two types of prokaryotic "immune systems," known as restriction-modification (R-M) and CRISPR-Cas (*C*lustered, *r*egularly *i*nterspaced *s*hort *p*alindromic *r*epeats-*C*RISPR-*a*ssociated *p*roteins) systems, resemble the mammalian immune system in their ability to actively and with high selectivity combat infectious elements (foreign DNA) (Goldberg and Marraffini 2015). Apart from their immune function, these systems significantly influence evolution and ecology of prokaryotes in a number of ways and have a range of applications in biotechnology (Ershova et al. 2015; Hille and Charpentier 2016).

In type II R-M systems, which are often found on plasmids, separate genes code for two main system components: a restriction enzyme, which cuts specific DNA sequences, and a methyltransferase, which methylates the same sequences and thereby protects them from cutting (Nagornykh et al. 2008). It is widely considered that the main condition for safely and efficiently establishing an R-M system in a naïve host cell, is a delayed beginning of expression of restriction enzyme with respect to methyltransferase. This delay provides enough time for a methyltransferase to protect a host genome, so that restriction enzyme later targets only invasive DNA. Apart from this constraint on their dynamics imposed by their function, we propose other potentially common R-M system dynamical properties, and ask if these can be achieved by a wide variety of R-M systems architectures and regulatory features (Rodic et al. 2017b). These hypotheses are tested by analyzing dynamical properties of different R-M systems, predicted by biophysical models including thermodynamically modeled transcription regulation and dynamically modeled transcript and protein expression.

Unlike R-M systems, which are considered rudimentary for their lack of ability to memorize past infections, CRISPR-Cas are advanced, adaptive prokaryotic immune systems, which store partial DNA sequences of former infectors as spacers flanked by direct repeats in a so-called CRISPR array (Hille and Charpentier 2016). Another constitutive part of a CRISPR-Cas system are genes coding for Cas proteins. In Type

I-E CRISPR-Cas system in *E. coli*, which is a model system for studying CRISPR-Cas regulation, CRISPR array is transcribed as a long pre-crRNA molecule which is further cut by Cas6e protein into small crRNAs, containing separate spacers. These crRNAs guide Cascade complexes constituted of Cas proteins to complementary foreign DNA, which is consequently destroyed. Somewhat surprisingly, while CRISPR-Cas is extensively used for designing various biotechnological tools, its native function and regulation in bacterial cells are not well understood. In particular, CRISPR-Cas is silenced in *E. coli* cells under standard conditions, which hinders observing its expression dynamics (Pul et al. 2010). However, transcription regulation of this system involves general features similar to those found in certain R-M systems, which can be used to predict the main features of CRISPR-Cas expression dynamics (Rodic et al. 2017a).

In this chapter, we aim to explain how a thermodynamic model of a given promoter regulation is formulated, by briefly describing a theoretical basis of thermodynamic modeling and showing how this approach is applied on examples of R-M systems, AhdI and EcoRV. Further, thermodynamic modeling of transcription is used as an input for dynamic modeling, predicting appropriate protein expression in a cell in time, which is discussed on the example of Esp1396I R-M system, for which protein expression dynamics were experimentally measured. We also show how measures for dynamical properties of interest were defined to compare expression dynamics of different R-M systems and to propose unifying principles that characterize their regulatory dynamics. To in silico predict the main qualitative properties of CRISPR-Cas dynamics, and to understand the significance of few characteristic regulatory features found in CRISPR-Cas, we introduce the idea of using a synthetic setup where R-M system transcription regulation with similar features is used as a proxy for not–well understood CRISPR-Cas transcription regulation. Based on the obtained results, we propose that regulatory dynamics of CRISPR-Cas and R-M systems may be governed by similar design principles imposed by their immune function.

2 Thermodynamic Modeling of Transcription Regulation

Thermodynamic modeling approach of gene transcription control is based on principles of statistical mechanics. As an input it takes levels of transcription factors, and patterns and affinities of their binding sites, while as an output it provides predictions of promoter transcription activity (Dresch et al. 2013).

As regulation of transcription initiation, which is a rate-limiting step in gene transcription, involves binding of protein molecules (RNA polymerase, transcription factors) to DNA (promoter region), let us start with a simple scenario in which one molecule of protein, present in some copy number in a cell, binds to one binding site on DNA. From a thermodynamics point of view, the cell interior can be approximated by a system exchanging energy with a much larger heat reservoir (its surroundings) (Phillips et al. 2012). Protein molecules in this system, among which energy is distributed, are approximated by noninteracting particles randomly moving in space confined to the cell volume. These particles can be arranged in a number of different ways, and every unique arrangement of particles corresponds to a particular *microstate* of the system. The probability of finding different microstates is given by the Boltzmann distribution, which we derive below.

2.1 Derivation of the Boltzmann Distribution

Consider a system (s) in contact with a thermal reservoir (r), which together constitute an isolated system with fixed total energy $E = E^{(s)} + E^{(r)}$. According to the second law of thermodynamics, such an isolated system evolves toward such partition of energy between the system and the reservoir, which corresponds to the largest number of microstates of the whole system (Phillips et al. 2012). Therefore, the probability that the system has energy $E_i^{(s)}$ is proportional to the number of the corresponding microstates of the overall system, $\Omega(E,E_i^{(s)}) = \Omega^{(s)}(E_i^{(s)}) \times \Omega^{(r)}(E-E_i^{(s)})$. System degeneracy is directly related to its entropy $S = k_B \ln(\Omega)$, where k_B is the Boltzmann constant, so the probability that the system has energy $E_i^{(s)}$ reads:

$$
\begin{aligned}
P\left(E_i^{(s)}\right) &\propto \exp\left(S^{(s)}\left(E_i^{(s)}\right)/k_B\right)\cdot \exp\left(S^{(r)}\left(E-E_i^{(s)}\right)/k_B\right)\\
&\approx \exp\left(S^{(s)}\left(E_i^{(s)}\right)/k_B\right)\cdot \exp\left(\left(S^{(r)}(E)-\frac{dS^{(r)}}{dE}\cdot E_i^{(s)}\right)/k_B\right)\\
&\propto \exp\left(S^{(s)}\left(E_i^{(s)}\right)/k_B\right)\cdot \exp\left(-E_i^{(s)}/\left(k_B\cdot T\right)\right),
\end{aligned}
\tag{1}
$$

where in the second step, the reservoir entropy is expended about $S^{(r)}(E)$ (note that this approximation is valid when a reservoir is much bigger than a system, so $E_i^{(s)} << E$), while in the third step the thermodynamic definition of temperature $(\partial S/\partial E)_{V,N} = 1/T$ is used. The first term in Eq. (1) gives the number of microstates of a system with energy $E_i^{(s)}$ (i.e., $\Omega^{(s)}(E_i^{(s)})$), while the second term is called the Boltzmann factor, and represents the unnormalized probability of selecting one particular system microstate at energy $E_i^{(s)}$, i.e. it represents a statistical weight of that microstate (Sneppen and Zocchi 2005).

2.2 Statistical Weights from Statistical Mechanics

In the problem of binding of a protein to its binding site considered above, all of the microstates can be grouped in one of the two system *macrostates*: the one in which the DNA binding site is occupied by the protein, or the one in which it is empty, where binding sites in these two states are characterized by the energies $\varepsilon_i^{(bs)}$ (so that *i* corresponds to *bound* or *unbound*). Thereby, the energy of the system ($E_i^{(s)}$) is a sum of the binding site energy and the kinetic energies of all unbound protein molecules. Since the probability of finding different microstates is given by the Boltzmann distribution, the weight associated with the macrostate with energy $E_i^{(s)}$ is proportional to the corresponding number of the system microstates ($\Omega^{(s)}$), multiplied by the Boltzmann factor (the numerator in the equation below):

$$P\left(E_i^{(s)}\right) = \frac{\Omega^{(s)}\left(E_i^{(s)}\right) \cdot \mathrm{e}^{-E_i^{(s)}/(k_B \cdot T)}}{\sum_i \left(\Omega^{(s)}\left(E_i^{(s)}\right) \cdot \mathrm{e}^{-E_i^{(s)}/(k_B \cdot T)}\right)}. \tag{2}$$

In the denominator of Eq. (2) is the so-called *partition function,* which represents a sum of statistical weights of all possible system microstates.

To determine $\Omega^{(s)}$ from Eq. (2), i.e. to count in how many ways protein molecules can be arranged, one needs to know how many states are available to one freely moving protein molecule with kinetic energy $\varepsilon_k = p^2/(2m)$ in a cell. According to the uncertainty principle from quantum mechanics, this question amounts to counting discrete cells of the size h (Planck's constant) in the phase-space containing three dimensions of particle position (r) and three dimensions of its momentum (p) (Stowe 2007; Sneppen and Zocchi 2005).

Therefore, the statistical weight of the system macrostate with binding site energy $\varepsilon_{bound}^{(bs)}$, where the protein binding site is occupied, is obtained by summing through all possible arrangements (permutations) of *N*−*1* indistinguishable protein molecules (because 1 is bound) in a cell phase-space, with that sum weighted by a corresponding Boltzmann factor (Phillips et al. 2012; Sneppen and Zocchi 2005):

$$\begin{aligned} Z_{ON} &= \frac{1}{(N-1)!}\left(\int_V \int \frac{\mathrm{d}^3 r \cdot \mathrm{d}^3 p}{h^3} \mathrm{e}^{-p^2/(2mk_B T)}\right)^{N-1} \mathrm{e}^{-\varepsilon_{bound}^{(bs)}/(k_B T)} \\ &\propto k^{N-1} \rho^{-(N-1)} \mathrm{e}^{-\varepsilon_{bound}^{(bs)}/(k_B T)}, \end{aligned} \tag{3}$$

where $k = (2mk_B T\pi/h^2)^{3/2}$ and $\rho = N/V$ (V is cell volume). Equivalently, a statistical weight of a system macrostate in which all protein molecules are free in a cell (with binding site energy $\varepsilon_{unbound}^{(bs)}$) reads:

$$\begin{aligned} Z_{OFF} &= \frac{1}{N!}\left(\int_V \int \frac{\mathrm{d}^3 r \cdot \mathrm{d}^3 p}{h^3} \mathrm{e}^{-p^2/(2mk_B T)}\right)^{N} \mathrm{e}^{-\varepsilon_{unbound}^{(bs)}/(k_B T)} \\ &\propto k^{N} \rho^{-N} \mathrm{e}^{-\varepsilon_{unbound}^{(bs)}/(k_B T)}. \end{aligned} \tag{4}$$

Taking into account that the total statistical weight (partition function) of this system is $Z = Z_{ON} + Z_{OFF}$, one can express the ratio of probabilities of finding a binding site in occupied and unoccupied state:

$$\frac{P_{ON}}{P_{OFF}} = \frac{Z_{ON}}{Z} \cdot \left(\frac{Z_{OFF}}{Z}\right)^{-1} = \frac{\rho}{k} \mathrm{e}^{-\Delta\varepsilon/(k_B T)}, \tag{5}$$

where $\Delta\varepsilon = \varepsilon_{bound}^{(bs)} - \varepsilon_{unbound}^{(bs)}$ is the binding energy. Conveniently, statistical weights are expressed in terms of $\Delta\varepsilon$ (i.e., normalized with Z_{OFF}). One should have in mind that binding of a protein to DNA induces significant conformational changes in both molecules, so $\Delta\varepsilon$ in the above equations corresponds to the (Gibbs) free energy of binding (often written as ΔG, which we will adopt below).

2.3 Statistical Weights from Equilibrium Biochemical Reactions

Binding of a protein present in a cell in concentration [P], to a binding site of concentration [BS] is, alternatively, described by the following chemical reaction:

$$[\mathrm{P}] + [\mathrm{BS}] \underset{K_d}{\rightleftarrows} [\mathrm{P} \sim \mathrm{BS}]\,, \tag{6}$$

characterized by the equilibrium dissociation constant $K_d = [\mathrm{P}]\cdot[\mathrm{BS}]/[\mathrm{P} \sim \mathrm{BS}]$. The ratio of probabilities of finding a binding site occupied and unoccupied is then

$$\frac{p_{ON}}{p_{OFF}} = \frac{[\mathrm{P} \sim \mathrm{BS}]}{[\mathrm{BS}]_{\mathrm{tot}}} \cdot \left(\frac{[\mathrm{BS}]}{[\mathrm{BS}]_{\mathrm{tot}}}\right)^{-1} = \frac{[\mathrm{P}]}{K_d}, \tag{7}$$

where $[\mathrm{BS}]_{\mathrm{tot}} = [\mathrm{BS}] + [\mathrm{P} \sim \mathrm{BS}]$ is a total binding site concentration. Equation (7) is equivalent to Eq. (5) obtained using statistical mechanics, where $[\mathrm{P}] = \rho$ and $K_d = k\cdot\exp(\Delta\varepsilon/(k_B T))$ (Sneppen and Zocchi 2005).

If a protein from the analyzed example is RNA polymerase (RNAP) binding to a promoter site, the promoter transcription activity can be approximated through a classical assumption that the transcription activity is proportional to equilibrium binding probability of RNAP to the promoter (Shea and Ackers 1985). Transcription from promoters with more complex regulation, including combinatorial binding of multiple transcription factors which results in more than two promoter configurations, can also be modeled in this way, as in the following example.

2.4 Modeling Transcription Regulation of AhdI R-M System

Thermodynamic modeling approach introduced above was applied in modeling transcription regulation of the R-M system AhdI, which belongs to a large group of R-M systems coding for an additional, control protein (C) which regulates transcription of system genes (Bogdanova et al. 2008). In this system, an operon containing control protein and restriction endonuclease genes (*c* and *res*), and a gene coding for methyltransferase (*met*) are oriented convergently and transcribed from

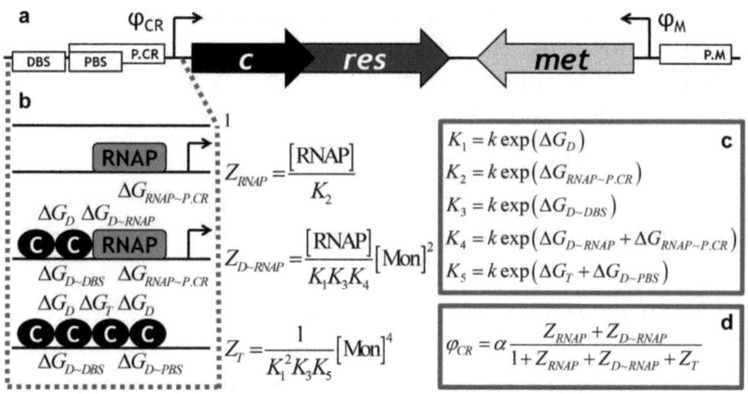

Fig. 1 Thermodynamic modeling of P.CR transcription regulation in AhdI R-M system. (**a**) Gene organization in AhdI system. P.CR, DBS, PBS, and P.M mark relative positions of the P.CR promoter, the distal and the proximal C protein binding site, and the P.M promoter, respectively. (**b**) Allowed P.CR configurations with their statistical weights denoted on the right, expressed in terms of the equilibrium dissociation constants (K) of reactions (8). Protein–DNA (below each configuration) and protein–protein (above interacting proteins) binding free energies (ΔG, in units of k_BT) are related to the appropriate equilibrium dissociation constants following the equations in (**c**). (**d**) P.CR transcription activity (φ_{CR}) is proportional to the fraction of statistical weights corresponding to transcriptionally active configurations (those containing an arrow in **b**)

the promoters denoted as P.CR and P.M, respectively (Fig. 1a). Methyltransferase methylates the P.M promoter, thereby repressing transcription of its own gene. On the other hand, transcription of the operon genes is regulated by binding of C protein dimers to the distal (DBS) and the proximal binding sites (PBS) in the P.CR promoter region.

Prior experiments of in vitro transcription from a wild type P.CR showed that transcription from this promoter is virtually inactive in the absence of C protein, and that it becomes first activated and then repressed with increasing C protein concentration (Bogdanova et al. 2008). This suggests that RNAP is presumably recruited to the promoter through a protein-protein contact with a bound C protein which, therefore, acts as a transcription activator. However, in the electrophoretic mobility shift assay experiments, only free DNA and complexes comprised of C protein tetramers bound to DNA were revealed in the whole range of varying C protein concentrations (Bogdanova et al. 2008; McGeehan et al. 2006). Furthermore, it was shown that DBS has a few orders of magnitude larger binding affinity than PBS, indicating that binding of C dimers to DNA is highly cooperative, i.e., a C dimer bound to DBS immediately recruits a second C dimer to PBS. As a bound C tetramer prevents RNAP from binding to the P.CR and thereby represses transcription of *c* and *res* genes, this raises a question of how transcription from the P.CR is activated. Therefore, quantitative modeling was used to test the proposed mechanism: that RNAP can passively outcompete a second C dimer from binding to PBS, which results in activation of transcription from the P.CR (Bogdanova et al. 2008).

The proposed thermodynamic model of the P.CR transcription regulation takes into account the following chemical reactions, characterized by the appropriate equilibrium dissociation constants (K):

$$\begin{aligned}
&[\text{Mon}] + [\text{Mon}] \underset{K_1}{\rightleftarrows} [\text{D}] \\
&[\text{RNAP}] + [\text{DNA}] \underset{K_2}{\rightleftarrows} [\text{RNAP} \sim \text{DNA}] \\
&[\text{D}] + [\text{DNA}] \underset{K_3}{\rightleftarrows} [\text{D} \sim \text{DNA}] \\
&[\text{D} \sim \text{DNA}] + [\text{RNAP}] \underset{K_4}{\rightleftarrows} [\text{D} \sim \text{RNAP} \sim \text{DNA}] \\
&[\text{D} \sim \text{DNA}] + [\text{D}] \underset{K_5}{\rightleftarrows} [\text{T} \sim \text{DNA}] ,
\end{aligned} \tag{8}$$

where [RNAP], [Mon], [D] and [DNA] stand for concentrations of RNA polymerase, C protein monomers and dimers, and DNA containing the P.CR promoter region, while [RNAP ~ DNA], [D ~ DNA], [D ~ RNAP ~ DNA] and [T ~ DNA] denote concentrations of established complexes of, respectively, RNAP bound to the P.CR, a C dimer bound to DBS, RNAP recruited to the promoter by a bound C dimer, and a bound C tetramer. This system of reactions describes establishing of the allowed P.CR equilibrium configurations characterized by the following statistical weights (Fig. 1b):

- 1—empty promoter;
- Z_{RNAP} = [RNAP ~ DNA]/[DNA]—only RNAP bound to the promoter, which corresponds to basal transcription of the operon genes;
- $Z_{D\text{-}RNAP}$ = [D ~ RNAP ~ DNA]/[DNA]—RNAP recruited to the promoter by a C dimer bound to DBS, resulting in transcription activation;
- Z_T = [T ~ DNA]/[DNA]—a second C dimer recruited to PBS by a C dimer bound to DBS, with obtained C tetramer repressing transcription.

Note that the configuration representing only a C dimer bound to PBS was not taken into account, as such a configuration was not observed in the experiments and has a very low probability due to a large cooperativity in C dimers binding. One should also note that this modeling approach involves the rapid equilibrium assumption applied to the binding reactions, which is justified by the fact that association and dissociation processes between a protein and a DNA molecule, or two protein molecules, are much faster compared to transcription, translation and protein/RNA degradation processes (Phillips et al. 2012). Consequently, the model considers only the frequency of different promoter configurations in equilibrium and cannot distinguish between different sequences of binding events leading to a given configuration—e.g., whether protein A binds to DNA first and prevents binding of protein B, or it displaces protein B when it is already bound to DNA.

The measured value of C protein dimerization constant (K_1) is by an order of magnitude larger than the range of C protein concentrations used in experiments, indicating that C protein is present in a cell in the form of monomers. Therefore, statistical weights of the corresponding configurations are expressed in terms of C monomer and RNAP concentrations and, either appropriate equilibrium dissociation constants (Fig. 1b), or binding free energies (Fig. 1c). According to the assumption introduced above, transcription activity of the P.CR is proportional to the fraction of statistical weights that correspond to bound RNAP (Fig. 1d). Absorbing all constants into few parameters (*x*, *y*, and *z*), P.CR transcription activity is obtained as a function of C protein monomer concentration:

$$\varphi_{CR}\,(\text{Mon}) = \alpha \frac{x + y[\text{Mon}]^2}{1 + x + y[\text{Mon}]^2 + z[\text{Mon}]^4}, \tag{9}$$

where α is a proportionality constant with units transcript amount over time. Equation (9) was fitted to the experimentally measured data, obtained for a wild type system (Fig. 2a), but also for systems in which mutations were introduced in the DNA sequences of DBS or/and PBS (Fig. 2b–d), which corresponds to changing $\Delta G_{D\text{-}DBS}$ or/and $\Delta G_{D\text{-}PBS}$ (see Fig. 1b) (Bogdanova et al. 2008). Fig. 2 shows that the proposed model, with only three free parameters (*x*, *y*, and *z*; α was given the value 1), is in very good agreement with the data for both the wild type and the mutated systems. Furthermore, when fitted to the mutants data, parameter values change as expected with respect to the wild-type case—e.g., decreasing the affinity of DBS strongly negatively affects parameters *y* and *z*, while it has no effect on parameter *x* (compare the Eq. (9) with statistical weights in Fig. 1b and c). All of the above indicates that the modeling can realistically explain in vitro measured transcription activities and, accordingly, that the proposed model appropriately describes the P.CR transcription regulation in AhdI system.

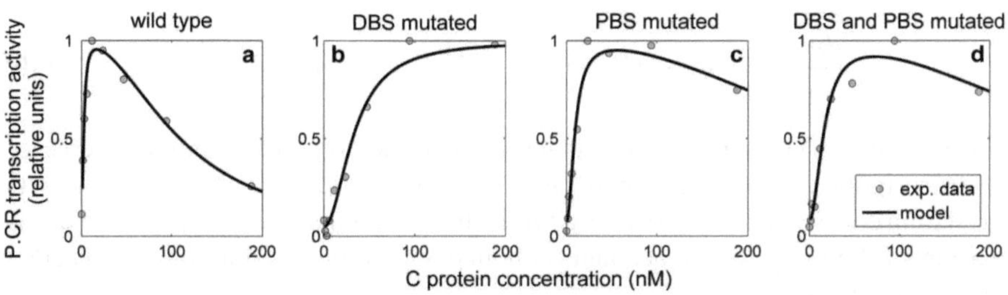

Fig. 2 Fitting experimentally measured dependence of P.CR transcription activity on C protein concentration in wild type and experimentally mutated systems, with a thermodynamic model of this promoter transcription regulation. Transcription activity was measured in arbitrary units and the values (grey circles) were normalized. Solid curves represent the fitted model Eq. (9). (**a**) Wild type system, (**b**) DBS affinity decreased, (**c**) PBS affinity decreased, (**d**) Decreased affinity of both DBS and PBS (Bogdanova et al. 2008)

3 Dynamic Modeling of Protein Expression

Dynamic modeling is the most common approach to model molecular networks and can be used to predict how protein amounts of interest—e.g. those of restriction enzyme and methyltransferase—change with time. State variables of the model represent concentrations (or numbers of molecules) of all mRNA and protein species in the system. These quantities dynamically depend on the combination of all processes that increase or decrease the corresponding amounts, characterized by appropriately defined rates (Le Novère 2015).

Experimentally observing dynamics of protein expression in a cell is, however, challenging due to a prerequirement for a synchronized cell population. Consequently, such measurements have been conducted on R-M systems in only two cases: for PvuII system, by introducing the system in a cell on a phage vector (Mruk and Blumenthal 2008), and for Esp1396I system, by monitoring fluorescently labeled R-M system proteins at the level of single cells (Morozova et al. 2016). In the latter case, experimental measurements were compared with predictions of a biophysical model of Esp1396I R-M system expression during its establishment in a newly transformed host (Morozova et al. 2016).

Similarly to AhdI system, Esp1396I system contains *c* and *res* genes in an operon, expressed from a promoter controlled by cooperative binding of two C dimers (see Fig. 1a and b). In contrast to an autoregulated *m* gene in AhdI system, in Esp1396I system, P.M is under control of C protein, where binding of one C dimer to its single binding site in this promoter region represses transcription of *m* gene (Bogdanova et al. 2009). P.CR and P.M regulation was thermodynamically modeled as explained above, to obtain relations for their transcription activities (φ_i) as functions of C protein concentration, which were further used as an input for a dynamic model describing how appropriate transcript (m_i) and protein (p_i) amounts change with time, for all three system components (i = *C, Res, Met* denoting C protein, restriction enzyme, and methyltransferase, respectively):

$$\frac{dm_i(t)}{dt} = \varphi_i - \lambda_i^m \cdot m_i, \qquad \frac{dp_i(t)}{dt} = \kappa_i \cdot m_i - \lambda_i^p \cdot p_i \tag{10}$$

Equation (10) takes into account that transcript and protein amounts are increased by transcription of the corresponding genes and translation of their transcripts (with translation constants κ_i), respectively, while these amounts are decreased with decay constants λ_i^m and λ_i^p, which account for both degradation and dilution of molecules due to cell division.

The proposed model of Esp1396I expression is minimal, in a sense that it takes into account only the experimentally established regulatory mechanisms, and that all model parameters are considered time-independent. Estimating the parameters by fitting this model to the data (Fig. 3a and b), is a difficult task due to the relatively large parameter space. This task is simplified by the fact that the parameters related to restriction enzyme expression can be estimated separately from those

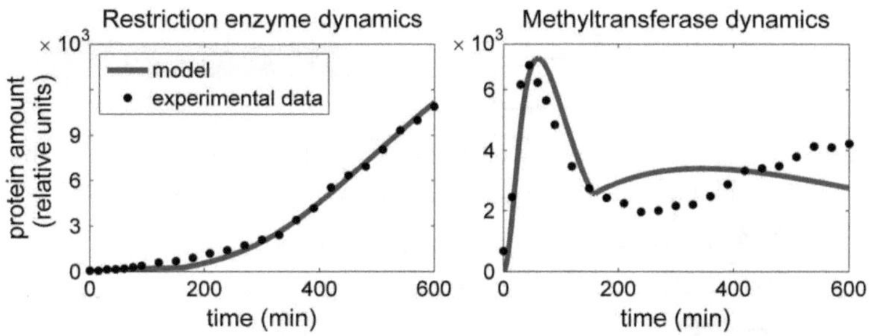

Fig. 3 Fitting experimentally measured data of single cell Esp1396I R-M system expression dynamics with a biophysical model. The zero time point corresponds to the plasmid entry in a naïve cell. (**a**) Restriction enzyme expression dynamics, (**b**) Methyltransferase expression dynamics (Morozova et al. 2016)

describing methytransferase expression, as methyltransferase does not control *c* and *res* expression. The observed good agreement of the model with the data is also supported by a subsequent experimental confirmation of very large restriction enzyme stability, which is consistent with inferred parameter values. Moreover, this minimal model can explain the main qualitative features of expression dynamics observed for Esp1396I system and proposed for R-M systems in general (Fig. 3a and b): a delayed beginning of restriction enzyme synthesis and high expression of methyltransferase early upon transforming a naïve cell. Improved quantitative agreement of the model with the data can likely be achieved by involving the dependence of at least some parameter values with time, imposed by changing conditions in a cell population or a desynchronization of cell and plasmid division. Specifically, during the first ~160 min cells in the culture divided with different (faster) rate compared to the rest of the experiment (Morozova et al. 2016), which is taken into account through decay parameters in the model, as previously explained. Therefore, it is plausible to assume that population dynamics also has significant effect on some other parameters of the model, which may be a subject of future modeling.

4 Modeling Expression of EcoRV R-M System

In contrast to AhdI and Esp1396I systems presented above, in EcoRV R-M system P.CR and P.M are oriented divergently and partially overlap causing mutually exclusive binding of RNAP to these promoters (Fig. 4a), which represents the most distinctive regulatory feature of EcoRV system (Semenova et al. 2005). Consequently, P.CR and P.M control is strongly coupled, making transcription regulation of this system more complex compared to AhdI system. Furthermore,

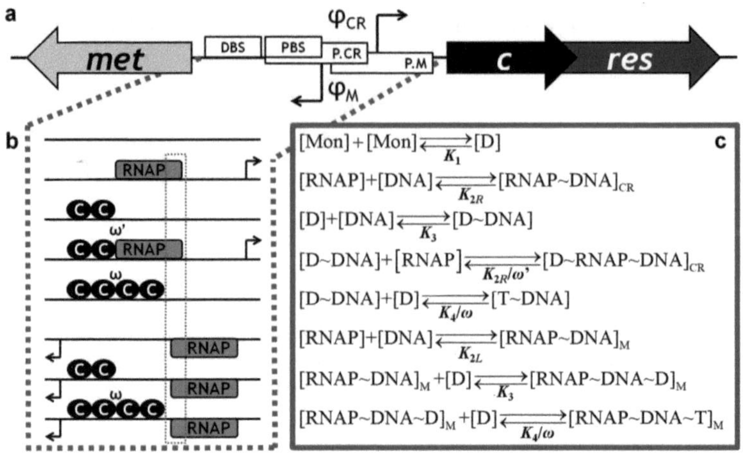

Fig. 4 Thermodynamic modeling of EcoRV R-M system transcription regulation. (**a**) Scheme of gene organization in EcoRV. Relative positions of operon and *met* promoters (P.CR and P.M) and distal and proximal C dimer binding sites (DBS and PBS) are denoted. (**b**) Allowed configurations of a DNA fragment separating *met* and *c* genes, with those transcriptionally active denoted with an arrow. Overlapping of P.CR and P.M is emphasized by framing their common fragment. (**c**) Chemical reactions in the model, with their equilibrium dissociation constants (K). Unlike in modeling AhdI transcription regulation (Fig. 1), cooperativities in binding of a second C dimer to PBS ($\omega \equiv \exp(-\Delta G_T)$) and of RNAP to P.CR ($\omega' \equiv \exp\left(-\Delta G^{CR}_{D\sim RNAP}\right)$) are here introduced as separate parameters, to enable perturbation of ω alone (see below in the text)

all characteristic AhdI regulatory features are absent from EcoRV. Namely: (1) no cooperativity in C dimers binding to DBS and PBS was experimentally found for EcoRV system, (2) *c* transcript is not leaderless in EcoRV, contrary to AhdI system whose leaderless *c* transcript is translated less efficiently than *res* and *met* transcripts, and (3) the equilibrium dissociation constant for a reaction of C protein dimerization is significantly lower than in AhdI system, leading to mostly C dimers in solution (Semenova et al. 2005).

To thermodynamically model EcoRV transcription regulation, one first needs to determine the allowed configurations of a DNA region separating the two divergent genes (Fig. 4b). Transcription regulation of the P.CR by C protein is similar to that found in AhdI system, except that now an additional configuration, consisting of only one C dimer bound to DBS, has to be included due to the absence of cooperativity. Regarding the P.M regulation, contrary to AhdI where it was C-independent, in EcoRV it is indirectly influenced by C protein, as it dictates when RNAP can bind to P.M due to overlapping promoters. From the equilibrium chemical reactions (Fig. 4c), which describe establishing of the allowed configurations, statistical weights can be determined and further used to obtain the equations for P.CR and P.M transcription activities:

$$\varphi_{CR}\,(\mathrm{Mon}) = \alpha \frac{\left(1 + \omega' \frac{[\mathrm{Mon}]^2}{K_1 K_3}\right)}{u\left(1 + \frac{[\mathrm{Mon}]^2}{K_1 K_3} + \omega \frac{[\mathrm{Mon}]^4}{5K_1^2 K_3^2}\right) + \left(1 + \omega' \frac{[\mathrm{Mon}]^2}{K_1 K_3}\right)}, \tag{11}$$

$$\varphi_{M}\,(\mathrm{Mon}) = \alpha \frac{u\left(1 + \frac{[\mathrm{Mon}]^2}{K_1 K_3} + \omega \frac{[\mathrm{Mon}]^4}{5K_1^2 K_3^2}\right)}{u\left(1 + \frac{[\mathrm{Mon}]^2}{K_1 K_3} + \omega \frac{[\mathrm{Mon}]^4}{5K_1^2 K_3^2}\right) + \left(1 + \omega' \frac{[\mathrm{Mon}]^2}{K_1 K_3}\right)}, \tag{12}$$

relying, again, on the assumption that promoter transcription activity is proportional to its equilibrium occupancy by RNAP. In deriving the above Eqs. (11) and (12), the following information from the experiments was used: a C dimer binds to DBS with approximately five times higher affinity compared to PBS, setting $K_4/K_3 = 5$, and the P.CR is considerably weaker than the P.M ($K_{2R} >> K_{2L}$, $u = K_{2R}/K_{2L}$) (Semenova et al. 2005). The thermodynamic model of EcoRV transcription regulation (Eqs. (11) and (12)) is incorporated in an appropriate dynamic model of transcript and protein expression, of the form given by Eq. (10). Furthermore, to estimate the model parameters, and since EcoRV expression dynamics has not been experimentally measured, it is useful to reduce their number by rescaling the appropriate variables. A detailed explanation of parameter estimation in the case of EcoRV is available in (Rodic et al. 2017b). Overall, this presents to our knowledge the first model of a divergent R-M system, which provides an opportunity to assess the effect of regulatory features found in such a system on its expression dynamics, by in silico introducing AhdI features in EcoRV system (see below).

5 Inferring Effects of R-M Systems Regulatory Features on Their Dynamical Properties

As all R-M systems share the same function, namely, efficiently destroying foreign DNA without harming the host cell, it is reasonable to hypothesize that their expression dynamics, constrained by their function, should exhibit some universal properties, regardless of the underlying regulation. Specifically, the following common dynamical properties of R-M system establishment in a naïve host cell have been proposed (Rodic et al. 2017b): (1) a time delay in expression of restriction enzyme with respect to methyltransferase, which provides time for genome protection, (2) a fast transition of restriction enzyme expression from the OFF to the ON state, to ensure rapid cell protection from incoming foreign DNA, and (3) a stable steady-state of the toxic molecule (restriction enzyme), as

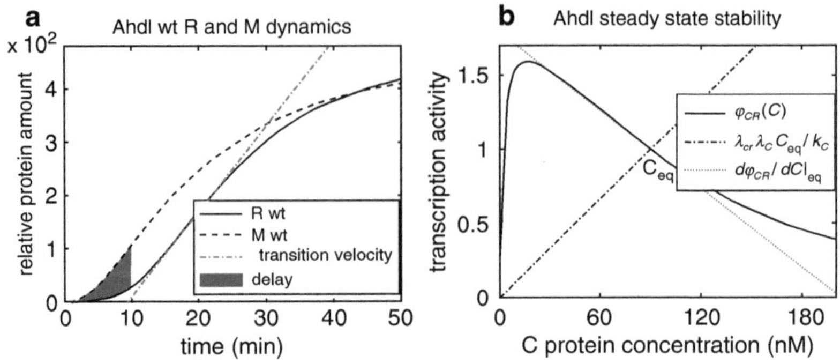

Fig. 5 Dynamical property observables. (**a**) Predicted restriction enzyme (R) and methyltransferase (M) expression dynamics upon system entry in a naïve bacterial host. Measures of R expression time delay and transition velocity are graphically represented, (**b**) Dependence of AhdI P.CR transcription activity on C protein concentration is provided by the full line, whose intersection with the dash-dotted line determines the equilibrium C protein concentration. Slope of the transcription activity curve at this equilibrium concentration (the dotted line) is related with the steady state stability (Rodic et al. 2017b)

fluctuations in restriction enzyme amount not matched by appropriate fluctuations in methyltransferase amount could lead to host cell death.

To quantify these properties, corresponding *dynamical property observables* were defined, which are graphically represented on the example of predicted AhdI wild type dynamics in Fig. 5 (Bogdanova et al. 2008; Rodic et al. 2017b). As a measure of the time delay, the ratio of the shaded areas in a perturbed and in a wild type system, spanning the first 10 min postinduction was introduced (Fig. 5a). A maximal slope of the sigmoidal restriction enzyme expression curve (the dash-dotted line in Fig. 5a) measures the transition velocity from the OFF (low restriction enzyme amount) to the ON (high restriction enzyme amount) state. Finally, the third dynamical property observable (Ω^2) related with the slopes of the dash-dotted and dotted curves shown in Fig. 5b quantifies stability of the restriction enzyme steady-state (Rodic et al. 2017b).

From Fig. 5 it can be readily inferred that AhdI exhibits all the listed dynamical properties. Moreover, perturbing characteristic AhdI regulatory features—i.e., large cooperativity in C dimers binding, high dissociation constant for C dimerization and low translation rate for the leaderless *c* transcript—abolishes these properties, leading to, presumably, less optimal AhdI expression dynamics (Rodic et al. 2017b). Thus, the requirement for the proposed dynamical properties might explain the existence of these characteristic AhdI regulatory features.

Despite missing all regulatory features inherent to AhdI system, and having a unique feature not present in AhdI (overlapping of P.CR and P.M), wild type EcoRV system also meets the same three dynamical properties (see the darkest R and M curves in Fig. 6a), arguing in favor of universality of these properties in different R-M systems. Therefore, the question emerges: why are AhdI regulatory features,

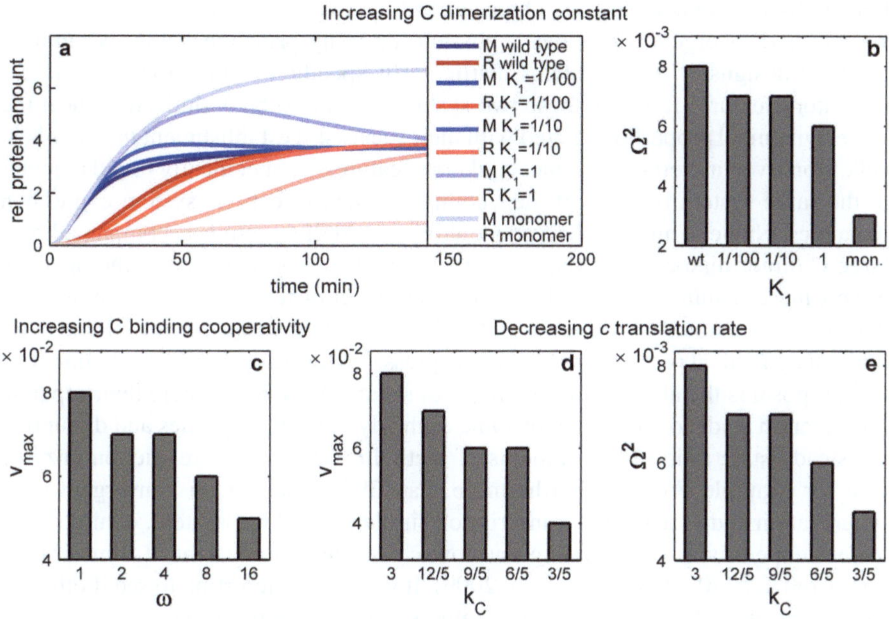

Fig. 6 In silico introducing AhdI-characteristic regulatory features in EcoRV system. Effect of increasing C dimerization constant K_1 on (**a**) dynamics of restriction enzyme (R) and methyltransferase (M) and (**b**) the steady-state stability. Effect of increasing cooperativity ω in C binding on (**c**) R OFF-ON transition velocity. Effect of decreasing the *c* translation constant k_C on (**d**) transition velocity and (**e**) steady-state stability (Rodic et al. 2017b)

apparently successful in optimizing this R-M system immune function, absent from EcoRV? This question can be addressed by in silico introducing characteristic AhdI regulatory features in EcoRV and observing their effect on the system dynamics (Rodic et al. 2017b).

To that end, the equilibrium dissociation constant of C dimerization (K_1 in Fig. 4c) was gradually increased toward an AhdI-characteristic value, which corresponds to a transition from the case where the solution contains mostly C dimers to the case where it contains mostly C monomers (Fig. 6a and b). This perturbation clearly has an adverse effect on two dynamical properties: on the OFF-ON transition velocity (note in Fig. 6a that the slope of R curves decreases as the dimerization constant is increased) and on the steady-state stability (Fig. 6b). Transition velocity of restriction enzyme expression is also decreased by introducing higher C binding cooperativity (increasing ω), as can be seen from Fig. 6c, and by decreasing *c* translation constant k_C (Fig. 6d). Less efficient *c* translation additionally leads to a less stable steady-state (Fig. 6e). Apparently, none of these three perturbations affect an extent of the time delay in restriction enzyme expression.

Overall, the observation that perturbing of the characteristic regulatory features abolishes one or more of the proposed dynamical properties for both AhdI and

EcoRV R-M systems (that are characterized by two different architectures, convergent and divergent), provides a possible unifying principle behind seemingly different designs of these systems. Particularly, specific combinations of different regulatory features found in these two systems appear to be optimized to meet the same dynamical properties, related with their successful establishment in a new host cell. Moreover, it seems that some regulatory features are specifically found together in the same system because of their complementary effects on system expression dynamics. Namely, high binding cooperativity in AhdI system is accompanied by a large C dimer dissociation constant, which may be a consequence of the opposite effects these features have on the steady-state level of restriction enzyme, thereby balancing the amount of this toxic molecule, while at the same time providing more optimal dynamical properties of system expression (Rodic et al. 2017b). In line with this proposal is the absence of both features in EcoRV system, where their separate introduction leads to abolishing of some of the dynamical properties and disrupting the steady-state ratio of the amounts of methyltransferase and restriction enzyme (see for example Fig. 6a). Furthermore, Esp1396I system with convergent gene organization and C regulated transcription similar as in AhdI system, exhibits both lower cooperativity in C binding and lower dissociation constant of dimerization compared to AhdI (Bogdanova et al. 2009). It would be interesting to see if other R-M systems, with different regulatory features, such as control by antisense RNAs or by translational coupling (Nagornykh et al. 2008), are similarly constrained by the proposed dynamical principles, and what are the roles of their regulatory features in achieving these principles.

6 Assessing the Significance of CRISPR-Cas Regulatory Features

Protection of a host bacterium by a CRISPR-Cas system requires its activation upon infection by foreign DNA, or upon setting the right environmental conditions (Ratner et al. 2015). However, expression dynamics of a native CRISPR-Cas have not been observed experimentally because this system (Type I-E) is silent in cells under standard conditions, even in the presence of an infecting phage, and signaling resulting in system activation is unknown (Pul et al. 2010). To date, experimental research on CRISPR-Cas transcription control in *E. coli* and *S. enterica* revealed some elements and regulatory features involved in system activation (Pul et al. 2010; Westra et al. 2010; Medina-Aparicio et al. 2011), specifically: (1) it is known that both CRISPR array and (Cascade) *cas* genes promoters are repressed by highly cooperative binding of global regulators, such as H-NS and LRP, (2) repressors can be outcompeted in binding by some global activators (e.g., LeuO), when present at elevated amounts. Therefore, highly cooperative repression, which is abolished by transcription activators, can be considered as one of the major Type I-E CRISPR-Cas regulatory features (Rodic et al. 2017a).

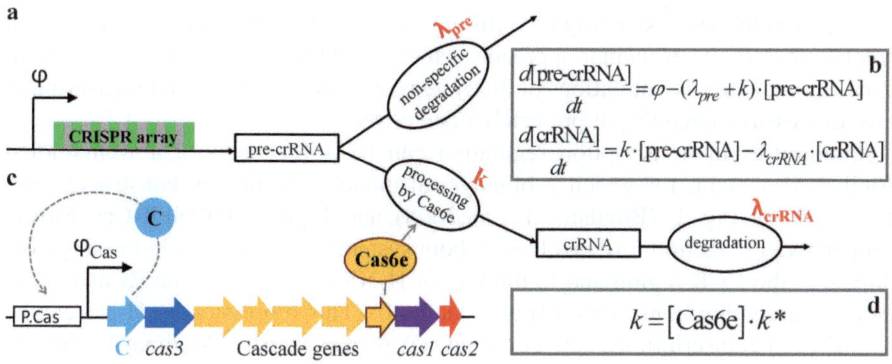

Fig. 7 Scheme of the proposed setup for CRISPR-Cas activation. (**a**) Pre-crRNA processing model scheme. Notation: φ—CRISPR array transcription rate (assumed constant in modeling), λ_{pre}, k and λ_{crRNA}—rates of the processes specified in the ovals (Djordjevic et al. 2012); (**b**) Equations which correspond to the scheme under **a** and which describe time dependence of pre-crRNA and crRNA amounts; (**c**) A schematic diagram of plasmid encoded *c* and *cas* genes under control of the Cas promoter (P.Cas), with a transcription rate φ_{Cas}. AhdI-like regulation of P.Cas by C protein, as denoted by a dashed arrow, is included in the cooperative model; (**d**) Dependence of a processing rate k on Cas6e amount is considered linear, in line with an experimentally determined very low amount of its substrate (pre-crRNA; k*—processing constant) (Rodic et al. 2017a)

Furthermore, another key regulatory feature of CRISPR-Cas expression is the fast nonspecific pre-crRNA degradation by an unidentified endonuclease (Djordjevic et al. 2012). Particularly, it was shown by modeling CRISPR transcript processing upon artificial overexpression of Cas proteins (for a scheme of a model and corresponding dynamic equations see Fig. 7a and b), that the main mechanism responsible for a large increase in crRNA amount from a small decrease in substrate (pre-crRNA) amount is the fast substrate degradation. Processing of pre-crRNA by an elevated amount of Cas proteins diverts the whole molecule from the path of nonspecific degradation, thereby amplifying the equilibrium crRNA amount for a few orders of magnitude (Djordjevic et al. 2012).

However, to more realistically predict CRISPR-Cas expression dynamics and to understand the significance of the established main regulatory features of this system, an appropriate mathematical description of gradual expression of a processing Cas6e protein upon system induction is also needed, as the pre-crRNA processing rate (k) directly depends on the level of Cas6e (Fig. 7d). As a detailed mechanism of Cas promoter (P.Cas) control is unknown, one possible approach involves replacing transcription control (while keeping the CRISPR-Cas pre-crRNA processing regulation) of a native P.Cas with that of a qualitatively and mechanistically similar, but better explored system, and in silico analyzing expression dynamics of the obtained construct (Rodic et al. 2017a). Such an approach would allow assessing the significance of CRISPR-Cas regulatory features, with a minimum of guessing (if a system used as a proxy was already well-studied so that all its major parameters are fixed). At the same time, in silico analysis would provide a much simpler and faster way of fulfilling the task of interest, in comparison to a complementary experimental approach, which would require creating an artificial gene circuit, extensive mutations of the system regulatory features and measuring in vivo expression dynamics of pre-crRNA and crRNA.

Similarities in transcription regulation can be noted between a well-studied AhdI R-M system, for which a biophysical model was already constructed and evaluated (see Fig. 1) (Bogdanova et al. 2008), and Type I-E CRISPR-Cas. Strong cooperative interactions are involved in both systems, in particular, in binding of C dimers to the P.CR region and in binding of H-NS molecules to the P.Cas region, thereby repressing transcription (Pul et al. 2010). This repression by H-NS can be abolished by transcriptional activator LeuO (Westra et al. 2010; Medina-Aparicio et al. 2011). Consequently, autoregulation (both positive and negative), similar to that exhibited by C protein in AhdI system, is likely found in CRISPR-Cas regulation, as LeuO activates, and also indirectly represses its own promoter (Chen et al. 2001; Stratmann et al. 2012). Thus, the main features of CRISPR-Cas transcription regulation, namely, gradual synthesis of Cas proteins, cooperativity in transcription regulation, and putative autoregulation, can be qualitatively mimicked by putting *cas* genes under transcription control found in AhdI. More precisely, the proposed setup for CRISPR-Cas activation analyzed in silico includes *cas* genes put under control of the P.CR promoter from AhdI (see Fig. 1b), which are introduced in a cell on a plasmid, marking the start of CRISPR-Cas activation (Fig. 7c); dynamics of crRNA generation due to the Cas6e processing activity is monitored.

To understand the effect of cooperative transcription regulation of *cas* genes, three (sub)models of *cas* genes regulation in the proposed setup are analyzed: (1) a baseline model, in which P.Cas transcription activity acquires its equilibrium value infinitely fast, (2) a constitutive model, with constant P.Cas transcription activity, and (3) a cooperative model, where P.Cas is cooperatively regulated by C protein in the same manner as the AhdI P.CR promoter. Figure 8 shows how the amount of crRNA, determined 20 min after the start of system activation, depends on the level of the processing rate k reached in equilibrium, in all three models. The crRNA

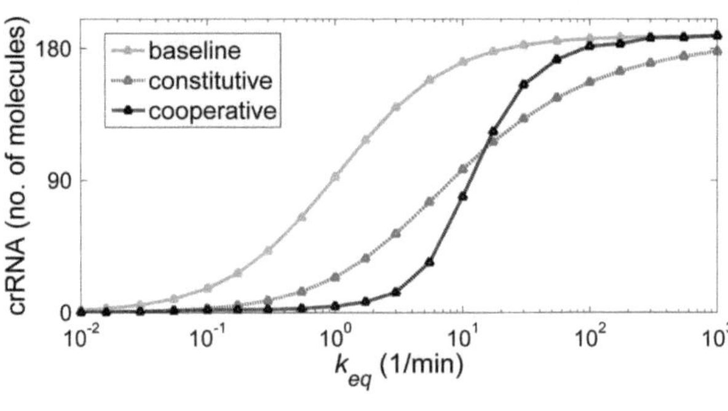

Fig. 8 Dependence of crRNA amount 20 min postinduction on the equilibrium value of a pre-crRNA processing rate k_{eq} in the three models of CRISPR-Cas activation (Rodic et al. 2017a)

amount at 20 min postinduction was specifically chosen, as this period is most relevant for a successful cell defense against an incoming virus.

The most prominent characteristic of the cooperative model is its switch-like behavior (Fig. 8), in contrast to much more gradual responses of the constitutive and the baseline models. This feature enables keeping the system in the OFF state in the case of possible leaks in P.Cas activity (corresponding to low k_{eq} values) and, on the other side, rapidly generating a sufficient amount of crRNAs once the induction signal is received, to efficiently combat viral infection. Furthermore, taking into account that the amount of crRNAs sufficient to negatively affect phage development was determined to be ~10 molecules (Pougach et al. 2010), the models predict that enough crRNAs is generated even at low (somewhat larger than 1 1/min) k_{eq} values, corresponding to the activated system regime. Therefore, CRISPR-Cas system expression regulation is probably mainly constrained by a requirement of rapidly producing a large amount of crRNAs (Rodic et al. 2017a).

In line with its switch-like behavior, at low k_{eq} values the cooperative model produces less crRNAs than the constitutive one. However, at high k_{eq} values an interesting cross-over behavior is observed, where the cooperative model approaches the limit of infinitely fast crRNA production (given by the baseline model). Even more crRNAs can be generated by jointly activating transcription of both *cas* genes and a CRISPR array, which relieves the saturation obtained when increasing only *cas* expression (Djordjevic et al. 2012; Rodic et al. 2017a). As k values around 100 1/min were encountered in experiments with artificial overexpression of *cas* genes (Pougach et al. 2010; Djordjevic et al. 2012), it is tempting to speculate that such high rates of pre-crRNA processing may also be reached in the native system, providing highly efficient protection to a bacterium.

Effect of abolishing the second major CRISPR-Cas regulatory feature, i.e., of decreasing the pre-crRNA nonspecific degradation rate (λ_{pre}) in the cooperative model, can be deduced from Fig. 9. Namely, the crRNA dynamics curve is gradually deformed with respect to the standard Hill (sigmoidal) shape, indicating that fast

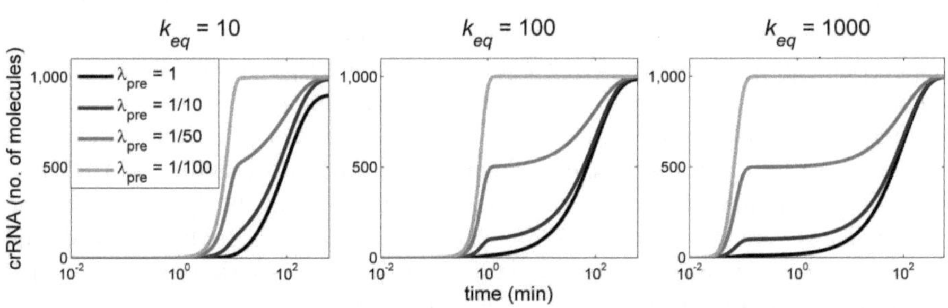

Fig. 9 Perturbing the fast nonspecific degradation of pre-crRNA. Effect of decreasing the pre-crRNA degradation rate λ_{pre} on crRNA expression dynamics at three different k_{eq} values can be seen in the figures under (**a**, **b**, and **c**). Native CRISPR-Cas is characterized by $\lambda_{pre} = 1$ 1/min. Zero time point corresponds to the start of system activation, i.e., to the moment of entrance of a plasmid carrying *cas* genes in a cell (Rodic et al. 2017a)

nonspecific pre-crRNA degradation is, together with cooperative transcription regulation, responsible for the system switch-like behavior. Another effect of decreasing λ_{pre}, which comes as a model prediction, is a decrease in the time delay of the onset of crRNA generation, most pronounced at high k_{eq} values. It has been shown previously that slow or delayed CRISPR interference (targeting of foreign DNA by Cascade-crRNA complexes) facilitates the *primed adaptation* (Künne et al. 2016; Musharova et al. 2017), i.e., the acquisition of invasive DNA fragments similar to already possessed spacers, to be incorporated as new spacers in the CRISPR array. This mechanism serves to minimize infection by phages with mutated genome sequences, which would otherwise evade the interference (Sternberg et al. 2016). The required delay could be achieved by a delay in crRNA generation, resulting from fast pre-crRNA degradation, as predicted by the model (Rodic et al. 2017a). Consequently, both main dynamical features, i.e., rapid transition of the system from OFF to ON state and the delay in the expression of the effector molecules (restriction enzyme and crRNAs), are observed in both R-M systems and mechanistically more complicated CRISPR-Cas systems.

7 Summary and Conclusion

Seemingly very different architectures and regulatory properties of AhdI and EcoRV systems can be explained by few common design principles, ensuring that expression dynamics of every R-M system is optimized to serve its purpose—namely, safe and efficient host cell protection from foreign DNA. Other R-M systems, representative of different regulatory mechanisms, should be investigated to test if unifying design principles can be defined at the level of the whole group of these prokaryotic immune systems. Moreover, having the same immune function, CRISPR-Cas systems may also obey similar design principles, as it was theoretically predicted by using a synthetic system to bypass the unknown CRISPR-Cas transcription control, while keeping the same transcript processing mechanism as in native CRISPR-Cas. Thereby, thermodynamic modeling proved to be an appropriate approach in describing R-M system transcription regulation, while in combination with a dynamic model of protein expression, it can be used to describe the main qualitative properties of R-M system dynamics of establishment in a single, naïve host cell. Further experimental and theoretical studies of CRISPR-Cas regulation may allow to more accurately understand its dynamics, in line with what is already done for R-M systems. Overall, the studies on bacterial immune systems summarized here underline a major goal of systems biology which is to discover common design principles in mechanistically otherwise different biological systems.

References

Bogdanova E, Djordjevic M, Papapanagiotou I et al (2008) Transcription regulation of the type II restriction-modification system AhdI. Nucleic Acids Res 36:1429–1442
Bogdanova E, Zakharova M, Streeter S et al (2009) Transcription regulation of restriction-modification system Esp1396I. Nucleic Acids Res 37:3354–3366
Chen C-C, Fang M, Majumder A et al (2001) A 72-base pair AT-rich DNA sequence element functions as a bacterial gene silencer. J Biol Chem 276:9478–9485
Djordjevic M, Djordjevic M, Severinov K (2012) CRISPR transcript processing: a mechanism for generating a large number of small interfering RNAs. Biol Direct 7:24–34
Dresch JM, Richards M, Ay A (2013) A primer on thermodynamic-based models for deciphering transcriptional regulatory logic. BBA-Gene Regul Mech 1829:946–953
Ershova A, Rusinov I, Spirin S et al (2015) Role of restriction-modification systems in prokaryotic evolution and ecology. Biochemistry-Moscow 80:1373–1386
Goldberg GW, Marraffini LA (2015) Resistance and tolerance to foreign elements by prokaryotic immune systems – curating the genome. Nat Rev Immunol 15:717–724
Hille F, Charpentier E (2016) CRISPR-Cas: biology, mechanisms and relevance. Philos T Roy Soc B 371:20150496
Künne T, Kieper SN, Bannenberg JW et al (2016) Cas3-derived target DNA degradation fragments fuel primed CRISPR adaptation. Mol Cell 63:852–864
Le Novère N (2015) Quantitative and logic modelling of molecular and gene networks. Nat Rev Genet 16:146–158
McGeehan J, Papapanagiotou I, Streeter S et al (2006) Cooperative binding of the C.AhdI controller protein to the C/R promoter and its role in endonuclease gene expression. J Mol Biol 358:523–531
Medina-Aparicio L, Rebollar-Flores J, Gallego-Hernández A et al (2011) The CRISPR/Cas immune system is an operon regulated by LeuO, H-NS, and leucine-responsive regulatory protein in Salmonella enterica serovar Typhi. J Bacteriol 193:2396–2407
Morozova N, Sabantsev A, Bogdanova E et al (2016) Temporal dynamics of methyltransferase and restriction endonuclease accumulation in individual cells after introducing a restriction-modification system. Nucleic Acids Res 44:790–800
Mruk I, Blumenthal RM (2008) Real-time kinetics of restriction–modification gene expression after entry into a new host cell. Nucleic Acids Res 36:2581–2593
Musharova O, Klimuk E, Datsenko KA et al (2017) Spacer-length DNA intermediates are associated with Cas1 in cells undergoing primed CRISPR adaptation. Nucleic Acids Res 45:3297–3307
Nagornykh M, Bogdanova E, Protsenko A et al (2008) Regulation of gene expression in a type II restriction-modification system. Russ J Genet 44:523–532
Phillips R, Kondev J, Theriot J et al (2012) Physical biology of the cell. Garland Science, New York
Pougach K, Semenova E, Bogdanova E et al (2010) Transcription, processing and function of CRISPR cassettes in Escherichia coli. Mol Microbiol 77:1367–1379
Pul Ü, Wurm R, Arslan Z et al (2010) Identification and characterization of E. coli CRISPR-cas promoters and their silencing by H-NS. Mol Microbiol 75:1495–1512
Ratner HK, Sampson TR, Weiss DS (2015) I can see CRISPR now, even when phage are gone: a view on alternative CRISPR-Cas functions from the prokaryotic envelope. Curr Opin Infect Dis 28:267–274
Rodic A, Blagojevic B, Djordjevic M et al (2017a) Features of CRISPR-Cas regulation key to highly efficient and temporally-specific crRNA production. Front Microbiol 8:2139
Rodic A, Blagojevic B, Zdobnov E et al (2017b) Understanding key features of bacterial restriction-modification systems through quantitative modeling. BMC Syst Biol 11:377–391
Semenova E, Minakhin L, Bogdanova E et al (2005) Transcription regulation of the EcoRV restriction–modification system. Nucleic Acids Res 33:6942–6951

Shea MA, Ackers GK (1985) The OR control system of bacteriophage lambda: a physical-chemical model for gene regulation. J Mol Biol 181:211–230

Sneppen K, Zocchi G (2005) Physics in molecular biology. Cambridge University Press, Cambridge

Sternberg SH, Richter H, Charpentier E et al (2016) Adaptation in CRISPR-Cas systems. Mol Cell 61:797–808

Stowe K (2007) An introduction to thermodynamics and statistical mechanics. Cambridge University Press, New York

Stratmann T, Pul Ü, Wurm R et al (2012) RcsB-BglJ activates the Escherichia coli leuO gene, encoding an H-NS antagonist and pleiotropic regulator of virulence determinants. Mol Microbiol 83:1109–1123

Westra ER, Pul Ü, Heidrich N et al (2010) H-NS-mediated repression of CRISPR-based immunity in Escherichia coli K12 can be relieved by the transcription activator LeuO. Mol Microbiol 77:1380–1393

Systems Biology of RNA-Binding Proteins in Amyotrophic Lateral Sclerosis

Tara Kashav and Vijay Kumar

Contents

Abstract Amyotrophic lateral sclerosis (ALS) is an adult-onset incurable neurodegenerative disease. Although the precise pathogenesis of ALS remains unknown, mutations in genes encoding RNA-binding proteins (RBPs) have been known as a major culprit. RBPs are involved in almost every aspect of RNA metabolism events from synthesis to degradation. Characteristic features of RBPs in neurodegeneration include misregulation of RNA processing, mislocalization of RBPs to the cytoplasm, and unusual aggregation of RBPs. Modern advancement in technology and computational capabilities suggests an optimistic future for deconvolution of the pathological changes associated with ALS to identify the pathomechanisms of ALS. Importantly, combination of highly multidimensional omic technologies involving proteomics, microarray, and mass spectrometry with computational systems biology approaches provides a systemic methodology to reveal novel mechanisms behind ALS. In this chapter, we begin by summarizing the ALS and involvement of

T. Kashav
Centre for Biological Sciences, Central University of South Bihar, Patna, India

V. Kumar (✉)
Centre for Interdisciplinary Research in Basic Sciences, Jamia Millia Islamia, New Delhi, India
e-mail: vijay9595@st.jmi.ac.in

N. Rajewsky et al. (eds.), *Systems Biology*, RNA Technologies,
https://doi.org/10.1007/978-3-319-92967-5_4

RBPs in ALS. Further, we provide a comprehensive overview of applications of systems biology to study ALS. We imagine that the integration of highly efficient computational tools with multiple omic analyses will help in the discovery of new therapeutic interventions in ALS.

Keywords Amyotrophic lateral sclerosis · RNA-binding proteins · Systems biology · Omics · Therapy

1 Introduction

Amyotrophic lateral sclerosis (ALS) is a serious neurodegenerative disease characterized by a progressive loss of motor neurons in the brain and spinal cord. According to the National ALS Registry, almost 12,000 people in the USA are diagnosed with ALS with a prevalence of 4.0 cases per 100,000 persons (Mehta et al. 2016). Majority of ALS cases (90%) are sporadic (sALS) having no family history of the disease, while the remaining 10% of cases are inherited and classified as familial ALS (fALS) (Shaw 2005; Robberecht and Philips 2013). The aetiology of ALS like other neurodegenerative diseases (NDs) is multifactorial, and the pathophysiology is mediated by various cellular pathways including glutamate excitotoxicity, oxidative stress, endoplasmic reticulum stress, neuroinflammation, mitochondrial dysfunction, axonal deregulation, protein misfolding and aggregation, proteasomal dysfunction, and RNA processing defects (Dunkel et al. 2012; Mancuso and Navarro 2015; Kumar et al. 2016a; Moujalled and White 2016). Targeting these different pathophysiological aberrations remains a challenge in the field of ALS (Pandya et al. 2013; Goyal and Mozaffar 2014; Bucchia et al. 2015; DeLoach et al. 2015; Nicholson et al. 2015).

Frontotemporal dementia (FTD) is characterized by the focal neuronal loss in the frontal and anterior temporal lobes of the brain. It is the second most common cause of dementia after Alzheimer's disease (AD). The prevalence of FTD in old age people is 10–20 per 100,000 peoples, and the rate of incidence is 3.5–4.2 per 100,000 peoples per year (Ratnavalli et al. 2002; Sieben et al. 2012; Perry and Miller 2013). FTD and ALS have overlapping clinical symptoms and molecular mechanism of pathological manifestation and are now being considered as representatives of a continuum of a broad neurodegenerative disorder (Lomen-Hoerth et al. 2002; Burrell et al. 2011; Kumar et al. 2016b). It has been estimated that approximately 15% of FTD patients meet ALS criteria and as much as 15% of ALS cases also display typical FTD symptoms like cognitive and behavioural impairment (Ringholz et al. 2005; Wheaton et al. 2007).

In this chapter, we will begin by reviewing the involvement of RNA-binding proteins (RBPs) in ALS. Finally, we will review some of the systems biology studies from human tissues, mouse, and different animal and cell culture models in ALS.

1.1 RNA-Binding Proteins in ALS

An increasing number of RBPs are linked to several NDs (Polymenidou et al. 2012; Belzil et al. 2013; Nussbacher et al. 2015; Kapeli et al. 2017; Coyne et al. 2017), which suggest the role of RBPs in preserving normal physiology of the nervous system. Mutations in genes encoding RBPs have been identified in patients with ALS, FTD, spinal muscular atrophy (SMA), and multisystem proteinopathy (MSP). ALS-linked RBPs have been successfully isolated from human cells or tissues as well as from model organisms like mouse, yeast, and drosophila that led to significant understanding of normal and pathological functions in ALS.

More than hundred genes have been linked with ALS in which a handful are RBPs that control RNA processing events (Wroe et al. 2008). Some examples are TAR DNA-binding protein 43 (TDP-43), fused in sarcoma/translocated in liposarcoma (FUS/TLS, referred to as FUS), chromosome 9 open reading frame 72 (C9orf72), heterogeneous nuclear ribonucleoprotein A1 (hnRNP A1), heterogeneous nuclear ribonucleoprotein A2/B1 (hnRNP A2/B1), Ewing's sarcoma breakpoint region 1 (EWSR1), and TATA-box binding protein associated factor 15 (TAF15) (Table 1).

These RBPs share many common structural features such as the presence of RNA recognition motifs (RRMs) and a disordered, Gly-rich prion-like domain (Fig. 1) (He and Smith 2009). They are mainly nuclear and multifunctional proteins that are present in the majority of cells and tissue types. The findings that complete loss of TDP-43 or hnRNP A1 in mice is embryonic lethal (Kraemer et al. 2010; Liu et al. 2017) and complete loss of FUS or EWSR1 in mice is postnatal lethal (Hicks et al. 2000; Li et al. 2007) demonstrated the significance of these proteins.

Moreover, TDP-43, FUS, and C9orf72 show association with stress granules under cellular stress condition (Bentmann et al. 2013; Daigle et al. 2013; Lee et al. 2016; Maharjan et al. 2017). These stress granules sequester proteins and RNAs leading to inhibition of translation of specific transcripts (Anderson and Kedersha 2009; Buchan 2014; Protter and Parker 2016). As a consequence, defects in RNA metabolism arise due to entrapment of proteins and mRNAs (Ramaswami et al. 2013), and subsequently protein aggregation occurs as well (Dewey et al. 2012; Aulas and Vande Velde 2015). As ALS develops, different cellular events are disrupted progressively leading to synaptic failure and muscle degeneration (Robberecht and Philips 2013). Thus, a clear understanding of the primary defects in RNA metabolism that triggers ALS pathogenesis is essential for our understanding of disease manifestation, progression, and treatment.

Table 1 RNA-binding proteins (RBPs) implicated in neurodegenerative diseases affecting RNA metabolism at various steps

Affected RNA metabolism events	RBPs	Associated diseases
Transcription	FUS, EWSR1, TAF15	ALS/FTD
Alternative splicing	TDP-43, FUS, TAF15, hnRNPA1, hnRNPA2/B1, EWSR1	ALS/FTD
	MBNL1/2, CUGBP	DM
	NOVA	POMA
	SMN	SMA
	RBFOX	Epilepsy, ataxia
Alternative polyadenylation	NOVA	POMA
	PABPN1	OPMD
	MBNL1/2	DM
Localization, transportation, and sequestration	TDP-43, FUS, hnRNPA2/B1, TAF15, EWSR1	ALS/FTD
	ATXN2	SCA2, ALS
	FMRP	FXS, FXTAS
	SMN	SMA
Degradation and turnover	hnRNPA1,hnRNPA2/B1	ALS/FTD
	MATR3	ALS
	FMRP, DGCR8, DROSHA	FXS, FXTAS
	MBNL1/2	DM
	PARK7	PD
	CELF4, HuR, ELAVL1	Epilepsy, PD

Abbreviations: DM, myotonic dystrophy; FXS, fragile X syndrome; FXTAS, fragile X-associated tremor/ataxia syndrome; OPMD, oculopharyngeal muscular dystrophy; PD, Parkinson's disease; POMA, paraneoplastic opsoclonus-myoclonus ataxia; SCA2, spinocerebellar ataxia type 2; SMA, spinal muscular atrophy

1.2 Application of Systems Biology to Unravel the Complexity of ALS

The field of systems biology provides a large collection of computational tools for investigating the mechanism of any biological processes based on modern, highly multidimensional omics datasets. The systems biology-directed perturbation analysis of in vitro or in vivo models helps to identify new pathomechanisms behind NDs (Diaz-Beltran et al. 2013; Wood et al. 2015). Systems biology can be considered as a research tool that utilizes biological, physical, chemical, mathematical, and computational methods to incorporate and analyse physiological and clinical information from laboratory experiment.

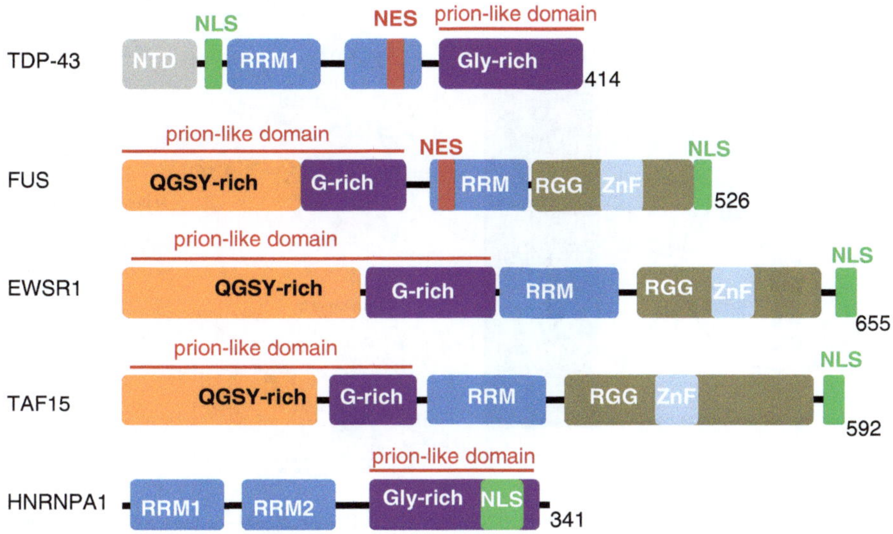

Fig. 1 ALS-associated RNA-binding proteins (RBPs)

To study ALS and other NDs, classical reductionist approaches have primarily focused on crucial genes and their products, and thus did not provide a complete understanding of these complex disorders. Therefore, although there are significant informations about the pathology coming from cellular and molecular studies, a systematic and comprehensive knowledge of pathomechanisms are still missing.

On the contrary, a systems biology perspective involves an integrative study of the key pathways involved in the physiological or pathophysiological state within the cells and organisms. A systems biology approach thus considered as a better and effective strategy to untangle the complex pathomechanisms of these multifactorial diseases.

Overall two different systems biology approaches were undertaken to investigate the events of ALS. The first one considered as descriptive encompasses system-wide analysis of biomolecular variations (mRNA, proteins, lipids, and metabolites) and identification of crucial molecular players in signalling pathways and disease process. The second one, which is more integrated and complex, identifies key molecules or networks which elaborate topological features at different levels. Through the analysis of structural properties and the network connectivity, we may gain important insights on their dynamic behaviour and illustrates the nature of healthy or diseased state (Fig. 2).

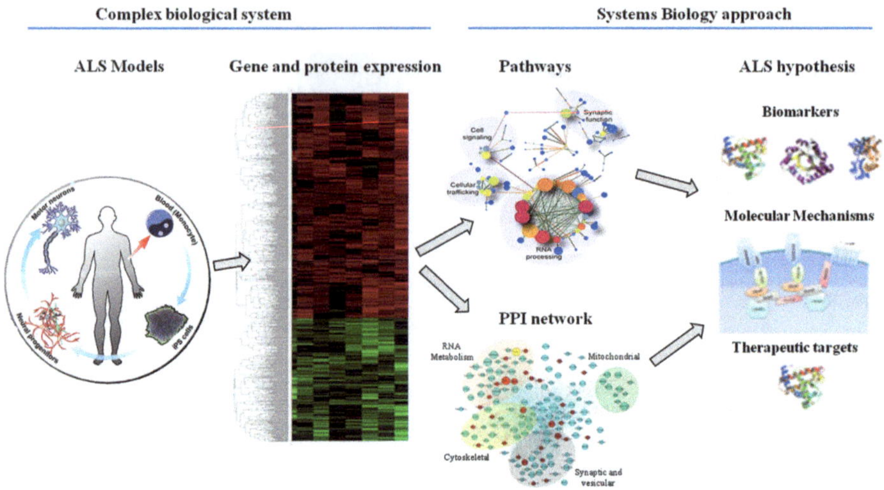

Fig. 2 Schematic representation of a systems biology approach to ALS. The altered pathways and protein–protein interaction (PPI) networks enable the integration of diverse information from high-throughput molecular data sets, extracting the complex molecular ALS response into testable hypotheses

2 Systems Biology of ALS

Structural and functional genomics, transcriptomics, proteomics, and metabolomics are effective tools to discover various mechanisms of ALS and FTD pathogenesis as well as for discovery of effective players in diagnosis and cure (Caballero-Hernandez et al. 2016).

2.1 Genomics of ALS

Genomics techniques have mostly led to the finding of genes contributing to two-thirds of fALS and almost 15% of sALS cases (Appel and Rowland 2012). The findings from genomic studies have demonstrated the complex nature of ALS at the genomic level and are now considered as a polygenic disease rather than a monogenic disease (van Blitterswijk et al. 2012). Mutations in the SOD1 gene were the first genetic abnormality associated with fALS and were identified by linkage mutation analysis in 1993 (Rosen et al. 1993). To date more than 180 gain-of-function mutations in SOD1 gene are described in ALS and opened up new research opportunities. Later, TARDBP gene was discovered by genome sequence analysis (Arai et al. 2006), and subsequently mutations in this gene were reported in both ALS and FTD cases (Kabashi et al. 2008; Sreedharan et al. 2008). Also, mutations in the FUS gene were recognized using the loss of heterozygosity mapping

(Kwiatkowski et al. 2009). Recently, novel technique such as exome sequencing has been successfully utilized for the identification of mutations in profilin 1 (PFN1) (Wu et al. 2012), ubiquilin 2 (Deng et al. 2011), Tank-binding kinase 1 (Freischmidt et al. 2015), hnRNPA1, and hnRNPA2B1 (Kim et al. 2013). An important success of ALS genomics has led to the identification of hexanucleotide repeat expansion (HRE) in the C9orf 72 gene, located in the non-coding region of chromosome 9 in ALS and FTD patient (DeJesus-Hernandez et al. 2011; Renton et al. 2011). The HRE is observed in 25 to 40% of fALS and in FTD cases and proved that ALS and FTD are inherited together and provide continuum.

2.2 *Transcriptomics of ALS*

Expression patterns of RNA in ALS have been investigated extensively in the search for biomarker of disease progression that will be helpful to gain information of ALS pathogenesis. Human tissues from patient such as the brain, spinal cord, and blood and animal models including SOD1 G93A mouse, TDP43 mouse, and rat model have been examined (Heath et al. 2013; Saris et al. 2013a, 2013b; Krokidis and Vlamos 2018). In recent years, many microarray-based studies to characterize the transcriptome alterations in ALS have been reported. These studies tried to investigate the role of novel genes in ALS and certain other neurodegenerative disorders. Using Mouse Genome 430 2.0 array from Affymetrix, oxidative stress-induced death of neurons by NMDA or hydrogen peroxide in SOD1 G93A cortical neurons has been reported (Boutahar et al. 2011). In another outstanding study (Heath et al. 2013), differential gene expression was reported in both fALS and sALS cases, elucidating the mechanism of motor neuronal death. The same research group studies human gene expression profile in mixed-cell and peripheral tissue samples, revealing the important genes in neuroinflammation, RNA splicing, and cytoskeletal involvement. The pharmacological targeting of the modified pathways and networks in ALS has also been studied using microarray-based transcriptomics (Paratore et al. 2012; Henriques and Gonzalez De Aguilar 2011).

Furthermore, using GeneChip Mouse Gene 1.0 ST (Affymetrix) arrays on C57BL/6J mouse brain, TDP-43 interacting genes involved in synaptic development and functions were identified (Narayanan et al. 2013). A detailed microarray studies on the brain of GMR-Gal4/UAS-TDP-43 transgenic drosophila model showed alterations in genes involved in oxidative homeostasis and cell cycle regulation (Zhan et al. 2013). Similarly, FUS-targeted genes were identified using GeneChip Mouse Exon 1.0. ST exon array (Affymetrix) in FUS-deficient motor neurons, cerebellum, cortical neurons, and glial cell (Fujioka et al. 2013). The same research group also examined alterations in gene expression and alternative splicing patterns of TDP-43-deficient primary cortical neurons (Honda et al. 2013) and compare with the transcriptome patterns obtained in FUS-deficient neurons. They showed that almost 25% of genes with altered expression levels and 10% of genes with differentially spliced exons were similar in the transcriptome profiles of TDP-43-deficient and

FUS-deficient primary cortical neurons. These results suggest that TDP-43 and FUS influence common downstream RNA-regulated events in ALS. The transcriptome profile of another RBP, TIA-1 depleting mice's spinal cord and cerebellum analysed using the Affymetrix GeneChip HT Mouse Genome 430 2.0, reveals that TIA-1 acts as a lipid homeostasis regulator (Heck et al. 2014).

Furthermore, sALS-linked epigenetic biomarkers were checked using Illumina Human Methylation DNA BeadChip array (Figueroa-Romero et al. 2012). This study provides an evidence of ALS-dependent methylation and misregulation of genes involved in neural development and differentiation and thus enhances our knowledge of disease pathogenesis and facilitates the finding of new targets. To extend microarray experiment focussing on pathophysiological mechanisms of ALS, next-generation sequencing technologies are employed to study the disease at the genomic, transcriptomic, and epigenetic levels (Reis-Filho 2009; Morozova and Marra 2008). RNA-Seq provides an extensive deep sequencing approach for the accurate quantification of the transcript levels (Wang and Xi 2013; Reis-Filho 2009). For instance, RNA-Seq analysis carried out by Illumina Genome Analyzer showed alternative splicing patterns associated with wild-type or mutated or knocked down FUS (van Blitterswijk et al. 2013). A significant number of differentially expressed ribosomal- and spliceosomal-related genes have been reported in R521G and R522S FUS mutations, indicating that misregulation of FUS might be responsible to the disease (van Blitterswijk et al. 2013). Utilizing HITS-CLIP technology together with RNA-Seq, the role of FUS as neuronal transcriptome regulators has been revealed (Nakaya et al. 2013). RNA-Seq approach has also demonstrated the significant role of TDP-43 in ion channel regulation and in synaptic transmission (Hazelett et al. 2012).

2.3 *Metabolomics of ALS*

Metabolomics can be viewed as the final result of transcribed genome fine-regulated by functional proteins. In ALS, brain metabolites have been non-invasively measured by the magnetic response spectroscopy (MRS) (Jones et al. 1995; Gredal et al. 1997). The researcher observed decreased N-acetylaspartate (NAA) levels in the motor cortex (Gredal et al. 1997), brainstem (Cwik et al. 1998), thalamus, and cerebellum (Ikeda et al. 1998) of ALS patients.

First global metabolomics study utilizing high-performance liquid chromatography on blood plasma of ALS patients was achieved in 2005 (Rozen et al. 2005). Here, the authors measured almost 300 metabolites and illustrated that ALS is connected to downregulation rather than the upregulation of metabolites level. They also reported 12 metabolites that appreciably increased in the patients kept on the riluzole medication. A similar study was performed recently in ALS patients categorized as 'possible', 'probable', or 'definite' and showed that 32 metabolites were considerably changed in the plasma of ‘definite’ ALS patients (Lawton et al. 2012). These metabolites were involved in neuronal loss, oxidative damage, hypermetabolism, and mitochondrial dysfunction.

Blasco et al. (2010) measured 17 metabolites employing high-resolution ^{1}H-MRS and found lower concentration of acetate and higher concentration of ascorbate and pyruvate in ALS patients. Wuolikainen et al. (2011) measured approximately 120 metabolites using gas chromatography combined to mass spectrometry (GS-TOFMS) and classified CSF samples with different ALS subtypes. They found noteworthy differences in metabolites’ profile among fALS, sALS, and SOD1-ALS patients, where fALS patients showed distinct and homogenous metabolomic profile than sALS patients. Similarly, a different metabolomic pattern has also been reported in ALS patients with a C9orf72 mutation when compared to sALS or FTD patients (Cistaro et al. 2014). By using ^{18}F-fluorodeoxyglucose-positron emission tomography (PET) in specific areas of the brain, the authors showed that C9orf72 mutation carriers showed a larger involvement of CNS than that of sALS or FTD patients (Cistaro et al. 2014).

Thus, metabolomic studies in human tissues and animal model have shown neuronal alteration and degeneration. However, sufficient experimental evidences in support of specific metabolomic profile (fingerprint) associated with disease are lacking, although important efforts with hopeful results are being obtained (Lawton et al. 2014; Wuolikainen et al. 2009).

2.4 Secretomics of ALS

Besides proteomics and metabolomics, secretomics could also tell pathological pathway and biomarkers in ALS. Secretomics facilitate the study of cellular secretion products from a variety of sources such as plasma, serum, urine, CSF, and saliva plus relevant cell culture media (Dowling and Clynes 2011). The ALS secretomes have been analysed by fewer studies where the role of mutant SOD1 and other ALS genes in various cell types involves non-cell autonomous mechanism for neuronal defects (Boillee et al. 2006; Ilieva et al. 2009). Non-neuronal cells such as astrocytes from either sALS or fALS cases have shown the secretion of inflammatory mediators, revealing a general mechanism of non-cell-mediated toxicity (Haidet-Phillips et al. 2011). Moreover, Gomes et al. (2010) demonstrated that NSC-34 motor neuronal cells overexpressing wild-type and mutant SOD1 secrete exosomes containing SOD1, and this secretion could be enhanced by chromogranin (Urushitani et al. 2006). Although in vivo role of SOD1 exosome secretion is still unclear (Grad et al. 2014), it would be highly interesting to further investigate the nature of exosome secretion in ALS.

3 C9orf72 'Omics' in FTD-ALS

In 2011, hexanucleotide repeat expansions of the G4C2 (HRE) in the non-coding region of the *C9orf72* gene (refereed here to C9) were found to be the common familial cause of ALS and FTD (referred as C9-FTD/ALS) (Renton et al. 2011; DeJesus-Hernandez et al. 2011) and were also found in sporadic cases later (Ling et al. 2013). The size of the repeat in ALS and FTD cases ranges between 700 to 1600 as compared to 2–23 in controls (DeJesus-Hernandez et al. 2011).

Currently, the pathological mechanisms associated with C9-FTD/ALS includes haploinsufficiency, RNA gain of toxicity, and dipeptide repeat (DPR) accumulation via repeat-associated non-ATG (RAN) translation (Stepto et al. 2014; Gendron et al. 2014; Todd and Petrucelli 2016). The repeat expansions have been shown to form RNA foci as well as five DPR species (poly-GA, poly-GR, poly-GP, ploy-PA, and poly-PR), which contribute to toxicity by sequestering other proteins and RNAs (Wen et al. 2017; Gendron and Petrucelli 2017). Since its discovery, significant effort has been made towards the understanding of pathomechanisms associated with C9 and a number of studies pointing towards dysregulation of RNA metabolism as a major contributor to C9 pathogenesis (Kumar et al. 2016c). As an example of how systems biology brings out important insights about the dynamics of ALS, in the following section, we point out some recent achievement in the case of C9-mediated ALS.

RNA-Seq experiments have revealed that C9 RNA foci sequester several members of the hnRNP family of splicing factors (Lee et al. 2013; Mori et al. 2013; Sareen et al. 2013; Cooper-Knock et al. 2014), resulting in altered splicing patterns of their RNA targets. Cooper-Knock et al. (2015) reported the upregulation of 'RNA splicing' genes in motor neurons and lymphoblastoid cell lines of patients with C9-ALS. Upregulation of these genes is consistent with an effort to compensate for sequestration of these proteins by RNA foci. Many of the differentially expressed genes have been independently identified as repeat-binding proteins, including hnRNPA3 and hnRNPH (Lee et al. 2013; Mori et al. 2013; Conlon et al. 2016). These results suggest that the splicing defects in repeat expansion carriers are due to RBP sequestration into foci.

To identify biologically relevant pathways, Satoh et al. (2014) used different pathway analysis tools including Kyoto Encyclopedia of Genes and Genomes (KEGG:www.kegg.jp), Ingenuity Pathways Analysis (IPA:www.ingenuity.com), and KeyMolnet (www.km-data.jp/keymolnet) to study molecular networks engaged in C9ALS by utilizing three different *C9orf72* omics datasets. These data sets were (i) proteome of *C9orf72* HRE RBPs which provides the most important biochemical information of C9-ALS (Haeusler et al. 2014; Cooper-Knock et al. 2014), (ii) transcriptome of iPSNs of patients with C9-ALS which represents the most effective cell culture model (Sareen et al. 2013), and (iii) transcriptome of motor neurons of C9-ALS patients acting as the most clinically appropriate in vivo source (Highley et al. 2014). The results of this study reveal *C9orf72* HREs involvement in the ribosome, spliceosome, and post-transcriptional modification of

RNA. Essentially, the proteome is enriched of RBPs having RNA recognition motifs and prion-like domain. Similarly, network analysis of differentially expressed genes in iPSNs of patients with C9-ALS shows that the majority of genes identified were underexpressed, namely, the genes encoding for extracellular matrix proteins and matrix metalloproteinases. Moreover, the authors did not observed any significant differences in splicing patterns of C9-ALS patients and controls. In addition, the authors also reported that the post-transcriptional RNA processing, cytoskeletal dynamics, and intracellular molecular transport have been affected in C9-ALS patients.

Recently, Petrucelli group (Prudencio et al. 2015) has also reported transcriptome alterations in the frontal cortex and cerebellum of C9-ALS, sALS, and controls. Their findings showed that a number of misregulated genes in C9-ALS were approximately double than in sALS, demonstrating the differences between these two forms of ALS. Gene ontology (GO) analyses showed that unfolded protein response (UPR), intracellular protein transport, and localization pathway are mainly affected in C9-ALS, while cytoskeleton organization, defence response, and synaptic transmission pathways are affected in sALS.

Furthermore, in induced pluripotent stem cell (iPSC) models of C9-ALS, TDP-43 redistribution from the nucleus to the cytoplasm was reported, suggesting the alteration of TDP-43 function. Also, mass spectrometry results indicating the sequestration of RBPs involved in splicing, translation, and nuclear export in C9RNA foci (Cooper-Knock et al. 2014; Rossi et al. 2015). Thus, C9 involvement in disease could be due to either alteration in gene transcription and splicing events or via RNA foci-mediated sequestration of proteins involved in RNA metabolism. Table 2 summarizes the transcriptome alterations in C9-ALS pathology.

4 Concluding Remarks

During the last 25 years, research on ALS is flourishing, and the amount of data has significantly increased. Defects in RNA metabolism have emerged as a crucial event in the pathogenesis of ALS/FTD. RBPs such as TDP-43, FUS, and C9orf72 pathology share common alterations in gene expression and post-transcriptional regulation. Given the significance of RNA and RBPs in cellular physiology, RNA processing can be considered as a crucial target for therapeutic intervention to patients not only with ALS/FTD but to a collection of neurodegenerative diseases.

Recent advances in microarray and next-generation sequencing technologies enable us to study the global analysis of genome, transcriptome, proteome, and metabolome, collectively termed as 'omics'. These omics study enables the characterization of the genome-wide molecular basis of diseases and identifies disease-specific molecular biomarkers. In particular, systems biology approaches will be

Table 2 Gene ontology (GO) terms based on analysis of up- or downregulated genes in C9-ALS cases

	Upregulated	Downregulated	References
	Extracellular matrix	Neuron differentiation	(Sareen et al. 2013)
	Cell adhesion	Cell–cell signalling	
	Cell–cell signalling	Synapse	
	Synaptic transmission		
	Neurological process		
Motor neurons	RNA Splicing	Cholesterol biosynthesis	(Cooper-Knock et al. 2015)
	Erythrocyte homeostasis	Regulation of glucose metabolism	
	Male sex differentiation	Regulation of nuclear division	
Lymphoblastoid cell lines	RNA splicing	Inflammatory response	
	Protein catabolic process	Regulation of action potential in neuron	
	Synaptic transmission	Striated muscle tissue development	
	Positive regulation of apoptosis		
Cerebellum	Pattern specification process	G-protein coupled receptor protein signalling pathway	(Prudencio et al. 2015)
	Skeletal system development	Cognition	
	Embryonic morphogenesis	Regulation of nucleotide biosynthetic process	
	Response to unfolded protein	Immune response	
	Inflammatory response	Regulation of nucleotide metabolic process	
Frontal cortex	Inflammatory response	Gas transport	
	Response to wounding	Oxygen transport	
	Defence response	Haemoglobin metabolic process	
	Response to unfolded protein		
	Digestion		

indispensable in understanding of the pathomechanisms and rational drug designing for the debilitating diseases. In future, it will be exciting to see how the systems biology data can be clinically translated.

Acknowledgement TK thanks the Department of Biotechnology, India, for providing DBT Biocare fellowship (BT/Bio-CARe/07/351/2016–2018). VK thanks the Department of Science of Technology, India, for the award of DST fast track fellowship (SB/YS/LS-161/2014).

Conflict of Interest The authors have declared that there is no conflict of interest.

References

Anderson P, Kedersha N (2009) RNA granules: post-transcriptional and epigenetic modulators of gene expression. Nat Rev Mol Cell Biol 10:430–436

Appel SH, Rowland LP (2012) Amyotrophic lateral sclerosis, frontotemporal lobar dementia, and p62: a functional convergence? Neurology 79:1526–1527

Arai T, Hasegawa M, Akiyama H et al (2006) TDP-43 is a component of ubiquitin-positive tau-negative inclusions in frontotemporal lobar degeneration and amyotrophic lateral sclerosis. Biochem Biophys Res Commun 351:602–611

Aulas A, Vande Velde C (2015) Alterations in stress granule dynamics driven by TDP-43 and FUS: a link to pathological inclusions in ALS? Front Cell Neurosci 9:423

Belzil VV, Gendron TF, Petrucelli L (2013) RNA-mediated toxicity in neurodegenerative disease. Mol Cell Neurosci 56:406–419

Bentmann E, Haass C, Dormann D (2013) Stress granules in neurodegeneration—lessons learnt from TAR DNA binding protein of 43 kDa and fused in sarcoma. FEBS J 280:4348–4370

Blasco H, Corcia P, Moreau C et al (2010) 1H-NMR-based metabolomic profiling of CSF in early amyotrophic lateral sclerosis. PLoS One 5:e13223

Boillee S, Yamanaka K, Lobsiger CS et al (2006) Onset and progression in inherited ALS determined by motor neurons and microglia. Science 312:1389–1392

Boutahar N, Wierinckx A, Camdessanche JP et al (2011) Differential effect of oxidative or excitotoxic stress on the transcriptional profile of amyotrophic lateral sclerosis-linked mutant SOD1 cultured neurons. J Neurosci Res 89:1439–1450

Bucchia M, Ramirez A, Parente V et al (2015) Therapeutic development in amyotrophic lateral sclerosis. Clin Ther 37:668–680

Buchan JR (2014) mRNP granules. Assembly, function, and connections with disease. RNA Biol 11:1019–1030

Burrell JR, Kiernan MC, Vucic S et al (2011) Motor neuron dysfunction in frontotemporal dementia. Brain 134:2582–2594

Caballero-Hernandez D, Toscano MG, Cejudo-Guillen M et al (2016) The 'Omics' of amyotrophic lateral sclerosis. Trends Mol Med 22:53–67

Cistaro A, Pagani M, Montuschi A et al (2014) The metabolic signature of C9ORF72-related ALS: FDG PET comparison with nonmutated patients. Eur J Nucl Med Mol Imaging 41:844–852

Conlon EG, Lu L, Sharma A et al (2016) The C9ORF72 GGGGCC expansion forms RNA G-quadruplex inclusions and sequesters hnRNP H to disrupt splicing in ALS brains. Elife 5:e17820

Cooper-Knock J, Walsh MJ, Higginbottom A et al (2014) Sequestration of multiple RNA recognition motif-containing proteins by C9orf72 repeat expansions. Brain 137:2040–2051

Cooper-Knock J, Bury JJ, Heath PR et al (2015) C9ORF72 GGGGCC expanded repeats produce splicing dysregulation which correlates with disease severity in amyotrophic lateral sclerosis. PLoS One 10:e0127376

Coyne AN, Zaepfel BL, Zarnescu DC (2017) Failure to deliver and translate-new insights into RNA dysregulation in ALS. Front Cell Neurosci 11:243

Cwik VA, Hanstock CC, Allen PS et al (1998) Estimation of brainstem neuronal loss in amyotrophic lateral sclerosis with in vivo proton magnetic resonance spectroscopy. Neurology 50:72–77

Daigle JG, Lanson NA Jr, Smith RB et al (2013) RNA-binding ability of FUS regulates neurodegeneration, cytoplasmic mislocalization and incorporation into stress granules associated with FUS carrying ALS-linked mutations. Hum Mol Genet 22:1193–1205
DeJesus-Hernandez M, Mackenzie IR, Boeve BF et al (2011) Expanded GGGGCC hexanucleotide repeat in noncoding region of C9ORF72 causes chromosome 9p-linked FTD and ALS. Neuron 72:245–256
DeLoach A, Cozart M, Kiaei A et al (2015) A retrospective review of the progress in amyotrophic lateral sclerosis drug discovery over the last decade and a look at the latest strategies. Expert Opin Drug Discov 10:1099–1118
Deng HX, Chen W, Hong ST et al (2011) Mutations in UBQLN2 cause dominant X-linked juvenile and adult-onset ALS and ALS/dementia. Nature 477:211–215
Dewey CM, Cenik B, Sephton CF et al (2012) TDP-43 aggregation in neurodegeneration: are stress granules the key? Brain Res 1462:16–25
Diaz-Beltran L, Cano C, Wall DP et al (2013) Systems biology as a comparative approach to understand complex gene expression in neurological diseases. Behav Sci (Basel) 3:253–272
Dowling P, Clynes M (2011) Conditioned media from cell lines: a complementary model to clinical specimens for the discovery of disease-specific biomarkers. Proteomics 11:794–804
Dunkel P, Chai CL, Sperlagh B et al (2012) Clinical utility of neuroprotective agents in neurodegenerative diseases: current status of drug development for Alzheimer's, Parkinson's and Huntington's diseases, and amyotrophic lateral sclerosis. Expert Opin Investig Drugs 21:1267–1308
Figueroa-Romero C, Hur J, Bender DE et al (2012) Identification of epigenetically altered genes in sporadic amyotrophic lateral sclerosis. PLoS One 7:e52672
Freischmidt A, Wieland T, Richter B et al (2015) Haploinsufficiency of TBK1 causes familial ALS and fronto-temporal dementia. Nat Neurosci 18:631–636
Fujioka Y, Ishigaki S, Masuda A et al (2013) FUS-regulated region- and cell-type-specific transcriptome is associated with cell selectivity in ALS/FTLD. Sci Rep 3:2388
Gendron TF, Petrucelli L (2017) Disease mechanisms of C9ORF72 repeat expansions. Cold Spring Harb Perspect Med 8:a024224
Gendron TF, Belzil VV, Zhang YJ et al (2014) Mechanisms of toxicity in C9FTLD/ALS. Acta Neuropathol 127:359–376
Gomes C, Escrevente C, Costa J (2010) Mutant superoxide dismutase 1 overexpression in NSC-34 cells: effect of trehalose on aggregation, TDP-43 localization and levels of co-expressed glycoproteins. Neurosci Lett 475:145–149
Goyal NA, Mozaffar T (2014) Experimental trials in amyotrophic lateral sclerosis: a review of recently completed, ongoing and planned trials using existing and novel drugs. Expert Opin Investig Drugs 23:1541–1551
Grad LI, Pokrishevsky E, Silverman JM et al (2014) Exosome-dependent and independent mechanisms are involved in prion-like transmission of propagated Cu/Zn superoxide dismutase misfolding. Prion 8:331–335
Gredal O, Rosenbaum S, Topp S et al (1997) Quantification of brain metabolites in amyotrophic lateral sclerosis by localized proton magnetic resonance spectroscopy. Neurology 48:878–881
Haeusler AR, Donnelly CJ, Periz G et al (2014) C9orf72 nucleotide repeat structures initiate molecular cascades of disease. Nature 507:195–200
Haidet-Phillips AM, Hester ME, Miranda CJ et al (2011) Astrocytes from familial and sporadic ALS patients are toxic to motor neurons. Nat Biotechnol 29:824–828
Hazelett DJ, Chang JC, Lakeland DL et al (2012) Comparison of parallel high-throughput RNA sequencing between knockout of TDP-43 and its overexpression reveals primarily nonreciprocal and nonoverlapping gene expression changes in the central nervous system of Drosophila. G3 (Bethesda) 2:789–802
He Y, Smith R (2009) Nuclear functions of heterogeneous nuclear ribonucleoproteins A/B. Cell Mol Life Sci 66:1239–1256
Heath PR, Kirby J, Shaw PJ (2013) Investigating cell death mechanisms in amyotrophic lateral sclerosis using transcriptomics. Front Cell Neurosci 7:259

Heck MV, Azizov M, Stehning T et al (2014) Dysregulated expression of lipid storage and membrane dynamics factors in Tia1 knockout mouse nervous tissue. Neurogenetics 15:135–144

Henriques A, Gonzalez De Aguilar JL (2011) Can transcriptomics cut the gordian knot of amyotrophic lateral sclerosis? Curr Genomics 12:506–515

Hicks GG, Singh N, Nashabi A et al (2000) Fus deficiency in mice results in defective B-lymphocyte development and activation, high levels of chromosomal instability and perinatal death. Nat Genet 24:175–179

Highley JR, Kirby J, Jansweijer JA et al (2014) Loss of nuclear TDP-43 in amyotrophic lateral sclerosis (ALS) causes altered expression of splicing machinery and widespread dysregulation of RNA splicing in motor neurones. Neuropathol Appl Neurobiol 40:670–685

Honda D, Ishigaki S, Iguchi Y et al (2013) The ALS/FTLD-related RNA-binding proteins TDP-43 and FUS have common downstream RNA targets in cortical neurons. FEBS Open Bio 4:1–10

Ikeda K, Iwasaki Y, Kinoshita M et al (1998) Quantification of brain metabolites in ALS by localized proton magnetic spectroscopy. Neurology 50:576–577

Ilieva H, Polymenidou M, Cleveland DW (2009) Non-cell autonomous toxicity in neurodegenerative disorders: ALS and beyond. J Cell Biol 187:761–772

Jones AP, Gunawardena WJ, Coutinho CM et al (1995) Preliminary results of proton magnetic resonance spectroscopy in motor neurone disease (amyotrophic lateral sclerosis). J Neurol Sci 129(Suppl):85–89

Kabashi E, Valdmanis PN, Dion P et al (2008) TARDBP mutations in individuals with sporadic and familial amyotrophic lateral sclerosis. Nat Genet 40:572–574

Kapeli K, Martinez FJ, Yeo GW (2017) Genetic mutations in RNA-binding proteins and their roles in ALS. Hum Genet 136:1193–1214

Kim HJ, Kim NC, Wang YD et al (2013) Mutations in prion-like domains in hnRNPA2B1 and hnRNPA1 cause multisystem proteinopathy and ALS. Nature 495:467–473

Kraemer BC, Schuck T, Wheeler JM et al (2010) Loss of murine TDP-43 disrupts motor function and plays an essential role in embryogenesis. Acta Neuropathol 119:409–419

Krokidis MG, Vlamos P (2018) Transcriptomics in amyotrophic lateral sclerosis. Front Biosci (Elite Ed) 10:103–121

Kumar V, Islam A, Hassan MI et al (2016a) Therapeutic progress in amyotrophic lateral sclerosis-beginning to learning. Eur J Med Chem 121:903–917

Kumar V, Islam A, Hassan MI et al (2016b) Delineating the relationship between amyotrophic lateral sclerosis and frontotemporal dementia: sequence and structure-based predictions. Biochim Biophys Acta 1862:1742–1754

Kumar V, Kashav T, Islam A et al (2016c) Structural insight into C9orf72 hexanucleotide repeat expansions: towards new therapeutic targets in FTD-ALS. Neurochem Int 100:11–20

Kwiatkowski TJ Jr, Bosco DA, Leclerc AL et al (2009) Mutations in the FUS/TLS gene on chromosome 16 cause familial amyotrophic lateral sclerosis. Science 323:1205–1208

Lawton KA, Cudkowicz ME, Brown MV et al (2012) Biochemical alterations associated with ALS. Amyotroph Lateral Scler 13:110–118

Lawton KA, Brown MV, Alexander D et al (2014) Plasma metabolomic biomarker panel to distinguish patients with amyotrophic lateral sclerosis from disease mimics. Amyotroph Lateral Scler Frontotemporal Degener 15:362–370

Lee KH, Zhang P, Kim HJ et al (2016) C9orf72 dipeptide repeats impair the assembly, dynamics, and function of membrane-less organelles. Cell 167:774–788.e717

Lee YB, Chen HJ, Peres JN et al (2013) Hexanucleotide repeats in ALS/FTD form length-dependent RNA foci, sequester RNA binding proteins, and are neurotoxic. Cell Rep 5:1178–1186

Li H, Watford W, Li C et al (2007) Ewing sarcoma gene EWS is essential for meiosis and B lymphocyte development. J Clin Invest 117:1314–1323

Ling SC, Polymenidou M, Cleveland DW (2013) Converging mechanisms in ALS and FTD: disrupted RNA and protein homeostasis. Neuron 79:416–438

Liu TY, Chen YC, Jong YJ et al (2017) Muscle developmental defects in heterogeneous nuclear Ribonucleoprotein A1 knockout mice. Open Biol 7:160303
Lomen-Hoerth C, Anderson T, Miller B (2002) The overlap of amyotrophic lateral sclerosis and frontotemporal dementia. Neurology 59:1077–1079
Maharjan N, Kunzli C, Buthey K et al (2017) C9ORF72 regulates stress granule formation and its deficiency impairs stress granule assembly, hypersensitizing cells to stress. Mol Neurobiol 54:3062–3077
Mancuso R, Navarro X (2015) Amyotrophic lateral sclerosis: current perspectives from basic research to the clinic. Prog Neurobiol 133:1–26
Mehta P, Kaye W, Bryan L et al (2016) Prevalence of amyotrophic lateral sclerosis – United States, 2012–2013. MMWR Surveill Summ 65:1–12
Mori K, Lammich S, Mackenzie IR et al (2013) hnRNP A3 binds to GGGGCC repeats and is a constituent of p62-positive/TDP43-negative inclusions in the hippocampus of patients with C9orf72 mutations. Acta Neuropathol 125:413–423
Morozova O, Marra MA (2008) Applications of next-generation sequencing technologies in functional genomics. Genomics 92:255–264
Moujalled D, White AR (2016) Advances in the development of disease-modifying treatments for amyotrophic lateral sclerosis. CNS Drugs 30:227–243
Nakaya T, Alexiou P, Maragkakis M et al (2013) FUS regulates genes coding for RNA-binding proteins in neurons by binding to their highly conserved introns. RNA 19:498–509
Narayanan RK, Mangelsdorf M, Panwar A et al (2013) Identification of RNA bound to the TDP-43 ribonucleoprotein complex in the adult mouse brain. Amyotroph Lateral Scler Frontotemporal Degener 14:252–260
Nicholson KA, Cudkowicz ME, Berry JD (2015) Clinical trial designs in amyotrophic lateral sclerosis: does one design fit all? Neurotherapeutics 12:376–383
Nussbacher JK, Batra R, Lagier-Tourenne C et al (2015) RNA-binding proteins in neurodegeneration: Seq and you shall receive. Trends Neurosci 38:226–236
Pandya RS, Zhu H, Li W et al (2013) Therapeutic neuroprotective agents for amyotrophic lateral sclerosis. Cell Mol Life Sci 70:4729–4745
Paratore S, Pezzino S, Cavallaro S (2012) Identification of pharmacological targets in amyotrophic lateral sclerosis through genomic analysis of deregulated genes and pathways. Curr Genomics 13:321–333
Perry DC, Miller BL (2013) Frontotemporal dementia. Semin Neurol 33:336–341
Polymenidou M, Lagier-Tourenne C, Hutt KR et al (2012) Misregulated RNA processing in amyotrophic lateral sclerosis. Brain Res 1462:3–15
Protter DS, Parker R (2016) Principles and properties of stress granules. Trends Cell Biol 26:668–679
Prudencio M, Belzil VV, Batra R et al (2015) Distinct brain transcriptome profiles in C9orf72-associated and sporadic ALS. Nat Neurosci 18:1175–1182
Ramaswami M, Taylor JP, Parker R (2013) Altered ribostasis: RNA-protein granules in degenerative disorders. Cell 154:727–736
Ratnavalli E, Brayne C, Dawson K et al (2002) The prevalence of frontotemporal dementia. Neurology 58:1615–1621
Reis-Filho JS (2009) Next-generation sequencing. Breast Cancer Res 11(Suppl 3):S12
Renton AE, Majounie E, Waite A et al (2011) A hexanucleotide repeat expansion in C9ORF72 is the cause of chromosome 9p21-linked ALS-FTD. Neuron 72:257–268
Ringholz GM, Appel SH, Bradshaw M et al (2005) Prevalence and patterns of cognitive impairment in sporadic ALS. Neurology 65:586–590
Robberecht W, Philips T (2013) The changing scene of amyotrophic lateral sclerosis. Nat Rev Neurosci 14:248–264
Rosen DR, Siddique T, Patterson D et al (1993) Mutations in Cu/Zn superoxide dismutase gene are associated with familial amyotrophic lateral sclerosis. Nature 362:59–62

Rossi S, Serrano A, Gerbino V et al (2015) Nuclear accumulation of mRNAs underlies G4C2-repeat-induced translational repression in a cellular model of C9orf72 ALS. J Cell Sci 128:1787–1799
Rozen S, Cudkowicz ME, Bogdanov M et al (2005) Metabolomic analysis and signatures in motor neuron disease. Metabolomics 1:101–108
Sareen D, O'Rourke JG, Meera P et al (2013) Targeting RNA foci in iPSC-derived motor neurons from ALS patients with a C9ORF72 repeat expansion. Sci Transl Med 5:208ra149
Saris CG, Groen EJ, Koekkoek JA et al (2013a) Meta-analysis of gene expression profiling in amyotrophic lateral sclerosis: a comparison between transgenic mouse models and human patients. Amyotroph Lateral Scler Frontotemporal Degener 14:177–189
Saris CG, Groen EJ, van Vught PW et al (2013b) Gene expression profile of SOD1-G93A mouse spinal cord, blood and muscle. Amyotroph Lateral Scler Frontotemporal Degener 14:190–198
Satoh J, Yamamoto Y, Kitano S et al (2014) Molecular network analysis suggests a logical hypothesis for the pathological role of c9orf72 in amyotrophic lateral sclerosis/frontotemporal dementia. J Cent Nerv Syst Dis 6:69–78
Shaw PJ (2005) Molecular and cellular pathways of neurodegeneration in motor neurone disease. J Neurol Neurosurg Psychiatry 76:1046–1057
Sieben A, Van Langenhove T, Engelborghs S et al (2012) The genetics and neuropathology of frontotemporal lobar degeneration. Acta Neuropathol 124:353–372
Sreedharan J, Blair IP, Tripathi VB et al (2008) TDP-43 mutations in familial and sporadic amyotrophic lateral sclerosis. Science 319:1668–1672
Stepto A, Gallo JM, Shaw CE et al (2014) Modelling C9ORF72 hexanucleotide repeat expansion in amyotrophic lateral sclerosis and frontotemporal dementia. Acta Neuropathol 127:377–389
Todd TW, Petrucelli L (2016) Insights into the pathogenic mechanisms of Chromosome 9 open reading frame 72 (C9orf72) repeat expansions. J Neurochem 138:145–162
Urushitani M, Sik A, Sakurai T et al (2006) Chromogranin-mediated secretion of mutant superoxide dismutase proteins linked to amyotrophic lateral sclerosis. Nat Neurosci 9:108–118
van Blitterswijk M, van Es MA, Hennekam EA et al (2012) Evidence for an oligogenic basis of amyotrophic lateral sclerosis. Hum Mol Genet 21:3776–3784
van Blitterswijk M, Wang ET, Friedman BA et al (2013) Characterization of FUS mutations in amyotrophic lateral sclerosis using RNA-Seq. PLoS One 8:e60788
Wang B, Xi Y (2013) Challenges for microRNA microarray data analysis. Microarrays (Basel) 2:34–50
Wen X, Westergard T, Pasinelli P et al (2017) Pathogenic determinants and mechanisms of ALS/FTD linked to hexanucleotide repeat expansions in the C9orf72 gene. Neurosci Lett 636:16–26
Wheaton MW, Salamone AR, Mosnik DM et al (2007) Cognitive impairment in familial ALS. Neurology 69:1411–1417
Wood LB, Winslow AR, Strasser SD (2015) Systems biology of neurodegenerative diseases. Integr Biol (Camb) 7:758–775
Wroe R, Wai-Ling Butler A, Andersen PM et al (2008) ALSOD: the amyotrophic lateral sclerosis online database. Amyotroph Lateral Scler 9:249–250
Wu CH, Fallini C, Ticozzi N et al (2012) Mutations in the profilin 1 gene cause familial amyotrophic lateral sclerosis. Nature 488:499–503
Wuolikainen A, Hedenstrom M, Moritz T et al (2009) Optimization of procedures for collecting and storing of CSF for studying the metabolome in ALS. Amyotroph Lateral Scler 10:229–236
Wuolikainen A, Moritz T, Marklund SL et al (2011) Disease-related changes in the cerebrospinal fluid metabolome in amyotrophic lateral sclerosis detected by GC/TOFMS. PLoS One 6:e17947
Zhan L, Hanson KA, Kim SH et al (2013) Identification of genetic modifiers of TDP-43 neurotoxicity in Drosophila. PLoS One 8:e57214

Systems Approaches to Map In Vivo RNA–Protein Interactions in *Arabidopsis thaliana*

Martin Lewinski and Tino Köster

Contents

Abstract Proteins that specifically interact with mRNAs orchestrate mRNA processing steps all the way from transcription to decay. Thus, these RNA-binding proteins represent an important control mechanism to double check which proportion of nascent pre-mRNAs is ultimately available for translation into distinct proteins. Here, we discuss recent progress to obtain a systems-level understanding of in vivo RNA–protein interactions in the reference plant *Arabidopsis thaliana* using protein-centric and RNA-centric methods as well as combined protein binding site and structure probing.

Keywords CLIP · iCLIP · mRNA interactome · RNA-binding protein · RNA immunoprecipitation · RNA–protein interaction

---

M. Lewinski · T. Köster (✉)
RNA Biology and Molecular Physiology, Bielefeld University, Bielefeld, Germany
e-mail: tino.koester@uni-bielefeld.de

N. Rajewsky et al. (eds.), *Systems Biology*, RNA Technologies,
https://doi.org/10.1007/978-3-319-92967-5_5

1 Introduction

RNA-binding proteins (RBPs) are a diverse class of proteins that control every step of RNA processing and RNA function in the cell. They are characterized by dedicated domains involved in RNA binding and can have accessory domains engaged in protein-protein interactions or enzymatic activities.

In higher plants, RBP function so far has been best studied in the reference plant *Arabidopsis thaliana*. Among the RBPs present in the *Arabidopsis* genome are 197 proteins with an RNA recognition motif (RRM), the most abundant type of RNA-binding domain, and 28 K homology (KH) domain proteins first identified in mammalian heterogeneous nuclear protein hnRNP K (Silverman et al. 2013). In addition, 26 Pumilio (PUM) domain proteins, nine DEAD-box helicases as well as five proteins with cold shock domains (CSDs) have been identified (Silverman et al. 2013). Another 450 proteins harbor pentatricopeptide repeat (PPR) domains. PPR domains consist of multiple 35-amino acid repeats of which two are known to be engaged in specific RNA recognition (Barkan and Small 2014). These proteins are imported into mitochondria or chloroplasts and regulate all aspects of RNA metabolism, e.g., RNA editing, splicing, RNA cleavage, and translation in organelles (Schmitz-Linneweber and Small 2008; Barkan and Small 2014).

A suite of Arabidopsis RBPs have been experimentally characterized, mainly through loss-of-function mutants and transgenic plants ectopically overexpressing RBPs. These approaches revealed a crucial role for RBPs in development (Kalyna et al. 2003; Ripoll et al. 2006; Kupsch et al. 2012; Völz et al. 2012; Ferrari et al. 2017; Foley et al. 2017; Teubner et al. 2017), timing of plant reproduction (Macknight et al. 1997; Streitner et al. 2008; Hornyik et al. 2010), responses to abiotic stress (Kim et al. 2007b, c, 2008, 2010; Park et al. 2009), pathogen defense (Fu et al. 2007; Qi et al. 2010; Jeong et al. 2011; Lyons et al. 2013; Nicaise et al. 2013), responses to phytohormones (Lu and Fedoroff 2000; Hugouvieux et al. 2001; Riera et al. 2006; Carvalho et al. 2010; Hackmann et al. 2014; Löhr et al. 2014), and circadian timekeeping (Heintzen et al. 1994; Staiger 2001; Jones et al. 2012; Schmal et al. 2013; Perez-Santángelo et al. 2014). At the biochemical level, an impact of defined RBPs on RNA processing including pre-mRNA splicing, 3′ end processing, processing of microRNA precursors, and translation has been described (Lopato et al. 1999; Simpson et al. 2003; Vazquez et al. 2004; Dong et al. 2008; Stauffer et al. 2010; Ren et al. 2012; Rühl et al. 2012; Juntawong et al. 2013; Sorenson and Bailey-Serres 2014; Staiger 2015; Carvalho et al. 2016). Recent attempts to comprehensively identify RBPs, summarized in Sect. 2, provided experimental evidence for RNA binding for most of the previously identified Arabidopsis RBPs and identified a plethora of proteins with noncanonical RBDs.

Systems approaches to describe RNA–protein interactions globally come in two main flavors (Fig. 1). In RNA-centric approaches, proteins associated with mRNAs are recovered by RNA pull-down and identified by mass spectrometry, a technique referred to as mRNA interactome capture (Baltz et al. 2012; Castello et al. 2012) (Fig. 1a). In protein-centric approaches, the focus is laid on a particular RBP. The

RNA complement associated with the RBP of interest, the ribonome, is identified via immunoprecipitation of the RBP from cell lysates and identification of the bound target RNAs, initially by microarrays (Tenenbaum et al. 2000; Galgano and Gerber 2011; Guerreiro et al. 2014) or more recently via high throughput sequencing (Licatalosi et al. 2008; König et al. 2010; Rossbach et al. 2014; Müller-McNicoll et al. 2016) (Fig. 1b).

2 The Arabidopsis RBPome

Of all predicted RBPs in Arabidopsis, RNA binding has only been experimentally confirmed for a limited number of them. A first attempt to globally identify proteins based on their ability to interact with mRNAs in vivo was made for cultured Arabidopsis cells (Schmidt et al. 2010). In this study, mRNAs and interactors were recovered under native conditions by affinity chromatography on an oligo(dT) cellulose column followed by two-dimensional gel electrophoresis. The protein components were identified via Maldi-TOF. In the RNA-bound proteome were a suite of RRM proteins including members of the family of glycine-rich RNA-binding proteins like *At*GRP2 (*Arabidopsis thaliana* glycine rich RNA-binding protein 2), *At*GRP7 and *At*GRP8 (Lewinski et al. 2016), the two oligouridylate-specific RBP45 and RBP47 proteins (Lorkovic et al. 2000), and CSD proteins.

In 2012, mRNA interactome capture was reported to comprehensively identify proteins interacting with mRNAs in mammalian cells (Baltz et al. 2012; Castello et al. 2012). This technique employs in vivo cross-linking of mRNA and bound proteins by UV light irradiation. The RNA–protein complexes are recovered by pull-down of polyadenylated RNAs using magnetic beads coated with oligo(dT). Proteins are released by RNase treatment, subjected to tryptic digest and identified via mass spectrometry (Fig. 1a). Following these pioneering studies, this technique was applied to a wide range of organisms including yeast, *Drosophila melanogaster*, *Caenorhabditis elegans*, Leishmania, trypanosomes, and Plasmodium (Mitchell et al. 2013; Beckmann et al. 2015; Matia-Gonzalez et al. 2015; Bunnik et al. 2016; Lueong et al. 2016; Sysoev et al. 2016; Wessels et al. 2016; Nandan et al. 2017). A minimal core mRNA bound proteome occurring in both human and yeast was defined by Beckmann and coworkers (Beckmann et al. 2015). Lately, mRNA interactome capture has also been successfully applied to Arabidopsis (Marondedze et al. 2016; Reichel et al. 2016; Zhang et al. 2016).

2.1 The mRNA Interactome of Arabidopsis Protoplasts

The first mRNA interactome capture experiments in Arabidopsis employed widely differing tissues to catalog RBPs. Gueten and coworkers chose protoplasts, cells without a cell wall, assuming that UV cross-linking should occur as efficiently as

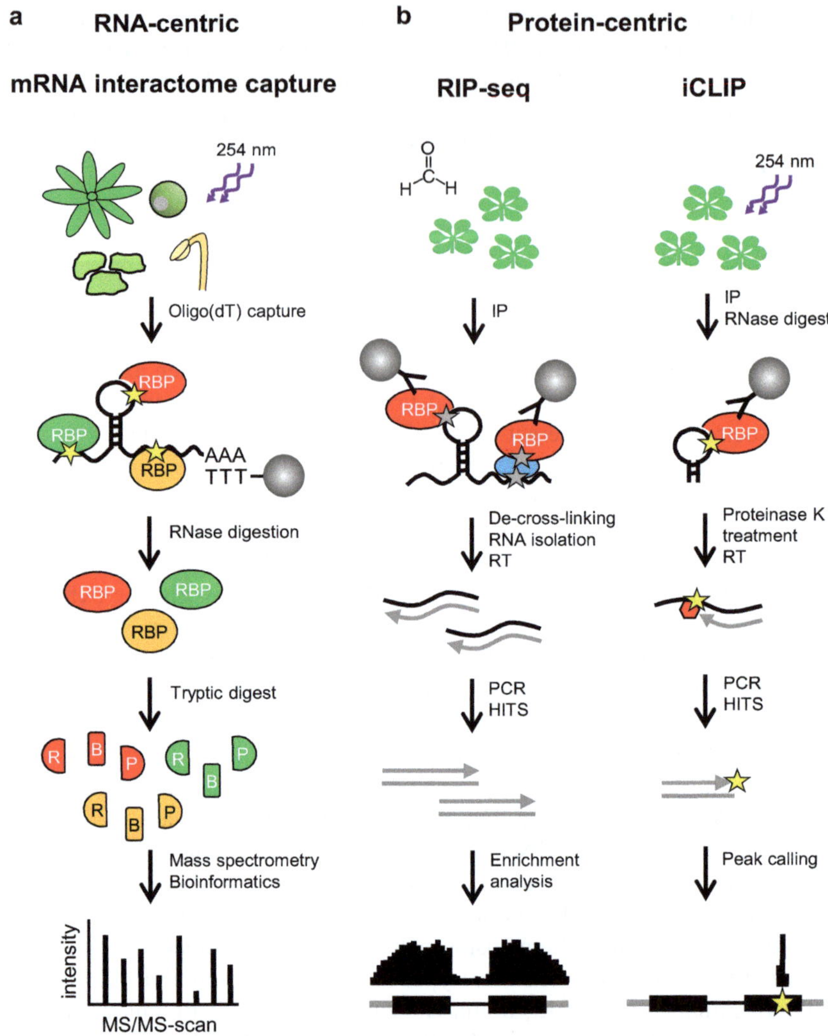

Fig. 1 Strategies to globally identify in vivo RNA–protein interaction in Arabidopsis. (**a**) RNA-centric strategies such as mRNA interactome capture employ oligo(dT) affinity capture. RNA and bound proteins are covalently linked in planta through UV irradiation. RNA–protein complexes are recovered by oligo(dT) pull-down. Proteins are released by RNase treatment, subjected to tryptic digest and identified via mass spectrometry. (**b**) Protein-centric methods focus on a particular RBP and aim at identifying its in vivo RNA targets. Based on the cross-linking agent, RIP using formaldehyde or CLIP technique using UV light are distinguished. Proteins are immunoprecipitated. In RIP-seq cross-links are reversed by heat treatment, RNA is isolated, subjected to reverse transcription and PCR amplification for HITS. Targets enriched upon RIP are determined relative to mock IP controls, e.g., Xing et al. 2015, or relative to polyadenylated RNA, e.g., Meyer et al. 2017. In iCLIP (König et al. 2010), RNA–protein complexes are subjected to RNase treatment. Bound proteins are digested with proteinase, leaving a polypeptide at the cross-link site. Reverse transcriptase stops there, allowing the detection of the cross-link site at the −1 position of the processed sequencing reads

in mammalian cell monolayers (Zhang et al. 2016). Leaf mesophyll protoplasts are also widely used in transient assays to study the regulation of gene expression.

A mesophyll protoplast mRNA interactome was defined with a total of 325 proteins based on enrichment in cross-linked samples vs. non-cross-linked controls with a $\log_2$ fold change above 2 (Zhang et al. 2016). Of these, one class was represented by 123 ribosomal proteins of which 52 were also present in the core mRNA-bound proteome of human and yeast cells (Beckmann et al. 2015). The second class comprised 70 proteins with a known RBD. For 41 of them, a role in mRNA binding and RNA biology had already been described while the remaining proteins had a potential role in mRNA processing. Moreover, 12 of the RBPs in the second class overlapped with the RBPs identified in the native oligo(dT) affinity chromatography approach (Schmidt et al. 2010). The third class comprised 132 candidate RBPs. Of these, 49 were metabolic enzymes, mainly oxidoreductases. Moreover, numerous proteins related to photosynthesis were found. As these are generally strongly expressed, their RNA binding activity and the domains involved beg for an independent validation. One of the enzymes was the Arabidopsis ortholog of phosphoglycerate kinase whose RNA binding capacity has previously been validated in yeast and human cells (Beckmann et al. 2015).

2.2 *The mRNA Interactome of Etiolated Arabidopsis Seedlings*

Another mRNA interactome capture experiment employed 4-days-old etiolated Arabidopsis seedlings (Reichel et al. 2016). This was based on the rationale that UV-absorbing pigments present in green plant tissue may interfere with UV cross-linking *in planta* and their absence in etiolated tissue may allow more efficient UV cross-linking.

Around 300 of the 746 proteins identified altogether were significantly enriched in UV cross-linked samples vs. non-cross-linked controls with a false discovery rate below 1% and designated the "At-RBP set." Eighty percent of these have a known RBD, and 75% have been linked to RNA biology. More than 400 additional proteins did not meet the significance criteria applied for the "At-RBP set" and were classified as "candidate RBPs."

Notably, of the 197 computationally predicted RRM proteins in Arabidopsis 160 were detected in the input fraction in etiolated seedlings (Silverman et al. 2013). Half of these were recovered in the "At-RBD set" and another 50 were present among the "candidate RBPs." Similarly, seven of the predicted KH proteins were present in the "At-RBD set" and 12 were among the "candidate RBPs." Of the predicted 450 members of the PPR protein family only 60 were detected in the input fraction, likely due to low abundance (Schmitz-Linneweber and Small 2008; Reichel et al. 2016). Only six PPR proteins were found in the "At-RBP set" and another twelve in the "candidate RBPs," likely because most RNAs in the organelles lack poly(A) tails. A comparison of the identified proteins to the mRNA interactome in other model organisms revealed that 52 were present in the interactomes of humans

(Baltz et al. 2012; Beckmann et al. 2015), mice (Kwon et al. 2013; Liao et al. 2016), and yeast (Beckmann et al. 2015) and were assigned to basic functions in RNA metabolism such as translation, splicing, and RNA unwinding.

In addition to RBPs with known RBDs many Arabidopsis proteins emerged that have not been linked to RNA binding so far. Among novel RBPs were proteins harboring a YT521-B homology (YTH) domain (Li et al. 2014). YTH domain proteins have been shown to bind N^6 methyladenosine and thus serve as readers of the m^6A mark in mammals (Wang et al. 2014). In addition, Alba domain containing proteins have been identified. Alba domain proteins are well characterized in archaebacteria where they act as transcriptional repressors and in other eukaryotes where they control translation (Goyal et al. 2016). In plants, they have not yet been functionally characterized. The only observation pointing to RNA binding is the recovery of an Arabidopsis Alba domain protein by RNA-affinity chromatography (Gosai et al. 2015). WHIRLY domain containing proteins have been characterized as single-stranded DNA binding proteins in organelles (Krause et al. 2009) and in maize, association of a WHIRLY protein with chloroplast transcripts has been observed (Prikryl et al. 2008). The identification of three WHIRLY proteins in the etiolated seedling interactome (Reichel et al. 2016) and of WHIRLY1 upon oligo(dT) affinity chromatography in Arabidopsis cells (Schmidt et al. 2010) now provides evidence for global in vivo RNA binding.

In addition, a plethora of proteins with potential RNA binding activity have been detected. To substantiate their RNA-binding properties, independent replication is desirable. Among those are proteins with the Domain of unknown function 1296, cytoskeletal proteins, and photoreceptors. The identification of plasma membrane intrinsic proteins has led to the speculation that aquaporins may be involved in transport of RNAs between cells (Reichel et al. 2016).

2.3 *The mRNA Interactome of Arabidopsis Cultured Cells and Leaves of Adult Plants*

Another mRNA interactome capture experiment was performed on cell suspension cultures generated from roots of the Arabidopsis accessions Col-0 and Landsberg *erecta*. In parallel, leaves of four-weeks-old Arabidopsis Col-0 plants were investigated (Marondedze et al. 2016). Of 1145 proteins identified altogether in these three samples, 914 appeared only in UV cross-linked samples, and 233 proteins were significantly enriched upon UV cross-linking relative to non-cross-linked samples. More than 350 proteins were known RBDs whereas 736 were novel candidate RBPs not previously assigned an RNA-related function or known RBD, including many enzymes of intermediary metabolism, and thus await further experimental proof (Marondedze et al. 2016).

The discovery of many novel RBPs begs for further investigation of the RNA-binding properties of these proteins. Accordingly, methods to define RNA targets of candidate RBPs genome wide using protein-centric methods have recently been adapted for the use in Arabidopsis, as discussed below.

3 Toward Arabidopsis Ribonomes

Approaches to globally identify in vivo targets of an RBP in Arabidopsis mostly rely on transgenic plants expressing an epitope-tagged version of the RBP. Immunopurification is performed via an antibody directed against the epitope tag. To mirror-image the endogenous expression pattern, authentic promoters are used and the constructs are introduced into a loss-of-function mutant (Köster and Staiger 2014). Alternatively, endogenous RBPs can be recovered with dedicated antibodies.

To freeze the in vivo RNA–protein interactions before cell lysis, cross-linking is performed by exposing plants to formaldehyde in RNA immunoprecipitation (RIP) or by UV irradiation in UV cross-linking and immunoprecipitation (CLIP) (Fig. 1b). Formaldehyde efficiently cross-links nucleic acids and proteins in vivo but also cross-links proteins. Thus, not only direct targets are recovered. This is circumvented by using 254 nm UV light that cross-links proteins directly binding to nucleic acids in the neighborhood of the excited nucleobase but does not cross-link proteins.

To date, a comprehensive determination of in vivo targets, the ribonome, has been performed for only a few Arabidopsis RBPs, both nucleocytoplasmic proteins and chloroplast-localized proteins with different tasks in posttranscriptional regulation. In the subsequent sections, selected examples are presented.

3.1 HLP1, An hnRNP A/B-Like Protein Involved in Alternative Polyadenylation

HLP1 is an Arabidopsis RBP resembling mammalian hnRNP A/B-like proteins (Zhang et al. 2015). High throughput sequencing (HITS)-CLIP of HLP1 fused to GFP and expressed under control of the strong, constitutive Cauliflower Mosaic Virus 35S RNA promoter identified above 5500 transcripts bound in vivo (Zhang et al. 2015). When endogenous HLP1 protein was precipitated by a specific antibody, 6850 transcripts bound in vivo were detected with an overlap of above 3000 transcripts to the HLP1-GFP precipitation. The prevalence of cross-linked regions near polyadenylation sites provoked the hypothesis that HLP1 may control polyadenylation. Indeed, in more than 2000 transcripts the distal polyadenylation site was preferred over the proximal polyadenylation site in *hlp1* mutant plants. Around 19% of these transcripts were also recovered by HLP1 HITS-CLIP, pointing to a role for HLP1 in the control of alternative polyadenylation, at least partly by direct binding. In line with this, MEME motifs overrepresented in the cross-link regions, namely A-rich (5′-AGAAAA-3′) and U-rich (5′-UUUUCU-3′) motifs, resembled motifs enriched in the vicinity of the poly(A) site, 5′-AAAGAAAA-3′ and 5′-UGUUUC-3′. The presence of cross-link regions in other parts of the transcripts apart from the 3′ untranslated region (UTR) suggests that HLP1 may also affect other aspects of pre-mRNA processing in addition to polyadenylation.

3.2 The Glycine-Rich RBP *At*GRP7

*At*GRP7 (*Arabidopsis thaliana* glycine rich RNA-binding protein 7) is another hnRNP-like protein with an N-terminal RRM and a C-terminus enriched in contiguous glycine residues. *At*GRP7 is regulated by the circadian clock and negatively autoregulates its own oscillations by alternative splicing and Nonsense-mediated decay (Staiger et al. 2003; Schmal et al. 2013). Additionally, it is involved in several steps of posttranscriptional regulation including alternative splicing, nucleic acid chaperone function, and pri-miRNA processing (Kim et al. 2007a; Streitner et al. 2012; Köster et al. 2014). To gain insights into the breadth of its in vivo targets, individual nucleotide resolution cross-linking and immunoprecipitation (iCLIP) and RIP-seq were performed (Meyer et al. 2017). *At*GRP7 fused to GFP was expressed from its own promoter including all regulatory elements (5′ UTR, intron, and 3′ UTR) in the *atgrp7-1* loss-of-function mutant. In parallel, transgenic plants expressing GFP alone or an RNA-binding dead variant of *At*GRP7 with a single conserved arginine in the RRM mutated to glutamine (*At*GRP7 $R^{49}Q$) were used as negative controls.

iCLIP identified 858 transcripts with significant iCLIP hits in four out of five biological replicates for *At*GRP7-GFP that were not present in the controls. RIP-seq identified 2453 transcripts enriched by *At*GRP7-GFP relative to total polyadenylated RNA. The higher number may be due to the higher cross-linking efficiency of formaldehyde compared to UV light, and the recovery of many indirect targets. 452 transcripts were common in both data sets, suggesting that they represent a set of high confidence binders. The iCLIP cross-link sites were observed in all transcript regions, the UTRs, coding sequence and introns. After correcting for the length of the feature in the genome, cross-link sites in the 3′ UTR prevailed. Conserved motifs in the vicinity of the cross-link sites generally were U/C rich.

To determine how *AtGRP7* may impact its downstream targets, the binding targets were cross-referenced against transcriptome data from *At*GRP7 overexpressing plants or loss-of-function mutants. In both, the *At*GRP7 overexpressors or the mutant, a similar number of transcripts was expressed at elevated or reduced levels compared to wild-type plants. Notably, significantly more differentially expressed iCLIP targets were downregulated in *At*GRP7-overexpressors than upregulated. In turn, more of the differentially expressed *At*GRP7 iCLIP targets were expressed at elevated in the mutant than at reduced levels. This indicates a predominantly negative effect of *At*GRP7 on its targets. Among the targets were more circadianly regulated transcripts than expected. In particular, elevated *At*GRP7 levels lead to damping of circadian oscillations of target transcripts including *DORMANCY/AUXIN ASSOCIATED FAMILY PROTEIN2* and *CCR-LIKE*. This conforms with the idea that the circadian clock regulated *At*GRP7 functions as a molecular slave oscillator, conveying temporal information from the core circadian clock within the cell (Rudolf et al. 2004). In addition, changes in splicing patterns were observed for iCLIP and RIP-seq targets upon misexpression of *At*GRP7, confirming a role for *At*GRP7 in the control of alternative splicing.

3.3 The Splicing Regulator SR45

Arabidopsis thaliana serine/arginine rich (SR)-like protein SR45, the counterpart of metazoan RNPS1, is an SR-like protein with two RS domains, flanking either side of the RRM (Badolato et al. 1995; Golovkin and Reddy 1999). Notably, recombinant Arabidopsis SR45 can activate splicing of a β-globin splicing reporter in HeLa cell S100 extracts (Ali et al. 2007). SR45 occurs in two splice isoforms that arise through differential usage of a 3′ splice site in intron 6. This leads to two protein isoforms that differ by seven amino acid residues and in their function: SR45.1 is involved in petal development in flowers, whereas SR45.2 is important for root growth (Zhang and Mount 2009). Genome-wide targets for SR45.1 were determined during early seedling development (Xing et al. 2015) and in inflorescences (Zhang et al. 2017), respectively.

In seedlings, RIP-seq identified 4361 transcripts from 4262 genes that were enriched upon precipitation of SR45.1-GFP from nuclei of transgenic plants compared to mock precipitation from wild type plants (Xing et al. 2015). These were designated SARs, for SR45 associated RNAs. A Gene Ontology term analysis showed that 43 of 147 abscisic acid (ABA) signaling genes (30%) were among the SARs, in line with a function for SR45 in the ABA signaling pathway (Carvalho et al. 2010). Hundred and forty-eight of the SARs had an altered expression in the *sr45-1* mutant, suggesting that binding of SR45 has functional consequences.

A MEME search for SR45 binding motifs revealed four overrepresented motifs within SAR genes. Two G/A rich motifs are largely positioned within exons and show strong similarity to the binding motifs of two metazoan splicing regulators Transformer 2 (Tra2) and serine/arginine-rich splicing factor 10 (SRSF10). Furthermore, one G/A rich motif closely resembles the GAAG motif, a known *cis*-regulatory element in regulating alternative splicing in plants. In contrast, two U/C rich motifs peak within intronic regions near 5′ and 3′ splice sites, in line with the observation that the majority of SARs were from intron-containing genes and the known role as a splicing regulator (Xing et al. 2015).

To gain insights into a potential role of SR45 in flower development, RIP-seq was performed for SR45.1-GFP in inflorescence tissue (Zhang et al. 2017). The resulting reads were analyzed by two different bioinformatics pipelines, one based on mapping reads to the genome and one directly quantifying annotated transcripts. SARs in inflorescence were defined based on a twofold enrichment compared to GFP only controls and the identification by both pipelines. Of 1812 SARs in inflorescence, 677 overlapped with the SARs in seedlings.

Notably, 19 transcripts encoding splicing factors were among the SARs including SR45 itself, the three SR proteins SR30, SR34, and SCL35, the pre-mRNA processing factors PRP39, PRP40A, PRP40B, and PRP2, and the RNA helicase RH42, pointing to a hierarchical regulation of posttranscriptional regulators (Keene 2007). Genes upregulated in the *sr45-1* mutant are enriched for defense response genes. Indeed, the *sr45-1* mutant was more resistant to bacterial and fungal pathogens. Of 68 upregulated defense response genes in *sr45-1*, 10 were SARs. Thus, SR45 has an additional role as a negative regulator of plant immunity.

Furthermore, 81 of the inflorescence SARs were aberrantly spliced in the *sr45-1* mutant. Determination of potential SR45 binding sites in inflorescence SARs uncovered an overrepresentation of the purine-rich motifs GGNGG, GNGGA, and GNGGNNG. Importantly, GGNGG and related motifs are enriched in introns and exons that are alternatively spliced in the *sr45-1* mutant, irrespective of the splicing event is favored or suppressed by SR45. This led to the suggestion that SR45 identifies regions for alternative splicing and acts as a facilitator for other splicing factors. However, the identified binding motifs for SR45 in inflorescences differ from that in seedlings, which might be in part due to the different bioinformatic tools used for motif determination. Both RIP-seq data sets nevertheless strengthen SR45's key role as an important splicing factor in Arabidopsis. However, in both RIP-seq experiments intron-less transcripts were identified in addition to intron-containing transcripts, pointing to functions of SR45 beyond its known role in pre-mRNA splicing.

Interestingly, a comparison between the U/C-rich motifs of *At*GRP7 and the U/C-rich motifs of SR45 identified by MEME in seedlings revealed a high degree of similarity (Meyer et al. 2017). The functional significance remains to be tested.

3.4 Cold Shock Protein 1

In bacteria, CSPs are upregulated upon cold stress and destabilize RNA secondary structure at low temperatures (Sommerville 1999). To elucidate a potential involvement of Arabidopsis CSPs in the regulation of cold responsive genes, RIP followed by gene chip analysis was performed for CSP1 (Juntawong et al. 2013).

More than 6000 mRNAs were identified. Comparison of these CSP1-associated transcripts in total RNA and RNA loaded onto polysomes revealed an enrichment of mRNAs associated with ribosome biogenesis in the pool of actively translating RNAs. The high GC content in 5′ UTRs of these mRNAs suggested that CSP1 is involved in removing secondary structures in the 5′ UTR to facilitate their translation. Accordingly, these mRNAs were less efficiently loaded onto polysomes at low temperature in the *atcsp1-1* mutant compared to wild type plants or CSP1 overexpressing plants (Juntawong et al. 2013).

3.5 The cpRNP Family

The highly abundant chloroplast ribonucleoproteins (cpRNPs) have been well characterized for their role in regulating chloroplast transcripts (Ohta et al. 1995). The cpRNPs comprise an acidic domain and two RRMs. They are encoded in the nucleus and imported into chloroplasts. Mutants in distinct cpRNPs are widely affected in processing of transcripts in the chloroplast, leading to defects in chloroplast development and, consequently, plant performance owing to the essential role of the chloroplast in photosynthetic energy (Ruwe et al. 2011). For example, mutants deficient in CP29A (29 kDa chloroplast protein A) and CP31A (31 kDa chloroplast protein A) showed gross defects at low ambient temperature. RIP performed with antibodies against the endogenous proteins and subsequent hybridization of coprecipitated RNAs on tiling arrays covering the Arabidopsis chloroplast genome (RIP-Chip) showed that CP29A and CP31A associate with large overlapping sets of chloroplast transcripts including strong enrichment for *psbB, psbD, psaA/B, atpB, ndhB* and intermediate enrichment for almost all chloroplast mRNAs (Kupsch et al. 2012). Both CP29A and CP31A are required for accumulation of chloroplast mRNAs under cold stress. Furthermore, binding of CP31A to 3′ ends of certain transcripts serves to protect these transcripts against 3′ exonuclease activity (Kupsch et al. 2012). Together with the known role of CP31A in RNA (Tillich et al. 2009) this points to multiple functions in posttranscriptional regulation in chloroplasts.

For CP33A (33 kDa chloroplast protein A), RIP-chip revealed an association with a large body of chloroplast mRNAs (Teubner et al. 2017). A global reduction in mRNAs and proteins making up the photosynthetic apparatus was found in the *cp33a* mutant. In line with a crucial role for CP33A in the development of the photosynthetic apparatus, *cp33a* null mutants have an albino phenotype and are not able to survive without external sucrose supply (Teubner et al. 2017).

3.6 The PPR Protein AtCPR1

In contrast to the broad substrate specificity of the cpRNPs, a very narrow substrate specificity was found for a representative of the PPR class of nuclear-encoded RBPs that are imported into organelles. *At*CPR1 (*Arabidopsis thaliana* CHLOROPLAST RNA PROCESSING 1) is important for the production of subunits of the thylakoid protein complexes (Ferrari et al. 2017). *Atcpr1* mutants are yellow-white because the subunits of the photosynthetic apparatus do not accumulate.

RIP-chip was performed for *At*CPR1 under native conditions. Hybridization of bound targets to chloroplast tiling arrays revealed specific binding of *At*CPR1 to only few transcripts, the *psaC* transcript encoding a photosystem I subunit, *petB-petD* encoding Cytochrome b_6 and the subunit IV of the cytochrome b_6/f complex. Because during RIP RNase was used to digest unprotected RNA, it was possible to delineate the binding regions. Binding to the *petB-petD* intergenic region correlated with a requirement for processing of the polycistronic transcript comprising *petB* and *petD* (Ferrari et al. 2017), thus providing proof for the functional relevance of the observed in vivo binding.

4 Combined Analysis of RNA–Protein Interaction and RNA Secondary Structure Landscapes

In addition to RNA sequence, RNA secondary structure also strongly influences the interaction of RBPs with their cognate RNA binding motifs (Cruz and Westhof 2009; Vandivier et al. 2016). RNA structure may facilitate binding of RBDs with a preference for double-stranded RNA or inhibit binding of RBPs with a preference for single-stranded RNA. Protein interaction profile sequencing (PIP-seq) allows simultaneous delineation of in vivo RNA secondary structure and protein-protected sites (PPSs) (Fig. 2) (Gosai et al. 2015). To identify PPSs, samples are treated with a single-strand specific or double-strand specific RNase. Proteins are then denatured before library preparation. To determine the RNA secondary structure, proteins are denatured by SDS and removed by protease digestion to make sites protected by proteins in vivo accessible for RNases. Collectively, motifs that are enriched in the samples used to determine protein protected sites compared to the samples used for structure determination are in vivo target sites of RBDs.

Gregory and coworkers applied PIP-seq to the nuclei of two specific cell types in the Arabidopsis roots that derive from epidermal cells through distinct differentiation, those cells bearing root hairs and those that do not (Foley et al. 2017). Distinct protein binding patterns were detected, and binding motifs either specific to hair cells, non-hair cells or common to both cell types were determined. To identify candidate proteins, RNA affinity chromatography was performed on immobilized oligonucleotides derived from enriched motifs. A GGN repeat motif enriched in sites protected in both hair cells and non-hair cells recovered SERRATE (SE) from root lysates, a zinc finger containing RBP involved in processing of miRNA precursors. A TG rich motif enriched in hair cell-specific protected sites identified AtGRP2, *At*GRP7 and *At*GRP8. Subsequently, *At*GRP8 was shown to regulate root hair development at the posttranscriptional level.

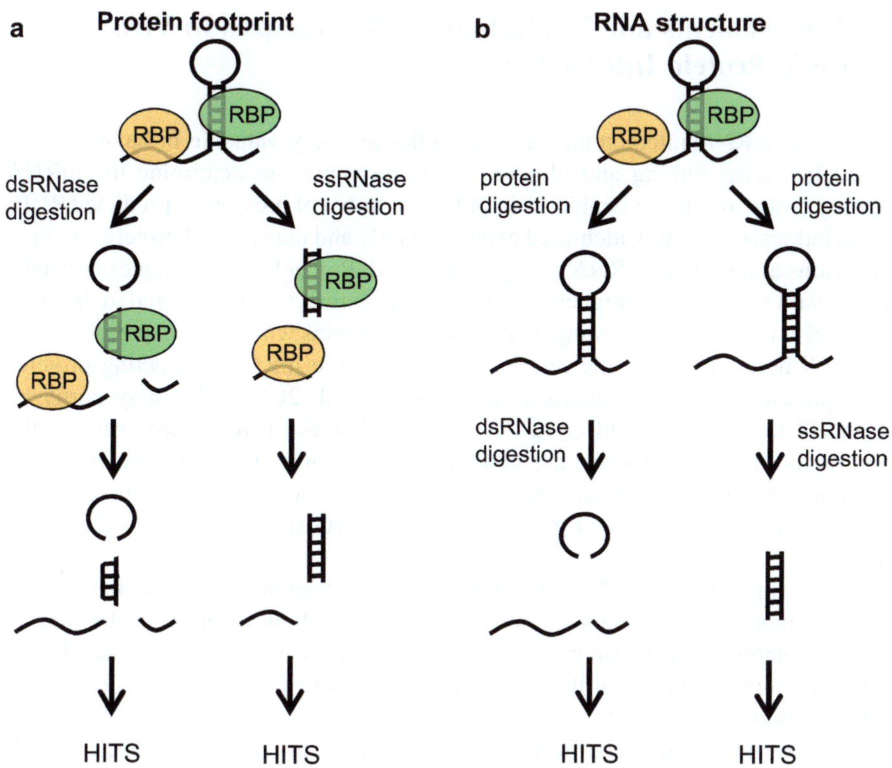

Fig. 2 Protein interaction profile sequencing (PIP-seq). (**a**) To identify protein binding sites, i.e., sites that are protected from RNase digestion by interacting proteins (PPS), samples are treated with an RNase specific for double-stranded RNA (left) or for single-stranded RNA (right). Subsequently, proteins are denatured, leaving either target sites for proteins with a preference for single-stranded regions (left), or target sites for proteins with a preference for double-stranded regions (right). These sequences are used to generate libraries for HITS. (**b**) To determine the RNA secondary structure, proteins are denatured in a first step. Subsequently, samples are treated with RNase specific for double-stranded RNA (left) or for single-stranded RNA (right). Again, libraries for HITS are prepared. Collectively, motifs that are enriched in the samples used to determine protein binding sites compared to the samples used for structure determination are in vivo target sites of RBDs

An advantage of PIP-seq is that it does not rely on an antibody to identify target sites within bound transcripts. In contrast, subsequent identification of the cognate binding proteins requires in vitro binding techniques. Thus, binding in vivo has to be confirmed by independent means.

5 Achievements and Limitations of Arabidopsis In Vivo RNA–Protein Interaction

The recent mRNA interactome capture studies are very valuable in having established UV cross-linking and oligo(dT) affinity capture to determine the mRNA binding proteome also in Arabidopsis. A large number of previously predicted RBPs in Arabidopsis were now identified experimentally and many novel proteins without a previous assignment to RNA biology unearthed. Reichel and colleagues noticed a bias toward proteins with higher abundance in the interactome compared to the input (Reichel et al. 2016), suggesting that additional proteins with lower expression level may still be identified in the future. Only few of the mRNA interacting proteins were present in all three interactomes (Köster et al. 2017). This may partly be attributed to the widely differing developmental stages investigated. Among the commonly identified proteins are numerous cytoplasmic ribosomal proteins from the small and large ribosomal subunits, likely due to their high abundance, as well as the ubiquitously expressed glycine-rich RBPs *At*GRP7 and *At*GRP8 (Köster et al. 2017).

Future applications are the dynamics of posttranscriptional networks in response to endogenous and exogenous stimuli cues by describing changes in the mRNA bound proteomes. Furthermore, as proteins binding to nonpolyadenylated RNAs obviously remain elusive in these approaches, transcript-specific approaches have to be developed.

Transcriptome-wide identification of target transcripts bound by selected RBPs in vivo has overcome a major limitation in research on plant RNA-based regulation. Nevertheless, except for the PPR proteins, we are still far from understanding the exact binding specificity of most proteins and the consequences in vivo binding has for the targets. To correlate in vivo binding with function, the impact of mutated candidate binding motifs on RBP binding and target gene expression has to be determined.

Most bioinformatics pipelines today discussing motif discovery are limited to sequence data. Current efforts focus on developing bioinformatics pipelines for identifying conserved motifs taking RNA structure context into consideration (Maticzka et al. 2014). Molecular dynamics of RNA molecules are still compute intensive but can shed light on possible interaction sites and three dimensional structures (Tuszynska et al. 2015; Boniecki et al. 2016). Finally, heterogeneous datasets and analyses, fusing several kinds of sources, can improve meta-analysis with in silico and in vivo datasets. This is yet limited in Arabidopsis but will improve the information quality in the near future. Additionally, it will be important to have comprehensive databases on RBP target sites linked to the Arabidopsis information portal (The International Arabidopsis Informatics Consortium 2012). Such resources will be of great value to improve a systems understanding of RNA–protein interaction.

Acknowledgments The work in Tino Köster's lab is supported by the DFG through grant KO 5364/1-1. Martin Lewinski is supported by the DFG through grant STA653/6-1 to Dorothee Staiger.

References

Ali GS, Palusa SG, Golovkin M et al (2007) Regulation of plant developmental processes by a novel splicing factor. PLoS One 2:e471

Badolato J, Gardiner E, Morrison N et al (1995) Identification and characterisation of a novel human RNA-binding protein. Gene 166:323–327

Baltz AG, Munschauer M, Schwanhäusser B et al (2012) The mRNA-bound proteome and its global occupancy profile on protein-coding transcripts. Mol Cell 46:674–690

Barkan A, Small I (2014) Pentatricopeptide repeat proteins in plants. Annu Rev Plant Biol 65:415–442

Beckmann BM, Horos R, Fischer B et al (2015) The RNA-binding proteomes from yeast to man harbour conserved enigmRBPs. Nat Commun 6:10127

Boniecki MJ, Lach G, Dawson WK et al (2016) SimRNA: a coarse-grained method for RNA folding simulations and 3D structure prediction. Nucleic Acids Res 44:e63–e63

Bunnik EM, Batugedara G, Saraf A et al (2016) The mRNA-bound proteome of the human malaria parasite Plasmodium falciparum. Genome Biol 17:147

Carvalho RF, Carvalho SD, Duque P (2010) The plant-specific SR45 protein negatively regulates glucose and ABA signaling during early seedling development in Arabidopsis. Plant Physiol 154:772–783

Carvalho RF, Szakonyi D, Simpson CG et al (2016) The Arabidopsis SR45 splicing factor, a negative regulator of sugar signaling, modulates SNF1-related protein kinase 1 stability. Plant Cell 28:1910–1925

Castello A, Fischer B, Eichelbaum K et al (2012) Insights into RNA biology from an atlas of mammalian mRNA-binding proteins. Cell 149:1393–1406

Cruz JA, Westhof E (2009) The dynamic landscapes of RNA architecture. Cell 136:604–609

Dong Z, Han MH, Fedoroff N (2008) The RNA-binding proteins HYL1 and SE promote accurate in vitro processing of pri-miRNA by DCL1. Proc Natl Acad Sci USA 105:9970–9975

Ferrari R, Tadini L, Moratti F et al (2017) CRP1 Protein: (dis)similarities between *Arabidopsis thaliana* and Zea mays. Front Plant Sci 8:163

Foley SW, Gosai SJ, Wang D et al (2017) A global view of RNA-protein interactions identifies post-transcriptional regulators of root hair cell fate. Dev Cell 41:204–220

Fu ZQ, Guo M, Jeong BR et al (2007) A type III effector ADP-ribosylates RNA-binding proteins and quells plant immunity. Nature 447:284–288

Galgano A, Gerber AP (2011) RNA-binding protein immunopurification-microarray (RIP-Chip) analysis to profile localized RNAs. Methods Mol Biol 714:369–385

Golovkin M, Reddy AS (1999) An SC35-like protein and a novel serine/arginine-rich protein interact with *Arabidopsis* U1-70K protein. J Biol Chem 274:36428–36438

Gosai S, Foley Shawn W, Wang D et al (2015) Global analysis of the RNA-protein interaction and RNA secondary structure landscapes of the *Arabidopsis* nucleus. Mol Cell 57:829–845

Goyal M, Banerjee C, Nag S et al (2016) The Alba protein family: structure and function. Biochim Biophys Acta Proteins Proteomics 1864:570–583

Guerreiro A, Deligianni E, Santos J et al (2014) Genome-wide RIP-chip analysis of translational repressor-bound mRNAs in the Plasmodium gametocyte. Genome Biol 15:493

Hackmann C, Korneli C, Kutyniok M et al (2014) Salicylic acid-dependent and -independent impact of an RNA-binding protein on plant immunity. Plant Cell Environ 37:696–706

Heintzen C, Melzer S, Fischer R et al (1994) A light- and temperature-entrained circadian clock controls expression of transcripts encoding nuclear proteins with homology to RNA-binding proteins in meristematic tissue. Plant J 5:799–813
Hornyik C, Terzi LC, Simpson GG (2010) The spen family protein FPA controls alternative cleavage and polyadenylation of RNA. Dev Cell 18:203–213
Hugouvieux V, Kwak JM, Schroeder JI (2001) An mRNA cap binding protein, ABH1, modulates early abscisic acid signal transduction in *Arabidopsis*. Cell 106:477–487
Jeong B, Lin Y, Joe A et al (2011) Structure function analysis of an ADP-ribosyltransferase type III effector and its RNA-binding target in plant immunity. J Biol Chem 286:43272–43281
Jones MA, Williams BA, McNicol J et al (2012) Mutation of *Arabidopsis* SPLICEOSOMAL TIMEKEEPER LOCUS1 causes circadian clock defects. Plant Cell 24:4907–4916
Juntawong P, Sorenson R, Bailey-Serres J (2013) Cold shock protein 1 chaperones mRNAs during translation in *Arabidopsis thaliana*. Plant J 74:1016–1028
Kalyna M, Lopato S, Barta A (2003) Ectopic expression of at RSZ33 reveals its function in splicing and causes pleiotropic changes in development. Mol Biol Cell 14:3565–3577
Keene JD (2007) RNA regulons: coordination of post-transcriptional events. Nat Rev Genet 8:533–543
Kim JS, Park SJ, Kwak KJ et al (2007a) Cold shock domain proteins and glycine-rich RNA-binding proteins from *Arabidopsis thaliana* can promote the cold adaptation process in Escherichia coli. Nucleic Acids Res 35:506–516
Kim JY, Park SJ, Jang B et al (2007b) Functional characterization of a glycine-rich RNA-binding protein 2 in *Arabidopsis thaliana* under abiotic stress conditions. Plant J 50:439–451
Kim YO, Pan S, Jung CH et al (2007c) A zinc finger-containing glycine-rich RNA-binding protein, at RZ-1a, has a negative impact on seed germination and seedling growth of *Arabidopsis thaliana* under salt or drought stress conditions. Plant Cell Physiol 48:1170–1181
Kim JS, Kim KA, Oh TR et al (2008) Functional characterization of DEAD-box RNA helicases in *Arabidopsis thaliana* under abiotic stress conditions. Plant Cell Physiol 49:1563–1571
Kim JY, Kim WY, Kwak KJ et al (2010) Glycine-rich RNA-binding proteins are functionally conserved in *Arabidopsis thaliana* and Oryza sativa during cold adaptation process. J Exp Bot 61:2317–2325
König J, Zarnack K, Rot G et al (2010) iCLIP reveals the function of hnRNP particles in splicing at individual nucleotide resolution. Nat Struct Mol Biol 17:909–915
Köster T, Staiger D (2014) RNA-binding protein Immunoprecipitation from whole-cell extracts. Methods Mol Biol 1062:679–695
Köster T, Meyer K, Weinholdt C et al (2014) Regulation of pri-miRNA processing by the hnRNP-like protein *At*GRP7 in Arabidopsis. Nucleic Acids Res 42:9925–9936
Köster T, Marondedze C, Meyer K et al (2017) RNA-binding proteins revisited: the emerging Arabidopsis mRNA interactome. Trends Plant Sci 22:512–526
Krause K, Herrmann U, Fuß J et al (2009) Whirly proteins as communicators between plant organelles and the nucleus? Endocytobiosis Cell Res 19:51–62
Kupsch C, Ruwe H, Gusewski S et al (2012) *Arabidopsis* chloroplast RNA binding proteins CP31A and CP29A associate with large transcript pools and confer cold stress tolerance by influencing multiple chloroplast RNA processing steps. Plant Cell 24:4266–4280
Kwon SC, Yi H, Eichelbaum K et al (2013) The RNA-binding protein repertoire of embryonic stem cells. Nat Struct Mol Biol 20:1122–1130
Lewinski M, Hallmann A, Staiger D (2016) Genome-wide identification and phylogenetic analysis of plant RNA binding proteins comprising both RNA recognition motifs and contiguous glycine residues. Mol Gen Genomics 291:763–773
Li D, Zhang H, Hong Y et al (2014) Genome-wide identification, biochemical characterization, and expression analyses of the YTH domain-containing RNA-binding protein family in *Arabidopsis* and Rice. Plant Mol Biol Report 32:1169–1186
Liao Y, Castello A, Fischer B et al (2016) The cardiomyocyte RNA-binding proteome: links to intermediary metabolism and heart disease. Cell Rep 16:1456–1469

Licatalosi DD, Mele A, Fak JJ et al (2008) HITS-CLIP yields genome-wide insights into brain alternative RNA processing. Nature 456:464–469

Löhr B, Streitner C, Steffen A et al (2014) A glycine-rich RNA-binding protein affects gibberellin biosynthesis in *Arabidopsis*. Mol Biol Rep 41:439–445

Lopato S, Kalyna M, Dorner S et al (1999) atSRp30, one of two SF2/ASF-like proteins from *Arabidopsis thaliana*, regulates splicing of specific plant genes. Genes Dev 13:987–1001

Lorkovic ZJ, Wieczorek Kirk DA, Klahre U et al (2000) RBP45 and RBP47, two oligouridylate-specific hnRNP-like proteins interacting with poly(A)+ RNA in nuclei of plant cells. RNA 6:1610–1624

Lu C, Fedoroff N (2000) A mutation in the Arabidopsis HYL1 gene encoding a dsRNA binding protein affects responses to abscisic acid, auxin, and cytokinin. Plant Cell 12:2351–2366

Lueong S, Merce C, Fischer B et al (2016) Gene expression regulatory networks in *Trypanosoma brucei*: insights into the role of the mRNA-binding proteome. Mol Microbiol 100:457–471

Lyons R, Iwase A, Gänsewig T et al (2013) The RNA-binding protein FPA regulates flg22-triggered defense responses and transcription factor activity by alternative polyadenylation. Sci Rep 3:2866

Macknight R, Bancroft I, Page T et al (1997) FCA, a gene controlling flowering time in Arabidopsis, encodes a protein containing RNA-binding domains. Cell 89:737–745

Marondedze C, Thomas L, Serrano NL et al (2016) The RNA-binding protein repertoire of *Arabidopsis thaliana*. Sci Rep 6:29766

Matia-Gonzalez AM, Laing EE, Gerber AP (2015) Conserved mRNA-binding proteomes in eukaryotic organisms. Nat Struct Mol Biol 22:1027–1033

Maticzka D, Lange S, Costa F et al (2014) GraphProt: modeling binding preferences of RNA-binding proteins. Genome Biol 15:R17

Meyer K, Köster T, Nolte C et al (2017) Adaptation of iCLIP to plants determines the binding landscape of the clock-regulated RNA-binding protein AtGRP7. Genome Biol 18:204

Mitchell SF, Jain S, She M et al (2013) Global analysis of yeast mRNPs. Nat Struct Mol Biol 20:127–133

Müller-McNicoll M, Botti V, de Jesus Domingues AM et al (2016) SR proteins are NXF1 adaptors that link alternative RNA processing to mRNA export. Genes Dev 30:553–566

Nandan D, Thomas SA, Nguyen A et al (2017) Comprehensive identification of mRNA-binding proteins of *Leishmania donovani* by interactome capture. PLoS One 12:e0170068

Nicaise V, Joe A, Jeong B et al (2013) Pseudomonas HopU1 affects interaction of plant immune receptor mRNAs to the RNA-binding protein GRP7. EMBO J 32:701–712

Ohta M, Sugita M, Sugiura M (1995) Three types of nuclear genes encoding chloroplast RNA-binding proteins (cp29, cp31 and cp33) are present in *Arabidopsis thaliana*: presence of cp31 in chloroplasts and its homologue in nuclei/cytoplasms. Plant Mol Biol 27:529–539

Park SJ, Kwak KJ, Oh TR et al (2009) Cold shock domain proteins affect seed germination and growth of *Arabidopsis thaliana* under abiotic stress conditions. Plant Cell Physiol 50:869–878

Perez-Santángelo S, Mancini E, Francey LJ et al (2014) Role for LSM genes in the regulation of circadian rhythms. Proc Natl Acad Sci USA 111:15166–15171

Prikryl J, Watkins KP, Friso G et al (2008) A member of the Whirly family is a multifunctional RNA- and DNA-binding protein that is essential for chloroplast biogenesis. Nucleic Acids Res 36:5152–5165

Qi Y, Tsuda K, Joe A et al (2010) A putative RNA-binding protein positively regulates salicylic acid-mediated immunity in *Arabidopsis*. Mol Plant Microbe Interact 23:1573–1583

Reichel M, Liao Y, Rettel M et al (2016) In planta determination of the mRNA-binding proteome of Arabidopsis etiolated seedlings. Plant Cell 28:2435–2452

Ren G, Xie M, Dou Y et al (2012) Regulation of miRNA abundance by RNA binding protein TOUGH in *Arabidopsis*. Proc Natl Acad Sci USA 109:12817–12821

Riera M, Redko Y, Leung J (2006) *Arabidopsis* RNA-binding protein UBA2a relocalizes into nuclear speckles in response to abscisic acid. FEBS Lett 580:4160–4165

Ripoll JJ, Ferrandiz C, Martinez-Laborda A et al (2006) PEPPER, a novel K-homology domain gene, regulates vegetative and gynoecium development in *Arabidopsis*. Dev Biol 289:346–359

Rossbach O, Hung L-H, Khrameeva E et al (2014) Crosslinking-immunoprecipitation (iCLIP) analysis reveals global regulatory roles of hnRNP L. RNA Biol 11:146–155

Rudolf F, Wehrle F, Staiger D (2004) Slave to the rhythm. Biochemist 26:11–13

Rühl C, Stauffer E, Kahles A et al (2012) Polypyrimidine tract binding protein homologs from *Arabidopsis* are key regulators of alternative splicing with implications in fundamental developmental processes. Plant Cell 24:4360–4375

Ruwe H, Kupsch C, Teubner M et al (2011) The RNA-recognition motif in chloroplasts. J Plant Physiol 168:1361–1371

Schmal C, Reimann P, Staiger D (2013) A circadian clock-regulated toggle switch explains AtGRP7 and AtGRP8 oscillations in *Arabidopsis thaliana*. PLoS Comput Biol 9:e1002986

Schmidt F, Marnef A, Cheung M-K et al (2010) A proteomic analysis of oligo(dT)-bound mRNP containing oxidative stress-induced *Arabidopsis thaliana* RNA-binding proteins *AT*GRP7 and *AT*GRP8. Mol Biol Rep 37:839–845

Schmitz-Linneweber C, Small I (2008) Pentatricopeptide repeat proteins: a socket set for organelle gene expression. Trends Plant Sci 13:663–670

Silverman IM, Li F, Gregory BD (2013) Genomic era analyses of RNA secondary structure and RNA-binding proteins reveal their significance to post-transcriptional regulation in plants. Plant Sci 205-206:55–62

Simpson GG, Dijkwel PP, Quesada V et al (2003) FY is an RNA 3′ end-processing factor that interacts with FCA to control the *Arabidopsis* floral transition. Cell 113:777–787

Sommerville J (1999) Activities of cold-shock domain proteins in translation control. BioEssays 21:319–325

Sorenson R, Bailey-Serres J (2014) Selective mRNA sequestration by OLIGOURIDYLATE-BINDING PROTEIN 1 contributes to translational control during hypoxia in *Arabidopsis*. Proc Natl Acad Sci USA 111:2373–2378

Staiger D (2001) RNA-binding proteins and circadian rhythms in *Arabidopsis thaliana*. Philos Trans R Soc Lond Ser B Biol Sci 356:1755–1759

Staiger D (2015) Shaping the Arabidopsis transcriptome through alternative splicing. Adv Bot 2015:13

Staiger D, Zecca L, Wieczorek Kirk DA et al (2003) The circadian clock regulated RNA-binding protein AtGRP7 autoregulates its expression by influencing alternative splicing of its own pre-mRNA. Plant J 33:361–371

Stauffer E, Westermann A, Wagner G et al (2010) Polypyrimidine tract-binding protein homologues from Arabidopsis underlie regulatory circuits based on alternative splicing and downstream control. Plant J 64:243–255

Streitner C, Danisman S, Wehrle F et al (2008) The small glycine-rich RNA-binding protein *At*GRP7 promotes floral transition in *Arabidopsis thaliana*. Plant J 56:239–250

Streitner C, Köster T, Simpson CG et al (2012) An hnRNP-like RNA-binding protein affects alternative splicing by in vivo interaction with target transcripts in *Arabidopsis thaliana*. Nucleic Acids Res 40:11240–11255

Sysoev VO, Fischer B, Frese CK et al (2016) Global changes of the RNA-bound proteome during the maternal-to-zygotic transition in Drosophila. Nat Commun 7:12128

Tenenbaum SA, Carson CC, Lager PJ et al (2000) Identifying mRNA subsets in messenger ribonucleoprotein complexes by using cDNA arrays. Proc Natl Acad Sci USA 97:14085–14090

Teubner M, Fuß J, Kühn K et al (2017) The RRM protein CP33A is a global ligand of chloroplast mRNAs and is essential for plastid biogenesis and plant development. Plant J 89:472–485

The International Arabidopsis Informatics Consortium (2012) Taking the next step: building an Arabidopsis information portal. Plant Cell 24:2248–2256

Tillich M, Hardel SL, Kupsch C et al (2009) Chloroplast ribonucleoprotein CP31A is required for editing and stability of specific chloroplast mRNAs. Proc Natl Acad Sci USA 106:6002–6007

Tuszynska I, Magnus M, Jonak K et al (2015) NPDock: a web server for protein–nucleic acid docking. Nucleic Acids Res 43:W425–W430

Vandivier LE, Anderson SJ, Foley SW et al (2016) The conservation and function of RNA secondary structure in plants. Annu Rev Plant Biol 67:463–488

Vazquez F, Gasciolli V, Crete P et al (2004) The nuclear dsRNA binding protein HYL1 is required for microRNA accumulation and plant development, but not posttranscriptional transgene silencing. Curr Biol 14:346–351

Völz R, von Lyncker L, Baumann N et al (2012) LACHESIS-dependent egg-cell signaling regulates the development of female gametophytic cells. Development 139:498–502

Wang X, Lu Z, Gomez A et al (2014) N^6-methyladenosine-dependent regulation of messenger RNA stability. Nature 505:117–120

Wessels H-H, Imami K, Baltz AG et al (2016) The mRNA-bound proteome of the early fly embryo. Genome Res 26:1000–1009

Xing D, Wang Y, Hamilton M et al (2015) Transcriptome-wide identification of RNA targets of Arabidopsis SERINE/ARGININE-RICH45 uncovers the unexpected roles of this RNA binding protein in RNA processing. Plant Cell 27:3294–3308

Zhang X-N, Mount SM (2009) Two alternatively spliced isoforms of the *Arabidopsis thaliana* SR45 protein have distinct roles during normal plant development. Plant Physiol 150:1450–1458

Zhang Y, Gu L, Hou Y et al (2015) Integrative genome-wide analysis reveals HLP1, a novel RNA-binding protein, regulates plant flowering by targeting alternative polyadenylation. Cell Res 25:864–876

Zhang Z, Boonen K, Ferrari P et al (2016) UV crosslinked mRNA-binding proteins captured from leaf mesophyll protoplasts. Plant Methods 12:42

Zhang X-N, Shi Y, Powers JJ et al (2017) Transcriptome analyses reveal SR45 to be a neutral splicing regulator and a suppressor of innate immunity in *Arabidopsis thaliana*. BMC Genomics 18:772

Systems-Level Analysis of Bacterial Regulatory Small RNA Networks

Julia Wong, Ignatius Pang, Marc Wilkins, and Jai J. Tree

Contents

Abstract The RNA landscape of all sequenced bacteria is littered with regulatory noncoding small RNAs (sRNA). Understanding the functions of these sRNAs has lagged behind their identification, as few high-throughput approaches existed to capture sRNA interactions in vivo. Recently, methodologies have been described that allow for profiling of the sRNA interaction network facilitating systems-level analysis sRNA regulation. This chapter discusses recent advances in our understanding of sRNA function, technical advances that allow us to capture sRNA interactions in vivo, and the computational tools that allow meaningful conclusions to be drawn from these data.

Keywords Small regulatory RNAs · Bacteria · RNase E · Hfq · Networks · sRNA interactomes

J. Wong · I. Pang · M. Wilkins · J. J. Tree (✉)
School of Biotechnology and Biomolecular Sciences, University of New South Wales, Sydney, New South Wales, Australia
e-mail: j.tree@unsw.edu.au

N. Rajewsky et al. (eds.), *Systems Biology*, RNA Technologies,
https://doi.org/10.1007/978-3-319-92967-5_6

1 Introduction

Our understanding of the diverse functions of regulatory RNAs has accelerated with recent advances in high throughput sequencing technologies, but were less apparent when the first bacterial noncoding RNA was identified nearly half a century ago (Brownlee 1971). Progress was initially gradual and the first example of a *trans*-acting regulatory small RNA (sRNA) was not described until 13 years after the initial discovery of 6S RNA, when MicF was proposed to regulate OmpF translation by base-pairing with the *ompF* ribosomal binding site (Mizuno et al. 1984). The general utility of regulatory small RNA regulation was recognized at the time and these authors stated that an artificial *trans*-acting small RNA could repress Lpp translation, shown 30 years later to be the target of an endogenous small RNA generated from the 3′ UTR of *cutC* (MicL) (Guo et al. 2014). At first, examples of *trans*-acting small RNAs were limited, with the next tranche of sRNAs (DsrA, DicF, OxyS, GcvB, Spot42 and RprA) identified over the following 6–17 years (Altuvia et al. 1997; Argaman et al. 2001). These studies predicted the existence of many more small RNAs and an initial *in silico* analysis identified another 24 novel small RNAs in the model prokaryote *E. coli* (Argaman et al. 2001). With the advent of tiled DNA microarrays and RNA sequencing technologies it has become apparent that small RNAs are abundant and ubiquitous, with well-studied bacteria like *Salmonella* and *E. coli* transcribing hundreds of sRNAs (Barquist and Vogel 2015; Huang et al. 2009). The bottleneck has now become understanding the diverse functions of small RNAs that have been shown to act at the level of transcription termination, transcript stability/processing, and translation (Gottesman and Storz 2011; Sedlyarova et al. 2016).

Accumulating evidence indicates that transcriptional and posttranscriptional gene regulatory networks are highly interleaved and small regulatory RNAs participate in mixed transcriptional and posttranscriptional regulatory circuits. Mixed circuits can have simple advantages like switching positive regulatory signals for negative (Gogol et al. 2011). They may also have special properties, for example two positive transcriptional regulators acting on a gene will create an OR logic gate with respect to their stimuli. Because of the natural order enforced by transcription and translation the equivalent circuit with a positive transcription factor and small RNA establish an AND logic gate (Papenfort et al. 2015). Posttranscriptional regulation by sRNAs may also have less noise and sharper kinetics that will be particularly important for responding to acute stress (Feng et al. 2015; Levine et al. 2007; Schu et al. 2015). Significantly, they expand the regulatory space for signal input for gene expression and this will undoubtedly be exploited by bacterial pathogens to adapt to complex environments like host tissues (Barquist et al. 2016).

Over the last 5 years, RNA proximity-dependent ligation protocols have provided high-throughput snapshots of noncoding RNA interaction networks in yeast, *C. elegans*, and cultured cell lines. More recently, these have been applied to bacterial small RNA networks and have begun to map out the structure of the small RNA regulatory network (sRNA interactome). These analyses identify interaction sites between sRNA and target RNAs and because the RNA fragments recovered are relatively short, predictions of base complementarity can be made with a reasonable degree of confidence. These studies are not without sampling biases, but have so far supported small RNA interactions with the ribosomal binding site as the major, canonical pathway for regulation although a significant proportion of sRNA interactions fall outside of the RBS and many interact with noncoding RNAs suggesting that sponging interactions are prevalent, hinting at as yet uncharacterized functions.

Within a given small RNA network each node represents a small RNA or mRNA target and each edge represents an RNA–RNA interaction. It is already apparent that an interaction may have many different regulatory outcomes depending on the sequence, structure, and protein-binding context. In this sense, the regulatory outcomes for many edges within the sRNA interactome are unknown and provide fertile ground for exploration using emerging technologies for transcriptome-wide analysis of RNA processing, termination, and ribosome recruitment. In this chapter, we aim to provide an outline of recent work on systems-level analyses of sRNA regulatory pathways. The mechanisms of sRNA regulation are increasingly diverse and we review what the interaction edges may mean in the context of an interactome and discuss methodologies for capturing these interactions. Statistical analysis of high-throughput RNA–RNA interaction data is in its infancy and we review statistical filters used to extract meaningful networks from interactome datasets. Finally, we discuss some of the features of the sRNA–RNA interaction network when viewed at a systems-level.

2 Functional Consequences of Small RNA–mRNA Interactions

Transcription and translation of an mRNA is controlled by the secondary structure of the RNA, codon usage, stability of the transcript, and interactions with ribosomes and RNA binding proteins. Increasing evidence suggests that sRNA interactions can modulate many of these processes and also participate in convoluted anti-antisense derepression pathways. To complicate the matter, the sources of regulatory small RNAs are expanding and it is now appreciated that these species can be encoded as independent transcripts or be processed from larger RNAs including mRNAs (reviewed in Kavita et al. 2017). In this section, we discuss some of the varied functional consequences of sRNA–mRNA interactions with attention to sequence context that may be used to identify them in sRNA interactome data, and how these may generate different outputs from an sRNA regulatory circuit.

2.1 Mechanisms of Small RNA Repression of mRNA Expression

Canonically, sRNA–mRNA interactions lead to translational repression and mRNA degradation. From the initial work on MicF-*ompF* repression, it was evident that sRNAs may be binding at the RBS and occluding 30S ribosomal subunit association (Coleman et al. 1984). This activity is widely observed and was formalized in a study demonstrating that sRNA binding within a five-codon window—correlating well with the footprint of a ribosome—was able to repress 30S association (Bouvier et al. 2008). More recently the five-codon window has been extended with the discovery of translation activating stem loops starting at position +20 nt within the CDS (Jagodnik et al. 2017). These stem loops appear to stabilize 30S association at suboptimal ribosomal binding sites and may prevent 30S sliding along the mRNA before an initiation complex can be assembled. The sRNA, OmrA, was shown to prevent stem loop formation within the *fepA* mRNA and represses translation by binding at +28 nt downstream of the start codon, extending the sRNA repression window to nine codons.

Both initiating and elongating ribosomes are known to stabilize transcripts by steric interference of RNase interactions with mRNA 5′ ends and internal cleavage sites (Deana and Belasco 2005). This initially suggested that sRNA inhibition of translation initiation might be a prerequisite for mRNA degradation simply by displacing ribosomes and increasing RNase access to the transcript. Later work demonstrated that the C-terminal domain of the sRNA chaperone Hfq interacts with the major endoribonuclease RNase E and work on the sRNA–mRNA pairs RyhB-*sodB* and MicC-*ompD* demonstrated that sRNAs can specifically direct RNase E-dependent mRNA cleavage (Bandyra et al. 2012; Prévost et al. 2011). It is now evident that, independent of translation inhibition, sRNAs can direct the cleavage of a transcript 6–13 nt downstream of the sRNA–mRNA duplex (Bandyra et al. 2012; Waters et al. 2017) and also at more distal sites that promote mRNA turnover (Lalaouna et al. 2015). Small RNAs themselves are also susceptible to RNase E cleavage and may be coupled to degradation of their target mRNA (Massé et al. 2003). Importantly, processing of an sRNA by RNase E can change the kinetics of its posttranscriptional regulation, by generating an active form that exposes the sRNA seed region for base pairing with target mRNAs (Chao et al. 2017), and by generating a 5′ monophosphorylated end that stimulates RNase E activity toward the duplexed sRNA and mRNA (Bandyra et al. 2012; Chao et al. 2017). From these studies, it is apparent that RNase E cleavage has highly context dependent effects on sRNA regulation—being used to mature sRNAs, inactivate and degrade unpaired sRNAs, cleave mRNAs 3′ of sRNA duplexes, and cleave mRNAs at more distal sites. Each of these processes will have different effects on the kinetics of sRNA regulation.

Subtle differences in base pairing have also been shown to determine some of the differences in sRNA repression. *Vibrio cholerae* encodes five homologous sRNAs, Qrr1-5, that regulate gene expression in response to quorum sensing signals. Using

Qrr3 as a model, Feng et al. (2015) demonstrated that the base pairing pattern of Qrr3 with it target mRNA can determine whether the sRNA is degraded with its mRNA target (coupled degradation), released from the degraded mRNA (catalytic degradation), or sequestered without triggering degradation (sequestration). This differential regulation of mRNA targets seems at odds with the relative simplicity of the 109 nt sRNA and suggests that the sRNA contains sequence or structures that modulate the fate of the sRNA and mRNA. Indeed, Qrr3 is stabilized by a 5′ stem loop (SL1) and mRNA base pairing that destabilizes this stem leads to coupled degradation of Qrr3 and its target mRNA. Base paring that does not destabilize SL1, does not lead to degradation of Qrr3, allowing recycling of the sRNA and catalytic degradation of the mRNA target. Extensive base pairing between Qrr3 and a target mRNA that retains the SL1 structure also prevents degradation of the mRNA target, leading to sequestration of both RNAs. Notably, each of these mechanisms results in distinct regulatory kinetics with catalytic degradation providing potent suppression of mRNA expression, coupled degradation leading to threshold-linear response kinetics that are highly sensitive to the concentration of competing RNA targets, and sequestration leading to more moderate control (Feng et al. 2015). This study highlights that for an sRNA–mRNA pair, structures that stabilize each RNA can be modulated by base pairing with significant impacts on expression kinetics. Additional rules for predicting sRNA regulatory kinetics and whether such careful tuning of sRNA–mRNA interactions is true of all sRNAs, will become apparent as the sRNA interactome is explored further.

2.2 *Mechanisms of Small RNA Activation of mRNA Expression*

Regulatory interactions (edges) between sRNAs and mRNAs can also positively affect gene expression and may occur through a number of mechanisms. The most common examples are dependent on anti-antisense interactions that involve repressive structures encoded in *cis* or sponging repressive sRNA interactions (reviewed in Papenfort and Vanderpool 2015). Posttranscriptional activation is best characterized for the stationary phase sigma factor RpoS (σ^S) where the unusually long 5′ UTR (576 nt) folds into an inhibitory stem loop structure that occludes the ribosomal binding site (Soper et al. 2011). Hfq facilitates interactions with the closed complex and allows at least 3 sRNAs, DsrA, RprA, and ArcZ, to base pair with the leader at position +94 to +119 and destabilize the inhibitory stem loop. This in turn allows 30S ribosomal subunits access to the ribosomal binding site and translation of RpoS (Majdalani et al. 2002; Soper et al. 2010; Soper and Woodson 2008). Surprisingly, recent work has demonstrated these sRNAs also promote *rpoS* transcription by inhibiting Rho terminator binding or translocation along the *rpoS* leader (Sedlyarova et al. 2016). Small RNA interactions had previously been shown to promote Rho interactions with the *chiP* mRNA by blocking translation, exposing a Rho utilization site (*rut*) within *chiP*, and leading to transcription termination. In the case of *rpoS*, sRNA interactions have the opposite effect, displacing Rho and allowing transcription. Whether sRNAs prevent initial binding or displace translocating Rho remains to be determined, but clearly the position of sRNA binding relative the *rut* site is critical for determining whether sRNA interactions promote or prevent transcription termination. For *rpoS*, sRNAs activate both transcription and translation and RNA-Seq analysis of transcripts similarly controlled by sRNAs and the small molecule inhibitor of Rho, bicyclomycin, suggests that hundreds of transcripts may be activated by a similar sRNA-dependent anti-termination mechanism. Rho utilization sites consist of minimal sequence features: repeated YC motifs separated by >6 nt of unstructured RNA spanning 60–90 nt, that make in silico predictions of this mode of sRNA activation challenging. RNA-Seq analysis of bicyclomycin treated cells or direct immunoprecipitation of Rho–RNA complexes (RIP-Seq, CLIP-Seq, or CRAC) will likely provide the context that is critical for predicting this regulation in a broad range of bacteria.

Small RNA recruitment of RNases facilitates degradation of a large number of sRNA–mRNAs pairs; however, RNase interactions with an mRNA are also sensitive to local RNA sequence and structure making them susceptible to sRNA regulation through modulation of the RNA structure. The major *E. coli* endoribonuclease RNase E is stimulated by interactions with unstructured monophosphorylated 5′ ends of transcripts and cleaves RNA at unstructured RNWUU motifs with a strong preference for U at the +2 position (Chao et al. 2017). Small RNA interactions with the 5′ end of a transcript—which can be mapped using a growing list of high-throughput sequencing techniques—inhibits RNase E recruitment and RNase E stimulation. This mechanism was demonstrated for the long cyclopropane fatty acid synthase-encoding transcript *cfa* in *Salmonella enterica* where RydC inhibits RNase E recruitment to the 210 nt *cfa* leader by binding between –109 and –99 nt relative to the *cfa* start codon (Fröhlich et al. 2013). Slightly more striking is the finding that recruitment of an sRNA–Hfq complex to the bicistronic transcript *pldB-yigL* does not lead to cleavage and degradation of the transcript by RNase E but inhibits processing. The critical determinant that decides the fate of the transcript on interactions with sRNA–Hfq complex appears to be the relative proximity of RNase E cleavage sites and, in this case the sRNA SgrS, base pairs with the preferred RNase E cleavage site and blocks RNase E access (Papenfort et al. 2013).

A variant of the anti-antisense derepression pathways described for RpoS are sRNA sponging interactions that positively activate mRNA translation by blocking sRNA function. The genome of enterohemorrhagic *E. coli* is littered with cryptic prophage elements and these were found to carry sRNAs that target sRNAs rather than mRNAs (termed anti-sRNAs or sRNA sponges) (Tree et al. 2014). The sRNA sponge, AgvB, represses the activity of GcvB through base-pairing with the R1 seed region. AgvB binding appears to sequester GcvB rather than trigger degradation, consistent with its binding site within an internal unstructured region (R1 seed) with only minimal interactions with the long 5′ stem loop of GcvB. This is in contrast to a 3′UTR derived sRNA, SroC, that also sponges GcvB and bases pairs with the

5′ end (nts 13–19), at least partly destabilizing the 5′ stem loop and triggering degradation of GcvB (Miyakoshi et al. 2015). Many Hfq-dependent sRNAs also carry intrinsic terminators that protect the 3′ end from exonucleolytic attack. Sponging interactions between the sRNA, ChiX (MicM), and an intergenic region (IGR) within the *chbBC* transcript act to alleviate ChiX repression of *chiP*. Base-pairing between the IGR and ChiX extends into the GC-rich stem of the intrinsic terminator and destablizes ChiX, indicating that sRNA degradation can also be targeted to stabilizing structures within 3′ end of an sRNA (Plumbridge et al. 2014). The result of these sponging interactions is the derepression of sRNA repressed transcripts (activation of mRNA translation) and demonstrates that recently defined rules governing catalytic, coupled, or sequestered kinetics of sRNA regulation of mRNAs also apply to sRNA sponging interactions.

An exciting development is the realization that stably transcribed RNA species like tRNA spacers (that are presumably not regulated in response to environmental signals) can also participate in sponging interactions with sRNAs and activate gene expression. The 3′ external transcribed spacer (ETS) of tRNA^{leuZ} interacts with the sRNAs RybB and RyhB, and sets minimum criteria for sRNA and target mRNA interactions—the interaction is subject to a stringency filter that prevents low strength interactions, and the sRNA is subject to a concentration threshold set by the abundance of the ETS and interaction strength of the target mRNA. Small RNA sponging interactions with sRNA or pre-tRNA fragments appear exceedingly common in sRNA interactome datasets (Helwak et al. 2013; Melamed et al. 2016; Waters et al. 2017), suggesting that these interactions play an important role in buffering the sRNA response.

Since the initial observations suggesting MicF is able to repress *ompF* translation by occluding the ribosomal binding site, a complex picture of sRNA-dependent regulation has emerged. Small RNA interactions are often heavily contextualized by the local sequence, structure, and protein binding of both the sRNA and target RNA. This presents unique challenges for deciphering the regulatory network from high-throughput datasets as interactions may represent varying degrees of positive, negative, sequestered, or sponged interaction. Combining sRNA interactome data with many of the emerging techniques for transcriptome-wide profiling of RNA structure and protein binding will likely provide the context required to interpret the functions of sRNA interactions. In the next sections, we focus in approaches that allow sRNA interactions to be predicted in silico or captured and sequenced in vivo.

3 Computational Prediction of sRNA–Target Interactions

Trans-encoded sRNAs interact with their targets through short >6 nucleotide regions of complementarity and this short seed sequence forms the basis for the computational prediction of sRNA targets. Additional sequence features, like conservation and accessibility of the sRNA and mRNA seed sequences, have improved predictions of sRNA targets; however, they still yield a high rate of false positives due to the relatively common occurrence of a given >6 nucleotide sequence in the genome. Whether a complementary seed sequence is targeted in vivo depends on the local mRNA structure and position of RNA chaperone binding sites, like the ARN_x motif used by Hfq. Programs such as IntaRNA (Busch et al. 2008) and TargetRNA2 (Kery et al. 2017) predict targets by looking for regions complementary to the sRNA seed, but also account for the RNA structure and accessibility of mRNA and sRNA seed sequences to calculate a minimal hybridization energy for the interaction. The free energy is a composite of the hybridization energy for the RNA duplex combined with the associated energy required to denature intramolecular duplexes of each RNA species (Busch et al. 2008).

The seed regions in sRNAs are more highly conserved than the rest of the sRNA and are often relatively free of secondary structure allowing them to remain accessible for regulatory interactions with targets (Peer and Margalit 2011). Two programs, CopraRNA and TargetRNA2, consider not only the sequence complementarity and hybridization energy between a specified sRNA and its potential targets but also the conservation of the target sequence amongst user-specified bacterial species (Kery et al. 2017; Wright et al. 2013). TargetRNA2 searches for sequences with homology to the sRNA within the GenBank database and defines highly conserved regions that are more likely to form the sRNA seed sequence. The sRNA and mRNA are independently folded using RNAplfold (Bernhart et al. 2006) and the regions of highest accessibility are used to predict sRNA–mRNA interactions. TargetRNA2 ranks to predicted mRNA targets using sRNA sequence conservation, sRNA and mRNA accessibility, and hybridization strength. It is possible for the user to filter for a subset of predicted targets, such as searching only for coexpressed genes in a user-supplied set of RNA-seq data that can reduce the false-positive rate by as much as half in *E. coli* (Kery et al. 2017).

Small RNAs are often expressed in response to discrete environmental stimuli and their regulons form biologically coherent responses to the input. RNAPredator (Eggenhofer et al. 2011) considers the accessibility of putative target sites, the hybridization energy of the interaction, and includes the ability to search for enrichment of gene ontologies in the predicted targets. The software uses RNAPlex to search for putative seed sequences and target interaction sites and computes the potential hybridization energies with an associated Z-score. RNAPlex does not account for intramolecular interactions or their associated free energies and is therefore faster than programs that account for these duplexes (Tafer and Hofacker 2008). Under the default settings, RNAPlex will include the coding regions of potential targets across the genome, as well as the 5′ and 3′ UTRs (set as 200 bp upstream or downstream of the coding regions, respectively). The user may specify genomic coordinates and filter for hybrids with a specific region of potential targets. Users can also select specific interactions of interest for postprocessing, such as searching for gene ontology enrichments in the predicted targets or accessing graphical depictions of structure around the RBS. The output of the program is the top 100 interactions ranked according to the computed interaction energies or prioritized according to user-specified parameters (Eggenhofer et al. 2011).

Regardless of what program is selected to predict sRNA interactions, all of the in silico target predictions must still be validated experimentally and even when incorporating seed accessibility and conservation, the false positive rates (FPRs) may be as high as 50% (Wright et al. 2013). One problem with computational approaches is the limited amount of conserved sequences and structures required for sRNA interactions to occur. Additionally, the algorithms have a limited ability to predict new interactions that mediate new modes of regulation because their algorithms use parameters that reflect known types of regulatory interactions. To address this, high-throughput RNA-seq methodologies have been developed to discover new regulatory RNA interactions. These datasets will likely identify conserved features and rules that can be used to refine the power and accuracy of *in silico* predictions.

4 Experimental Approaches for Identifying sRNA–Target Interactions

One of the first transcriptome-wide experimental approaches for finding the targets of a specific sRNA involves a short "pulse" of expression and microarray analysis to identify differentially expressed transcripts. This was based on the observation the sRNA targets are often rapidly degraded by recruitment of RNases, but has also identified many transcripts that are stabilized by sRNAs. This approach successfully identified multiple transcripts for the iron sparing response in *E. coli* as targets of the bacterial sRNA RyhB (Massé et al. 2005). RyhB was overexpressed from an arabinose-inducible promoter in cells lacking the chromosomal copy of RyhB for 15 min to prevent recovery of indirect effects of RyhB expression on the transcriptome. RNA was collected, reverse transcribed, labeled, and hybridized to whole genome microarrays. Multiple transcripts related to iron metabolism, including the Fe-S cluster containing succinate dehydrogenase subunit C (*sdhC*) and superoxide dismutase (*sodB*) were identified as direct targets of RyhB (Massé et al. 2005). These experiments provided evidence of a role of the sRNA RyhB in regulating the expression of transcripts involved in cellular adaptation to iron starvation and evidence that bacterial sRNAs control regulons that are biologically and ontologically coherent. One disadvantage of this approach is that it depends on (de)stabilization of the sRNA target and interactions that do not facilitate degradation (e.g., sequestered interactions) will not be detected. Methodologies that detect sRNA–mRNA interactions directly, rather than downstream sequelae have recently been developed, although pulsed sRNA expression studies remain a staple of sRNA research as they are comparatively inexpensive and easy to perform.

MS2-affinity purification and sequencing, or MAPS, can be used to identify direct interactions between an sRNA and its target even if the target RNA is not (de)stabilized. In MAPS, an sRNA is fused to the MS2 RNA aptamer at the 5′ end and is expressed under the control of an arabinose-inducible promoter (Fig. 1). Cell lysates from the sRNA-MS2 expressing cells are bound to an affinity column coated with the MS2 phage coat protein. The MS2 aptamer has a high affinity for the MS2 phage coat protein and can be used to affinity purify MS2 tagged sRNA–target RNA complexes. Bound sRNA–mRNA duplexes are washed extensively, eluted from the column and sequenced on an Illumina platform (Lalaouna et al. 2015). MAPS analysis has been used to capture sRNA–mRNA complexes in *E. coli* and *Staphylococcus aureus*, demonstrating the broad utility of the technique (Lalaouna et al. 2013, 2015; Tomasini et al. 2017). Notably, MAPS was used to capture the interactome of the well-studied sRNA RyhB and recovered known and new targets including the 3′ external transcribed spacer (ETS) of $tRNA^{leuZ}$. The $tRNA^{leuZ}$ ETS is stabilized by RyhB and acts as a sponge to buffer RyhB interactions with target mRNAs, providing a critical stringency filter that limits unwanted low affinity interactions (Lalaouna et al. 2015). Therefore, MAPS facilitates the discovery of targets directly bound by an MS2-tagged sRNA or target in an unbiased manner.

Recently, a number of techniques have been developed that profile sRNA–mRNA interactions in vivo using RNA proximity-dependent ligation. In these approaches, base-paired RNAs that have 5′ and 3′ ends in close proximity, can be ligated together using RNA ligase to form a single contiguous RNA that can be sequenced as a single read. The sequence reads that map to interacting RNAs are variously termed chimeric, hybrid, or chiastic read. For consistency, we refer to the ligated RNA–RNA molecule as a chimera, and the sequence reads that map discontiguously to a genome or transcriptome as hybrids.

Global sRNA target identification by ligation and sequencing (GRIL-seq) is an RNA proximity-dependent ligation protocol that identifies direct targets of an sRNA but does not require aptamer tagging of the RNA and ligates the sRNA and target in vivo before purification of the RNA duplexes and sequencing of the chimeras (Fig. 1). Small RNAs of interest are expressed from an inducible promoter and coexpressed with T4 RNA ligase 1 in order to ligate RNAs in close proximity. Primary transcripts are "capped" with a triphosphate group that is not a substrate for T4 RNA ligase; however, most Hfq-bound transcripts are enriched for 5′ monophosphorylated ends that can be ligated to free 3′-OH ends (Bandyra et al. 2012). These authors reasoned that T4 RNA ligase would preferentially ligate small RNA 5′P ends and the proximal 3′OH of mRNAs cleaved by recruited RNases. RNA–RNA chimeras are captured using polyadenylated, sRNA-specific oligos that facilitate recovery of sRNA–mRNA pairs using oligo d(T) conjugated magnetic beads, and sequenced on an Illumina platform (Han et al. 2016). The authors used IntaRNA to examine the predicted interaction sites on the targets of the *Pseudomonas aeruginosa* sRNA PrrF1. IntaRNA predicted interactions with the 5′ end of putative targets, close to the ribosome binding site, as expected for the majority of characterized sRNA–target interactions. GRIL-seq also recovered interactions with 3′ ends of mRNAs and,

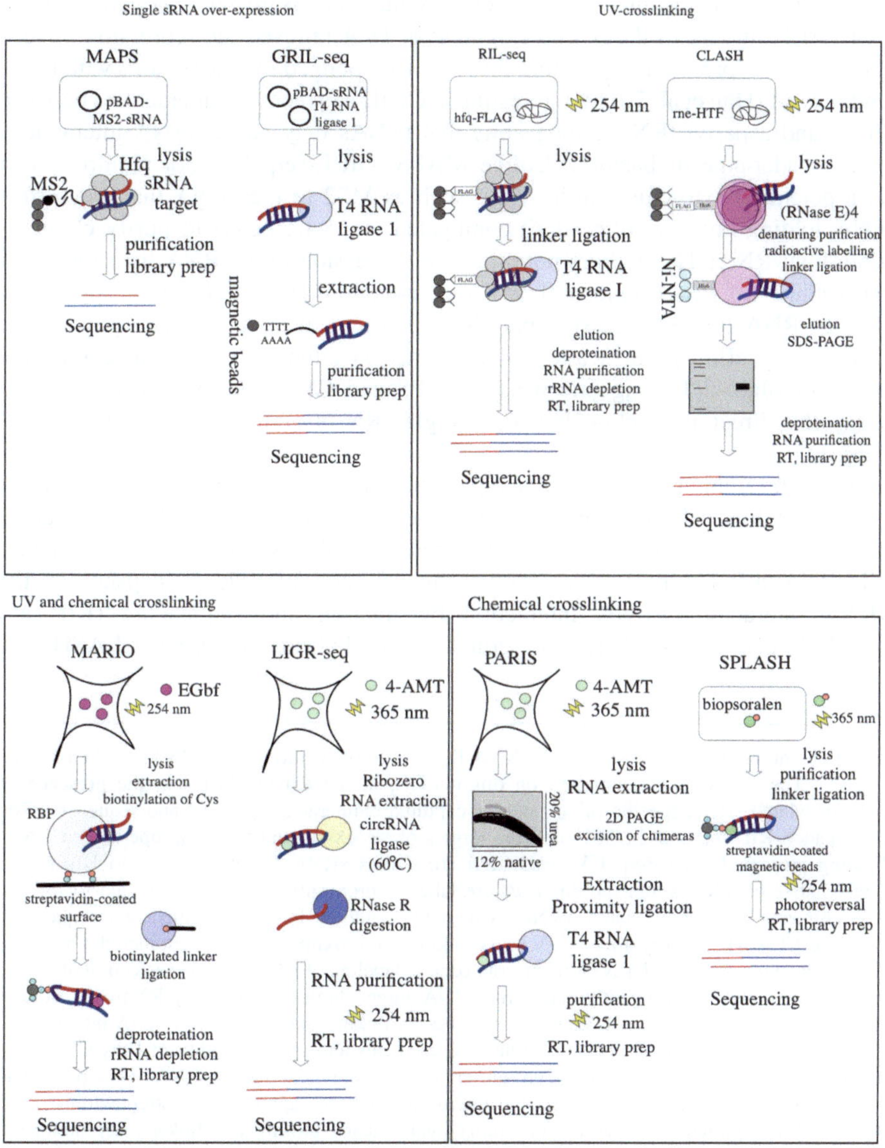

Fig. 1 Techniques to investigate sRNA–target interactions. Top left panel: MAPS and GRIL-seq find the targets of single overexpressed sRNAs. MAPS uses MS2-MBP-coated amylose resin to capture transcripts that interact with an MS2-tagged sRNA expressed from an inducible promoter. GRIL-seq also uses an inducible promoter to express a sRNA of interest but requires the

though IntaRNA was unable to predict these interactions, the authors demonstrated that PrrF1 binding to the 3′ end fragment of *katA* relieved the repression of full-length *katA* expression, suggesting that PrrF1 is sponged by interactions with the 3′ end of *katA* (Han et al. 2016). Interestingly, GRIL-seq recovers interactions between PrrF1 and aspartyl tRNAs, suggesting that tRNAs may buffer sRNA interactions in a broad range of bacteria. Unlike MAPS, GRIL-seq does not require fusion of a potentially structured affinity tag such as MS2 to the sRNA that may affect RNA folding but uses oligo-d(T) conjugated magnetic beads to purify chimeras from total RNA. The authors note that the expression of T4 RNA ligase affected the viability of *P. aeruginosa* after 1 h of induction (Han et al. 2016), suggesting that T4 RNA ligase expression may have detrimental effects on cell physiology. Thus, using affinity purification of chimeras on either MS2 protein conjugated resins or oligo-d(T) magnetic beads, MAPS and GRIL-seq provide a means to assay the direct interaction between single sRNAs and their respective targets (Fig. 1).

High-throughput RNA sequencing experiments have identified hundreds of new regulatory RNAs in bacteria and there is a need for tools that analyze sRNA function at a systems level. These techniques simultaneously capture all sRNA–mRNA interactions within the cell and provide a snapshot of the sRNA interactome. RNA interaction by ligation and sequencing (RIL-seq) and RNase E cross-linking and sequencing of hybrids (RNase E-CLASH) are

---

Fig. 1 (continued) coexpression of T4 RNA ligase 1 to proximity ligate sRNAs to their targets in vivo. The chimeras are enriched on oligo-d(T) coated magnetic beads in the presence of polyadenylated sRNA-specific oligos and the captured chimeras are purified and sequenced. Top right panel: RIL-seq and CLASH find the targets of sRNAs bound to an epitope-tagged RNA-binding protein. In RIL-seq, UV irradiation cross-links sRNA–target duplexes to Hfq-FLAG under native conditions and the duplexes are labeled, proximity ligated, and ligated to linkers. The chimeras are deproteinated, rRNA is depleted, and the RNA is purified and sequenced. RNase E-CLASH similarly utilizes UV cross-linking and proximity ligation, but includes a second denaturing purification on Ni-NTA resins, radioactive labeling of RNA, and size selection of RNA–protein complexes to stringently select for sRNA–target duplexes bound by RNase E. Bottom left panel: MARIO combines UV and chemical cross-linking to capture RNA–RNA interactions. During MARIO, UV irradiation cross-links nucleic acids and a combination of EthylGlycol bis and formaldehyde (EGbf) cross-links protein–protein complexes and RNA–RNA interactions. The cross-linked proteins are biotinylated and the complexes are captured on a streptavidin-coated surface. The captured chimeras are proximity-ligated using biotinylated linkers, deproteinated, and sequenced. Bottom right panel: LIGR-seq, PARIS, and SPLASH use psoralen derivatives to capture RNA–RNA interactions. LIGR-seq relies on the reversible UV-induced intercalating agent 4AMT to cross-link interacting RNAs together, the RNA is proximity ligated using a circular RNA ligase at high temperature, the RNA is purified, the cross-linking is photoreversed, and the RNA is sequenced. Similarly, PARIS utilizes reversible 4AMT-mediated cross-linking of RNA–RNA duplexes but selects for chimeras through a two-dimensional electrophoresis step. Excised duplexes are proximity ligated, purified and sequenced. SPLASH uses biotinylated psoralen (biopsoralen) to cross-link and affinity capture RNA duplexes on streptavidin-coated magnetic beads. The cross-linking is reversed, the RNA complexes are purified and sequenced. See the text for more details. "RT" denotes reverse transcription

techniques aimed at understanding the global regulatory network of sRNA interactions by cross-linking sRNA–target RNA duplexes to an RNA-binding protein (RBP) of interest (Fig. 1). Similar to GRIL-seq, both techniques rely on proximity-dependent ligation but capture interacting RNAs by UV cross-linking them to an affinity-tagged RBP expressed from its native chromosomal promoter.

During RIL-seq, bacteria expressing a FLAG-tagged Hfq are exposed to UV light to cross-link RNAs to Hfq. The Hfq–RNA complexes are immunoprecipitated on anti-FLAG magnetic beads, trimmed with RNase A/T1, and treated with polynucleotide kinase to convert 5′OH and 3′ P ends into 5′ P and 3′ OH ends that are substrates for T4 RNA ligase. sRNA and mRNA ends that are in close proximity are ligated using T4 RNA ligase, washed, and the ligated RNA duplexes released by protease K digestion of covalently cross-linked Hfq-FLAG. The released RNA is extracted, purified, and treated to deplete rRNA, before sequencing on an Illumina platform (Fig. 1). Using this method, Melamed and colleagues recovered 2817 unique, statistically significant RNA–RNA interactions on Hfq, including 1631 unique sRNA–mRNA interactions and 56% of previously confirmed interactions. RIL-seq analysis of Hfq-dependent sRNA interactions is performed under native conditions that recover a broad range of interactions with low stringency that are subsequently filtered for statistical significant *in silico* to reveal biologically meaningful interactions. This approach recovered a novel sponging interaction between PspH and Spot42, leading to derepression of Spot42 target transcripts (Melamed et al. 2016). While MAPS and GRIL-seq can identify the targets of one sRNA, RIL-seq can be used to profile Hfq-dependent sRNA–RNA interactions in the cell.

RNA–protein complexes that are cross-linked with UV are attached by a covalent bond that is stable under denaturing conditions. This allows stringent purification of the RNA–protein complex and is exploited in UV cross-linking and immunoprecipitation (CLIP-Seq) approaches like CRAC (UV cross-linking and analysis of cDNA) that use denaturing purification steps (SDS-PAGE separation of RNA–protein complexes and/or denaturing purification) to generate high confidence maps of RNA binding protein sites throughout the transcriptome. This stringent purification approach can also be used to capture RNA–RNA interactions and was initially demonstrated for miRNA–RNA interactions mediated by Argonaute 2 in human embryonic kidney cells (Helwak et al. 2013). Conceptually similar protocols that use stringent purification to reduce background have been reported for yeast (Kudla et al. 2011), *C. elegans* (Sugimoto et al. 2015), and human cell lines (Grosswendt et al. 2014). Recently, we have shown that RNase E forms an appropriate scaffold for capturing sRNA–target RNA interactions by CLASH in the human pathogen *E. coli* serotype O157:H7 (enterohemorrhagic *E. coli*). In RNase E-CLASH, the scaffold protein of the *E. coli* RNA degradosome, RNase E, is dual affinity-tagged at the C-terminus with a His_6 and 3X FLAG tag linked by a TEV protease cleavage site (HTF). The His_6 tag allows purification under stringent conditions. After UV cross-linking, RNA–RNase E–HTF complexes are initially immunoprecipitated under native conditions on M2 anti-FLAG resin. The

RNase E–HTF cross-linked to RNA is eluted from the M2 anti-FLAG resin by digestion with TEV protease and the eluted RNA–protein complex (RNP) is trimmed using an RNase A/T1 cocktail. To facilitate rapid washing and buffer changes during successive enzymatic treatments, the RNP is bound to Ni-NTA resin and washed under denaturing conditions. The RNA ends are then prepared for intermolecular ligation as RNase A and T1 leave 5′OH and 3′P ends that need to be prepared for intermolecular ligation by T4 RNA ligase 1. At this step RNase E-CLASH differs from Ago2-CLASH in that miRNAs bare a 5′-P close to the seed that is buried within Argonaute and is protected from trimming. This is used to enrich miRNAs as ligation substrates as longer RNAs will carry a cleaved 5′OH. Bacterial sRNAs are significantly longer than miRNAs and the 5′ ends require phosphorylation to generate ligation-competent ends. Hybrid reads are generated by ligating one end of the duplexed RNAs together and ligating the free 5′ and 3′ ends to RNA linkers that allow sequencing of the chimeric RNA. We find that this can be achieved during ligation of the 5′ and 3′ linkers without the need for a dedicated intramolecular ligation step. Both 5′ and 3′ ligations are performed with T4 RNA ligase 1 and can be adenylated by preadenylated linkers or can use ATP to energize ligation of the intraduplex RNA ends. The efficiency of intraduplex ligation of RNAs with 5′ and 3′ sequencing linkers is typically 1–2% (Helwak et al. 2013; Grosswendt et al. 2014; Sugimoto et al. 2015; Waters et al. 2017). The RNA–RNase E–His_6 complexes can then be eluted from the Ni-NTA resins using imidazole and the RNP complexes are size-selected on an SDS polyacrylamide gel that is transferred to nitrocellulose. The radioactive complexes of the appropriate size are excised from the membrane and RNase E digested by proteinase K. The released RNA is extracted, purified, and cDNA libraries are constructed and sequenced on an Illumina platform (Fig. 1, Waters et al. 2017). CLASH performed on enterohemorrhagic *E. coli* RNase E led to the identification of 1733 statistically significant, unique sRNA–mRNA interactions and 176 874 RNA–RNA interactions in total. One of the sRNA–mRNA interactions identified included the *E. coli*-specific sRNA Esr41 which binds and represses the expression of the *cirA* transcript encoding an outer membrane ferric iron receptor. The expression of *cirA* renders the cells susceptible to the bacterial colicin toxins involved in interspecific competition between *Salmonella* and *E. coli* in the gut (Nedialkova et al. 2014). Overexpression of Esr41 led to colicin resistance, providing evidence for the biological significance of the Esr41–*cirA* interaction recovered from RNaseE-CLASH (Waters et al. 2017). Both RIL-seq and RNaseE-CLASH rely on the use of ultraviolet light to cross-link RNA to protein; however, RNaseE-CLASH uses a second denaturing purification to stringently select for bound chimeras (Fig. 1). Both techniques provide a snapshot of sRNA interactions occurring the bait protein (either Hfq or RNase E) in the cell under specific conditions and a means to find RNA–RNA interactions that regulate gene expression at a systems level.

5 Chemical Cross-Linking of RNA Duplexes to Probe RNA–RNA Interactions

Recent studies have demonstrated that chemically cross-linked RNA–RNA interactions can be purified and sequenced providing transcriptome-wide maps of the RNA interactome that is independent of specific bait proteins. The cross-linking of RNA–RNA species can be achieved by chemical cross-linking, the most common being psoralen or its derivative 4-aminomethyltrioxsalen (4AMT). While psoralen cross-linking techniques have not been used to probe bacterial sRNA–RNA interactions, it has been used to probe RNA–RNA interactions in *E. coli* (Aw et al. 2016) and may represent an innovative methodology for capturing the sRNA interactome without the biases introduced by using a scaffolding protein (RNase E or Hfq) to purify the interacting RNAs. Psoralen analysis of RNA interactions and structures (PARIS) has been used to identify miRNAs and their targets, and can be used to probe the structure of longer RNAs by identifying intramolecular RNA interactions. The PARIS protocol was used to examine RNA–RNA interactions in human tissue culture and mouse embryonic stem cells and employs the intercalating agent 4AMT to reversibly cross-link RNA duplexes together after exposure to UV irradiation at 365 nm. The cross-linked RNAs can be resolved by treating with UV light at 254 nm. 4AMT preferentially cross-links RNA at staggered U residues and provides a means to fix the duplexed RNA–RNA pair during purification. The 4AMT cross-linked RNA–RNA complexes are purified, trimmed, and deproteinated before the RNA is separated by size on a native polyacrylamide gel. As cross-linked RNA–RNA complexes migrate differently from ssRNA, RNA interactions can be separated from non-cross-linked RNA using a second dimension of denaturing polyacrylamide gel electrophoresis. Up to 0.5% of the input RNA is recovered as dsRNA after the 2D gel electrophoresis step, suggesting that the elutions from the gel are enriched for chimeric RNA. The dsRNA is extracted from the gel, proximity ligated, and the 4AMT cross-links are reversed by irradiation at 254 nm. The RNA is purified and sequenced on an Illumina platform (Fig. 1). Unlike RIL-seq or RNase E-CLASH PARIS does not require a bait protein to capture RNA–RNA interactions and can identify complex RNA structures such as pseudoknots. Each sequence read corresponds to a single RNA–RNA interaction with less than 6% of the reads corresponding to background. PARIS facilitated the discovery of structures and interactions that spanned greater than 200 nucleotides and across multiple genomic features, including the structures of RBP binding sites or repetitive elements such as Alu, for which structures were previously unknown (Lu et al. 2016). In this manner, PARIS uses the chemical cross-linking agent 4AMT to provide a snapshot of abundant RNA interactions in the cell and the structures that comprise these interactions without requiring the selection of an RNA bait protein.

Ligation of interacting RNA followed by high-throughput sequencing (LIGR-seq) also uses 4-AMT to cross-link RNAs together upon irradiation with UV light at 365 nm. The cross-linked RNA is extracted from cells, treated to deplete rRNA, trimmed, and proximity ligated using circular RNA (circRNA) ligase that efficiently ligates proximal RNA ends and allows ligation at an elevated temperature (60°C) to prevent spurious hybridization and ligation events. Single-stranded RNA is removed by RNase R treatment to enrich for ligated RNA chimeras and the cross-linking is photoreversed. The RNA chimeras are washed, purified, and sequenced on an Illumina platform. The controls for LIGR-seq omitted the cross-linker 4-AMT or the proximity ligation step in order to identify and quantify any chimeras that occurred spuriously (Fig. 1). Similar to PARIS, LIGR-seq is capable of detecting interactions with highly structured RNA and employing LIGR-seq on 293T cells detected known structures that comprise the 80S and 5S ribosomal subunits as well as novel interactions been small nucleolar RNAs (snoRNAs) and other RNA classes (Sharma et al. 2016). One disadvantage of LIGR-seq is that, while it can detect RNA structure and long-range interactions with high specificity, it is unable to recover short RNA sequences such as miRNAs because these small fragments are inefficiently ligated by circRNA ligase (Sharma et al. 2016). Potentially, LIGR-seq will require modification in order to be applied to the study of bacterial sRNA interactions.

In contrast, sequencing of psoralen-cross-linked, ligated, and selected hybrids (SPLASH) utilizes biotinylated-psoralen (biopsoralen) to cross-link RNA–RNA interactions. Cross-linked RNAs from human tissue culture and yeast cells are enriched by extracting biotin-labeled chimeras with streptavidin-conjugated magnetic beads. The extracted RNAs are proximity ligated and sequenced on an Illumina platform (Fig. 1). The authors note that the entry of biopsoralen into human cells is inefficient and requires the treatment of cells with the detergent digitonin for a short period of time (5 min) prior to biopsoralen treatment. Additionally, ten times the concentration of biopsoralen was required to achieve a similar cross-linking efficiency in yeast and *E. coli* cells (Aw et al. 2016), suggesting that biopsoralen labeling may be limited by the entry of biopsoralen into the cells of interest. The authors also noted that RNA was damaged during the photoreversal of biopsoralen cross-linking and the length of time for photoreversal needs to be optimized in order to minimize RNA damage before sequencing. Despite these challenges, employing SPLASH led to the identification of more than 8000 intermolecular RNA interactions and more than 4000 intramolecular interactions across all of the cells sampled. SPLASH was used to identify mRNA–mRNA interaction networks that were correlated with similar subcellular locations and enrichment for similar biological functions (Aw et al. 2016). Thus, SPLASH uses a biotinylated form of the chemical cross-linker psoralen to cross-link and capture RNA–RNA interactions and represents another systems-level approach to identifying all of the RNA interactions in the cell.

RIL-seq and RNase E-CLASH require the dual-affinity tagging of the bait protein and, even if the recombinant protein is expressed from its endogenous chromosomal location, there may be unintended effects of tagging the protein on the cell. Additionally, the use of the aforementioned techniques requires selection of a single bait protein for each experiment. SPLASH depends on biopsoralen which, as previously discussed, requires tissue treatment or high concentrations in order to get efficient RNA labeling. The development of mapping RNA–RNA Interactome in vivo (MARIO) sought to address these issues by assaying protein-dependent RNA interactions in vivo without epitope-tagging proteins or adding an exogenous molecule like biopsoralen. MARIO requires a combination of UV cross-linking to cross-link RNA to protein followed by a second EthylGlycol bis–formaldehyde cross-linking to cross-link protein–protein complexes. UV cross-linking the RNA to protein stabilizes the RNA–protein ternary complexes and allows a stringent denaturing purification to be performed, thereby reducing background. The dually cross-linked RNA is partially digested then ligated to 5′ biotinylated linkers. The biotinylated chimeras are affinity purified on streptavidin magnetic beads and purified under denaturing conditions to ensure specific recovery of chimeric RNA–RNA complexes. A proximity ligation step is performed at a high volume overnight to avoid spurious in vitro intermolecular ligation, the complexes are washed and eluted from the magnetic beads, and sequenced on an Illumina platform (Fig. 1). The controls for MARIO were extensive and designed to help identify and eliminate background signal: a non-cross-linked control that was not treated with EthylGlycol bis–formaldehyde, a control that incorporated an unbiotinylated linker, and a control that performed ligation between *Drosophila* and mouse DNA to identify the amount of hybrids generated from RNA interactions occurring in vitro. Incorporating information from all of the controls, the false positive rate for MARIO was estimated to be 2.5–6.5% (Nguyen et al. 2016). MARIO performed in mouse embryonic stem cells provided evidence for the scale-free and hierarchical nature of the RNA networks, suggesting that RNA–RNA interactions are not random or entirely promiscuous in nature (Nguyen et al. 2016). By combining UV cross-linking and EthylGlycol bis–formaldehyde cross-linking, MARIO illuminates the entire network of protein-assisted RNA interactions in the cell without adding photoactivatable nucleosides or epitope-tagging proteins, providing insight into the shape and structure of the networks that underlie the transcriptome.

The techniques to assay RNA–RNA interactions are increasingly dependent on deep sequencing data. While the incorporation of biochemical steps to enrich for chimeric RNA–RNA interactions can help decrease background signal and increase the specificity of chimera recovery, these techniques increasingly rely on the computational tools to analyze the sequencing data to separate biologically meaningful interactions from background within these large datasets.

6 Statistical Analysis of sRNA Interaction Data

The computational analysis of high-throughput RNA interaction experiments involves two basic steps: identifying hybrid reads that represent chimeric RNA–RNA interactions and determining statistically significant RNA–RNA interactions. Because each technique requires the incorporation and design of specialized controls to minimize and aid in the identification of background signal, the computational pipelines for each technique are custom-built. Similarly, the statistical analysis for each dataset is customized to probe the variability of each experiment and the conditions dictated by each technique. The different computational approaches for identifying and calling hybrid reads, and determining statistical significance are explored here.

6.1 Sources of Background in sRNA Interaction Experiments

The task of identifying statistically significant hybrids depends on the ability to minimize sources of background signal without removing low-abundance RNA–RNA interactions. One of the most common sources of background is signal from the highly abundant rRNAs and tRNAs. Techniques such as RIL-seq (Melamed et al. 2016) and LIGR-seq (Sharma et al. 2016) deplete rRNA prior to sequencing in order to circumvent the accumulation of signal at these highly expressed loci. However, some of the interactions between rRNA and other classes of RNAs may represent biologically significant interactions. The vast majority of the rRNA–mRNA interactions captured by SPLASH, for example, is thought to be the capture of mRNAs during translation (Aw et al. 2016). While these interactions are not the regulatory interactions SPLASH was intended to capture, these interactions were identified as statistically significant and are likely representative of a true biological interaction.

In the case of RBP-mediated RNA–RNA interactions that are investigated by RNase E-CLASH or RIL-seq, nonspecific associations between either RNA and the RBP or RNA with the affinity resins can be sources of background. In RNase E-CLASH, a control strain that lacked the dual affinity-tagged copy of *rne* was used to control for nonspecific interactions (Waters et al. 2017). In RIL-seq, an untagged *E. coli* strain was also used to control for nonspecific binding of RNA to the resins. Melamed et al. (2016) also included a control where Hfq-HTF expressing *E. coli* were not exposed to UV light to control for nonspecific associations between RNA and Hfq, as well as a control where *E. coli* and *Saccharomyces cerevisiae* lysates were mixed together after cross-linking and lysis and the frequency of cross-species hybrids was calculated. Surprisingly, 3.2–3.8% of hybrids were recovered without UV cross-linking under native conditions suggesting UV cross-linking is dispensable. However, in the absence of UV cross-linking, control RNA spike-in experiments demonstrated that nearly 4% of chimeras were hybrids between *E. coli* and *S. cerevisiae* while less than 1% of the chimeras were cross-species hybrids in the presence of cross-linking indicating that UV cross-linking prevents in vitro RNA–RNA interactions during Hfq purification and library preparation (Melamed et al. 2016). Even in the absence of a bait RBP, nonspecific interactions with the affinity moieties can occur. In MARIO, a control experiment that omits the biotinylated linker accounts for any interactions that occur with the streptavidin magnetic beads (Nguyen et al. 2016). Even after accounting for both nonspecific interactions with the affinity resins and spurious ligation, the false positive rate for MARIO was estimated to be up to 6.5%, suggesting that there are as of yet unidentified sources of background in these experiments.

Finally, spurious ligation between proximal RNAs when ligated in small volumes can produce RNA–RNA chimeras that serve as background signal. If the RNA–RNA duplexes are free in solution (i.e., not bound to an affinity matrix), spurious proximity ligation of RNAs may occur. Some studies have diluted free RNA–RNA complexes to prevent random ligation of free RNAs to select only for those interactions that are duplexed in vivo (Helwak et al. 2013; Nguyen et al. 2016).

While experimental controls can be designed to help identify the level of background for a particular experiment, computationally accounting for these sources of background and filtering them before statistical analyses remains a significant challenge for the field. As yet, there is no consensus on the best practices to analyze RNA–RNA interaction data. The following section explores the computational pipelines used to filter for significant RNA interactions.

6.2 Computational Pipelines and Statistical Analyses to Identify Chimeras

The computational pipelines are custom-built to probe the particular RNA interaction of interest, and, due to the tendency for each technique to include unique controls and slightly different approaches, there are custom statistical analyses for each experimental design. However, the initial steps of computational analysis are the same: quality filtering and trimming to remove low quality sequence, demultiplexing of samples to parse the different experimental conditions or tissues analyzed, merging of paired-end sequence reads (if applicable) and collapsing of PCR duplicates. The sequence reads are then aligned to the transcriptome using annotated genome files or custom-built databases that include transcript boundaries, miRNA information, or other noncoding regulatory RNA. There are now a diversity of tools to quality filter, demultiplex, trim, clip, and collapse PCR duplicates in sequence reads. The *hyb* package (Travis et al. 2014) used in our RNase E-CLASH study (Waters et al. 2017) makes use of Flexbar or the Fastx toolkit, PARIS uses the Trimmomatic pipeline and removes PCR duplicates using readCollapse from the icSHAPE package, while MARIOtools uses custom Python and R scripts to perform these tasks (Nguyen et al. 2016). The subsequent steps involve alignment of reads to the genome or transcriptome and identification of reads that map discontiguously to the genome or a transcript database. Many read aligners have been used to call reads that map discontiguously including BLAST, BLAT, PBLAT, bwa, STAR, bowtie, and bowtie2. Comparison of quality filtering and read aligners using an Ago2-CLASH dataset suggests that blastn/blastall alignment may yield 5–10% more hybrid reads than bowtie2 but with a significantly longer run time, indicating that minor improvements can be made to hybrid recovery at the expense of processing time (Travis et al. 2014). After alignment, custom scripts are generally used to identify hybrid reads representing RNA–RNA interactions, and varies depending on whether "joining linkers" are used to connect the duplexed RNAs or the native 5′

and 3′ ends of the RNAs are ligated. Overlapping hybrid reads can then be clustered or merged to collate reads representing the same RNA–RNA interaction. The RNA–RNA interactions must then be filtered for statistical significance and the method for calculating statistical significance has yet to reach a consensus between protocols and software packages. Visualization of RNA–RNA interaction data is challenging and is often represented as links within circular interaction plots, network graphics, overlaid on specific RNA structures of interest, or custom tracks (arcs) in genome visualization tools like the UCSC or IGV genome browsers. Here we focus on the computation pipelines used by RNase E-CLASH, RIL-Seq, MARIO, PARIS, and LIGR-Seq as these more recent additions to the RNA interactome toolbox include methodologies for quantifying statistical confidence.

6.2.1 Identifying RNA–RNA Interactions with Sequencing Datasets

MARIOtools (https://mariotools.ucsd.edu/html/) is a Python and R-based pipeline that takes paired-end sequencing data and collapses PCR duplicates, demultiplexes sequence read data, and removes linker sequences using custom Python scripts. The resulting reads are aligned to the transcriptome using Bowtie and hybrid reads are identified. Because MARIO uses a biotinylated linker to join proximal RNA ends, mapping the linker sequence within the read easily identifies the hybrid read halves. A similar strategy is used for hiCLIP and simplifies the identification of independent RNAs. The software takes an input of a fasta file that contains all linker sequences and the output will split all of the fragments into classes that either contain a linker (Paired1 or Paired2), do not contain a linker (NoLinker), or contain only the linker. The Paired1 and Paired2 files containing fragments that are part of putative hybrids are then aligned to a user-supplied annotated transcriptome file using Bowtie and the mapped chimeras are written to an output file.

RIL-seq employs a different approach to identify hybrid reads. Paired-end sequencing is used the generate RIL-seq libraries and the last 25 nt of each paired end sequence read is mapped to the genome using bwa. Mate pair reads that align to the same transcript (concordantly or in reverse order), or within 1000 nt are called "single mapping" (nonhybrid read). If the two ends of a mate pair map at least 1000 bp apart, then that mate pair is identified as a hybrid.

Aligater (https://github.com/timbitz/Aligater) is a package written in Julia (v0.4) and Perl (v5) to identify hybrids in LIGR-seq data. To identify hybrid sequence reads, *aligater* recursively chains blocks of alignment generated by bowtie2 for each read and identifies regions of high quality alignment to determine whether it maps to a single position or contains multiple RNAs. Gaps between alignment blocks are penalized and the penalty can be adjusted for recovery of specific RNA species or for low quality libraries. After penalty scoring, each hybrid read is assigned a quality score (LIGQ). LIGR-Seq does not use a dedicated "joining linker" to ligate duplexed RNAs and the ligation site, like RNase E-CLASH and PARIS, is identified during alignment by discontiguous mapping. To verify that the ligation site is correctly called and represents the junction of two interacting RNAs, the ligation site is realigned to the transcriptome using BLAST (Sharma et al. 2016).

RNase E-CLASH identifies interactions using the *hyb* pipeline (Travis et al. 2014) originally developed for Ago2-CLASH data (Travis et al. 2014). *Hyb* trims, collapses, aligns, identifies hybrids, counts and clusters overlapping interactions, and provides information on interaction strength. *Hyb* can use BLAST, blat, or bowtie to map reads to the transcriptome (or genome) and identifies reads that map discontiguously and do not overlap or contain gaps greater than 4 bp. The putative hybrids are then filtered by selecting hybrids with high confidence mapping scores. The candidate hybrids are then folded *in silico* and the interaction sites are defined by base-pairing between the hybrids halves to illustrate the seed-target interaction. The corresponding hybridization energies are output into the resultant *hyb* file, as well as information about the RNAs in each hybrid, such as start/end, strand, and RNA subclass (Travis et al. 2014).

Hybrids from the PARIS (Lu et al. 2016) dataset are similarly called by filtering for reads that map discontiguously to human transcriptome coordinates. Reads are aligned to the genome using STAR and custom scripts cluster the mapped reads into duplex groups which are then further filtered using a connection score to identify hybrids with high confidence. Putative hybrids were also filtered for PCR duplication, intramolecular interactions, and potential splicing products (Lu et al. 2016).

RNA–RNA interactions are identified from high-throughput sequencing data after alignment to genomic or transcriptomic coordinates and successive filtering steps to identify high quality reads that map to two unique positions within the reference sequence. Once a candidate list of hybrid reads has been assembled, statistical analyses must be performed to determine which interactions are statistically overrepresented.

6.2.2 Determining the Statistical Significance of RNA–RNA Interactions in Sequencing Datasets

RNase E-CLASH, MARIO, and LIGR-seq utilize conceptually similar analyses to identify statistically significant RNA–RNA interactions. These approaches estimate the probability of recovering the hybrid based on hypergeometric or multinomial distributions of the RNA's relative abundance. To calculate the statistical significance of RNA–RNA interactions recovered by LIGR-Seq, *aligater* assumes that any two RNA fragments in the sequencing dataset can ligate and that random ligation will be a function of RNA abundance. True RNA–RNA interactions should be represented by significantly more hybrid reads than predicted for random ligation of RNAs of the same abundance. The authors first accounted for the differences in read recovery in each experiment by normalizing the transcript levels to reads per million (RPM). The probability of recovering the observed number of hybrids (given the abundance of each RNA in the dataset) is modeled as two independent draws from a multinomial distribution. The p-values are adjusted for multiple testing (number of RNA–RNA interactions in the dataset) using a Bonferroni correction. These authors also calculate a ratio of observed/expected interactions using AMT-treated versus untreated samples. The final dataset is filtered for RNA–RNA interactions represented by >10 hybrid reads, FDR < 0.1, and observed/expected ratio >1.1. A false positive dataset can be generated from high quality interactions with observed/expected ratios <0.9. For highly expressed transcripts, the estimated false positive rate was up to 4.4% while the false positive rate for transcripts with low levels of expression was up to 25% (Sharma et al. 2016) indicating that the false positive rate is negatively correlated with the expression level of the respective transcripts in the interaction.

RNase E-CLASH and LIGR-seq use essentially the same approach to determine the statistical significance of an RNA–RNA interaction. For RNase E-CLASH the number of single mapping, and hybrid read-halves, that map to a position within the transcriptome are taken as the relative abundance of the hybrid read-half within the pool of RNAs (again expressed as reads per million) and the probability of recovering two interacting RNAs within the distribution is modeled as two draws from a multinomial distribution. The *p*-values are corrected for multiple testing using the Benjamini–Hochberg correction, and *p*-values from replicate experiments are combined (Waters et al. 2017).

Conceptually, MARIO employs a similar method to determine the probability that an RNA–RNA interaction was recovered by chance. The total number of RNAs that map to a hybrid read-half is used to determine the abundance of the RNA in the dataset and a hypergeometric distribution is used to model the probability of randomly recovering an RNA–RNA interaction (Nguyen et al. 2016). The *p*-value is then corrected for multiple testing using a Benjamini–Hochberg procedure. Though these techniques differ in their approach to purifying chimeras from cells, the statistical analyses to identify biologically significant RNA–RNA interactions are similar and represent a conceptually simple way to identify statistically significant RNA–RNA interactions from sequencing data.

An alternative approach is taken in RIL-Seq to identify statistically significant RNA–RNA interactions. RIL-seq maps hybrid read-halves to genomic coordinates that are divided into 100 nucleotide windows. Hybrids within pairs of windows represented by more than 5 reads are counted. A Fisher's exact test is used to determine whether the window pairs occur more frequently than expected by random (referred to as S-chimeras). The interaction window is extended up to 500 nt and lowest *p*-value is taken as the true value and corrected for multiple testing using a Bonferroni procedure, and also by dividing by the number windows with >10 hybrid reads. An odds ratio is calculated to determine the magnitude of overrepresentation of a hybrid in the data (Melamed et al. 2016). Because RIL-seq is performed under native purification conditions, the authors have applied stringent statistical filters to identify statistically significant hybrids. RIL-Seq does not include random nucleotides within the ligated RNA linkers, so PCR duplicates cannot be differentiated from independent RNA–RNA interactions that occurred in vivo. Nevertheless, this approach identifies a high number of experimental verified interactions and provides a valuable tool for profiling Hfq-dependent sRNA–mRNA interactions.

The statistical analysis of PARIS data ranks interactions based using a connection score. The connection score for the two hybrid-halves is calculated by dividing the number of reads that connect the hybrid halves by the geometric mean of reads mapping to each hybrid half. Thus, a high connection score indicates a large number of reads that connect those RNAs relative to transcript abundance, and any hybrids with a connection score of less than 0.01 are discarded ensuring that the dataset is not dominated by low abundance hybrids in high abundance transcripts (Lu et al. 2016). The PARIS protocol also provides a method to view hybrids in the Integrative Genomics Viewer whereby the interacting arms of the hybrids are connected by arcs (Lu et al. 2016). Thus, PARIS relies on a unique connection score to normalize hybrid read counts to transcript abundance and to assess the overrepresentation of a putative interaction.

As the technologies to identify and analyze RNA regulatory circuits advance, so do the computational pipelines to extract biological significance from these datasets. For example, MARIO provides packages to dissect and visualize RNA structure and understand the role of secondary structure in RNA–RNA interactions (Nguyen et al. 2016). Once the RNA interactions have been identified and filtered for statistical significance, the user can focus on particular interactions and their contribution to a biological process. The ability to analyze these networks at a systems level and understand the contribution of all of the sRNAs and their interactions with target RNAs to bacterial adaptation and phenotypes remains a significant challenge.

7 sRNA–RNA Interaction Networks

RNA–RNA interactions can be represented as an interaction network similar to the analysis of pairwise protein–protein interaction networks (Vidal et al. 2011). This makes them amenable to existing methods and tools which have been used to analyze protein–protein and other biological networks. Network visualization provides a powerful means for users to represent and interact with the network. In these networks, each RNA is represented as a "node," while pairwise RNA–RNA interactions are referred to as an "edge," represented by lines connecting nodes. The network acts as a scaffold for which multiple sources of data can be integrated and displayed using visual cues.

The overall characteristics of the interaction network can be described by a number of network statistics (Barabasi and Oltvai 2004). The most commonly used network statistic is the number of immediate interaction partners of each node, also called the "node degree," and is equivalent to the number of RNA binding partners of each node (first neighbors) within the RNA–RNA interaction network. Nodes with high degree are often represented as "hubs" within the network and are likely to have important functions (Keller 2005). Nodes may also be centrally or peripherally placed within the network, termed the "node betweenness centrality." The node betweenness centrality measure is commonly used to describe the degree of connectivity in the network and is based on the concepts of shortest paths. Deletion of nodes with high betweenness centrality is likely to disconnect parts of the network and can be used to identify bottlenecks in the flow of signals in a signaling network (Breitkreutz et al. 2012). To calculate the "betweenness" score, the shortest paths between every pair of nodes within a connected network component is enumerated. The proportion of shortest paths that transverse the query node out of the total number of shortest paths is the "betweenness" score. As may be expected, the top 20 highest "betweenness" RNAs in our RNase E-CLASH data are regulatory small RNAs, but unexpectedly this also includes eight pathogen-specific sRNAs, suggesting that these play key roles in posttranscriptional regulation in the pathogen.

It has been suggested that the distribution of node degrees within biological networks follows a heavy tail distribution (Keller 2005). This led to the hypothesis that many biological networks have node degree distributions that fit the "scale-free power law" distribution (Barabasi and Albert 1999). Scale-free networks are thought to be more robust as random deletion of a node through gene mutation is more likely to affect nodes with a small number of interaction partners and thus will have less impact on overall network connectivity (Barabasi and Albert 1999). The node degree distribution of the RNase E-CLASH RNA–RNA interaction network is presented in Fig. 2a and can be fitted to a power-law distribution suggesting that the network is scale-free (Kolmogorov–Smirnov test of equality of distribution, p-value $= 1$). Although the distribution does not span two orders of magnitude in both the *x*- and *y-axes* required for strong statistical support, it has been suggested that many biological networks do not pass this filter for being a scale-free network (Barabasi and Albert 1999).

Barabasi and Albert (1999) hypothesized that preferential attachment is an important mechanism for generating scale-free networks (Barabasi and Albert 1999). Preferential attachment describes the process where nodes with high number of interaction partners tend to gain more interactions as the network continuously expands. This general idea of network growth and preferential attachment may be related to the evolutionary age of a sRNA within the sRNA interactome dataset (Kacharia et al. 2017). Interestingly, analysis of the evolutionary age of sRNAs within our interactome dataset (Ghadie et al. 2018) showed that relatively "young" and "middle-aged" sRNAs have less interaction partners than "old" sRNAs (Bonferroni adjusted p-value < 0.05, Fig. 2b). The above suggests that mRNAs could be preferentially attached to older sRNAs, which have a higher number of interaction partners. A power-law node degree distribution can arise from a number of generative processes besides preferential attachment (Barabasi and Albert 1999; Keller 2005) or by a combination of mechanisms (Ghadie et al. 2018). However, the high node degree of evolutionary old sRNAs suggests that preferential attachment contributes to growth of the scale-free sRNA interaction network.

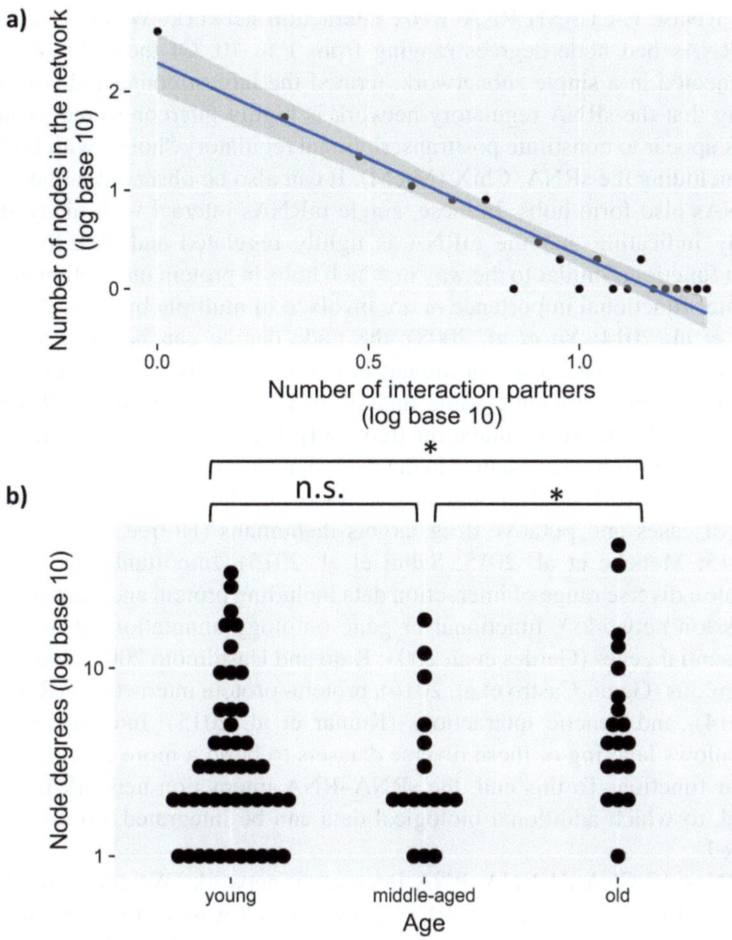

Fig. 2 Network analysis of the RNase E-CLASH RNA–RNA interaction network. (**a**) The frequency of nodes versus the number of interaction partners (node degrees), represented with a scatter plot in log-log scale. The node degree distribution could be fitted to a power-law distribution with an exponent of 2.2 (blue line), the 95% confidence interval shaded in dark grey. The Kolmogorov–Smirnov test did not reject the null hypothesis that the observed distribution is equivalent to a power-law (p-value = 1). The above analyses were performed using the "fit_power_law" function of the "igraph" library (Csardi and Nepusz 2006) of the R statistical analysis software (Team 2013). (**b**) A dot plot comparing the node degrees of the sRNA versus the evolutionary age of the sRNA (Ghadie et al. 2018). The node degrees is shown in log scale. The "old" sRNA had significantly more interaction partners than young or middle-aged sRNA (*Wilcoxon rank-sum test, Bonferroni adjusted p-value < 0.05). The number of interaction partners of young and middle-aged sRNA was not significantly different (denoted by n.s.). EcOnc are sRNAs that are specific to EHEC and therefore evolutionarily young. There were 12 sRNAs in the network with no age information and these were not included in this analysis

In the RNase E-CLASH RNA–RNA interaction network (Waters et al. 2017), the 94 sRNAs had node degrees ranging from 1 to 70. Of these 94 sRNAs, 56 were connected in a single subnetwork, termed the largest connected component, suggesting that the sRNA regulatory network is highly interconnected. A number of sRNAs appear to constitute posttranscriptional regulatory "hubs" with high node degrees including the sRNA, ChiX (MicM). It can also be observed in the network that mRNAs also form hubs. In these, single mRNAs interact with many sRNAs, potentially indicating that the mRNA is tightly regulated and therefore has an important function. Similar to the way in which hubs in protein interaction networks have greater functional importance or are involved in multiple biological functions (Rolland et al. 2014; Yu et al. 2008), the node degree can be used to identify sRNAs that likely have a disproportionate effect on the network. Examples of hubs with diagnostic or therapeutic value include p53 (Collavin et al. 2010) and Ras (Kauke et al. 2017), where mutation frequently leads to cancer, and interaction hubs linked to antifungal resistance in the pathogen *Cryptococcus neoformans* (Kim et al. 2015). Network analysis has also yielded insights into the mechanisms of complex diseases and putative drug targets in humans (Hofree et al. 2013; Isik et al. 2015; Menche et al. 2015; Sahni et al. 2015). Importantly networks can incorporate a diverse range of interaction data including protein and transcript levels (coexpression networks), functional or gene ontology annotations (Carbon et al. 2017), essential genes (Gerdes et al. 2003; Kato and Hashimoto 2007), transcription factor regulons (Gama-Castro et al. 2016), protein–protein interactions (Rajagopala et al. 2014), and genetic interactions (Kumar et al. 2015). Integrative network analysis allows layering of these diverse datasets to build a more complete picture of cellular function. To this end, the sRNA–RNA interaction network may act as a scaffold, to which additional biological data can be integrated, covisualized, or coanalyzed.

The RNase E-CLASH RNA–RNA interaction network (Waters et al. 2017) is visualized in Fig. 3 using the network visualization software Cytoscape (Kohl et al. 2011). Figure 3a visualizes all interactions that involve sRNAs, EcOncs (sRNAs specific to enterohemorrhagic *E. coli*), and their mRNA targets (false discovery rate < 0.05). Cytoscape enables the user to filter the network by node degree and display subsets of the network that have two or more interaction partners (Fig. 3b). This highlights a core subnetwork in which nodes are more tightly connected, including sRNAs which have more than one target, and mRNAs which are regulated by multiple sRNAs. Small RNA–sRNA, sRNA–EcOnc, and EcOnc–EcOnc interactions likely represent sRNA sponging interactions within the RNase E-CLASH dataset (Fig. 3c). Small RNAs that regulate essential genes can also be identified within the network and may represent biologically significant nodes. In Fig. 3d, essential genes have been manually curated from the EHEC str. Sakai genome using essential genes from *E. coli* str. K12 substr. MG1655 (Ecoliwiki 2017) and are controlled by both "core" genome encoded and pathogen-specific sRNAs.

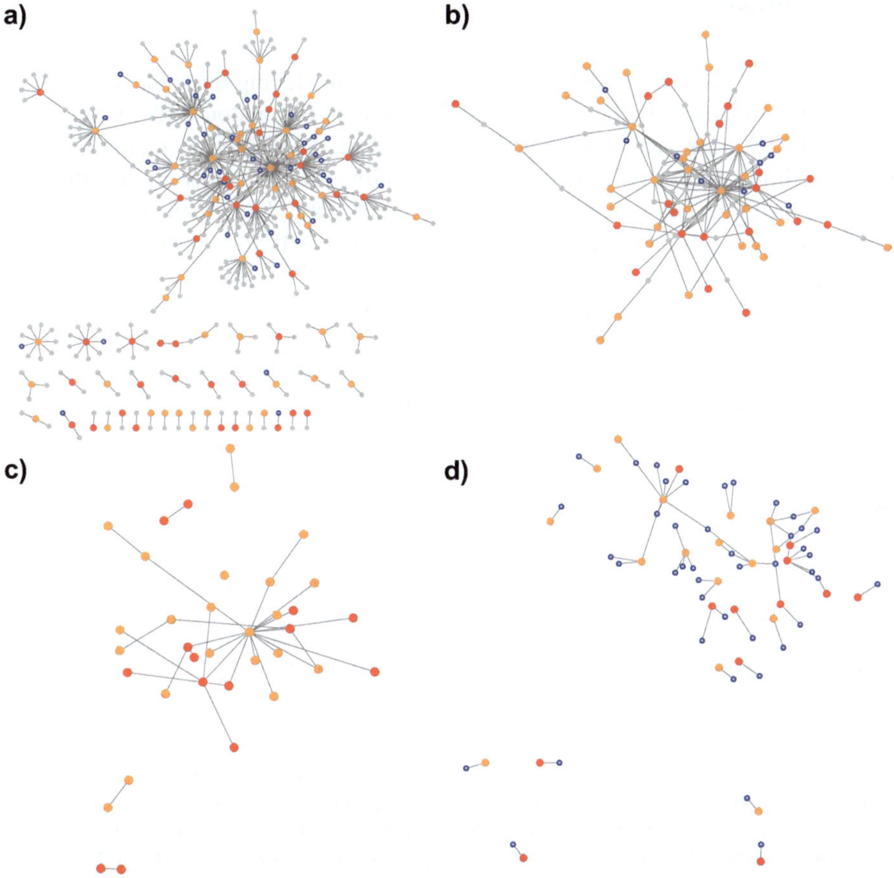

Fig. 3 Visualization of the RNase E-CLASH RNA–RNA interaction network using Cytoscape. (**a**) This network visualizes all interactions involving sRNAs, EcOncs (EHEC-specific sRNAs), and their mRNA targets (false discovery rate < 0.05). Subsequent panels show different subsets of this network, with the position of the nodes fixed in order to facilitate comparisons. The high abundance regulatory RNAs CsrB, tmRNA, and 6S RNA were removed from the network. The RNAs AgvB1 and AgvB2 have the same sequence and their nodes are merged. (**b**) The largest connected component showing only nodes which had 2 or more interaction partners in panel (**a**). Only nodes with >1 interaction partners are shown in this network. (**c**) Subnetworks of sRNA–sRNA, sRNA–EcOnc, and EcOnc–EcOnc interactions. (**d**) Subnetworks showing essential mRNA transcripts targeted by sRNAs and EcOncs. Essential genes were identified in *E. coli* str. K12 substr. MG1655 and mapped to *E. coli* O157:H7 str. Sakai orthologs using sequence homology. The sRNAs are represented as large orange nodes, the EcOncs are represented as red nodes, and mRNA are represented in grey nodes. Essential genes are represented by blue colored node borders

8 Conclusions

Systems-level analyses of sRNA interactions are broadening our understanding of small RNA function. Further technical developments will undoubtedly increase the sensitivity of these analyses beyond a snapshot of the sRNA interactome occurring on single RNAs or proteins, to transcriptome-wide chaperone-independent analyses. Approaches for processing RNA interactome data is still highly varied but is expected to reach a consensus as more datasets become available. Finally, small RNA interaction networks are interleaved with the transcriptional regulatory network conferring unique properties and kinetics to regulatory circuits. Adding additional layers of regulatory information to the sRNA interactome will likely reveal many of these "mixed" regulatory pathways and identify regulatory hubs or bottlenecks that may be exploited for therapeutic benefit.

References

A list of manually curated essential genes from the Ecoliwiki (2017) Retrieved December 31, 2017 from http://ecoliwiki.net/colipedia/index.php?title=Welcome_to_EcoliWiki&oldid=1491803

Altuvia S, Weinstein-Fischer D, Zhang A et al (1997) A small, stable RNA induced by oxidative stress: role as a pleiotropic regulator and antimutator. Cell 90:43–53

Argaman L, Argaman L, Hershberg R et al (2001) Novel small RNA-encoding genes in the intergenic regions of *Escherichia coli*. Curr Biol 11:941–950

Aw JGA, Shen Y, Wilm A et al (2016) In vivo mapping of eukaryotic RNA interactomes reveals principles of higher-order organization and regulation. Mol Cell 62:603–617

Bandyra KJ, Said N, Pfeiffer V et al (2012) The seed region of a small RNA drives the controlled destruction of the target mRNA by the endoribonuclease RNase E. Mol Cell 47:943–953

Barabasi AL, Albert R (1999) Emergence of scaling in random networks. Science 286:509–512

Barabasi AL, Oltvai ZN (2004) Network biology: understanding the cell's functional organization. Nat Rev Genet 5:101–113

Barquist L, Vogel J (2015) Accelerating discovery and functional analysis of small RNAs with new technologies. Annu Rev Genet 49:367–394

Barquist L, Westermann AJ, Vogel J (2016) Molecular phenotyping of infection-associated small non-coding RNAs. Philos Trans R Soc Lond 371:20160081

Bernhart SH, Hofacker IL, Stadler PF (2006) Local RNA base pairing probabilities in large sequences. Bioinformatics 22:614–615

Bouvier M, Sharma CM, Mika F et al (2008) Small RNA binding to 5′ mRNA coding region inhibits translational initiation. Mol Cell 32:827–837

Breitkreutz D, Hlatky L, Rietman E et al (2012) Molecular signaling network complexity is correlated with cancer patient survivability. Proc Natl Acad Sci USA 109:9209–9212

Brownlee GG (1971) Sequence of 6S RNA of *E. coli*. Nat New Biol 229:147–149

Busch A, Richter AS, Backofen R (2008) IntaRNA: efficient prediction of bacterial sRNA targets incorporating target site accessibility and seed regions. Bioinformatics 24:2849–2856

Carbon S, Dietze H, Lewis SE et al (2017) Expansion of the gene ontology knowledgebase and resources: the gene ontology consortium. Nucleic Acids Res 45:D331–D338

Chao Y, Li L, Girodat D et al (2017) In vivo cleavage map illuminates the central role of RNase E in coding and non-coding RNA pathways. Mol Cell 65:39–51

Coleman J, Green PJ, Inouye M (1984) The use of RNAs complementary to specific mRNAs to regulate the expression of individual bacterial genes. Cell 37:429–436

Collavin L, Lunardi A, Del Sal G (2010) p53-family proteins and their regulators: hubs and spokes in tumor suppression. Cell Death Differ 17:901–911
Csardi G, Nepusz T (2006) The igraph software package for complex network research. Int J Complex Syst 1695:1695
Deana A, Belasco JG (2005) Lost in translation: the influence of ribosomes on bacterial mRNA decay. Genes Dev 19:2526–2533
Eggenhofer F, Tafer H, Stadler PF et al (2011) RNApredator: fast accessibility-based prediction of sRNA targets. Nucleic Acids Res 39:W149–W154
Feng L, Rutherford ST, Papenfort K et al (2015) A Qrr noncoding RNA deploys four different regulatory mechanisms to optimize quorum-sensing dynamics. Cell 160:228–240
Fröhlich KS, Papenfort K, Fekete A et al (2013) A small RNA activates CFA synthase by isoform-specific mRNA stabilization. EMBO J 32:2963–2979
Gama-Castro S, Salgado H, Santos-Zavaleta A et al (2016) RegulonDB version 9.0: high-level integration of gene regulation, coexpression, motif clustering and beyond. Nucleic Acids Res 44:D133–D143
Gerdes SY, Scholle MD, Campbell JW et al (2003) Experimental determination and system level analysis of essential genes in *Escherichia coli* MG1655. J Bacteriol 185:5673–5684
Ghadie MA, Coulombe-Huntington J, Xia Y (2018) Interactome evolution: insights from genome-wide analyses of protein–protein interactions. Curr Opin Struct Biol 50:42–48
Gogol EB, Rhodius VA, Papenfort K et al (2011) Small RNAs endow a transcriptional activator with essential repressor functions for single-tier control of a global stress regulon. Proc Natl Acad Sci USA 108:12875–12880
Gottesman S, Storz G (2011) Bacterial small RNA regulators: versatile roles and rapidly evolving variations. Cold Spring Harb Perspect Biol 3:a003798
Grosswendt S, Filipchyk A, Manzano M et al (2014) Unambiguous Identification of miRNA: target site interactions by different types of ligation reactions. Mol Cell 54:1042–1054
Guo MS, Updegrove TB, Gogol EB et al (2014) MicL, a new σE-dependent sRNA, combats envelope stress by repressing synthesis of Lpp, the major outer membrane lipoprotein. Genes Dev 28:1620–1634
Han K, Tjaden B, Lory S (2016) GRIL-seq provides a method for identifying direct targets of bacterial small regulatory RNA by in vivo proximity ligation. Nat Microbiol 2:16239
Helwak A, Kudla G, Dudnakova T et al (2013) Mapping the human miRNA interactome by CLASH reveals frequent noncanonical binding. Cell 153:654–665
Hofree M, Shen JP, Carter H et al (2013) Network-based stratification of tumor mutations. Nat Methods 10:1108–1115
Huang HY, Chang HY, Chou CH et al (2009) sRNAMap: genomic maps for small non-coding RNAs, their regulators and their targets in microbial genomes. Nucleic Acids Res 37:D150–D154
Isik Z, Baldow C, Cannistraci CV et al (2015) Drug target prioritization by perturbed gene expression and network information. Sci Rep 5:17417
Jagodnik J, Chiaruttini C, Guillier M (2017) Stem-loop structures within mRNA coding sequences activate translation initiation and mediate control by small regulatory RNAs. Mol Cell 68:158–170
Kacharia FR, Millar JA, Raghavan R (2017) Emergence of new sRNAs in enteric bacteria is associated with low expression and rapid evolution. J Mol Evol 84:204–213
Kato J, Hashimoto M (2007) Construction of consecutive deletions of the *Escherichia coli* chromosome. Mol Sys Biol 3:132
Kauke MJ, Traxlmayr MW, Parker JA et al (2017) An engineered protein antagonist of K-Ras/B-Raf interaction. Sci Rep 42:5831
Kavita K, de Mets F, Gottesman S (2017) New aspects of RNA-based regulation by Hfq and its partner sRNAs. Curr Opin Microbiol 42:53–61
Keller EF (2005) Revisiting "scale-free" networks. BioEssays 27:1060–1068
Kery MB, Feldman M, Livny J et al (2017) TargetRNA2: identifying targets of small regulatory RNAs in bacteria. Nucleic Acids Res 42:124–129

Kim H, Jung K-W, Maeng S et al (2015) Network-assisted genetic dissection of pathogenicity and drug resistance in the opportunistic human pathogenic fungus *Cryptococcus neoformans*. Sci Rep 5:8767

Kohl M, Wiese S, Warscheid B (2011) Cytoscape: software for visualization and analysis of biological networks. In: Hamacher M, Eisenacher M, Stephan C (eds) Data mining in proteomics, vol 696. Humana, New York, pp 291–303

Kudla G, Granneman S, Hahn D et al (2011) Cross-linking, ligation, and sequencing of hybrids reveals RNA-RNA interactions in yeast. Proc Natl Acad Sci USA 108:10010–10015

Kumar A, Beloglazova N, Bundalovic-Torma C et al (2015) Conditional epistatic interaction maps reveal global functional rewiring of genome integrity pathways in *Escherichia coli*. Cell Rep 14:648–661

Lalaouna D, Simoneau-Roy M, Lafontaine D et al (2013) Regulatory RNAs and target mRNA decay in prokaryotes. Biochim Biophys Acta 1829:742–747

Lalaouna D, Carrier MC, Semsey S et al (2015) A 3′ external transcribed spacer in a tRNA transcript acts as a sponge for small RNAs to prevent transcriptional noise. Mol Cell 58:393–405

Levine E, Zhang Z, Kuhlman T et al (2007) Quantitative characteristics of gene regulation by small RNA. PLoS Biol 5:1998–2010

Lu Z, Zhang QC, Lee B et al (2016) RNA duplex map in living cells reveals higher-order transcriptome structure. Cell 165:1267–1279

Majdalani N, Hernandez D, Gottesman S (2002) Regulation and mode of action of the second small RNA activator of RpoS translation, RprA. Mol Microbiol 46:813–826

Massé E, Escorcia FE, Gottesman S (2003) Coupled degradation of a small regulatory RNA and its mRNA targets in *Escherichia coli*. Genes Dev 17:2374–2383

Massé E, Vanderpool CK, Gottesman S (2005) Effect of RyhB small RNA on global iron use in *Escherichia coli*. J Bacteriol 187:6962–6971

Melamed S, Peer A, Faigenbaum-Romm R et al (2016) Global mapping of small RNA-target interactions in bacteria. Mol Cell 63:884–897

Menche J, Sharma A, Kitsak M et al (2015) Uncovering disease-disease relationships through the incomplete interactome. Science 347:1257601

Miyakoshi M, Chao Y, Vogel J (2015) Cross talk between ABC transporter mRNAs via a target mRNA-derived sponge of the GcvB small RNA. EMBO J 34:1478–1492

Mizuno T, Chou MY, Inouye M (1984) A unique mechanism regulating gene expression: translational inhibition by a complementary RNA transcript (micRNA). Proc Natl Acad Sci USA 81:1966–1970

Nedialkova LP, Denzler R, Koeppel MB et al (2014) Inflammation fuels colicin Ib-dependent competition of *Salmonella serovar* typhimurium and *E. coli* in enterobacterial blooms. PLoS Pathogens 10:1003844

Nguyen TC, Cao X, Yu P et al (2016) Mapping RNA–RNA interactome and RNA structure in vivo by MARIO. Nat Commun 7:12023

Papenfort K, Vanderpool CK (2015) Target activation by regulatory RNAs in bacteria. FEMS Microbiol Rev 39:362–378

Papenfort K, Sun Y, Miyakoshi M et al (2013) Small RNA-mediated activation of sugar phosphatase mRNA regulates glucose homeostasis. Cell 153:426–437

Papenfort K, Espinosa E, Casadesús J (2015) Small RNA-based feedforward loop with AND-gate logic regulates extrachromosomal DNA transfer in *Salmonella*. Proc Natl Acad Sci USA 112:E4772–E4781

Peer A, Margalit H (2011) Accessibility and evolutionary conservation mark bacterial small-RNA target-binding regions. J Bacteriol 193:1690–1701

Plumbridge J, Bossi L, Oberto J et al (2014) Interplay of transcriptional and small RNA-dependent control mechanisms regulates chitosugar uptake in *Escherichia coli* and *Salmonella*. Mol Microbiol 92:648–658

Prévost K, Desnoyers G, Jacques JF et al (2011) Small RNA-induced mRNA degradation achieved through both translation block and activated cleavage. Genes Dev 25:385–396

Rajagopala SV, Sikorski P, Kumar A et al (2014) The binary protein-protein interaction landscape of *Escherichia coli*. Nat Biotechnol 32:285–290

Rolland T, Taşan M, Charloteaux B (2014) A proteome-scale map of the human interactome network. Cell 159:1212–1226

Sahni N, Yi S, Taipale M et al (2015) Widespread macromolecular interaction perturbations in human genetic disorders. Cell 161:647–660

Schu DJ, Zhang A, Gottesman S et al (2015) Alternative Hfq-sRNA interaction modes dictate alternative mRNA recognition. EMBO J 34:2557–2573

Sedlyarova N, Shamovsky I, Bharati BK et al (2016) sRNA-mediated control of transcription termination in *E. coli*. Cell 167:111–121

Sharma E, Sterne-Weiler T, O'Hanlon D et al (2016) Global mapping of human RNA-RNA interactions. Mol Cell 62:618–626

Soper TJ, Woodson SA (2008) The rpoS mRNA leader recruits Hfq to facilitate annealing with DsrA sRNA. RNA 14:1907–1917

Soper T, Mandin P, Majdalani N et al (2010) Positive regulation by small RNAs and the role of Hfq. Proc Natl Acad Sci USA 107:2–7

Soper TJ, Doxzen K, Woodson SA (2011) Major role for mRNA binding and restructuring in sRNA recruitment by Hfq. RNA 17:1544–1550

Sugimoto Y, Vigilante A, Darbo E et al (2015) hiCLIP reveals the in vivo atlas of mRNA secondary structures recognized by Staufen 1. Nature 519:491–494

Tafer H, Hofacker IL (2008) RNAplex: a fast tool for RNA-RNA interaction search. Bioinformatics 24:2657–2663

Team R (2013) R Development Core Team. R: a language and environment for statistical computing 55:275–286

Tomasini A, Moreau K, Chicher J et al (2017) The RNA targetome of *Staphylococcus aureus* non-coding RNA RsaA: impact on cell surface properties and defense mechanisms. Nucleic Acids Res 45:6746–6760

Travis AJ, Moody J, Helwak A et al (2014) Hyb: a bioinformatics pipeline for the analysis of CLASH (crosslinking, ligation and sequencing of hybrids) data. Methods 65:263–273

Tree JJ, Granneman S, McAteer SP et al (2014) Identification of bacteriophage-encoded anti-sRNAs in pathogenic *Escherichia coli*. Mol Cell 55:199–213

Vidal M, Cusick MEE, Barabási A-L (2011) Interactome networks and human disease. Cell 144:986–998

Waters SA, McAteer SP, Kudla G et al (2017) Small RNA interactome of pathogenic *E. coli* revealed through crosslinking of RNase E. EMBO J 36:374–387

Wright PR, Richter AS, Papenfort K et al (2013) Comparative genomics boosts target prediction for bacterial small RNAs. Proc Natl Acad Sci USA 110:E3487–E3496

Yu H, Braun P, Yildirim MA et al (2008) High-quality binary protein interaction map of the yeast interactome network. Science 322:104–110

Epioncogene Networks: Identification of Epigenomic and Transcriptomic Cooperation by Multi-omics Integration of ChIP-Seq and RNA-Seq Data

Fabian Volker Filipp

Contents

F. V. Filipp (✉)
Systems Biology and Cancer Metabolism, Program for Quantitative Systems Biology, University of California Merced, Merced, CA, USA
e-mail: filipp@ucmerced.edu;
https://systemsbiology.ucmerced.edu; https://orcid.org/0000-0001-9889-5727

N. Rajewsky et al. (eds.), *Systems Biology*, RNA Technologies,
https://doi.org/10.1007/978-3-319-92967-5_7

Abstract Next generation sequencing and systems biology have changed our understanding of oncogenesis. Master regulators at the transcriptional and epigenomic level have the ability to affect how an entire network of cancer genes behaves, and thereby taking on an oncogenic role. Further, epigenetic factors cooperate and team up with transcription factors to control specific gene target networks. Transcriptomics in combination with epigenomic profiling and measurement of chromatin accessibility enables global detection of epigenetic modifications and characterization of transcriptional and epigenetic footprints. Chromatin remodelers and transcription factors are in close communication via recognition of posttranslational histone modifications, DNA methylation marks, and sequence motifs to coordinate dynamic exchange of chromatin between open transcriptionally active conformations and compacted silences ones. Integration of complementary high-throughput sequencing platforms (HiC, DNAseI-Seq, MNase-Seq, FAIRE-Seq, ATAC-Seq, ChIP-Seq, ChIA-PET, TBS-Seq, WGBS-Seq, RNA-Seq, GRID-Seq) including chromatin higher-order structures, DNase hypersensitive sites, chromatin accessibility, histone modification, chromatin binding, and DNA methylation enables identification of cooperation and gene target networks. In cancer, due to the ability to team up with transcription factors, epigenetic factors concert mitogenic and metabolic gene networks claiming the role of a cancer master regulators or epioncogenes.

Keywords ATAC-Seq · Cancer systems biology · ChIA-PET · ChIP-Seq · Chromatin accessibility · Coactivation · Cooperation · CpG · DNAseI-Seq · Epigenetics · Epigenome · Epigenomics · Epioncogene · FAIRE-Seq · Gene set enrichment · GRID-Seq · HAT · HDAC · HiC · Histone modification · KDM · KMT · Master regulator · MNase-Seq · Motif enrichment · Multi-omics · Omics · Precision medicine · PRMT · Regulome · Resistance · Rewiring · RNA-Seq · Target gene · TBS-Seq · Transcription factor target · Transcriptomics · Upstream regulator · WGBS-Seq

1 Introduction

1.1 The Importance of the Epigenome in Development and Disease

An altered epigenome is a novel hallmark of cancer influencing transcriptional changes and contributing to oncogenic progression. The dynamic nature of the epigenome represents a fascinating layer of information to gain mechanistic understanding of human development and disease patterns. Genome-wide studies have identified chromatin and histone regulators as one of the most frequently dysregulated functional classes in a wide range of cancer types (Timp and Feinberg 2013; Filipp 2017a). Rightly so, clinical efforts are not ignorant to the informative potential of the epigenome. Diagnostic and prognostic successes have proven power for disease stratification and molecular targeting in precision medicine (Filipp 2017b). However, gene target networks of epigenomic factors remain poorly characterized and require multi-level, big-data-type analyses demanding in data volume, variety, and complexity. Here, we provide a logical framework on how to interrogate epigenomic and transcriptomic regulation at a systems biology level in order to gain insight into functional networks.

1.2 From Targeted Approaches and Biomarker Discovery to Epigenome-Wide Profiling

Pioneering studies of epigenetic enzymes provided biochemical insights into key factors and established a molecular code or language that governs regulatory principles aside from information stored in the order of nucleotides (Turner 1993; Strahl and Allis 2000). Initially, discovery of epigenetic biomarkers had to rely on targeted approaches using individual gene loci known or suspected to be involved in the etiology or progression of the disease or other phenotype under study (Issa et al. 1994). Despite challenges due cell-specific nature of epigenomic states and how these can vary with developmental stage and in response to environmental factors, targeted approaches yielded a number of important epigenomic biomarkers and effector genes (Sharrard et al. 1992; Hiltunen et al. 1997; Kondo et al. 2008). However, targeted approaches require *a priori* knowledge for the selection of candidate biomarkers.

2 High-throughput Sequencing Platforms for Epigenomics and Chromatin Accessibility

2.1 Overview of Next Generation Sequencing Platforms

With the advent of next generation sequencing technology, DNA and chromatin associated changes could be studied (Fig. 1). Genome-wide approaches including chromatin immunoprecipitation (ChIP) in combination with next generation sequencing (ChIP-Seq) (Barski et al. 2007; Johnson et al. 2007; Zheng et al. 2010), chromosome conformation capture in combination with high-throughput sequencing (HiC-Seq) (Lieberman-Aiden et al. 2009), assay for transposase-accessible chromatin using sequencing (ATAC-Seq) (Buenrostro et al. 2013, 2015; Corces et al. 2017), or whole genome bisulfite sequencing (WGBS-Seq) (Maunakea et al. 2010) and others to characterize epigenomic states. Epigenomic profiling focused on chromatin interactions, nucleosome accessibility, or DNA marks has the ability to capture different aspects including chromatin modification and chromatin binding factors, the spatial organization of chromosomes, open chromatin states, and DNA

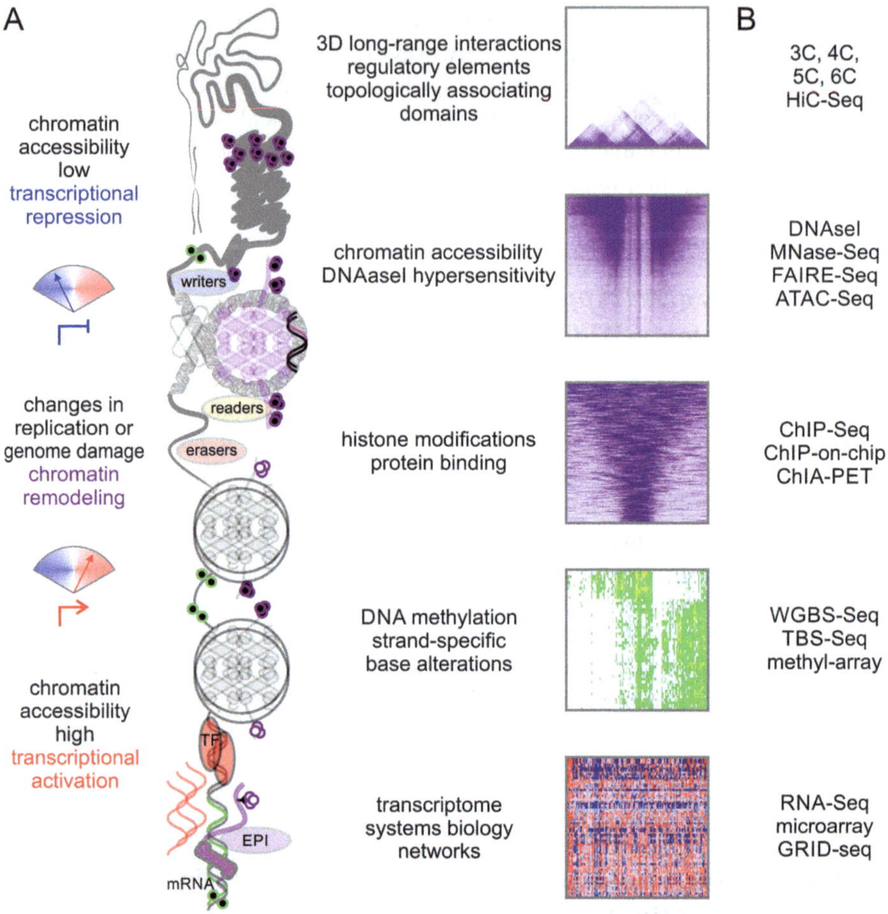

Fig. 1 Emerging high-throughput sequencing technologies have the ability to generate complementary epigenomic profiles. (**a**) The epigenomic and transcriptional landscape can be navigated by interrogating the dynamic structure of chromatin and its impact on the functional realization of the genome. (**b**) 3D chromatin interactions can be accessed by chromosome conformation capture in combination with high-throughput sequencing (HiC-Seq). DNase I hypersensitive sites sequencing (DNaseI-Seq) or assay for transposase-accessible chromatin using sequencing (ATAC-Seq) are examples of platforms allowing for insights into chromatin accessibility and transcription factor site occupancy. Chromatin immunoprecipitation in combination with next generation sequencing (ChIP-Seq) offers a platform to understand the intracellular regulatory landscape of the epigenome by determining signals from binding factors or chemical modifications of histones via specific antibodies. DNA methylation can be quantified by whole genome bisulfite sequencing (WGBS-Seq), targeted bisulfite sequencing (TBS-Seq), or methyl-arrays. The functional impact of the epigenomic landscape is validated at the transcriptional level by high-throughput gene expression experiments including RNA sequencing (RNA-Seq), microarrays, or global RNA interactions with DNA by deep sequencing (GRID-Seq)

methylation at base-pair resolution. In chromatin interaction analysis by paired-end tag sequencing (ChIA-PET), by combining HiC with ChIP it is possible to capture all chromatin interactions and map topologically associated domains (Fullwood et al. 2009; Li et al. 2017a). Methodology focused on global RNA interactions with DNA by deep sequencing (GRID-Seq) enables the comprehensive identification of the entire repertoire of chromatin interacting RNAs and their respective binding sites (Li et al. 2017b). Recent advancements include trimming of sample requirements and pooling of patient samples to identify disease associations and establish prognostic signatures compatible with clinical samples (Tehranchi et al. 2016; Qu et al. 2017).

2.2 Identification of Transcriptional and Regulatory Networks

The field of genome-wide expression profiling, or transcriptomics, is a powerful approach to capture the functionally realized genome at a certain time. The general goal of high-throughput transcriptomics studies by complementary DNA (cDNA) microarrays or RNA sequencing (RNA-Seq) is to quantify and compare gene expression profiles in order to detect differentially expressed genes (Sultan et al. 2008; Pan et al. 2008). Historically, next-generation sequencing technology and the availability of a comprehensive human genome annotation (Lander et al. 2001; Venter et al. 2001) has greatly accelerated scope and resolution of transcriptome-wide regulatory studies. Transcriptomics studies demonstrate the functional genome and present an ideal complementary data platform to epigenome-wide sequencing assays.

2.3 The Concept of Epigenomic and Transcriptomic Cooperation

The epigenome and the transcriptome are closely intertwined. Therefore, monitoring gene expression by RNA-Seq is a useful and necessary complement to epigenomic profiling. Further, transcription factors sample their corresponding cognate sites in a stochastic manner throughout the genome. Transcription factor activity, transcription factor motif enrichment, and transcriptional networks provide key insights into effects of epigenomic landscapes on gene expression. The concept of epigenomic and transcriptomic cooperation was introduced (Wilson et al. 2017; Wilson and Filipp 2018). A self-reinforcing, positive feedback enables a close teamwork of the transcriptional and epigenomic machinery, where one component opens the chromatin, another recognizes gene-specific DNA motifs, others scaffold between histones, cofactors, and the transcriptional complex (Qi and Filipp 2017). This highlights a close connection between the epigenomic and transcriptomic machinery, albeit much of the underlying principles remain to be discovered.

3 A Dynamic Epigenetic Code

3.1 Posttranslational Modifications Impacting Chromatin Structure and Function

Distinct patterns of reversible, covalent histone marks introduced the idea of an epigenetic language or code, a language edited and read by proteins and communicated in addition to four-letter base-code of DNA (Turner 1993; Strahl and Allis 2000). Posttranslational modification of histone proteins is part of the central epigenetic code. It includes methylation, acetylation, phosphorylation, citrullination, propionylation, butyrylation, formylation, crotonylation, proline isomerization, ubiquitination, sumoylation, glycosylation, and adenosine diphosphate-ribosylation of solvent accessible lysine, arginine, serine, and threonine residues at the termini of core histone proteins. Some of these modifications are understood to play important roles in the regulation of chromatin structure and function (Zentner and Henikoff 2013). Others remain to be deciphered and integrated into the epigenetic code.

3.2 Close Collaboration of Chromatin Remodeling with the Transcriptional Machinery

Coordination of the epigenetic program with transcription factors is key to successful tissue formation. Modulation of gene expression is accomplished by controlling initiation of transcription by assembly of the RNA polymerase II complex (POLR2, in human the holoenzyme consists of gene products encoded by POLR2A-POLR2L, GeneID: 5430-5441), general transcription factors (GTF2s, GeneID: 2957-2969), TATA-box binding protein associated factor 1-15 (TAFs, GeneID: 6872-6883, 8148, 83860, 129685), promoter recognition, replacement of components of stalled polymerase complexes by tissue-specific transcription factors, chromatin accessibility, ATP-dependent chromatin remodeling factors, posttranslational histone modifications, and DNA methylation.

3.3 Cross-talk Between Interconnected Epigenetic Forces

Histone modifiers are complementary but interconnected forces (Filipp 2017a) in the network of different histone editing enzymes that write, erase, and read epigenetic marks (Fig. 1). Hydrophobic modifications of distinct epigenetic marks may lead to condensed, transcriptionally silent heterochromatin, whereas distinct polar epigenetic marks may cause local formation of transcriptionally active euchromatin. For example, lysine histone methylation can compact chromatin, while acetylation or histone lysine demethylation can open chromatin, despite many exceptions to such simplified rules. Extensive data on chromatin accessibility and transcript abundance are available, allowing for organizing, analyzing, and modeling regulatory relationships of transcriptional control by epigenetic mechanisms (Kundaje et al. 2015; Fernandez et al. 2016).

3.4 Homeostasis Between Histone Writers, Erasers, and Readers

The dynamic process of histone modifications is mediated by the balance between opposing sets of enzymes in healthy cells (Fig. 1a). Histone writers chemically modify solvent accessible amino acids of histone tails, while histone erasers counteract. Further, reader domains determine state and type of posttranslational histone modifications allowing to distinguish for example between acetylation, monomethylation vs trimethylation of specific lysine residues (Taverna et al. 2007; Musselman et al. 2012). Together, they provide a fine-tuned clockwork for regulating chromatin structure and dynamics (Fig. 1a). Histone methylation is operated by lysine methyltransferases (KMTs), protein arginine methyltransferases (PRMTs) (Rea et al. 2000), and demethylases (KDMs) (Shi et al. 2004; Yamane et al. 2006; Whetstine et al. 2006), which were discovered to take key roles in gene expression (Filipp 2017a). Histone methylation increases the hydrophobicity of altered nucleosomes and promotes compaction of chromatin. Histone acetyltransferases (HATs) and deacetylases (HDACs) govern acetylation of histone lysine residues, influence the plasticity of chromatin structure by changing the electrical properties of histones, and improve the stability of many non-histone proteins by covering ubiquitination sites. Non-degradative monoubiquitylation alters nucleosome stability, nucleosome reassembly, and higher order compaction of the chromatin. Sumoylation of the core histones is associated with transcriptional silencing, and transcription factor sumoylation can decrease gene expression by promoting recruitment of chromatin modifying enzymes. Histone phosphorylation is a transient histone modification associated with local chromatin opening and transcriptional activation. Histone phosphorylation marks are important for regulation of the DNA damage response. Homeostasis between histone writers, erasers, and readers is vital for development and maintenance of healthy tissue and—if lost—can lead to developmental defects, autoimmunity, and uncontrolled proliferation.

4 Discovery of Epigenomic and Transcriptomic Cooperation Networks

4.1 Deciphering the Regulome by Complementary Functional Genomics Technologies

The combination of both, transcriptomic and epigenomic sequencing platforms, can resolve different levels of gene regulation, transcription factor binding motifs, DNA and chromatin modifications, and how each component is coupled to a functional output (Fig. 1b). Together, transcriptomic and epigenomic readouts generate comprehensive data on regulatory interactions, the so-called regulome. The regulome describes the interplay between genes and their products and defines the control network of cellular factors determining the functional outcome of a genomic component. The reconstruction of regulatory gene networks is stated as one of the main objectives of systems biology (Filipp 2013a, b). Regulatory networks in biology are intrinsically hierarchical and governed by interactions and chemical modifications (Cheng et al. 2015; Ay et al. 2015). The hierarchical nature can be accounted to the predominantly linear flow of information according to the central dogma of biology (Crick 1970). However, an accurate description of the regulome is a difficult task due to the dynamical nature of epigenetic, transcriptional, and signaling networks. Systems biology has the ability to integrate genome-wide epigenomic data recorded by ChIP-Seq, ATAC-Seq, WGBS-Seq, and RNA-Seq to identify gene targets of a regulatory event (Zecena et al. 2018; Wilson and Filipp 2018). The integrated analysis of such data—on the one hand based on gene networks, on the other hand based on sequence features of high-resolution sequencing data—captures cooperation among regulators (Wilson and Filipp 2018). Effective experimental design and data analysis of complementary epigenomic and transcriptomic platforms are required to decipher such epigenomic and transcriptional cooperation that has profound impact in development and disease.

4.2 Design of Coordinated Multi-omics Regulome Experiments

Gene expression is regulated by binding and modification of DNA and chromatin influencing RNA polymerase activity. Coordinated ChIP-Seq and RNA-Seq experiments are well-equipped to capture different epigenomic and transcriptomic levels governing the circuitry of a regulatory network (Fig. 2). A well-designed experimental setup pinpoints upstream master regulators as well as the effector network, streamlines data processing, facilitates prioritization of recorded information, and generates hypotheses for follow-up studies. ChIP-Seq is the most direct way to detect chromatin binding events and chemical modifications of histones. In regulome studies, experimental goals of ChIP-Seq assays may focus on in vivo

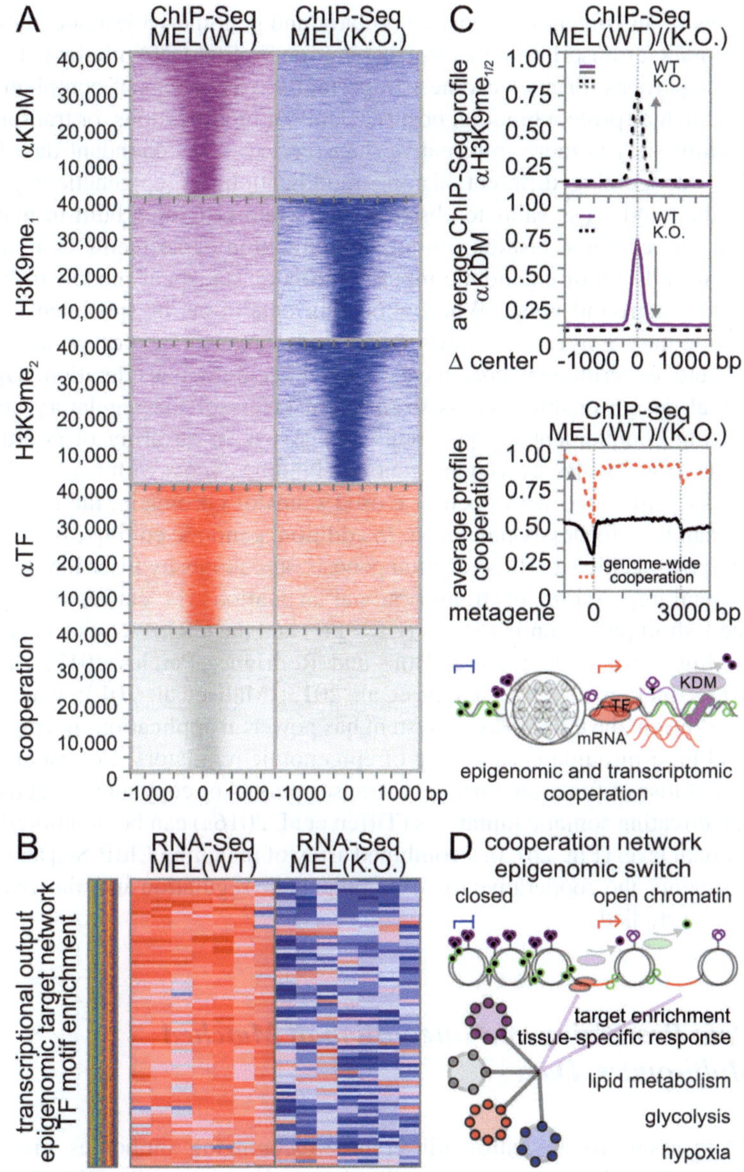

Fig. 2 The experimental design of regulome studies comprises multi-level controls to refine cooperation networks between epigenomic remodelers and transcription factors. Matched experiments of (**a**) epigenomic and (**b**) transcriptomic profiles elucidate synergistic forces across different regulatory levels and facilitate identification of effector networks. (**c**) Division of labor ascertains target specificity at genome-wide reach. In an integrated but modular setting, enzymatic domains of epigenomic factors control chromatin accessibility via histone modifications, while cooperating transcription factors navigate accessible promotor motifs. (**d**) Epigenomic and transcriptomic cooperation reinforces a specific effector network. Epigenomic master regulators have the ability to coactivate gene targets by collaborating with tissue-specific transcription factors. In tumors, epioncogenes support cell autonomy by preadaptation and promotion of cancer metabolism

binding sites of an epigenomic master regulator and genome-wide association with chromatin marks, transcription factors, recognition motifs, DNA-binding enzymes, histones, chaperones, or nucleosomes. Importantly, assessment of complementary data levels such as protein binding, posttranslational modifications, or transcription factor binding sites is more informative than replicates of redundant data levels. Antibodies recognizing different histone modifications or epigenetic regulators can strengthen and cross-validate observations, while providing built-in matching controls in an economic fashion. Immunoprecipitation experiments with specific antibodies usually require genomic input controls or statistical models to be able to subtract background noise. When allocating conditions of specimens, wealth of complementary epigenomic information and statistical power vs number of replicates and experimental cost have to be taken into consideration. Optimal experimental design enables access to different layers of the regulatory network by organizing complementary specimens and assays in an array of coordinated experiments (Fig. 2a). By overlaying genomic binding events with tracks of epigenomic marks, associated with open or closed states of chromatin, the epigenomic landscape can be effortlessly navigated. In addition, genomic editing offers tools to conduct target-specific site-directed mutagenesis, knockouts by insertion-deletions, transcriptional repression, or transcriptional activation via clustered, regularly interspaced, short palindromic repeats (CRISPR) and the CRISPR-associated (Cas) system (Ishino et al. 1987; Torres-Ruiz and Rodriguez-Perales 2017; Gasiunas et al. 2012; Jinek et al. 2012; Cong et al. 2013; Mali et al. 2013; Wang et al. 2013). In addition, the CRISPR/Cas system has powerful applications in epigenome editing and high-throughput screening of epigenomic regulators. Correspondingly, conditions of loss of function, oncogenic transcriptional or copy number activation, and hyperactivating somatic mutations (Tiffen et al. 2016a) can be monitored at the transcriptional level (Fig. 2b). In a combined array of matching ChIP-Seq and RNA-Seq experiments, the cooperative forces of epigenetic regulation and transcriptional output can be studied.

4.3 Data Processing and Integration of Matched Multi-omics Data

A ChIP-Seq assay utilizes chromatin epitope-recognizing antibodies, cross-links chromatin via formaldehyde, fragments the chromatin, captures the DNA fragments bound to chromatin using an antibody specific to it, and sequences the ends of the captured fragments (Barski et al. 2007). The processing workflow (Kharchenko et al. 2008) covers quality controls, genome alignment (Langmead et al. 2009), data normalization, assessment of reproducibility, and peak calling (Zhang et al. 2008; Feng et al. 2011). ChIP-Seq data is information-rich and contains sequence coordinates, genomic location, peak high, peak shape, event features, and significance of the detected event. Such multidimensional genomic and epigenomic data is conveniently stored in browser extensible data (BED) format (Quinlan and Hall 2010) providing flexible ways to display wiggle (WIG) track formats in genome browsers (Kent et al. 2002; Robinson et al. 2011). Similarly rich and complex—despite often diminished in mainstream applications—RNA-Seq offers multiple different perspectives on transcriptomic diversity such as alternative splicing, allele-specific expression, or noncoding transcripts (Sultan et al. 2008; Pan et al. 2008). Quantification and comparison of gene expression levels across conditions requires alignment and mapping of read counts onto the genome, normalization of expression levels, and identification of differentially expressed transcripts (Fonseca et al. 2014). During the conversion from raw next-generation sequencing data in the file format for sequences with quality scores (FASTQ) (Pearson and Lipman 1988; Cock et al. 2010) to binary version of sequence alignment map (BAM) files, the output of the differential expression analysis is consolidated into a single table file in text format with gene ID, symbol, transcript ID, expression level, p value, and q value. The aligned reads are passed to a quantification method to obtain gene expression values, normalization, and in a comparative setting fold change, test vs control expression (Mortazavi et al. 2008; Trapnell et al. 2012, 2013). Notably, such platforms often do not output read counts but produce instead normalized reads per kilobase of transcript per million fragments mapped (RPKM) (Mortazavi et al. 2008) in single-ended sequencing experiments and the corresponding fragments per kilobase of transcript per million fragments mapped (FPKM) in paired-end RNA-Seq, where two reads can correspond to a single fragment (Trapnell et al. 2010). In an alternative workflow that facilitates a more direct comparison between ChIP-Seq and RNA-Seq data, gene expression, differential expression, and differential binding data values are quantified as read counts scaled via the median of the geometric means of counts across all libraries. Such read counts are compatible with open tools for statistical testing, differential gene expression, and binding analysis (Robinson and Smyth 2007; Robinson et al. 2010; Anders and Huber 2010). The regulome analysis workflow combines ChIP-Seq and/or ATAC-Seq and RNA-Seq data with annotation of functional genomic context, differential binding analysis, differential gene expression analysis, pathway enrichment, and motif analysis (Wilson et al. 2016, 2017; Wilson and Filipp 2018). A multi-omics precision medicine profile of a malignant melanoma patient illustrates how multiple matched omics datasets can be integrated and visualized for clinical diagnostics (Filipp 2017b). Conceptually, data mapping in multi-omics medicine profiles mirrors the data flow according to the central dogma of biology (Filipp 2013a).

4.4 Understanding Epigenomic Regulome Networks

The regulome study of KMTs and KDMs in cancer rationalizes a streamlined workflow to connect epigenomic factors with transcriptional effector networks (Fig. 2). ChIP-Seq experiments with antibodies against epigenomic remodelers demonstrate absence of binding upon loss of function of the epigenetic target enzyme. Coherently, binding events of a lysine demethylase are associated with gain of histone lysine methylation upon loss of function of the epigenetic eraser enzyme (Fig. 2a). A comprehensive array of coordinated ChIP-Seq and RNA-Seq experiments comprises epigenomic transitions, change of binding upon loss of function, and differential expression of associated target genes in matching specimens. Such unbiased design reveals functional implications of effector networks but also shows non-enzymatic binding of epigenetic factors and multivalent scaffolding functions (Wilson et al. 2017). In addition, both ChIP-Seq and RNA-Seq data can be interrogated for enrichment of transcription factor motifs (Fig. 2b). Via genome-wide annotation and integration of sequencing reads, it becomes apparent that corresponding profiles of histone modifications are reversed upon loss of function mirroring the enzymatic function of the epigenetic modifier (Wilson et al. 2017). Cooperative epigenomic and transcription factor binding coincides with promoter sites on meta-gene coordinates enriched for histone lysine demethylation—overall indicators of transcriptionally activating epigenetic remodeling (Fig. 2c). Enriched transcription factor motifs can be cross-examined in the ChIP-Seq data using position site specific matrix models (Wilson et al. 2016). Conversely, direct transcription factor binding can be assessed using specific antibodies and validate overlapping sites. Detected motif enrichment and overlapping transcription factor binding sites coinciding with epigenomic remodeling highlight a close teamwork of the transcriptional and epigenomic machinery (Fig. 2d). Once more, the epigenomic factor facilitates chromatin accessibility, while the transcription factor guides motif recognition. Notably, a functional epigenetic protein complex constitutes several individual recognition modules to interpret the epigenetic language. Efficiency of protein domain architecture and gain of binding enthalpy via multi-component complexes explain why multivalent interactions of writer, reader, and reader domains emerge as a prevalent mechanism of epigenomic recognition. Rigid experimental design and stringent data processing allow for unbiased genome-wide associations of epigenomic networks. Thereby, the dynamic nature of chromatin remodeling, biochemical hurdles of elusive complexes, and complexity of data structure, can be overcome. In the field of epigenomics, simple, binary protein-protein interactions may be intangible. This emphasizes how the phenomenon of chromatin accessibility coordinates factors, facilitates multivalent, low-affinity associations, bridges temporal disconnect, and makes direct contact needless. Sequencing experiments focused on chromatin states are able to report on diverse aspects including chromatin modifications, binding, cooperation, accessibility, or occupancy. By incorporating complementary data levels into a hierarchical experimental layout, regulatory instances, and the flow of biological information in normal and transformed cells is revealed.

4.5 Detecting Motif Enrichment of Epigenomic Cooperation

The detection of regulatory events can be accomplished by genomic or functional genomic profiling (Guan et al. 2015; Edwards et al. 2016; Lanning et al. 2017). The footprints of cooperating transcription factors are found in cognate sequence motifs specific to their DNA binding domains (Wilson et al. 2016; Zecena et al. 2018). Such sequence motifs are pronounced in events of cooperating epigenomic activity. Detected motif enrichment highlights modularity, versatility, and efficacy of epigenomic cooperation providing target specificity at genome-wide reach. The number of detected events in genome-wide epigenomic binding studies provides statistical power for sequence motif discovery and gene target enrichment. As a consequence, high-resolution epigenomic studies often arrive at multiple plausible solutions, while each suggested interaction or association carries statistical significance. Enrichment scores reflect the degree to which a set is overrepresented of a ranked list. By walking down the list the score increases when encounter an element is encountered in the set but decreases if missing. Importantly, computations of enrichment scores can be performed at the gene or sequence level. In addition, transcriptome studies provide directionality of regulation, transcriptional activation or repression upon epigenomic activity—an important aspect lacking in coordinate-based ATAC-Seq or ChIP-Seq experiments. Integration of different sequence, gene, or network-based approaches priorities high-fidelity cooperation partners in epigenomic regulation. By intersecting epigenomic and transcriptomic data followed by analysis of motif enrichment (AME) and transcription factor target (TFT), and upstream regulator analysis (URA) approaches, it is possible to gain insights into gene networks associated with epigenomic regulators (Wilson and Filipp 2018) (Fig. 3). Computational response element searching algorithms are able to estimate a sequence's likelihood in belonging to the response element of the query transcription factor using position site specific matrices where each position in a query transcription factor model gives each of the four letters in the DNA alphabet a score based on the probability of that nucleotide being found at that position (Bailey and Gribskov 1998). Motif discovery, motif enrichment, and motif scanning used the multiple expectation maximization for motif elicitation (MEME) and discriminative regular expression motif elicitation (DREME) suite software toolkits from a set of user supplied unaligned sequences for ChIP-Seq regions (Bailey 2002; Bailey et al. 2015). De novo motif analysis programs MEME and DREME identify frequently detected DNA sequences patterns and similarity matches of recurring ATAC-Seq or ChIP-Seq sequences with DNA motifs of deposited studies in genomic sequence databases (Grant et al. 2011). After a motif of interest is discovered the genomic sequences of the ChIP sequenced data is scanned using the MEME suite software find individual motif occurrences (FIMO) (Grant et al. 2011) for individual motif occurrences using a position specific matrix to compute a log-likelihood ratio score for each submitted sequence. The position specific matrix is used further to analyze the sequenced data for motif enrichment for identifying potential coactivators within the data (Wilson et al. 2016, 2017; McLeay and Bailey 2010). An important branch

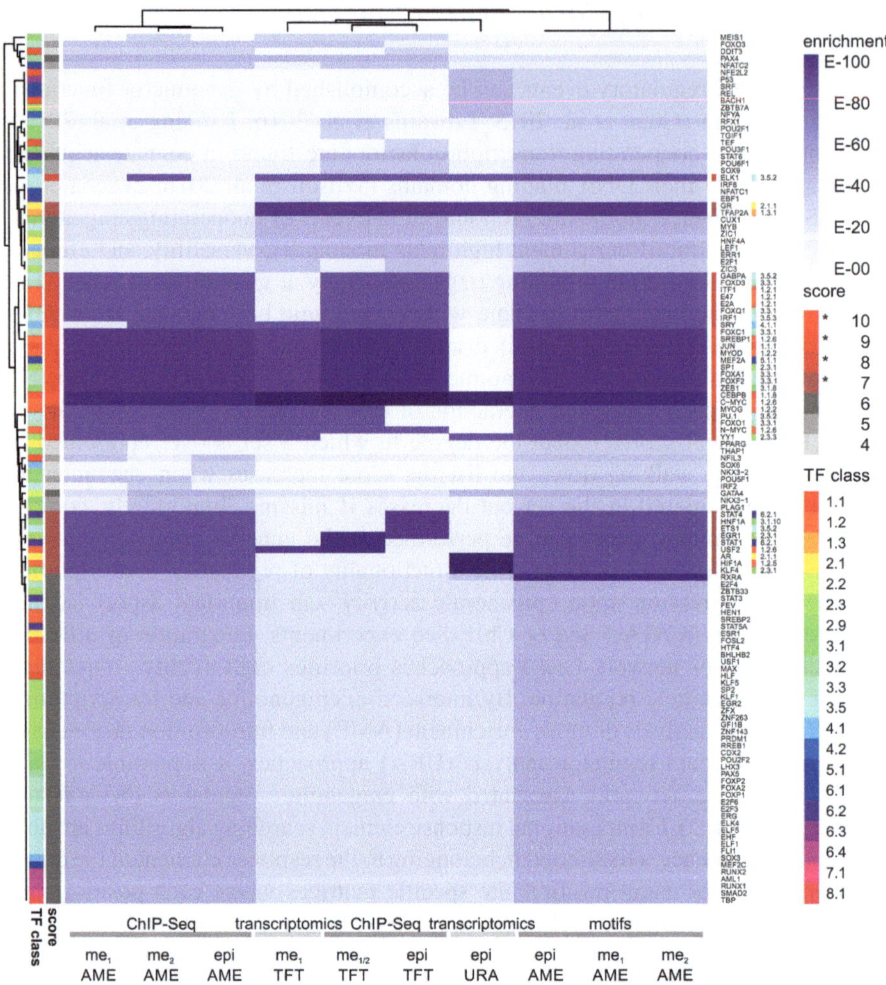

Fig. 3 Identification of transcriptional and epigenomic coactivation by complementary data platforms and omics analysis techniques. Complementary epigenomic and transcriptomic data serves as input for four different analysis platforms. Importantly, such genome-wide information can be accessed at the sequence or gene level providing different level of depth and resolution. Analysis of motif enrichment (AME) and transcription factor target (TFT), and upstream regulator analysis (URA) approaches provide insights into gene networks associated with epigenomic regulators, epioncogenes, and epigenomic cooperation

of epigenomic research is focused on the structural basis of chromatin interactions and dynamics. Epigenetic factors contain structural domains that have evolved to carry enzymatic chromatin remodeling functions. These include the royal family of folds and plant homeodomain zinc fingers (Taverna et al. 2007). In contrast, genomic approaches highlight DNA sequence associations and emphasize that chromatin is organized around our genetic material. Integrative, data-driven science has the opportunity to overcome disciplinary boundaries and merge seemingly opposite approaches from atomic observations to network insights.

5 Epigenomic Dysregulation in Cancer

5.1 *Cross-talk of Chromatin-Associated Events*

If the tightly controlled balance between histone writers and erasers is dysregulated, cells respond with transcriptional changes and pathological features manifest. Aberrant methylation of histones, in particular hypermethylation, is thought to influence the pathobiology of cancer by disrupting the same pathways as are affected by deleterious mutations and promoter cytosine-phosphate-guanine (CpG) site DNA hypermethylation (Baylin et al. 1991; Laird et al. 1995). Additionally, posttranslational modifications of non-histone chromatin remodeling enzymes can influence interactions with other transcriptional regulators, and alter their enzymatic activity. Importantly, posttranslational modification of histone proteins does lead not only to the binding of specific reader proteins but also to changes in the affinity for writers, erasers, or readers of other histone modifications. This induces a cross-talk between different chromatin modifiers that allows a spatiotemporal control of chromatin-associated events. Such a cross-talk is the focus of current investigations contributing to our understanding of epigenomic and transcriptional cooperation.

5.2 *Epigenomic Master Regulators in Cancer*

Within the regulome, epigenetic master regulators (Filipp 2017a) position themselves at the top of cellular hierarchies and control distinct phenotypic programs via chromatin modification without altering the core DNA sequence. Epigenetic oncogenes or tumor suppressors can arise when an epigenetic master regulators is somatically activated or lost, and contributes to cancer initiation and progression. Epigenetic master regulators utilize reversible chemical modifications of chromatin, histone or nucleotide marks, and affect gene activity without altering the core DNA sequence. In cancer, such epigenetic master regulator are found at the top of regulatory hierarchies, particularly in pathways related to cellular proliferation, survival, fate, and differentiation. For the manifestation of a genomic or non-genomic aberration of an epigenetic master regulator, it is a necessity that its own activity is affected by somatic mutation, copy number alteration, expression levels, protein cofactors, or methylation status. Epigenetic master regulators often accomplish target specificity of their phenotypic program by cooperation with members of the transcriptional machinery and therefore may depend on tissue-specific expression of such auxiliary factors. In cancer, an epigenetic master regulator populates an extreme state and is either permanently switched on or off. An epigenetic master regulator will become a cancer driver, if it is not functionally neutral but contributes to tumorigenesis or disease progression in its hyperactive or deactivated state. Genomic profiling of cancer patients has the ability to identify coincidence or mutual exclusivity of somatic alterations of epigenomic and transcription factors. Extreme states of epigenetic master regulators by somatic loss or gain of function in cancer may emphasize preexisting cooperative interactions with transcription factors, which may be subtle and difficult to detect under normal circumstances. A defined challenge in the field of epigenetic master regulators is to identify cancer-specific vulnerabilities in gene targets and biological pathways that are frequently and consistently perturbed under the control of an epigenetic driver.

6 The Power of Cancer Systems Biology

6.1 Multi-omics Support of Cooperation Networks

In a comprehensive genomic survey, the distribution of somatic alterations of epigenetic modifiers in cancer was established (Filipp 2017a). Captivatingly, neither activation nor loss of function dominated the landscape of epigenetic enzymes. It was found that chromatin decondensation can cause transcriptional activation of oncogenes but also histone hypermethylation can cause transcriptional repression of tumor suppressor genes. Histone methylation and DNA methylation are tightly linked and rely mechanistically on each other. Lysine methylation initiates, targets, or maintains DNA methylation, and vice versa (Tiffen et al. 2016a). In addition, there is a strong cooperation of epigenetic factors with the transcriptional complex. Cooperation with transcription factors or other members of the epigenetic machinery can target, amplify, or mute specific transcriptional responses. High-throughput technology in combination with multi-omics systems biology is necessary to decipher the dynamic interplay between epigenomics and functional output in biological and biomedical settings (Filipp 2013b). In particular, solid bridges between complementary next generation sequencing platforms have not yet been established and remain a future opportunity to elucidate mechanism of epigenomic cooperation.

6.2 Epioncogenes Are Cancer Drivers Concerting Mitogenic Gene Networks

The ability of epigenomic regulators to team up and synergize with transcription factors, facilitates control over specific gene networks and highlights their role as epigenomic cancer driver, master regulator, or epioncogenes. Recent multi-omics data has shown that the H3K9-JMJD family member, KDM3A (GeneID: 55818), is hyperactivated in epithelial and neuroektodermal cancers with poor prognosis due to activation of cellular metabolism (Wilson et al. 2017). KDM3 and KMT1 family members focused on remodeling of H3K9 marks display a great range of dysregulation in cancer. In melanoma, lung cancer, prostate cancer, and sarcoma copy number and transcriptional upregulation of KDM3A enables a predominant role as amplifier and epioncogene by transcriptionally activating oncogenic target genes (Yamane et al. 2006; Wilson et al. 2017). Genome-wide assessment of how epioncogenes control a mitogenic output requires matched epigenomic, chromatin binding, and transcriptomic profiles paired with CRISPR/Cas9 genome editing or stable hairpin RNA knockdown experiments (Fig. 2). Presence of ChIP-Seq binding vs genomic input and knockout but loss of epigenomic mark due to demethylation activity ensures bona fide identification of the detected signal. Loss of the epigenetic modifier results in reduced histone lysine demethylase activity and increases epigenomic H3K9 methylation in promotor region of target genes. In addition, transcriptomics validates gene targets in glycolysis, lipid metabolism, hypoxia, and anaplerosis that respond by differential expression upon targeted genome editing (Filipp et al. 2012a, 2012b). Overlap of complementary data platforms and analysis techniques identifies and cross-validates key transcriptional regulators focused metabolic functions (Fig. 3). For cancer patients with specific somatic activation of KDM3A, epigenomic rewiring is a profound contributor to oncogenic progression and a rational therapeutic target. Taken together, precision medicine efforts in combination with cancer systems biology have the ability to elucidate genome- and epigenome-wide alterations and identify molecular pathways suitable for drug targeting.

6.3 Gene Networks by Multi-omics Integration of Complementary Data Platforms

Interestingly, KDMs cooperate with a network of transcription factors rather than an isolated partner, while maintaining and accomplishing gene target and DNA sequence specificity (Fig. 4). Removal of repressive histone marks results in increased chromatin accessibility for transcription factors to recognize their response elements and implement regulation of gene expression. Conversely, the ability of transcription factors to recognize specific DNA motifs in promoter regions of target genes is an attractive feature for histone modifiers to associate with, since they often lack DNA specificity domains and depend on ternary complexes. Such indirect, transient, and dynamic interactions are intrinsic to epigenomic cooperation but can be narrowed down by acquisition of matching ChIP-Seq, ATAC-Seq, or RNA-Seq datasets. Individual epigenomic and transcriptomic sequencing experiments following enrichment analyses are able to delineate association of transcription factors with epigenomic events. However, the degeneracy of motif recognition mediated by structural domains of transcription factor families leaves ambiguity despite base-level resolution and depth of sequencing data. Strategies that

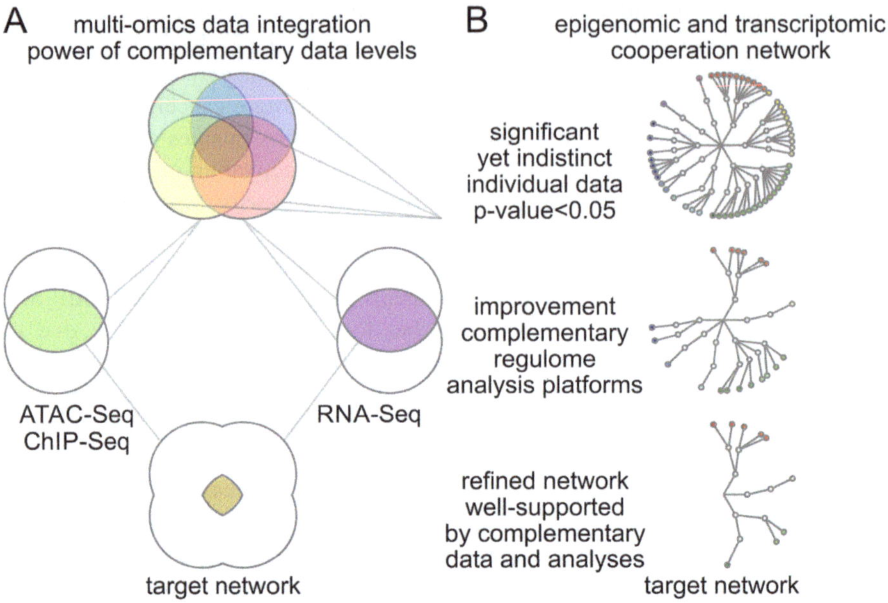

Fig. 4 Multi-omics integration of complementary data platforms and analyses yields well-refined target networks. Prioritization of genome-wide profiles is challenging, since hundreds of transcription factors are significantly associated with each individual data analysis platform or high-throughput sequencing technology. (**a**) Big data challenges can be overcome by systems biology analysis and integration of complementary but mutually supporting multi-omics data. (**b**) Motif similarity is visualized by transcription factor family trees classifying superclass, class, and family of transcription factors (from inward to outward) based on the characteristics of their DNA-binding domains. Single epigenomic or transcriptomic datasets examined by different analysis tools result in improved resolution but leave ambiguities. Gene target networks involved in epigenomic and transcriptomic cooperation can be identified by intersecting complementary data platforms and analysis techniques

take advantage of complementary point of views and integrate epigenomic coordinates, motif enrichment of chromatin accessibility, and transcriptional changes create a high-level mutual support network (Fig. 4a). By intersecting epigenomic and transcriptomic data followed by regulome and enrichment analyses it is possible to insights into transcriptional gene networks associated with epigenomic regulators (Fig. 4b). In particular the direction of regulation based on the transcriptional response is of complementary value to the coordinate-based epigenomic data.

6.4 *Durable Genome-Wide Rewiring and Target Specificity by Cooperative Networks*

For the oncogenic nature and target specificity of epigenomic regulators cooperation with transcription factors is key. KDMs cooperate with mitogenic basic helix-loop-helix factors including MYC proto-oncogene (MYC, Gene ID: 4609), hypoxia inducible factor 1 alpha subunit (HIF1A, Gene ID: 3091), and sterol regulatory element binding transcription factor 1 (SREBF1, Gene ID: 6720) and derive their lipogenic program from association with nuclear receptors. By overlaying the sequence and genomic data produced through matched experiment epigenomic events can be correlated with the transcriptomic effect of histone remodelers and transcription factors. Exploration into the cooperative roles of epigenetic histone modifiers and transcription factor families in gene regulatory networks contribute to our understanding of how promiscuous epigenomic and transcriptional programs assist in oncogenesis.

6.5 *Targeting Epigenomic Networks in Cancer*

Epigenetic dysregulation contributes to cancer and is recognized as important factor in controlling immune surveillance. Epigenomic master regulators carry not only vital roles in tissue development and differentiation of the immune systems, they also have the ability to down-regulate tumor suppressor genes, trigger uncontrolled proliferation, and evade immune recognition by eliminating chemokines, cytokines, and corresponding receptors involved in immune system activation (Tiffen et al. 2015, 2016a, b). Furthermore, epigenetic modifiers can activate many silenced genes. Some of them are immune checkpoints regulators that control immune responses. Drug inhibition studies in combination with epigenomic experiments suggest that epigenetic drugs prime the immune response by increasing expression of tumor-associated antigens and immune-related genes, as well as modulating membrane surface receptors involved in immune system activation. Oncogenic changes of the epigenome have profound regulatory consequences. Cooperative interaction partners are recruited and recapitulated from healthy settings or may be combined in novel ways that contribute to oncogenic progression. Therefore, enhancing our understanding of regulatory chromatin landscapes is vital. Scientific discoveries of epigenomic cooperation and rewiring have the potential to improve the outcomes of cancer immunotherapy by combining epigenetic-targeting drugs with immune checkpoint inhibitors.

7 Conclusion

Functional interactions between epigenomic and transcriptomic effector proteins are generally complex, frequently transient, and often require the association of additional scaffolding factors. Insights from complementary multi-omics platforms across different biological level include chromatin accessibility, binding and modifications of chromatin, DNA methylation, and transcriptional activity. Therefore, the high-resolution mapping of dynamic chromatin features such as nucleosome positioning, histone modifications and histone variant composition are ideally complemented by mapping the transcriptional machinery, histone modifying enzymes and histone modification binding proteins. Specific histone modification patterns are commonly associated with open or closed chromatin states and genomic elements, and are linked to distinct biological outcomes such as transcription activation or repression. Disruption of patterns of histone modifications is associated with loss of proliferative control and cancer. Therefore there is tremendous therapeutic potential in understanding and targeting histone modification pathways. Thus, investigating cooperation of chromatin remodelers and the transcriptional machinery is not only important for elucidating fundamental mechanisms of chromatin regulation, but also necessary for the design of targeted therapeutics.

Acknowledgements The power of systems biology comprises that information content of a network is greater than the sum of all individual nodes. The same principle applies to a team and highlights why teamwork is so valuable. To all students of the Systems Biology and Cancer Metabolism Laboratory, who taught me about friendship, selflessness, and—above all—unwavering loyalty. This work on cancer systems biology is generously supported by the National Institutes of Health and the National Science Foundation. F.V.F. is grateful for the support of grants CA154887 from the National Institutes of Health, National Cancer Institute, CRN-17-427258 by the University of California, Office of the President, Cancer Research Coordinating Committee, the Goethe Institute, Washington, DC, USA, and the Federal Foreign Office, Berlin, Germany.

References

Anders S, Huber W (2010) Differential expression analysis for sequence count data. Genome Biol 11:R106

Ay A, Gong D, Kahveci T (2015) Hierarchical decomposition of dynamically evolving regulatory networks. BMC Bioinf 16:161

Bailey TL (2002) Discovering novel sequence motifs with MEME. Curr Protoc Bioinformatics Chapter 2:Unit 2.4. doi:10.1002/0471250953.bi0204s00

Bailey TL, Gribskov M (1998) Combining evidence using p-values: application to sequence homology searches. Bioinformatics 14:48–54

Bailey TL, Johnson J, Grant CE et al (2015) The MEME suite. Nucleic Acids Res 43:W39–W49

Barski A, Cuddapah S, Cui K et al (2007) High-resolution profiling of histone methylations in the human genome. Cell 129:823–837

Baylin SB, Makos M, Wu JJ et al (1991) Abnormal patterns of DNA methylation in human neoplasia: potential consequences for tumor progression. Cancer Cells 3:383–390

Buenrostro JD, Giresi PG, Zaba LC et al (2013) Transposition of native chromatin for fast and sensitive epigenomic profiling of open chromatin, DNA-binding proteins and nucleosome position. Nat Methods 10:1213–1218

Buenrostro JD, Wu B, Litzenburger UM et al (2015) Single-cell chromatin accessibility reveals principles of regulatory variation. Nature 523:486–490

Cheng C, Andrews E, Yan KK et al (2015) An approach for determining and measuring network hierarchy applied to comparing the phosphorylome and the regulome. Genome Biol 16:63

Cock PJ, Fields CJ, Goto N et al (2010) The Sanger FASTQ file format for sequences with quality scores, and the Solexa/Illumina FASTQ variants. Nucleic Acids Res 38:1767–1771

Cong L, Ran FA, Cox D et al (2013) Multiplex genome engineering using CRISPR/Cas systems. Science 339:819–823

Corces MR, Trevino AE, Hamilton EG et al (2017) An improved ATAC-seq protocol reduces background and enables interrogation of frozen tissues. Nat Methods 14:959–962

Crick F (1970) Central dogma of molecular biology. Nature 227:561–563

Edwards L, Gupta R, Filipp FV (2016) Hypermutation of DPYD deregulates pyrimidine metabolism and promotes malignant progression. Mol Cancer Res 14:196–206. https://doi.org/10.1158/1541-7786.MCR-15-0403
Feng J, Liu T, Zhang Y (2011) Using MACS to identify peaks from ChIP-Seq data. Curr Protoc Bioinf Chapter 2:Unit 2 14
Fernandez JM, de la Torre V, Richardson D et al (2016) The BLUEPRINT data analysis portal. Cell Syst 3(491–495):e495
Filipp FV, Ratnikov B, De Ingeniis J, Smith JW, Osterman AL, Scott DA (2012a) Glutamine-fueled mitochondrial metabolism is decoupled from glycolysis in melanoma. Pigment Cell Melanoma Res 25(6):732–739. https://doi.org/10.1111/pcmr.12000
Filipp FV, Scott DA, Ronai ZA, Osterman AL, Smith JW (2012b) Reverse TCA cycle flux through isocitrate dehydrogenases 1 and 2 is required for lipogenesis in hypoxic melanoma cells. Pigment Cell Melanoma Res 25(3):375–383. https://doi.org/10.1111/j.1755-148X.2012.00989.x
Filipp FV (2013a) Cancer metabolism meets systems biology: pyruvate kinase isoform PKM2 is a metabolic master regulator. J Carcinog 12:14. https://doi.org/10.4103/1477-3163.115423
Filipp FV (2013b) A gateway between omics data and systems biology. J Metabolomics Syst Biol 1:1. https://doi.org/10.13188/2329-1583.1000003
Filipp FV (2017a) Crosstalk between epigenetics and metabolism—Yin and Yang of histone demethylases and methyltransferases in cancer. Brief Funct Genomics 16:320–325. https://doi.org/10.1093/bfgp/elx001
Filipp FV (2017b) Precision medicine driven by cancer systems biology. Cancer Metastasis Rev 36:91–108. https://doi.org/10.1007/s10555-017-9662-4
Fonseca NA, Marioni J, Brazma A (2014) RNA-Seq gene profiling: a systematic empirical comparison. PLoS One 9:e107026
Fullwood MJ, Liu MH, Pan YF et al (2009) An oestrogen-receptor-alpha-bound human chromatin interactome. Nature 462:58–64
Gasiunas G, Barrangou R, Horvath P et al (2012) Cas9-crRNA ribonucleoprotein complex mediates specific DNA cleavage for adaptive immunity in bacteria. Proc Natl Acad Sci USA 109:E2579–E2586
Grant CE, Bailey TL, Noble WS (2011) FIMO: scanning for occurrences of a given motif. Bioinformatics 27:1017–1018
Guan J, Gupta R, Filipp FV (2015) Cancer systems biology of TCGA SKCM: efficient detection of genomic drivers in melanoma. Sci Rep 5:7857. https://doi.org/10.1038/srep07857
Hiltunen MO, Alhonen L, Koistinaho J et al (1997) Hypermethylation of the APC (adenomatous polyposis coli) gene promoter region in human colorectal carcinoma. Int J Cancer 70:644–648
Ishino Y, Shinagawa H, Makino K et al (1987) Nucleotide sequence of the iap gene, responsible for alkaline phosphatase isozyme conversion in Escherichia coli, and identification of the gene product. J Bacteriol 169:5429–5433
Issa JP, Ottaviano YL, Celano P et al (1994) Methylation of the oestrogen receptor CpG island links ageing and neoplasia in human colon. Nat Genet 7:536–540
Jinek M, Chylinski K, Fonfara I et al (2012) A programmable dual-RNA-guided DNA endonuclease in adaptive bacterial immunity. Science 337:816–821
Johnson DS, Mortazavi A, Myers RM et al (2007) Genome-wide mapping of in vivo protein-DNA interactions. Science 316:1497–1502
Kent WJ, Sugnet CW, Furey TS et al (2002) The human genome browser at UCSC. Genome Res 12:996–1006
Kharchenko PV, Tolstorukov MY, Park PJ (2008) Design and analysis of ChIP-seq experiments for DNA-binding proteins. Nat Biotechnol 26:1351–1359
Kondo Y, Shen L, Cheng AS et al (2008) Gene silencing in cancer by histone H3 lysine 27 trimethylation independent of promoter DNA methylation. Nat Genet 40:741–750
Kundaje A, Meuleman W, Ernst J et al (2015) Integrative analysis of 111 reference human epigenomes. Nature 518:317–330
Laird PW, Jackson-Grusby L, Fazeli A et al (1995) Suppression of intestinal neoplasia by DNA hypomethylation. Cell 81:197–205

Lander ES, Linton LM, Birren B et al (2001) Initial sequencing and analysis of the human genome. Nature 409:860–921
Langmead B, Trapnell C, Pop M et al (2009) Ultrafast and memory-efficient alignment of short DNA sequences to the human genome. Genome Biol 10:R25
Lanning NJ, Castle JP, Singh SJ et al (2017) Metabolic profiling of triple-negative breast cancer cells reveals metabolic vulnerabilities. Cancer Metab 5:6. https://doi.org/10.1186/s40170-017-0168-x
Li G, Chen Y, Snyder MP et al (2017a) ChIA-PET2: a versatile and flexible pipeline for ChIA-PET data analysis. Nucleic Acids Res 45:e4
Li X, Zhou B, Chen L et al (2017b) GRID-seq reveals the global RNA–chromatin interactome. Nat Biotechnol 35:940–950
Lieberman-Aiden E, van Berkum NL, Williams L et al (2009) Comprehensive mapping of long-range interactions reveals folding principles of the human genome. Science 326:289–293
Mali P, Yang L, Esvelt KM et al (2013) RNA-guided human genome engineering via Cas9. Science 339:823–826
Maunakea AK, Nagarajan RP, Bilenky M et al (2010) Conserved role of intragenic DNA methylation in regulating alternative promoters. Nature 466:253–257
McLeay RC, Bailey TL (2010) Motif enrichment analysis: a unified framework and an evaluation on ChIP data. BMC Bioinf 11:165
Mortazavi A, Williams BA, McCue K et al (2008) Mapping and quantifying mammalian transcriptomes by RNA-Seq. Nat Methods 5:621–628
Musselman CA, Lalonde ME, Cote J et al (2012) Perceiving the epigenetic landscape through histone readers. Nat Struct Mol Biol 19:1218–1227
Pan Q, Shai O, Lee LJ et al (2008) Deep surveying of alternative splicing complexity in the human transcriptome by high-throughput sequencing. Nat Genet 40:1413–1415
Pearson WR, Lipman DJ (1988) Improved tools for biological sequence comparison. Proc Natl Acad Sci USA 85:2444–2448
Qi J, Filipp FV (2017) An epigenetic master regulator teams up to become an epioncogene. Oncotarget 8:29538–29539. https://doi.org/10.18632/oncotarget.16484
Qu K, Zaba LC, Satpathy AT et al (2017) Chromatin accessibility landscape of Cutaneous T cell lymphoma and dynamic response to HDAC inhibitors. Cancer Cell 32(27-41):e24
Quinlan AR, Hall IM (2010) BEDTools: a flexible suite of utilities for comparing genomic features. Bioinformatics 26:841–842
Rea S, Eisenhaber F, O'Carroll D et al (2000) Regulation of chromatin structure by site-specific histone H3 methyltransferases. Nature 406:593–599
Robinson MD, Smyth GK (2007) Moderated statistical tests for assessing differences in tag abundance. Bioinformatics 23:2881–2887
Robinson MD, McCarthy DJ, Smyth GK (2010) edgeR: a Bioconductor package for differential expression analysis of digital gene expression data. Bioinformatics 26:139–140
Robinson JT, Thorvaldsdottir H, Winckler W et al (2011) Integrative genomics viewer. Nat Biotechnol 29:24–26
Sharrard RM, Royds JA, Rogers S et al (1992) Patterns of methylation of the c-myc gene in human colorectal cancer progression. Br J Cancer 65:667–672
Shi Y, Lan F, Matson C et al (2004) Histone demethylation mediated by the nuclear amine oxidase homolog LSD1. Cell 119:941–953
Strahl BD, Allis CD (2000) The language of covalent histone modifications. Nature 403:41–45
Sultan M, Schulz MH, Richard H et al (2008) A global view of gene activity and alternative splicing by deep sequencing of the human transcriptome. Science 321:956–960
Taverna SD, Li H, Ruthenburg AJ et al (2007) How chromatin-binding modules interpret histone modifications: lessons from professional pocket pickers. Nat Struct Mol Biol 14:1025–1040
Tehranchi AK, Myrthil M, Martin T et al (2016) Pooled ChIP-Seq links variation in transcription factor binding to complex disease risk. Cell 165:730–741

Tiffen JC, Gunatilake D, Gallagher SJ et al (2015) Targeting activating mutations of EZH2 leads to potent cell growth inhibition in human melanoma by derepression of tumor suppressor genes. Oncotarget 6:27023-27036. https://doi.org/10.18632/oncotarget.4809

Tiffen J, Wilson S, Gallagher SJ et al (2016a) Somatic copy number amplification and hyperactivating somatic mutations of EZH2 correlate with DNA methylation and drive epigenetic silencing of genes involved in tumor suppression and immune responses in melanoma. Neoplasia 18:121–132. https://doi.org/10.1016/j.neo.2016.01.003

Tiffen JC, Gallagher SJ, Tseng HY et al (2016b) EZH2 as a mediator of treatment resistance in melanoma. Pigment Cell Melanoma Res 29:500–507. https://doi.org/10.1111/pcmr.12481

Timp W, Feinberg AP (2013) Cancer as a dysregulated epigenome allowing cellular growth advantage at the expense of the host. Nat Rev Cancer 13:497–510

Torres-Ruiz R, Rodriguez-Perales S (2017) CRISPR-Cas9 technology: applications and human disease modelling. Brief Funct Genomics 16:4–12

Trapnell C, Williams BA, Pertea G et al (2010) Transcript assembly and quantification by RNA-Seq reveals unannotated transcripts and isoform switching during cell differentiation. Nat Biotechnol 28:511–515

Trapnell C, Roberts A, Goff L et al (2012) Differential gene and transcript expression analysis of RNA-seq experiments with TopHat and Cufflinks. Nat Protoc 7:562–578

Trapnell C, Hendrickson DG, Sauvageau M et al (2013) Differential analysis of gene regulation at transcript resolution with RNA-seq. Nat Biotechnol 31:46–53

Turner BM (1993) Decoding the nucleosome. Cell 75:5–8

Venter JC, Adams MD, Myers EW et al (2001) The sequence of the human genome. Science 291:1304–1351

Wang H, Yang H, Shivalila CS et al (2013) One-step generation of mice carrying mutations in multiple genes by CRISPR/Cas-mediated genome engineering. Cell 153:910–918

Whetstine JR, Nottke A, Lan F et al (2006) Reversal of histone lysine trimethylation by the JMJD2 family of histone demethylases. Cell 125:467–481

Wilson S, Qi J, Filipp FV (2016) Refinement of the androgen response element based on ChIP-Seq in androgen-insensitive and androgen-responsive prostate cancer cell lines. Sci Rep 6:32611. https://doi.org/10.1038/srep32611

Wilson S, Fan L, Sahgal N et al (2017) The histone demethylase KDM3A regulates the transcriptional program of the androgen receptor in prostate cancer cells. Oncotarget 8:30328–30343. https://doi.org/10.18632/oncotarget.15681

Wilson S, Filipp FV (2018) A network of epigenomic and transcriptional cooperation encompassing an epigenomic master regulator in cancer. NPJ Syst Biol Appl 4:24. https://doi.org/10.1038/s41540-018-0061-4

Yamane K, Toumazou C, Tsukada Y et al (2006) JHDM2A, a JmjC-containing H3K9 demethylase, facilitates transcription activation by androgen receptor. Cell 125:483–495

Zecena H, Tveit D, Wang Z, Farhat A, Panchal P, Liu J, Singh SJ, Sanghera A, Bainiwal A, Teo SY, Meyskens FL Jr, Liu-Smith F, Filipp FV (2018) Systems biology analysis of mitogen activated protein kinase inhibitor resistance in malignant melanoma. BMC Syst Biol 12 (1):33. https://doi.org/10.1186/s12918-018-0554-1

Zentner GE, Henikoff S (2013) Regulation of nucleosome dynamics by histone modifications. Nat Struct Mol Biol 20:259–266

Zhang Y, Liu T, Meyer CA et al (2008) Model-based analysis of ChIP-Seq (MACS). Genome Biol 9:R137

Zheng W, Zhao H, Mancera E et al (2010) Genetic analysis of variation in transcription factor binding in yeast. Nature 464:1187–1191

Coupling Large-Scale Omics Data for Deciphering Systems Complexity

Ali Nehme, Zahraa Awada, Firas Kobeissy, Frédéric Mazurier, and Kazem Zibara

Contents

Abstract Recent development in high-throughput experiments has provided great amount of data that is being used in translational personalized medicine. Data

---

A. Nehme
Universite de Tours, CNRS UMR 7292, LNOx Team, Tours, France

ER045, PRASE, DSST, Lebanese University, Beirut, Lebanon

Z. Awada
ER045, PRASE, DSST, Lebanese University, Beirut, Lebanon

F. Kobeissy
Department of Biochemistry and Molecular Genetics, Faculty of Medicine, American University of Beirut, Beirut, Lebanon

F. Mazurier
Universite de Tours, CNRS UMR 7292, LNOx Team, Tours, France

K. Zibara (✉)
ER045, PRASE, DSST, Lebanese University, Beirut, Lebanon

Biology Department, Faculty of Sciences-I, Lebanese University, Beirut, Lebanon
e-mail: kzibara@ul.edu.lb

N. Rajewsky et al. (eds.), *Systems Biology*, RNA Technologies,
https://doi.org/10.1007/978-3-319-92967-5_8

available in public databases is increasing exponentially as a result of the progress in omics technologies including genomics, epigenomics, transcriptomics, proteomics, and metabolomics. Advancements in computing power and machine intelligence are affecting large-scale data analysis and integration. Two types of data integration are often considered: horizontal and vertical meta-analysis. The former integrates multiple studies of the same type, while the latter integrates data at different biological levels. This integrative approach provides a better understanding of systems complexity as a result of the global view that it offers from a biological point of view. This chapter describes the different types of omics analysis and discusses the methods of integrating multi-omics data using a case study.

Keywords Omics · Integration · High-throughput · Large-scale · Systems biology

1 Introduction

Omics is a rapidly developing, multidisciplinary, emerging field covering among others genomics, transcriptomics, proteomics, epigenomics, and metabolomics. High-throughput technologies permit simultaneous assessment of multiple cellular components, providing functional outputs of key cellular pathways at different hierarchical levels. The cellular components from which these omics data are derived act as one unified system in vivo; thus, it is evident to integrate omics data. No single omic analysis can fully unravel the complexities of fundamental biology, as the regulation of the system certainly occurs at many levels (Tomescu et al. 2014). Thus, incorporation of multi-omics information can provide a more comprehensive analysis from a biological point of view. Integration aims to bridge the gaps between vast amounts of data generated for systematic understanding of biology. Indeed, information that is thoroughly gathered, but does not have significant findings on its own, may find great value when combined. Studying biological phenomena at omics and multi-omics levels will probably lead to significant progress in personalized medicine (Chen et al. 2012). The revolution of omics profiling technologies will considerably benefit health care, especially in disease mechanism elucidation, molecular diagnosis, and personalized treatment (Fig. 1).

2 Genomics

Genomics is the first milestone in the omics era and the most established omics field. Our genome encodes all the information needed to develop from a single cell into a highly complex functional organism. Therefore, decoding the DNA sequence is vital for almost all branches of biology. The breakthrough started in 1977 by Frederick Sanger (Sanger and Coulson 1975) who developed first DNA sequencing

	1950s-1960s	1970s	1980s	1990s	2000s
Genomics	**1952** Electrophoresis **1953** DNA double helix	**1975** Southern blot **1977** DNA sequencing	**1982** shotgun sequencing		**2000** NGS **2005** GWAS
Epigenomics	**1952** Electrophoresis		**1988** ChIP		**2006** DNA methylation maps **2007** ChIP-Seq **2008** Dnase-Seq
Transcriptomics	**1952** Electrophoresis			**1995** DNA Microarrays	**2008** RNA-Seq
Proteomics and Metabolomics	**1952** Electrophoresis **1956** Protein sequencing **1959** GC-MS **1966** Tandem MS	**1970** SDS-PAGE **1973** HPLC **1975** 2DE **1979** Gas chromatography	**1986** LC-MS **1987** MALADI-MS **1989** ESI-MS	**1990** Liqiud chromatpgraphy **1992** Capillary electrophoresis **1996** LC-MS/MS **1999** Protein microarray	**2003** MALADI-TOF **2015** AE-MS

Fig. 1 Time line of omics technologies

method, which became the basis for modern innovations (Shendure and Ji 2008). Development of the polymerase chain reaction, the availability of high quality nucleic acid modifying enzymes, and progress in automated DNA sequencing enabled the Human Genome Project to be completed in 2003 (International Human Genome Sequencing Consortium 2004). Since then, the landscape of genomics has evolved significantly and outstanding progress has been made in genome sequencing technologies. Further developments have led to the second generation sequencing that overcame Sanger's restrictions, allowing to sequence millions of bases for several genomes in a relatively short time, which helped to accomplish the 1000 Genomes Project (1000 Genomes Project Consortium et al. 2015). This allowed easier detection of inter-individual variations in the genome, to understand evolution, and most importantly, to identify genomic causes of disease development (Eid et al. 2009). However, a relatively long time was still required to run the machines to completion, which resulted in low read lengths. In less than a decade, increased throughput of sequencing and dramatic reduction in costs have led to third generation sequencing, also known as next generation sequencing (NGS), which rely on highly advanced technologies such as Single Molecule Real-Time (SMRT) and nanopore sequencing (Shendure and Ji 2008). Third generation sequencing offers the advantage ofcost- and time-efficient generation of long reads with high accuracy. NGS technologies with faster, cheaper, and more accurate sequencing represent a principal shift in measuring genomic variants and interactions in the entire genome (Table 1).

Table 1 History of some omics projects

Year	Projects
1990	Human Genome Project initiated
2001	Draft of Human Genome Project
2002	HapMap Project launched
2003	Completion of Human Genome Project
2003	ENCODE Project launched
2004	ENSEMBL—an example of a gene annotation tool
2005	HapMap Project results
2007	Human metabolome draft
2008	1000 Genomes Project launched
2010	Phase 1 of 1000 Genomes Project completed
2012	1000 Genomes Project published 1092 genomes
2012	ENCODE Project initial results published

2.1 *Single Nucleotide Polymorphisms (SNPs)*

The current revolution in genomics makes it possible, not only to determine our entire DNA sequence but also to understand how our specific genome can inform our health. As high-throughput sequencing technologies provide a resolution at the single nucleotide level, genetic variations can now be identified including rare single nucleotide polymorphisms (SNPs) and copy number variations (CNVs) (Buermans and den Dunnen 2014). SNPs, which correspond to changes in a single nucleotide among a population, are the most common type of genetic variation, accounting for 90% of the differences in the human genome. Some SNPs are associated with predisposition to disease, how fast a disease progresses, the response of a disease to a given treatment, and drug side effects. SNPs are among the most useful methods for predicting the risk of disease development, which may help taking measures to avoid certain conditions. In fact, knowing the specific genetic features of a patient's cancer will allow physicians to better match the treatment to the individual's profile, perhaps by increasing the effectiveness of therapy and minimizing serious side effects.

2.2 *Genome-Wide Association Studies (GWAS)*

Genome-wide association studies (GWAS) were made possible owing to the availability of microarrays and genotyping technologies. Millions of SNPs have been mapped and linked to diseases (Klein et al. 2005). The first remarkable success of GWAS arrived in 2005, with the identification of a variant in complement factor H gene, which associated with age-related macular degeneration (Maraganore et al. 2005). To date, more than 2000 loci of common human diseases have been identified by GWAS, which made it possible to identify genetic traits from

SNPs without sequencing the entire genome (Manolio and Collins 2009). GWAS studies have clearly provided numerous recognized associations, which is helpful in understanding complex diseases and identifying new potential drug targets. However, GWAS did not significantly help in predicting the risk of common diseases, drug efficacy or toxicity. In addition, false positive or false negative findings are produced by GWAS studies due to participants' heterogeneity (Ng et al. 2009). Besides, common variants are not responsible for all disease risks and thus GWAS will not detect all variants involved.

2.3 Next Generation Sequencing (NGS) and Whole Genome Sequencing (WGS)

The arrival of high-throughput NGS has accelerated the discovery of genetic variants and genome-wide profiling of expressed sequences and epigenetic marks, allowing for systems-based analyses of diseases. Researchers can now obtain, through this technology, the most comprehensive view of genomic information and associated biological implications (Qin et al. 2010). The first proof of concept that NGS technology could be used to detect genetic disorders was provided by Shendure's group in 2009 (Byron et al. 2016). Indeed, Whole Genome Sequencing (WGS) is becoming one of the most widely used applications in NGS and has quickly gained broad applicability in medicine. It is increasingly being applied in clinical diagnosis, as it can identify genetic variations associated with diseases, determine genes that cause cancer, and detect pathogens in patient samples or isolates (Caskey et al. 2014). Moreover, WGS has the potential to accelerate the early detection of disorders and the identification of pharmacogenetic markers to customize treatments (DePristo et al. 2011). This technology is reshaping medicine and is expected to improve health until reaching personalized medicine. Clinical application of WGS revealed huge genetic differences among individuals including DNA variations in coding and/or regulatory regions of genes implicated in drug metabolizing enzymes, transporters, receptors, or drug targets (DePristo et al. 2011).

2.4 Whole Exome Sequencing (WES)

Whole exome sequencing (WES) has become a popular choice for genetic studies, primarily for identification of disease-associated gene variants involved in disease development and clinical diagnosis. The human exome consists of 1% of the human genome but harbors 85% of disease-related variants (Stadler et al. 2010). Sequencing of the complete coding regions could potentially uncover mutations causing rare genetic disorders as well as variants predisposing to common diseases.

WGS offers exciting new opportunities for the diagnosis and treatment of diseases leading to a new era of personalized medicine (Cantor et al. 2010).

In summary, NGS technologies, in recent years, have generated tremendous and complex genomic data sets that accumulated in the public domain, representing a major breakthrough in data acquisition (Anderson 1981). Human genetic variations produced by the International Hap Map Project has provided a genome-wide database that represents a radically new approach for searching for genetic variants associated with complex diseases (Consortium 2010). HapMap and the 1000 Genomes Project produced extensive catalogs of human genetic variations (ENCODE Project Consortium 2012) making it possible to investigate complex phenotypes and multifactorial diseases using GWAS. Of importance, we are moving forward in identifying all functional genomic elements through the ENCODE Project (Carithers and Moore 2015) and in understanding the role of noncoding variants in tissue-specific contexts through the GTEx Project (GTEx Consortium 2013). Moreover, NGS studies brought significant discoveries of new mutations in most common cancers and contributed to the identification of key genetic variants in oncology. Indeed, the Cancer Genome Atlas (TCGA) and International Cancer Genome Consortium (ICGC) have performed genome and exome sequencing on thousands of tumor–normal pairs. These studies have described the mutational landscapes for over 20 cancer types, demonstrating that tumors can vary dramatically in both the type and number of mutations.

The future of genomic technologies holds great promise; however, we should make an effort to link and integrate the information that is being generated in order for genomic data to have a more meaningful impact on our understanding of biology and diseases.

3 Epigenomics

In 1942, Waddington used the term epigenetics to denote changes in phenotype without changes in genotype, in order to explain aspects of development for which there was little mechanistic understanding (Waddington 1942). Chemical modifications of DNA were detected as early as 1948 (Hotchkiss 1948). In the mid-1960s, pioneering work of Allfrey (Allfrey et al. 1964) on histone modifications, in particular histone acetylation, led to the hypothesis that acetylation is closely linked to gene activity (Verdin and Ott 2014). The role of DNA methylation in gene regulation was proposed in the mid-1970s by Holliday and Pugh (Holliday and Pugh 1975), among others. By 1980, the functional connection between DNA methylation and gene repression was established, as was the existence of CpG islands (Bird et al. 1985). Beginning 2003, Encyclopedia of DNA Elements (ENCODE) project was the first international project to describe all the functional elements encoded in the human genome by mapping epigenetic modifications and by using large-scale epigenome profiling to identify regulatory elements in the human genome. ENCODE became a member of the International Human Epigenome Consortium

(IHEC), a project that was launched in 2010 that aimed to generate 1000 reference epigenomes in primary tissues and cell types.

The DNA sequence was the core of genomic research until the emergence of epigenomics when chemical compounds surrounding the DNA were shown to direct the function of genome as a whole. Given its physical association with the genome, the epigenome has been proposed to perform key roles in dictating genome structure and function, including the timing, strength, and memory of gene expression (Kouzarides 2007). Epigenomics has progressed over the past decade and has been acknowledged as an explanation for interindividual and intraindividual diversity, as well as a source of hidden information beyond genes, which can be influenced by intrinsic and extrinsic factors. Epigenetic modifications are fundamental to cellular differentiation and help determine cellular identity; for instance, what distinguishes a skin cell from a brain or other cell types (Lister et al. 2011). Epigenomic profiling was crucial in discovering many significant associations between chromatin features and genomic function at the level of gene expression, gene regulation, cell identity, ageing, and even disease development (Clark et al. 2016). Errors in epigenetic modifications can cause abnormal gene activity or inactivity, leading to genetic disorders. Conditions including cancer, metabolic and degenerative disorders have all been found to be related to epigenetic errors. Their alterations have been associated to early stages of cellular transformation in tumors (Kulis and Esteller 2010).

3.1 Histone and DNA Modifications

Epigenomes include both histone and DNA modifications, layered on top of the genome, which are responsible for providing information to genes during particular events. The epigenetic condition of a cell is affected by both developmental and environmental factors including nutrients, toxins, infection and drugs. Thus, epigenetics exhibits a close relationship between the environment and the genome. The chromatin can be altered in different ways; however, only few chromatin features have been shown to be functionally involved in gene expression. The first chromatin mark to gain attention, in the late 1940s, was DNA methylation due to its uneven distribution in the genome and its heritability (Hotchkiss 1948). DNA methylation is vital to cellular processes including X chromosome inactivation, gene suppression, genomic imprinting, and disease development. On the other hand, certain proteins can indirectly alter the genome's accessibility for transcription factors by binding to histone proteins. Changes in these proteins influence distinctive processes in the cell, including the activation or inactivation of transcription, chromosome packaging, and DNA damage and repair. Histone modification is an important posttranslational process that plays a key role in gene expression and represents by far the largest category among known chromatin marks. To date, a total of 12 chemical modifications have been described, which can occur at more than 130 posttranslational modification sites. The theoretical number of combinatorial possibilities is truly astronomical (Khare et al. 2012) and, consequently, our knowledge of their functional roles is still limited (Tessarz and Kouzarides 2014).

3.2 Epigenetics and Diseases

For decades, research has focused on isolating genes that contribute to a particular phenotype. Approaches such as genome-wide association studies (GWAS) identify locations in the human genome at which variations in DNA sequence are linked to specific phenotypes. However, variants located in a DNA region that does not encode a protein will not provide insights into the regulatory mechanisms underlying the association. In these cases, comprehensive epigenomic analyses can provide the missing link between genomic variation and cellular phenotype (Welter et al. 2014). A variety of diseases are triggered by alterations of epigenetic patterns, including changes in DNA methylation, posttranslational histone modifications, or chromatin structure. These changes represent an exceptionally interesting layer of information for disease stratification and precision medicine (Dirks et al. 2016). Some health conditions, e.g., Beckwith Wiedemann syndrome, Prader–Willi syndrome, and Angelman syndrome, are thought to be associated with a change in genome imprinting. Accordingly, the nature of epigenomic imprinting and its impact on the human genome could be connected in medical practice, i.e., in diagnosis and treatment of related conditions. Indeed, Polak et al. (Polak et al. 2015) investigated the distribution of cancer-associated genetic mutations in a set of diverse cancers and correlated them with specific epigenomic features. Indeed, the mutation profile of each cancer could be predicted from the epigenomic signature of the cell type from which that cancer was originated. Remarkably, epigenomic signatures of cancer cell lines were poor predictors of this profile; hence, the authors concluded that the density and distribution of cancer mutations are strongly linked to a cell-type-specific epigenomic signature. Compared to normal cells, the genome in tumor cells has been shown to be hypomethylated, with hypermethylation happening onlyin the genes engaged with tumor cell invasion, cell cycle control, DNA repair, and processes for which silencing would induce the spread of cancer. For instance, in colon cancer, hypermethylation acts as a biomarker in the progress of disease. Epigenetic modifications of chromatin hold considerable promise for therapies since most of them are reversible, owing to the adaptive nature of epigenetic control. Researchers are now studying the link between DNA methylation and human diseases such as malignancy, muscular dystrophy, and different congenital defects. Their discoveries could be fundamental in helping the development of therapies and in understanding conditions that develop during embryonic growth because of abnormal methylation of the X chromosome and gene imprinting. Epigenetics has been and will continue to be one of the most innovative research areas in modern biology and medicine.

4 Transcriptomics

Transcriptomics is one of the functional genomic approaches while the transcriptome is a collection of all ribonucleic acids (RNA) present in a cell. RNAs are macromolecules with diverse cellular and biological functions; for instance, they are templates for protein synthesis and assist in gene regulation. The transcriptome is dynamic, as the levels of RNA transcripts vary during developmental stages and in response to certain conditions. The goal oftranscriptome analysis is to identify differentially expressed genes in different conditions, thus, indicating the genes or pathways associated with these conditions. Over the years, different methods have been developed to study the transcriptional activity of genes from semiquantitative methods such as northern blotting and quantitative-PCR (Weis et al. 1992), to high-throughput methods like microarrays and RNA sequencing (Schena et al. 1995). The first attempts to study the whole transcriptome began in the early 1990s (Adams et al. 1991). In 1995, Serial Analysis of Gene Expression (SAGE) was developed, based on Sanger sequencing (Velculescu et al. 1995). SAGE was instrumental in profiling novel and known transcripts but was labor-intensive and had limited scope and quantitative capability.

Transcriptomics offers a global view of the molecular transcriptional activity in cells, which affects both human physiology and pathology. Predisease prediction is not possible through transcriptomics because gene expression levels vary considerably before disease initiation. However, it is very suitable for either early identification of a disease or classifying patients into subgroups to predict their health outcomes. Deregulation of long noncoding RNAs (lncRNA) has been implicated in various diseases (Schmitz et al. 2016) such as myocardial infarction (Ishii et al. 2006), diabetes (Arnes et al. 2016), and cancer (Gupta et al. 2010). Gene expression profiling using microarrays is a swift approach to diagnose, via biomarker genes, a wide range of diseases such as cancer, diabetes, arthritis, and rheumatoid arthritis (Heller 2002). A significant promise was held with the emergence of RNA-Seq technology for its application in diagnosis, prognosis, and treatment of various disorders, including cancers and infectious diseases. RNA-Seq analysis compares transcriptomes across different developmental stages, diseases, or specific conditions. For example, comparing the transcriptomes of tumor and normal cells and looking for copy number alterations or alternative spliced variants are valuable methodologies that are missing in microarrays. Moreover, transcriptome analysis may provide a comprehensive snapshot of active genes during various stages of development. Besides, RNA-Seq approaches have allowed for the large-scale identification of transcriptional start sites (TSS) in addition to uncovering alternative promoter usage and novel splicing alterations. These regulatory elements are important in human disease, and therefore, defining such variants is crucial to the interpretation of disease-association studies. In addition, RNA-Seq can identify disease-associated SNPs, allele-specific expressions, and gene fusions contributing to our understanding of disease causal variants (Khurana et al. 2016). Moreover, RNA-Seq could be used in the diagnosis and management of infectious diseases

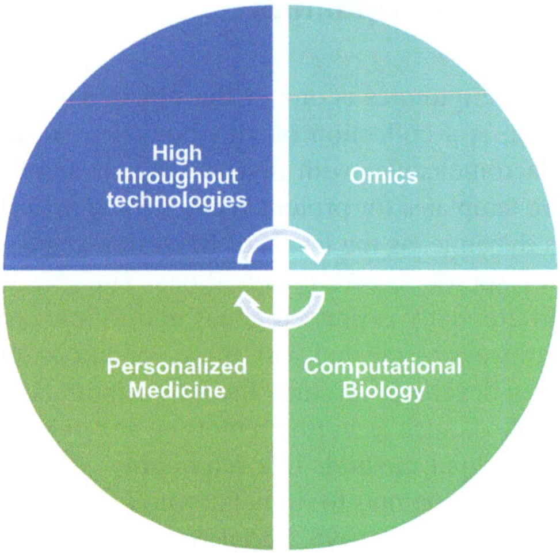

Fig. 2 Omics integration core units

such as monitoring for drug-resistant populations during therapy and tracking the origin and spread of the Ebola virus (Shabman et al. 2014).

Some of the accessible databases for transcriptomics is ArrayExpress (Parkinson et al. 2007), hosted by the European Bioinformatics Institute (EBI) site, which enables researchers to submit array data and conduct analysis. In addition, Gene Expression Omnibus (GEO) is a public repository of microarray and RNA-Seq data hosted by the NCBI site. These websites publish raw data to accompany a research publication for public access (Barrett et al. 2013). A variety of cancer studies and their profiling were performed using microarrays. A large fraction of this data can be found integrated in Oncomine, a cancer microarray database and Web-based data-mining platform (Rhodes et al. 2004, 2007).

Transcriptomics has revolutionized our understanding of how genomes are expressed. Over the last three decades, new technologies have redefined what is possible to investigate while integration with other omics technologies is giving an increasingly combined view of the complexities of cellular systems (Fig. 2).

5 Proteomics

With the mapping of the human genome, proteomics has rapidly emerged as a new field of research. Proteomics is complementary to genomics but encompasses functional analysis. The twenty first century has been designated as the postgenomic or proteomic era. Proteomics is the study, under defined set of conditions, of the whole protein set and their interactions in a cell, tissue, or organism (Consortium 2001). Proteins within the cell provide structure, produce energy, as well as allow for communication and movement. Highly specialized proteins regulate most biochemical reactions in a cell. Hence, the identification, quantification, and characterization of all cellular proteins is of extreme significance in order to understand molecular processes that mediate cellular functions.

The term proteoform, introduced in 2013, is a relatively new terminology to designate all different molecular forms in which the protein product of a single gene can be found (Smith et al. 2013). Human cells apparently have as few as 19,000 protein-encoding genes, representing less than 2% of the genome. However, these genes specify more than 1,000,000 final products of proteoforms, including the products of alternatively spliced RNAs (Roy et al. 2013), nonsynonymous SNPs (Schaefer et al. 2012; Wu and Zeng 2012), and extensive posttranslational modifications that add chemical moieties on amino acids or remove residues from the protein (Farley and Link 2009; Abou-Abbass et al. 2016). Besides, proteins change drastically as genes are turned on or off due to environmental changes. Proteins are also pleotropic since they have numerous distinct functions. They are not autonomous; instead, they act in complexes with other proteins. Indeed, protein–protein interactions (PPIs) and multiprotein complexes underlie almost all of the vital biochemical processes occurring in almost all living cells and tissues (Alberts 1998). Consequently, elucidating the role of proteins, especially on a large scale, is more difficult than for nucleic acids.

Analysis of proteomes, performed through the highly accurate and sensitive technology of mass spectrometry (MS), is promising for the introduction of proteomics in precision medicine. MS can be used for screening and diagnosis of disease and metabolic disorders, monitoring drug therapy, identifying drug toxicity, and discovering new biomarkers. Protein sequencing by MS has increased due to its ability to tolerate protein complexes and the possibility of high throughput options. MS-based proteomics can reveal the proteome's quantity and hence allows understanding the biochemical state of cells or tissues. Platforms originating from MS are unique in their potential to detect and quantify known and unknown proteins on a large scale. This technology has significantly contributed to the unraveling of cellular signaling networks, elucidation of the dynamics of PPIs in different cellular states, and improved diagnosis and molecular understanding of disease mechanisms. Interestingly, MS can also be applied to investigate the posttranslational modifications of proteins. On the other hand, MS imaging (MSI) is performed to localize panels of biomolecules in tissues and it can visualize the spatial distribution of molecules such as biomarkers, proteins, and metabolites based on their molecular masses. MALDI mass spectrometry imaging was used for the analysis of whole body tissues. Indeed, the distribution of drugs and metabolites was detected following drug administration that was useful to analyze novel therapeutics and provide deeper insights into toxicological and therapeutic processes (Worrall et al. 2001). Combining MSI to histology enables the extraction of molecular profiles from specific tissue regions or histopathological entities. MSI can facilitate, with high sensitivity and specificity, the classification of tumors during surgery (Balog et al. 2013). MS has been implemented into clinical research since MALDI and ESI approaches were developed 30 years ago. The main hurdle for the clinical adoption of these assays is their complexity and cost.

Developments in clinical high throughput MS made possible the publication of the human proteome draft (Uhlén et al. 2015). By comparative clinical proteomics, scientists can associate proteomic changes of patients in comparison to known proteomes in the databases. Clinically, translation of these findings allows the possibility of mass spectral analysis of patient proteomes. Physicians will eventually be able to compare patients' proteomes with their own healthy archived records (Lindskog 2015). Today, the new field of clinical mass spectrometry proteomics (cMSP) seeks to unify disparate basic science approaches and validate them for clinical proteomics (Lehmann et al. 2017). The complete characterization of all proteins has been the ultimate aim since the introduction of proteomics (Farrah et al. 2014). The Human Proteome Project focuses on characterizing the human proteome, while the Human Protein Atlas Project attempts to produce antibodies for all human encoded proteins. On the other hand, the Proteome X change consortium gathers proteomic data (Vizcaíno et al. 2014). Finally, many public repositories for PPI information have been created such as the Database of Interacting Proteins (DIP) (Xenarios 2002), MIntAct (Orchard et al. 2014) and Molecular Interaction Database (MINT) (Licata et al. 2012).

6 Metabolomics

Metabolomics is the comprehensive study of the metabolome, the repertoire of biochemical moleculesfound in cells, tissues, or body fluids. It serves as bridging the gap between genotype and phenotype, giving a complete view of how cells function in addition to identifying new changes in certain metabolites (Fiehn 2002). The latter are the end products of the processes that occur within a cell and provide a molecular outline of cellular activity reflecting biochemical processes occurring in a particular phenotype. The metabolome reacts to environmental stimuli or disease long ahead the transcriptome or proteome. The study of metabolomics will help understanding the physiological state of an organism and the functional changes in metabolic pathways that drive a disease.

Metabolic processes are associated to several vital aspects of human health. Metabolomics gives clues about one's health status, which is encoded by the genome and altered by environmental factors. The metabolic profile gives quantifiable data of biochemical state from normal physiology to pathophysiology in a way that is frequently not evident from gene expression studies. Over the years, metabolomics has been commonly employed in the understanding of pathophysiological processes including cancer and diabetes for the identification of disease onset predictive biomarkers, prognosis, and treatment monitoring (Friedrich 2012; Kim et al. 2016). The metabolome is highly responsive to biological regulatory mechanisms such as epigenetics, transcription, and posttranslational modification, the analysis of which presents a unique approach to characterize the phenotype. However, metabolomics by itself may not be sufficient to fully characterize complex biological systems or pathologies like cancer. The significant connection between health and metabolome is powered by small alterations in biochemical pathways to produce drastic changes in cell metabolites. For instance, metabolomics has given crucial perceptions into disease pathogenesis in disorders with clear metabolic causes such as diabetes and heart disease (Palmer et al. 2015). Nevertheless, when metabolic perturbations do not cause the disease process, they can still stimulate modifications in the metabolome. For example, inflammatory conditions (Jung et al. 2013), neurodegenerative disease (Trushina and Mielke 2014), infections (Schoen et al. 2014), and cancer (Kim et al. 2009) could change cell metabolism, permitting to recognize and classifynovel biomarkers that assist in the prediction, diagnosis, and comprehension of disease. Recently, metabolomics has been applied to dermatology, where it was used to distinguish characteristic biomarker profiles in metastatic melanoma (Abaffy et al. 2013), basal cell carcinoma (Mun et al. 2016), intense intermittent porphyria (Carichon et al. 2014), and atopic dermatitis (Assfalg et al. 2012).

Metabolomics efficiently integrates genetic and environmental factors and has been applied in various studies including biomarker discovery (Crutchfield et al. 2016), disease mechanisms and drug activity (García-Cañaveras et al. 2015) and metabolism (Das et al. 2016). MS metabolomics offers the possibility to create novel noninvasive diagnostic and screening tests, and covers novel metabolic pathwaysthat may enhance diagnosis and comprehension of pathologies. Various reports have shown that metabolic phenotypes can give novel insights in the study of gene function and pathogenic pathways other than diagnosis and prognosis (Guo et al. 2015). Indeed, in order to identify appropriate therapies to particular subgroups, MS qualitative methods and their analysis are used to assess the complex metabolic phenotypes of patients. A proficient approach for understanding the pathogenic part of metabolite perturbations is to consolidate metabolomics datasets from different omics fields. Integrated omics permits analyzing and inferring the influences from genetics, environment and microbial factors helping to reach personalized medicine. In fact, few approaches have been established for the integration of various omics datasuch as transformation-based integration and model-based integration (Ritchie et al. 2015). For instance, Bayesian networks are a link based integration technique that can be utilized to show dependent conditions between gene expression, proteins, and metabolites (Li et al. 2016). Recently, Guo et al. have utilized an assortment of blood-based metabolomics to evaluate the biologic importance and penetrance of genetic mutations in a group of healthy volunteers who were diagnosed disease-free at the time of sampling (Guo et al. 2015). Interestingly, this study recognized early abnormalities in metabolites engaged with lipolysis, glycolysis, and amino acid metabolism in healthy subjects that possess genetic mutations related with diabetes. These subjectswere later identified to have the disease. Guo's study gives a model of how metabolomics data can give perception into the clinical implication of different omics data in early detection of disease (Guo et al. 2015).

Together with the other omics, metabolomics constitutes one of the building blocks of systems biology. This fast-growing domain generates huge amounts of valuable data that require integration with other omics data and comprehensive analysis in order to be fully interpreted.

7 Coupling Large-Scale Data: A Case Study on RAAS

The renin-angiotensin-aldosterone system (RAAS) is a “ubiquitous” system that is expressed locally in various tissues and exerts multiple paracrine/autocrine effects involved in tissue physiology and homeostasis. RAAS includes successive enzymatic reactions resulting in the conversion of the “inactive” substrate angiotensinogen (AGT) into various active peptides that elicits cellular effects by binding to specific membrane receptors (Atlas 2007). The system is considered a hallmark in cardiovascular homeostasis and pathophysiology. However, RAAS inhibitors that are currently used still can not reach their desired effects and hold certain drawbacks such as adverse side effects, incomplete blockage, and poor end-organ protection (Nehme and Zibara 2017a, b). This could be explained by the fact that the system includes different pathways with alternative and synonymous enzymes and receptors having tissue-specific expression and activity. Therefore, treatments targeting RAAS components should be achieved in a disease- and tissue-specific manner.

We recently integrated transcriptomic, genomics, proteomics, and metabolomics data to reveal the tissue-specific organization of RAAS in atherosclerotic lesions. Horizontal integration of multiple microarray datasets was first used to decipher the RNA coexpression patterns of 37 RAAS genes in atheroma lesions, including enzymes and receptors (Nehme et al. 2015, 2016a). Expression analysis and hierarchical clustering was done to reveal the transcriptional map of extRAAS in atheroma using six microarray datasets, available on gene expression omnibus (GEO) database, containing 839 human atheroma samples. The map revealed highly reproducible coexpression modules of extRAAS components displaying favored pathways in atheroma. Interestingly, three main modules were identified showing intramodule functional relationships, where one module included mostly receptor-coding genes, while the other two included enzyme-coding genes. Interestingly, similar results were obtained from GEO datasets containing mouse atherosclerotic samples, but different from normal human arterial tissues (Nehme et al. 2016a). Promoter and transcription factor (TF) enrichment analyses were then done to identify candidate transcription regulators of the system in atherosclerotic lesions. A total of 21 transcription factors with enriched binding sites in the promoters of coordinated genes were extracted, showing specific positive and negative correlations with each of the identified RAAS modules. Atheroma showed specific correlations between RAAS and the identified TFs although some similarities in atheroma RAAS organization were shared with kidney and adipose tissues (Atlas 2007; Nehme et al. 2016a). Since RNA is not functional on its own, we performed a global proteomics study to identify expression levels of RAAS components at the proteome level. Unfortunately, only five proteins from RAAS were among the list of detectable and measured proteins. This could be either due to the fact that measurement was not possible on the rest of the proteins because of the low specificity achieved through global analysis, or that some of the unmeasured proteins required special extraction protocols, such as receptors for example.

Therefore, targeted study, taking into account special extraction procedures, should be done in the future. Nonetheless, two of the RAAS measured proteins that were upregulated at the RNA level were also highly upregulated at the protein level, including the substrate of the system, angiotensinogen. On the other hand, the system's activity depends on the local levels of angiotensin peptides metabolites, which in turn affects their rates of production and degradation. Therefore, we used mass spectrometry to reveal the kinetics of labeled spiked-in angiotensin-I metabolite (Ang-I) in paired early and advanced atherosclerotic lesions (Nehme et al. 2016b). Our results suggested that progression of atherosclerosis could be related to the increased production of the metabolite angiotensin-II peptide (Ang-II) along with the decreased production of the atheroprotective angiotensin-(1-7) peptide (Ang-(1-7)). However, Ang-II may exert proatherogenic and atheroprotective effects depending on the available expressed receptor. Going back to transcriptomics data, only the proatherogenic angiotensin type-1 receptor (AT1R) is expressed in atheroma, both in human and mouse, whereas the atheroprotective angiotensin type-2 receptor AT2R is not present (Nehme et al. 2016a), thus favoring proatherogenic effects by Ang-II peptide. Interestingly, these results were in line with a previous in vivo study that showed the benefits of using Ang-(1-7) along with blockers. Overall, our study deciphered the organization of RAAS at the transcriptome and proteome levels, showing its functionality at the metabolomic level and proposed candidate transcription factors that can be used as therapeutic targets for the treatment of atherosclerosis.

8 Conclusion and Perspectives

Genomics data provide important information of the biological identity of cells and individuals. In fact, such data can be useful in hereditary, prognostic and evolutionary, studies. However, they offer only little information on the molecular mechanisms implicated in cellular physiology and pathophysiology. In fact, a single gene mutation may lead to thousands of changes at the RNA, protein, metabolic and signaling levels, which constitutes the basis for disease development. Transcriptomic studies are usually the first line of evidence used to understand the mechanisms of cellular pathophysiology. However, several studies have recently demonstrated discrepancies in the expression levels between different molecular techniques, which raises the importance of vertical integration of data (Maier et al. 2009; Koussounadis et al. 2015). Interestingly, differentially expressed mRNAs were found to correlate better with their protein product than nondifferentially expressed mRNAs, which may raise the confidence in the use of differential mRNA expression for biological discovery (Koussounadis et al. 2015). Nonetheless, posttranslational modifications are indispensable for the function of a large array of proteins, including transcription factors and signaling molecules. Therefore, protein expression measurement remains very powerful for assessing the functionality of the system. On the other hand, global cellular activity depends on the physicochemical status of the cell, including oxygen, glucose, and ions levels. In addition, enzymatic efficiency depends on the concentrations of substrates and products, and the affinity between the substrate and the enzyme. A substrate could be a protein or a nonprotein molecule, considered nonbiological, such as glucose, reactive oxygen species (ROS), and ionic metals. These molecules could be measured by high-throughput metabolomics techniques that are providing insights for the treatment of various cancers. Finally, the physicochemical status of the cell takes us back to the genomic level by affecting epigenomic markers that in turn defines gene expression patterns.

Studies in genomics, transcriptomics, epigenomics, proteomics, and metabolomics have shaped our understanding of cellular complexity and heterogeneity. Each of them provides a one-dimensional insight view of cellular function. It is now evident that single omics analysis does not provide enough information for the understanding of a biological system; however, a complete view of a complex biological system can be achieved by a unified and global integrative analysis. Compared to single omics interrogations, multi-omics can offer researchers a more noteworthy understanding of the flow of information, from the cause of disease to the functional outcomes or relevant interactions (Civelek and Lusis 2013). Multidimensional analysis is being considered crucial for completely understanding the extent of traits architecture.

References

1000 Genomes Project Consortium, Auton A, Brooks LD et al (2015) A global reference for human genetic variation. Nature 526:68–74

Abaffy T, Möller MG, Riemer DD et al (2013) Comparative analysis of volatile metabolomics signals from melanoma and benign skin: a pilot study. Metabolomics 9:998–1008

Abou-Abbass H, Abou-El-Hassan H, Bahmad H et al (2016) Glycosylation and other PTMs alterations in neurodegenerative diseases: current status and future role in neurotrauma. Electrophoresis 37:1549–1561

Adams MD, Kelley JM, Gocayne JD et al (1991) Complementary DNA sequencing: expressed sequence tags and human genome project. Science 252:1651–1656

Alberts B (1998) The cell as a collection of protein machines: preparing the next generation of molecular biologists. Cell 92:291–294

Allfrey VG, Faulkner R, Mirsky AE (1964) Acetylation and methylation of histones and their possible role in the regulation of RNA synthesis. Proc Natl Acad Sci USA 51:786–794

Anderson S (1981) Shotgun DNA sequencing using cloned DNase I-generated fragments. Nucleic Acids Res 9:3015–3027

Arnes L, Akerman I, Balderes DA et al (2016) betalinc1 encodes a long noncoding RNA that regulates islet beta-cell formation and function. Genes Dev 30:502–507

Assfalg M, Bortoletti E, D'Onofrio M et al (2012) An exploratory 1H-nuclear magnetic resonance metabolomics study reveals altered urine spectral profiles in infants with atopic dermatitis. Br J Dermatol 166:1123–1125

Atlas SA (2007) The renin-angiotensin aldosterone system: pathophysiological role and pharmacologic inhibition. J Manag Care Pharm JMCP 13:9–20

Balog J, Sasi-Szabo L, Kinross J et al (2013) Intraoperative tissue identification using rapid evaporative ionization mass spectrometry. Sci Transl Med 5:194ra93–194ra93

Barrett T, Wilhite SE, Ledoux P et al (2013) NCBI GEO: archive for functional genomics data sets—update. Nucleic Acids Res 41:D991–D995
Bird A, Taggart M, Frommer M et al (1985) A fraction of the mouse genome that is derived from islands of nonmethylated, CpG-rich DNA. Cell 40:91–99
Buermans HPJ, den Dunnen JT (2014) Next generation sequencing technology: advances and applications. Biochim Biophys Acta 1842:1932–1941
Byron SA, Van Keuren-Jensen KR, Engelthaler DM et al (2016) Translating RNA sequencing into clinical diagnostics: opportunities and challenges. Nat Rev Genet 17:257–271
Cantor RM, Lange K, Sinsheimer JS (2010) Prioritizing GWAS results: a review of statistical methods and recommendations for their application. Am J Hum Genet 86:6–22
Carichon M, Pallet N, Schmitt C et al (2014) Urinary metabolic fingerprint of acute intermittent porphyria analyzed by 1H NMR spectroscopy. Anal Chem 86:2166–2174
Carithers LJ, Moore HM (2015) The Genotype-Tissue Expression (GTEx) Project. Biopreserv Biobank 13:307–308
Caskey CT, Gonzalez-Garay ML, Pereira S, McGuire AL (2014) Adult genetic risk screening. Annu Rev Med 65:1–17
Chen R, Mias GI, Li-Pook-Than J et al (2012) Personal omics profiling reveals dynamic molecular and medical phenotypes. Cell 148:1293–1307
Civelek M, Lusis AJ (2013) Systems genetics approaches to understand complex traits. Nat Rev Genet 15:34–48
Clark SJ, Lee HJ, Smallwood SA et al (2016) Single-cell epigenomics: powerful new methods for understanding gene regulation and cell identity. Genome Biol 17:72
Consortium IH 3 (2010) Integrating common and rare genetic variation in diverse human populations. Nature 467:52–58
Consortium IHGS (2001) Initial sequencing and analysis of the human genome. Nature 409:860–921
Crutchfield CA, Thomas SN, Sokoll LJ, Chan DW (2016) Advances in mass spectrometry-based clinical biomarker discovery. Clin Proteomics 13:1
Das MK, Arya R, Debnath S et al (2016) Global urine metabolomics in patients treated with first-line tuberculosis drugs and identification of a novel metabolite of ethambutol. Antimicrob Agents Chemother 60:2257–2264
DePristo MA, Banks E, Poplin R et al (2011) A framework for variation discovery and genotyping using next-generation DNA sequencing data. Nat Genet 43:491–498
Dirks RAM, Stunnenberg HG, Marks H (2016) Genome-wide epigenomic profiling for biomarker discovery. Clin Epigenetics 8:122
Eid J, Fehr A, Gray J et al (2009) Real-time DNA sequencing from single polymerase molecules. Science 323:133–138
ENCODE Project Consortium {fname} (2012) An integrated encyclopedia of DNA elements in the human genome. Nature 489:57–74
Farley AR, Link AJ (2009) Identification and quantification of protein posttranslational modifications. Methods Enzymol 463:725–763
Farrah T, Deutsch EW, Omenn GS et al (2014) State of the human proteome in 2013 as viewed through peptideatlas: comparing the kidney, urine, and plasma proteomes for the biology- and disease-driven human proteome project. J Proteome Res 13:60–75
Fiehn O (2002) Metabolomics—the link between genotyopes and phenotypes. Plant Mol Biol 48:155–171
Friedrich N (2012) Metabolomics in diabetes research. J Endocrinol 215:29–42
García-Cañaveras JC, Jiménez N, Gómez-Lechón MJ et al (2015) LC-MS untargeted metabolomic analysis of drug-induced hepatotoxicity in HepG2 cells. Electrophoresis 36:2294–2302
GTEx Consortium TGte (2013) The Genotype-Tissue Expression (GTEx) project. Nat Genet 45:580–585
Guo L, Milburn MV, Ryals JA et al (2015) Plasma metabolomic profiles enhance precision medicine for volunteers of normal health. Proc Natl Acad Sci USA 112:E4901–E4910

Gupta RA, Shah N, Wang KC et al (2010) Long non-coding RNA HOTAIR reprograms chromatin state to promote cancer metastasis. Nature 464:1071–1076

Heller MJ (2002) DNA microarray technology: devices, systems, and applications. Annu Rev Biomed Eng 4:129–153

Holliday R, Pugh JE (1975) DNA modification mechanisms and gene activity during development. Science 187:226–232

Hotchkiss RD (1948) The quantitative separation of purines, pyrimidines, and nucleosides by paper chromatography. J Biol Chem 175:315–332

International Human Genome Sequencing Consortium (2004) Finishing the euchromatic sequence of the human genome. Nature 431:931–945

Ishii N, Ozaki K, Sato H et al (2006) Identification of a novel non-coding RNA, MIAT, that confers risk of myocardial infarction. J Hum Genet 51:1087–1099

Jung J, Kim SH, Lee HS et al (2013) Serum metabolomics reveals pathways and biomarkers associated with asthma pathogenesis. Clin Exp Allergy 43:425–433

Khare SP, Habib F, Sharma R et al (2012) HIstome—a relational knowledgebase of human histone proteins and histone modifying enzymes. Nucleic Acids Res 40:D337–D342

Khurana E, Fu Y, Chakravarty D et al (2016) Role of non-coding sequence variants in cancer. Nat Rev Genet 17:93–108

Kim K, Aronov P, Zakharkin SO et al (2009) Urine metabolomics analysis for kidney cancer detection and biomarker discovery. Mol Cell Proteomics MCP 8:558–570

Kim Y, Jeon J, Mejia S et al (2016) Targeted proteomics identifies liquid-biopsy signatures for extracapsular prostate cancer. Nat Commun 7:11906

Klein RJ, Zeiss C, Chew EY et al (2005) Complement factor H polymorphism in age-related macular degeneration. Science 308:385–389

Koussounadis A, Langdon SP, Um IH et al (2015) Relationship between differentially expressed mRNA and mRNA-protein correlations in a xenograft model system. Sci Rep 5:10775

Kouzarides T (2007) Chromatin modifications and their function. Cell 128:693–705

Kulis M, Esteller M (2010) DNA methylation and cancer. Adv Genet 70:27–56

Lehmann S, Brede C, Lescuyer P et al (2017) Clinical mass spectrometry proteomics (cMSP) for medical laboratory: what does the future hold? Clin Chim Acta 467:51–58

Li S, Todor A, Luo R (2016) Blood transcriptomics and metabolomics for personalized medicine. Comput Struct Biotechnol J 14:1–7

Licata L, Briganti L, Peluso D et al (2012) MINT, the molecular interaction database: 2012 Update. Nucleic Acids Res 40:D857–D861

Lindskog C (2015) The potential clinical impact of the tissue-based map of the human proteome. Expert Rev Proteomics 12:213–215

Lister R, Pelizzola M, Kida YS et al (2011) Hotspots of aberrant epigenomic reprogramming in human induced pluripotent stem cells. Nature 471:68–73

Maier T, Güell M, Serrano L (2009) Correlation of mRNA and protein in complex biological samples. FEBS Lett 583:3966–3973

Manolio TA, Collins FS (2009) The HapMap and genome-wide association studies in diagnosis and therapy. Annu Rev Med 60:443–456

Maraganore DM, de Andrade M, Lesnick TG et al (2005) High-resolution whole-genome association study of Parkinson disease. Am J Hum Genet 77:685–693

Mun J-H, Lee H, Yoon D et al (2016) Discrimination of basal cell carcinoma from normal skin tissue using high-resolution magic angle spinning 1H NMR spectroscopy. PLoS One 11:e0150328

Nehme A, Zibara K (2017a) Cellular distribution and interaction between extended renin-angiotensin-aldosterone system pathways in atheroma. Atherosclerosis 263:334–342

Nehme A, Zibara K (2017b) Efficiency and specificity of RAAS inhibitors in cardiovascular diseases: how to achieve better end-organ protection? Hypertens Res 40:903–909

Nehme A, Cerutti C, Dhaouadi N et al (2015) Atlas of tissue renin-angiotensin-aldosterone system in human: a transcriptomic meta-analysis. Sci Rep 5:10035

Nehme A, Cerutti C, Zibara K (2016a) Transcriptomic analysis reveals novel transcription factors associated with renin–angiotensin–aldosterone system in human atheroma. Hypertension HYPERTENSIONAHA.116.08070

Nehme A, Marcelo P, Nasser R et al (2016b) The kinetics of angiotensin-I metabolism in human carotid atheroma: an emerging role for angiotensin (1-7). Vascul Pharmacol 85:50–56

Ng SB, Turner EH, Robertson PD et al (2009) Targeted capture and massively parallel sequencing of 12 human exomes. Nature 461:272–276

Orchard S, Ammari M, Aranda B et al (2014) The MIntAct project—IntAct as a common curation platform for 11 molecular interaction databases. Nucleic Acids Res 42:D358–D363

Palmer ND, Stevens RD, Antinozzi PA et al (2015) Metabolomic profile associated with insulin resistance and conversion to diabetes in the Insulin Resistance Atherosclerosis Study. J Clin Endocrinol Metab 100:E463–E468

Parkinson H, Kapushesky M, Shojatalab M et al (2007) ArrayExpress—a public database of microarray experiments and gene expression profiles. Nucleic Acids Res 35:D747–D750

Polak P, Karlić R, Koren A et al (2015) Cell-of-origin chromatin organization shapes the mutational landscape of cancer. Nature 518:360–364

Qin W, Kozlowski P, Taillon BE et al (2010) Ultra deep sequencing detects a low rate of mosaic mutations in tuberous sclerosis complex. Hum Genet 127:573–582

Rhodes DR, Yu J, Shanker K et al (2004) ONCOMINE: a cancer microarray database and integrated data-mining platform. Neoplasia 6, 6(1)

Rhodes DR, Kalyana-Sundaram S, Mahavisno V et al (2007) Oncomine 3.0: genes, pathways, and networks in a collection of 18,000 cancer gene expression profiles. Neoplasia 9:166–180

Ritchie MD, Holzinger ER, Li R et al (2015) Methods of integrating data to uncover genotype–phenotype interactions. Nat Rev Genet 16:85–97

Roy B, Haupt LM, Griffiths LR (2013) Review: alternative splicing (AS) of genes as an approach for generating protein complexity. Curr Genomics 14:182–194

Sanger F, Coulson AR (1975) A rapid method for determining sequences in DNA by primed synthesis with DNA polymerase. J Mol Biol 94:441–448

Schaefer C, Meier A, Rost B, Bromberg Y (2012) Snpdbe: constructing an nsSnp functional impacts database. Bioinformatics 28:601–602

Schena M, Shalon D, Davis RW, Brown PO (1995) Quantitative monitoring of gene expression patterns with a complementary DNA microarray. Science 270:467–470

Schmitz SU, Grote P, Herrmann BG (2016) Mechanisms of long noncoding RNA function in development and disease. Cell Mol Life Sci 73:2491–2509

Schoen C, Kischkies L, Elias J, Ampattu BJ (2014) Metabolism and virulence in Neisseria meningitidis. Front Cell Infect Microbiol 4:114

Shabman RS, Jabado OJ, Mire CE et al (2014) Deep sequencing identifies noncanonical editing of Ebola and Marburg virus RNAs in infected cells. mBio 5:e02011

Shendure J, Ji H (2008) Next-generation DNA sequencing. Nat Biotechnol 26:1135–1145

Smith LM, Kelleher NL, Linial M et al (2013) Proteoform: a single term describing protein complexity. Nat Methods 10:186–187

Stadler ZK, Thom P, Robson ME et al (2010) Genome-wide association studies of cancer. J Clin Oncol Off J Am Soc Clin Oncol 28:4255–4267

Tessarz P, Kouzarides T (2014) Histone core modifications regulating nucleosome structure and dynamics. Nat Rev Mol Cell Biol 15:703–708

Tomescu OA, Mattanovich D, Thallinger GG (2014) Integrative omics analysis. A study based on Plasmodium falciparum mRNA and protein data. BMC Syst Biol 8:S4

Trushina E, Mielke MM (2014) Recent advances in the application of metabolomics to Alzheimer's disease. Biochim Biophys Acta Mol Basis Dis 1842:1232–1239

Uhlén M, Fagerberg L, Hallström BM et al (2015) Proteomics. Tissue-based map of the human proteome. Science 347:1260419

Velculescu VE, Zhang L, Vogelstein B, Kinzler KW (1995) Serial analysis of gene expression. Science 270:484–487

Verdin E, Ott M (2014) 50 years of protein acetylation: from gene regulation to epigenetics, metabolism and beyond. Nat Rev Mol Cell Biol 16:258–264
Vizcaíno JA, Deutsch EW, Wang R et al (2014) ProteomeXchange provides globally coordinated proteomics data submission and dissemination. Nat Biotechnol 32:223–226
Waddington CH (1942) The epigenotype. Endeavour 1:18–20. https://doi.org/10.1093/ije/dyr184
Weis JH, Tan SS, Martin BK, Wittwer CT (1992) Detection of rare mRNAs via quantitative RT-PCR. Trends Genet 8:263–264. https://doi.org/10.1016/0168-9525(92)90242-V
Welter D, MacArthur J, Morales J et al (2014) The NHGRI GWAS Catalog, a curated resource of SNP-trait associations. Nucleic Acids Res 42:D1001–D1006
Worrall JA, Kolczak U, Canters GW, Ubbink M (2001) Interaction of yeast iso-1-cytochrome c with cytochrome c peroxidase investigated by [15N, 1H] heteronuclear NMR spectroscopy. Biochemistry (Mosc) 40:7069–7076
Wu JR, Zeng R (2012) Molecular basis for population variation: from SNPs to SAPs. FEBS Letters.:2841–2845
Xenarios I (2002) DIP, the Database of Interacting Proteins: a research tool for studying cellular networks of protein interactions. Nucleic Acids Res 30:303–305

Deciphering the Universe of RNA Structures and *trans* RNA–RNA Interactions of Transcriptomes In Vivo: From Experimental Protocols to Computational Analyses

Stefan R. Stefanov and Irmtraud M. Meyer

Contents

S. R. Stefanov · I. M. Meyer (✉)
Laboratory of Bioinformatics of RNA Structure and Transcriptome Regulation, Berlin Institute for Medical Systems Biology, Max Delbrück Center for Molecular Medicine, Berlin-Buch, Germany

Freie Universität, Department of Biology, Chemistry, and Pharmacy, Institute of Chemistry and Biochemistry, Berlin, Germany
e-mail: irmtraud.meyer@cantab.net

N. Rajewsky et al. (eds.), *Systems Biology*, RNA Technologies,
https://doi.org/10.1007/978-3-319-92967-5_9

Abstract The last few years have seen an explosion of experimental and computational methods for investigating RNA structures of entire transcriptomes in vivo. Very recent experimental protocols now also allow *trans* RNA–RNA interactions to be probed in a transcriptome-wide manner. All of the experimental strategies require comprehensive computational pipelines for analysing the raw data and converting it back into actual RNA structure features or *trans* RNA–RNA interactions. The overall performance of these methods thus strongly depends on the experimental and the computational protocols employed. In order to get the best out of both worlds, both aspects need to be optimised simultaneously. This review introduced the methods and proposes ideas how they could be further improved.

Keywords RNA secondary structures · *trans* RNA–RNA interactions · RNA structure prediction · RNA–RNA interaction prediction · Transcriptomes · In vivo RNA structure probing · In vivo probing of *trans* RNA–RNA interaction · RNA structurome · RNA interactome

1 Introduction

The remarkable chemical properties of RNA allow transcripts in vivo to directly interact with themselves (via so-called RNA structure) or in *trans* with other transcripts, DNA and proteins. Many known RNA functions are expressed in terms of RNA structure. Substantial insight into the potential functional roles of any RNA can already be gained by studying its so-called RNA secondary structure, i.e. the set of base-paired sequence positions that form base pairs via hydrogen bonds (the consensus base pairs are $\{G, C\}$, $\{G, U\}$ and $\{A, U\}$). Obviously, the functional roles of any RNA can be encoded not only via RNA structure features, but also via sequence signals such as the sequence of codons defining a contiguous open-reading frame at messenger-RNA (mRNA) level or the sequence of nucleotides defining a protein-binding site. As it turns out, many ways of encoding functional information into a transcript are mutually compatible. For example, any given transcript may have RNA structure while simultaneously interacting with other molecules such as other transcripts, DNA or proteins. Or, one and the same stretch of RNA may encode a functional RNA structure as well as codon information on protein synthesis. Cases like these are not only found in viral genomes where space constraints force different layers of information to overlap (Pedersen et al. 2004b; Watts et al. 2009) but can also occur in otherwise perfectly ordinary coding transcripts of model organisms such as human, mouse and fruit fly. Luckily, overlapping layers of information can be detected in silico provided dedicated computational methods are employed that are capable of explicitly dis-entangling them (Pedersen et al. 2004a,b; Meyer and Miklos 2005). It is already known that RNA structure features can act as exquisite sensors of the complex in vivo environment and change according to sometimes subtle changes of intrinsic and extrinsic factors.

Examples of these factors range from single-nucleotide modifications of the primary RNA transcript (e.g. tRNAs and rRNAs require a range of well-defined chemical modifications at distinct sequence positions in order to become functionally active in vivo) and other changes of the primary transcript sequence (cleavage, splicing, tail-adding, A-to-I RNA editing, etc.) to changes of the surrounding temperature, changing *trans* interacting partners (ligands, other RNAs, proteins, DNA) and changes of the transcription speed. A wealth of recent evidence supports the notion of *alternative RNA structure expression* (Meyer 2017), i.e. that a single transcript can encode and express not just one, but several distinct RNA structures which are differentially expressed depending on the specific in vivo environment. Known cases do include examples not only from bacteria, but also from model organisms such as the fruit fly (Steif and Meyer 2012; Zhu et al. 2013; Zhu and Meyer 2015; Mazloomian and Meyer 2015). There is, for example, strong statistical evidence for differentially expressed, local RNA structure features near splice sites that define tissue-specific splice isoforms (Mazloomian and Meyer 2015). The corresponding RNA structure changes are mediated by tissue-specific A-to-I RNA editing of these structural features (Mazloomian and Meyer 2015). Alternative RNA structure expression allows one and the same (coding or non-coding) transcript to wear a series of distinct functional hats throughout its cellular life depending on its directly surrounding in vivo environment (extrinsic factors) and any modifications it undergoes itself (intrinsic changes). Taken together, the transcriptome thus offers exceptional potential to functionally link all layers of the central dogma of biology in a well-regulated manner that depends on the specific in vivo environment, thereby influencing gene expression and determining the organism's overall complexity. For better or worse, the days where we may silently assume the validity of the one-sequence-one-structure dogma are over. This has far-reaching implications on how we should experimentally probe RNA structures and *trans* RNA–RNA interactions in vivo and how we should model these features computationally.

Whereas protein–protein, DNA–protein and protein–RNA interactions have been the subject of intense experimental and computational research for a while, transcriptome-wide investigations of RNA structures and general methods for detecting *trans* RNA–RNA interactions in vivo have only emerged fairly recently. On the experimental side, one major step forward was made very recently (2016) via the publication of three experimental protocols that can directly probe both RNA structure features and *trans* RNA–RNA interactions in a transcriptome-wide fashion in vivo. On the computational side, *ab initio* methods for predicting truly novel *trans* RNA–RNA interactions based on primary sequence data are only just emerging (Lai and Meyer 2016). Even these most recent experimental methods rely heavily on the computational analysis of their raw data to infer any actual RNA structures or *trans* RNA–RNA interactions. Any biological insight gained from experimental in vivo studies is thus a complex function of the *combined experimental and computational strategies* employed. The purpose of this review is therefore to describe, highlight and discuss key features of these experimental and computational pipelines that contribute critically to the overall results. The focus here is thus almost exclusively on method development. We therefore refer the reader to the respective original papers and recent reviews, e.g. (Bevilacqua et al. 2016), regarding the biological insights gained.

2 Transcriptome-Wide Experimental Methods for Determining RNA Structures In Vivo in a Nucleotide-Specific Way

In vivo, RNAs are surrounded by aqueous solution. Any experimental investigation of RNA secondary structures and *trans* RNA–RNA interactions (RNA structures and RNA–RNA interactions in the following) with potential relevance to biological in vivo systems thus has to happen in solution (Ehresmann et al. 1987).

2.1 Brief Survey of Experimental In Vitro Methods

2.1.1 Physical Methods

Early experimental methods for RNA structure probing comprise physical methods such as X-ray crystallography and nuclear magnetic resonance spectroscopy (NMR) (Lengyel et al. 2014). Both methods take the RNA out of its cellular context, especially so X-ray crystallography, where the ability to crystallise implies the almost complete removal of the solvent. Even then, not all RNAs crystallise equally well (some not at all), so that database of RNA structures derived by X-ray crystallography has inherent biases. NMR imposes a considerable limitation on the length of the RNAs that can be investigated. Both in vitro methods are low-throughput in the sense that they typically investigate a single RNA at a time. Especially NMR requires considerable human expertise to design and interpret all experiments required to determine a RNA structure. Experiments for different RNAs are considered on a case by case basis. These general limitations none-withstanding, NMR and X-ray crystallography have generated a wealth of important insights on RNA structure properties in vitro. Discrepancies between the RNA structures derived from NMR and from X-ray crystallography experiments give an early indication that RNA structure features are fairly context-sensitive (Higgs 2000). Based on these early observations, differences between RNA structures in vitro and in vivo could thus be expected.

2.1.2 Enzymatic Methods

RNA structure features can also be probed using RNases. These ribonucleases correspond to naturally occurring proteins that cleave at specific paired (i.e.

double-stranded (ds)) or unpaired (i.e. single-stranded (ss)) nucleotides. Each type of RNase comes with distinct specificities (e.g. RNAse T1 (ssG), RNAse A (ssC/U), RNAse S1 (ssRNA) and RNase V1 (dsRNA)). Probing the same RNA with different RNases in separate experiments is a good way to independently assess complementary RNA structure features (and to also estimate the corresponding false positive rates via consistency checks). The size of these proteins ($> 10{,}000$ Da) (Ehresmann et al. 1987), however, prevents them from easily crossing cellular membranes and from resolving smaller RNA structure details, e.g. small bulges. Their use has thus been limited to in vitro studies so far (Ehresmann et al. 1987; Weeks 2010; Knapp 1989; Woese et al. 1980; Aultman and Chang 1982; Guerrier-Takada et al. 1983; Kertesz et al. 2010).

In vitro experiments have the advantage of allowing to examine select aspects of the complex in vivo environment in isolation, e.g. changes in the ion concentrations, temperature or interaction partners. In vivo, however, many such effects including those that cannot be easily replicated in vitro conspire to create a complex environment that cannot be readily replicated in vitro. This is mostly due to the fact in vivo, intrinsic and extrinsic changes to the transcript happen in a space-wise and time-wise carefully orchestrated way which is often impossible to replicate in vitro. Several experimental and theoretical studies have, for example, confirmed that RNA structure formation in vivo can happen co-transcriptionally and that this yields functional RNA structures that can differ significantly from the so-called minimum-free-energy (MFE) RNA structures predicted for already synthesised transcripts assuming thermodynamic equilibrium (Morgan and Higgs 1996; Meyer and Miklos 2004; Wiebe and Meyer 2010; Lai et al. 2013; Proctor and Meyer 2013). This effect is particularly pronounced for transcripts longer than around 200 nt (Morgan and Higgs 1996), i.e. a significant portion of any transcriptome.

Overall, it should not come as a surprise that RNA structures in vitro have been found to differ from those in vivo (Kwok et al. 2013; Tyrrell et al. 2013; Lai et al. 2013). This has major implications for how we should computationally model RNA structures and RNA–RNA interactions that are functionally relevant in vivo. As we will see in the following, many well-known and commonly-used computational methods for predicting these features are based on the assumption that the RNA in question is in thermodynamic equilibrium (and already fully synthesised).

2.2 *Experimental Methods for Determining RNA Structures In Vivo*

Many existing experimental methods for RNA structure determination in vivo rely on small structure probing molecules (< 500 Da) that (a) can either be readily introduced into living cells via the cellular membrane (Kwok et al. 2013; Zaug and Cech 1995; Wells et al. 2000; Moazed et al. 1986; Harris et al. 1995; Merino et al. 2005;

Wilkinson et al. 2006; Mortimer and Weeks 2007; Watts et al. 2009; Steen et al. 2012; Rice et al. 2014; Spitale et al. 2015) or that (b) be generated directly inside the cell (e.g. hydroxyl radicals generated by the high-flux photon beam of a synchrotron source (Latham and Cech 1989; Sclavi et al. 1997)). One exception is RNA structure probing via cryo-electron-microscopy (cryo-EM) (Lengyel et al. 2014) which shall not be discussed here as it is a low-throughput. Similar to RNases, both strategies ((a) and (b)) can be used to probe many RNAs simultaneously, i.e. in a massively parallel fashion. Unlike RNases which act by cutting the transcripts into shorter sub-sequences, these strategies only *modify* individual nucleotides of the underlying transcripts chemically. Compared to RNases, these chemical RNA structure probing methods thus have the significant, strategic advantage of respecting the linear identity of the underlying transcript. One significant disadvantage of these chemical RNA structure probing methods, however, is that higher-dimensional information on secondary and tertiary RNA structure features is converted into position-specific information along the linear sequence of the transcript. This linearisation implies, in particular, that any direct information on actual base pairs is entirely lost.

The main task of the computational interpretation is thus to convert the experimental probing information for individual nucleotides back into RNA structures involving actual base pairs. It is important to note here that all of these experimental methods chemically modify single, individual nucleotides, but that the *reason* for each such modification typically extends well beyond the confines of the modified nucleotide itself. That is, the modified nucleotide captures its wider secondary and tertiary RNA structure context. It is thus not entirely appropriate to say that these chemical RNA structure probing methods have single-nucleotide resolution. We will see later on that this has important implications for the computational interpretation of the experimental structure probing data.

Depending on the chemical used for chemical RNA structure probing, these methods can be sub-divided into those that target *unpaired nucleotides in a nucleotide-specific way* and those that act in a *ribose-specific way*, see Table 1 for an overview. The first group comprising DMS and CMCT modifies distinct positions in a nucleotide-specific way, but *unpaired nucleotides only*, whereas reagents of the second group (so-called SHAPE reagents) alkylate the C2'-hydroxyl group of the ribose and thereby the group acts in a way which is *neither nucleotide-specific nor completely pairing-status-specific*. SHAPE stands for selective 2'-hydroxyl alkylation analysed by primer extension (McGinnis et al. 2012; Merino et al. 2005; Weeks 2010). SHAPE reagents assess the flexibility of the RNA backbone and thereby probe the local RNA structure environment of each type of nucleotide. Raw SHAPE reactivity values thus have the advantage of covering both paired and unpaired nucleotides in any given RNA. The downside, however, is that the distributions of SHAPE values for paired and unpaired nucleotides typically have a non-negligible overlap which requires carefully computational dis-entangling. An additional complication arises due to the fact that all SHAPE reagents also react with water. Different SHAPE reagents have different half-lives in water ($t_{1/2}$ hydrolysis at a specific temperature) spanning several orders of magnitude. These details have

Table 1 Overview of reagents used for transcriptome-wide in vivo probing of RNA structures (*cis*) and *trans* RNA–RNA interactions (*trans*)

	Chemical	Probing	Specificity	Sites of modification
(1)	DMS	*cis*	Nucleotide-specific	N_1A, N_3C, N_7G
(2)	CMCT	*cis*	Nucleotide-specific	N_3U, N_1G
(3)	NMIA	*cis*	Ribose-specific	$C_2'OH$
(4)	1M7	*cis*	Ribose-specific	$C_2'OH$
(5)	1M6	*cis*	Ribose-specific	$C_2'OH$
(6)	NAI-N3	*cis*	Ribose-specific	$C_2'OH$
(7)	Hyroxyl radical	*cis*	Ribose-specific	$C_4'H$
(8)	AMT	*cis, trans*	Nucleotide-specific	Base-pairing pyrimidine
(9)	Biopsoralen	*cis, trans*	Nucleotide-specific	Base-pairing pyrimidines

Chemical probing of transcriptome-wide RNA structure features (see *cis* above) in vivo has so far been done utilising both nucleotide-specific (DMS and CMCT) and ribose-specific reagents (NMIA, 1M7, 1M6, NAI-N3, hydroxyl radical (Latham and Cech 1989; Sclavi et al. 1997; Soper et al. 2013)). The nucleotide-specific reagents modify only *unpaired sequence positions* in a highly *nucleotide-specific way*. In contrast to this, most ribose-specific reagents act by alkylating the C2'-hydroxyl group of the ribose of an individual sequence position and thereby assesses the flexibility of the RNA backbone in the vicinity of the chemically modified nucleotide. In contrast to the nucleotide-specific reagents, these so-called SHAPE reagents thus yield chemical modifications of both, unpaired and base-paired nucleotides. These reagents ((1)–(6)) have been used in transcriptome-wide screens of RNA structure features in vivo, see Table 2 and the text for more information. AMT and biopsoralen are both psoralen-derivatives. They covalently cross-link base-pairing pyrimidines in conjunction with UV-light at 365 nm. This cross-linking can be reversed using UV-light at 254 nm. They have been used in recent, transcriptome-wide in vivo experiments to probe both RNA structure features (see *cis*) and *trans* RNA–RNA interactions (see *trans*), see Table 2 and the text for more information. Abbreviations used: DMS (dimethyl sulfate) (Kwok et al. 2013; Zaug and Cech 1995; Wells et al. 2000), CMCT (1-cyclohexyl-(2-morpholinoethyl)carbodiimide metho-p-toluene sulfonate) (Moazed et al. 1986; Harris et al. 1995), NMIA (N-methylisatoic anhydride) (Merino et al. 2005; Wilkinson et al. 2006), 1M7 (1-methyl-7-nitroisatoic anhydride) (Mortimer and Weeks 2007; Watts et al. 2009), 1M6 (1-methyl-6-nitroisatoic anhydride) (Steen et al. 2012; Rice et al. 2014), NAI-N3 (2-methylnicotinic acid imidazolide-azide) (Spitale et al. 2015), AMT (4'-aminomethyltrioxsalen) (Calvet and Pederson 1979; Sharma et al. 2016; Lu et al. 2016) and biopsoralen (biotinylated psoralen (psoralen-PEG$_3$-biotin)) (Aw et al. 2016)

to be carefully considered for making the correct choice for each specific research question, e.g. when trying to investigate RNA structure features as function of time.

In principle, it is also possible to probe RNA structure features with molecules that occur naturally to some extent in living cells, e.g. hydroxyl radicals (Latham and Cech 1989; Sclavi et al. 1997). Similar to SHAPE reagents, this chemical acts in a ribose-specific manner and acts both on paired and unpaired nucleotides. Unlike all SHAPE reagents, however, it modifies the C4'-H group (rather than the C2'-hydroxyl group) of the ribose and thereby tends to probe the tertiary RNA structure environment of individual sequence positions. In normal circumstances in vivo, the concentration of hydroxyl radicals is too low for RNA structure probing. In order to artificially increase the concentration for successful RNA structure probing in vivo, X-ray radiation can be used, e.g. generated by a synchrotron source which can generate photon beams of sufficiently high flux. This has already allowed RNA structure probing with high, time-wise resolution in vitro (Sclavi et al. 1997) and in vivo (Soper et al. 2013).

2.3 Experimental Methods for Transcriptome-Wide Probing RNA Structures In Vivo

The above methods for the chemical probing of RNA structures in vivo can naturally probe many RNAs simultaneously. The key achievement of the last few years was to realise that these methods can be combined with high-throughput transcriptome-wide next-generation sequencing (NGS). For this, RNA structure information is first converted into a linearised sequence signal. This is done for many transcripts in parallel. In a second step, these linearised sequence signals are efficiently read out using high-throughput sequencing (typically, NGS).

The corresponding experimental methods can be classified according to (a) the chemical used for RNA structure probing and (b) the protocol employed for converting structure probing information into sequence-based information that can be read in a parallelised fashion using NGS techniques. The second step can comprise a variety of different extraction, depletion and enrichment steps whose features are also key determinants of the overall sensitivity and specificity of the combined experimental protocol.

As the focus here is on in vivo methods, we review in vitro methods for transcriptome-wide RNA structure only briefly. Historically, PARS (parallel analysis of RNA structures) was the first to assess RNA structures in a massively parallel fashion using RNases for enzymatic RNA structure probing (Kertesz et al. 2010; Wan et al. 2012; Righetti et al. 2016; Wan et al. 2014, 2013; Del Campo et al. 2015). Other in vitro approaches have since included those based on enzymatic structure probing (DS/SSRNA-SEQ (Zheng et al. 2010; Li et al. 2012a,b) and FRAG-SEQ (Underwood et al. 2010) as well as approaches based on chemical probing (DMS-SEQ (Rouskin et al. 2014) and RING-MAP (Homan et al. 2014) using DMS, HRF-SEQ (Kielpinski and Vinther 2014) and MOHCA-SEQ (Cheng et al. 2015) using hydroxyl radicals, SHAPE-SEQ and SHAPE-SEQ 2.0 (Lucks et al. 2011; Loughrey et al. 2014; Watters et al. 2016b) (using SHAPE reagent 1M7) and SHAPES (Poulsen et al. 2015) (using SHAPE reagent NPIA)). Some in vitro methods employ two or more chemical reagents, e.g. CHEMMOD-SEQ (Hector et al. 2014) (DMS and SHAPE reagent 1M7), MAP-SEQ (Seetin et al. 2014) (DMS, CMCT and SHAPE reagent 1M7) and CIRS-SEQ (Incarnato et al. 2014) (DMS and CMCT). It is especially advantageous to combine nucleotide-specific with ribose-specific chemical modifications as these complement each other and enable valuable cross-checks. These in vitro methods are appropriate for RNA structure probing, if the artificial setting can be justified for addressing specific scientific questions. Care has to be taken, however, not to simply generalise these in vitro results to various in vivo settings.

All of the existing in vivo methods employ chemical probes for RNA structure probing. In all cases, the raw structure probing data consists of probing values for individual sequence positions, not base pairs. Most of the currently existing in vivo methods employ DMS as structure probing reagent, such as STRUCTURE-SEQ (Ding et al. 2014, 2015), DMS-SEQ (Rouskin et al. 2014), MOD-SEQ (Talkish et al. 2014; Lucks et al. 2011) and targeted STRUCTURE-SEQ (Fang et al. 2015). In addition, SHAPE-based approaches such as SHAPE-MAP (Smola et al. 2015a,b; Siegfried et al. 2014; Lavender et al. 2015; Mauger et al. 2015) (SHAPE reagents: 1M7, 1M6 and NMIA) and ICSHAPE (Spitale et al. 2015; Flynn et al. 2016) (SHAPE reagent: NAI-N3) now exist, as well as earlier in vitro approaches such as SHAPE-SEQ (Lucks et al. 2011; Mortimer et al. 2012) (SHAPE reagent: 1M7) that were extended to combine the earlier SHAPE-reagent with DMS-based probing in cell SHAPE-SEQ (Watters et al. 2016a,b), see Table 2 for an overview. The major steps of all currently existing in vivo RNA structure probing methods are

Table 2 Overview of methods used for transcriptome-wide in vivo probing of RNA structures (*cis*) and *trans* RNA–RNA interactions (*trans*)

	Name	Probing	Reagent
(a)	STRUCTURE-SEQ	*cis*	DMS
(b)	DMS-SEQ	*cis*	DMS
(c)	MOD-SEQ	*cis*	DMS
(d)	SHAPE-MAP	*cis*	1M7, 1M6, NMIA
(e)	ICSHAPE	*cis*	NAI-N3
(f)	In cell SHAPE-SEQ	*cis*	1M7, DMS
(g)	Targeted STRUCTURE-SEQ	*cis*	DMS
(h)	PARIS	*cis, trans*	AMT
(i)	SPLASH	*cis, trans*	Biopsoralen
(j)	LIGR-SEQ	*cis, trans*	AMT

The first few methods ((a)–(g)) probe RNA structure features by chemically modifying individual nucleotides, either using reagents that act in a nucleotide-specific way on *unpaired sequence positions only* (e.g. DMS) or using SHAPE-reagents that act in a ribose-specific way and thereby assess base-paired and unpaired sequence positions (e.g. 1M7, 1M6, NMIA, NAI-N3), see Table 1 for more information. All of these methods convert RNA structure probing information into a linearised sequence signal of position-specific chemical modifications that can be read out in a massively parallel fashion using next-generation sequencing methods. In particular, none of these methods retains direct information on specific base pairs. PARIS, SPLASH and LIGR-SEQ simultaneously probe RNA structure features and *trans* RNA–RNA interactions by covalently cross-linking individual duplexes, i.e. more or less contiguous stretches of base pairs involving the same or two different RNAs. These duplexes are subsequently trimmed and their ends ligated, thereby retaining information on both sub-sequences involved in a duplex, before the cross-linking is reversed and the linearised duplexes are sequenced using next-generation sequencing

2.3.1 Step 1: RNA Structure Probing

The goal of this step is to probe RNA structures using a reagent that induces chemical modifications into individual nucleotides.

The key aspect to consider is: Could any step of the protocol for RNA structure probing actually interfere with the in vivo RNA structures in a way which would alter them *before* they are probed?

This is perhaps the most important aspect to optimise. If this fails, no subsequent step in the experimental or computational analysis can fix it. (1) For this, the chemical properties of the probing reagents need to considered and their potential direct or indirect impact on RNA structure features be examined, e.g. in dedicated in vitro experiments prior to the in vivo ones. These experiments have to be conducted in a way that can distinguish reactions on different time-scales. (2) It is also important to consider the possibility that the chemical modifications induced during RNA structure probing alter the RNA structure while it is being probed. (3) Lastly, if RNA structure probing is done by more than a single probing reagent, this should happen in separate experiments keeping everything, but the probing reagent, unchanged.

In terms of future developments, it would be beneficial to have fast and efficient ways to stop RNA structure probing in vivo. This would help to conserve the RNA structure probing signal and allow detailed investigations of RNA structures as function of time.

2.3.2 Step 2: RNA Extraction, rRNA Depletion and RNA Enrichment

In this step, the pool of chemically modified transcripts of interest is extracted and enriched and unwanted transcripts are removed to prevent them from being sequenced (e.g. rRNAs which account for the majority of transcripts, yet are typically not the focus of the investigation).

The key challenge here is to ensure that extraction and enrichment are done with maximum specificity. Any true signal lost cannot be recovered later on.

For enrichment, a polyA RNA enrichment step is often applied. This implies, however, that non-polyA transcripts (e.g. non-coding RNAs, circular RNAs) are omitted from all subsequent steps of the analysis. The user needs to decide whether this is actually wanted and otherwise adapt the original protocol.

2.3.3 Step 3: Library Preparation for High-Throughput Sequencing

Different in vivo RNA structure probing protocols differ substantially in how the enriched pool of chemically modified transcripts is converted into a library for NGS sequencing. As soon as the library has been sequenced, the corresponding reads have to be mapped back to the underlying genome/transcriptome before the computational analysis of RNA structure features can start. As this mapping comes with its own significant challenges, it is imperative to optimise the experimental library preparation w.r.t. the subsequent computational analysis.

2.3.4 Key Aspects to Consider for Optimisation

(A) What is the expected average length of the final reads (excluding the length of any primers and/or adapters that are removed in silico prior to mapping the reads back to the genome/transcriptome)?

For those methods that detect RNA structure probing signals via chemical-induced reverse transcriptase halting, e.g. STRUCTURE-SEQ and ICSHAPE, this length is primarily determined by the average distance between the initiation site of reverse transcription (RT) and the first chemically modified nucleotide upstream. It thus depends both on the specificity of the chemical used for RNA structure probing as well as the mechanism used for RT initiation (example: DMS (which only probes unpaired nucleotides) and random hexamer primers for RT initiation in case of STRUCTURE-SEQ). Note that the mechanism used for RT initiation (e.g. random primers of different lengths may preferentially bind to single-stranded regions of the transcript) may introduce its own biases that may be relevant to the subsequent, computational RNA structure interpretation. The effective average read length may also be influenced by additional RNA fragmentation steps, e.g. random fragmentation by Mg^{2+}-mediated hydrolysis in ICSHAPE. For these methods, a well-chosen combination of probing reagent and RT initiation can thus optimise the expected average read length.

For those methods that detect RNA structure probing signals via chemical-induced reverse transcriptase read-through, e.g. SHAPE-MAP (Siegfried et al. 2014; Smola et al. 2015a,b; Lavender et al. 2015; Mauger et al. 2015), the natural average length of reads is primarily determined by the default fragmentation step of the corresponding library preparation protocol (Nextera in case of SHAPE-MAP) and *not* by the average distance between RT initiation and any nucleotides modified via chemical RNA structure probing. This is a significant conceptual advantage over methods that detect RNA structure probing signals via reverse transcriptase halting.

(B) How much RNA structure probing information is retained in a single read?

Ideally, we would like to retain structure probing information for entire, individual transcripts. If we loose this information, e.g. during library preparation, we cannot detect *RNA structure diversity*, i.e. the possibility that different copies of the same transcript assume different RNA structures in vivo. Also, in order to maximise the RNA structure information for each individual transcript, chemical RNA structure probing should happen in a way that saturates each transcript with structure probing signals (in a way which does not risk altering the underlying RNA structure itself).

For most of the existing protocols for RNA structure probing in vivo, however, the requirements for optimising the library preparation are not in line with the above requirements for optimising the RNA structure probing information. The library preparation of STRUCTURE-SEQ and ICSHAPE, for example, is set up to generate reads that correspond to one chemically modified nucleotide only, namely the chemically modified sequence position that is first encountered upstream of the RT initiation site (chosen by a hexamer primer in case of STRUCTURE-SEQ and chosen by Mg^{2+}-induced random fragmentation in case of ICSHAPE). Any correlations between RNA structure probing information from the same transcript are thereby lost. In addition, saturated RNA structure probing would have the tendency to further lower the average read length, making the subsequent mapping even harder.

The best way to circumvent this problem is to choose a library preparation protocol that does not rely on RT transcriptase halting for detecting the RNA structure probing signal. This can, for example, be done using Mn^{2+} mediated reverse transcriptase read-through of the modified nucleotide positions as in SHAPE-MAP. This strategy, however, has the undesired side effect of introducing a generally higher error rate for reverse transcription.

(C) What is the overall efficiency of all steps in the protocol?

Some protocols, e.g. ICSHAPE, incorporate a second enrichment step by chemically treating the RNA-structure-probed transcripts in a second step in vivo to prepare their subsequent biotinylation using click-chemistry (this happens after RNA extraction, rRNA depletion and RNA enrichment). This second biotin-based enrichment step has the advantage of further increasing the specificity.

Overall, protocols for in vivo RNA structure probing differ substantially in the number of steps required for library preparation. Any additional steps in the overall protocol, however, have the tendency of reducing the overall sensitivity and efficiency as the inefficiencies and biases of each step add up. Generally, it is thus advisable to minimise the total number of steps and to optimise each step in terms of specificity and sensitivity.

3 Interpreting the Experimental RNA Structure Probing Data In Silico

The above in vivo methods for transcriptome-wide RNA structure probing generate raw transcriptome sequencing data (reads) which must be computationally processed and interpreted for any actual RNA structures to be inferred.

Basically, any computational analysis has to achieve the reversal of the experimental protocol, namely to convert a purely sequence-based signal back into RNA structures involving base pairs. This is challenging due to a number of reasons:

(a) The sequence signals induced by chemically encoded RNA structure probing can be noisy, biased and/or incomplete. For example, any particular SHAPE values cannot be unambiguously interpreted as being derived from a paired or unpaired nucleotide.

(b) RNA structure probing information from any transcript is fragmented in the existing experimental protocols, i.e. the full sequence identity of the RNA structure probing signal is lost and cannot be retrieved later computationally. Correlated structure probing information is currently only retained within individual reads.
(c) Next-generation sequencing itself introduces errors and biases, e.g. sequencing errors whose rate depends on the position within each read.
(d) The mapping of sequenced reads to a reference genome/transcriptome is not straightforward and can induce different kinds of errors, biases and missing data. This is a particular concern for experimental protocols that encode RNA structure probing information in terms of nucleotide changes, e.g. SHAPE-MAP. There, sequenced reads cannot be readily mapped back to their original transcripts without carefully considering SNP-like discrepancies. This requires dedicated, probabilistic mapping methods such as those used in transcriptome-wide RNA editing studies, see e.g. (Mazloomian and Meyer 2015).
(e) Only once the sequenced reads have been mapped to a reference transcriptome, can the actual inference of RNA structures begin. This can be done using a range of conceptually different computational strategies. These are introduced in the following.

Most existing computational methods focus on utilising SHAPE reactivity values as input information to infer RNA structure information. The following describes different underlying conceptual strategies for converting raw SHAPE reactivity values along one linear transcript into distinct RNA structure(s). These approaches not only employ different strategies for RNA structure prediction, but also differ in the (implicit or explicit) assumptions they make in interpreting the raw structure probing data. Roughly, all existing computational approaches can be classified according to how they address three main aspects:

(a) How the raw, sequence-position-specific RNA structure probing is processed. Examples include re-scaling and normalisation procedures.
(b) How the raw, sequence-position-specific RNA structure probing is interpreted and integrated into RNA structure prediction.
(c) How RNA structures are captured in a predictive model that utilises experimental RNA structure probing data. All of these methods model RNA structures at secondary-structure level. These methods differ substantially in their implicit and explicit assumptions. Examples include thermodynamic methods that derive the thermodynamically most stable RNA secondary structure (so-called minimum-free energy (MFE) methods), methods that consider Boltzmann ensembles of RNA secondary structures in thermodynamic equilibrium and, most recently, probabilistic methods for RNA secondary structure prediction that predict the maximum likelihood RNA secondary structure, see Table 5.

As we will see in the following, early methods incorporate experimentally derived RNA structure probing information into thermodynamic methods for

RNA secondary structure prediction (MFE approach). More recently, RNA structure probing information has been integrated in a fully probabilistic manner into probabilistic methods for RNA secondary structure prediction. These new methods offer conceptually convincing ways of seamlessly combining experimental RNA structure probing data with RNA structure prediction.

3.1 Interpreting SHAPE Reactivity Values as Pseudo-Energies for Paired Sequence Positions

Many commonly used computational methods for RNA secondary structure prediction, e.g. MFOLD (Zuker 2003) and RNAFOLD (Zuker and Stiegler 1981), utilise a so-called thermodynamic model of RNA secondary structures. These methods decompose any (pseudo-knot-free) RNA secondary structure into a sum of Lego-like, structural RNA secondary structure building blocks and express the total free energy of the RNA structure as sum of the free-energy contributions of these structural building blocks. The underlying thermodynamic models, e.g. the well-known Turner model (Mathews et al. 1999) on which MFOLD and RNAFOLD are based, rely on many parameters that correspond to physical entities that have been determined experimentally. For a given input RNA sequence, these models employ efficient dynamic programming algorithms such as the Zuker–Stiegler algorithm (Zuker and Stiegler 1981) to derive the RNA secondary structure with the minimum overall free energy. The corresponding minimum-free-energy (MFE) structure is reported as output. For any given input sequence, these methods predict a single MFE RNA secondary structure. Thermodynamic methods for RNA secondary structure prediction such as MFOLD and RNAFOLD make the implicit assumptions that any given input sequence (a) is already fully synthesised and (b) that it will assume an MFE RNA secondary structure. In particular, these methods assume any input RNA to be in thermodynamic equilibrium and to be naked, i.e. without any *trans* interaction partners such as ligands, proteins or other RNAs. As we know, this assumption is generally not justified in in vivo settings.

Early efforts to integrate chemical RNA structure probing data into RNA structure prediction try to interpret these data as modifications to the default thermodynamic model used for RNA secondary structure prediction. For this, experimentally determined RNA structure probing values are somehow converted into free energy contributions assigned to individual sequence positions.

Deigan et al. (2009) were the first to interpret the position-specific SHAPE reactivity values α_i as position-specific free-energy corrections ΔG_i^D to the nominal free energy terms in the thermodynamic model for RNA structure prediction:

$$\Delta G_i^D = m \log(\alpha_i + 1) + b$$

Here, α_i denotes the experimentally determined SHAPE reactivity value for sequence position i in the transcript (i.e. $i \in \{1, \ldots, L\}$ for a transcript of L nucleotides length) and m and b are free parameters with default values $m = 2.6$ and $b = -0.8\,\text{kcal}\,\text{mol}^{-1}$, see Low et al. (2014), Qi et al. (2012) for other parametrisations. In the dynamic programming recursion which derives the most stable RNA secondary structures, these ΔG_i^D values are added to the nominal energy contribution for *each base-paired sequence position i*. Any contributions from SHAPE reactivity values from *un-paired sequence positions* are completely ignored.

This approach by Deigan was later extended to work on DMS input data (Cordero et al. 2012a); pseudo-energies are derived from a log-likelihood ratio of a nucleotide being unpaired versus paired. Eddy (2014) pointed out that base-pairing probabilities for individual sequence positions, p_i, can be linked to position-specific pseudo-energies *if one may assume that a naked, fully synthesised RNA is in thermodynamic equilibrium*. This can be achieved because $p_i(\pi_i = 1) \propto e^{-\Delta G_i/RT}$. That is, the probability that sequence position i is base-paired, i.e. $p_i(\pi_i = 1)$, is proportional to $e^{-\Delta G_i/RT}$, where ΔG_i is the pseudo-energy assigned to position i (here, R denotes the universal Gas constant and T the absolute temperature in degrees Kelvin).

3.2 *Interpreting SHAPE Reactivity Values as Pseudo-Energies for Paired and Unpaired Sequence Positions*

The above approach by Deigan introduces an unnatural bias into the interpretation of SHAPE reactivity values. Even though experimentally determined SHAPE reactivity values have a continuous spectrum, covering both paired and unpaired nucleotides, SHAPE-derived pseudo-energies are effectively only assigned to *paired sequence positions*.

Zarringhalam et al. (2012) propose a strategy which is symmetric w.r.t. paired and unpaired sequence positions. Similar to Deigan, they interpret SHAPE reactivity values α_i along the transcript as position-specific corrections ΔG_i^Z to the free energy terms of the underlying transcript position i:

$$\Delta G_i^Z = \beta|\pi_i - \alpha_i^r|$$

Here, α_i^r denotes the (rescaled version of the) experimentally determined SHAPE reactivity value and π_i is the corresponding pairing status of sequence position i, i.e. $\pi_i = 0$ for an un-paired and $\pi_i = 1$ for a paired sequence position. The rescaling of the original SHAPE reactivity values α_i is achieved via a piecewise-linear function which re-scales the values so that the resulting values satisfy $\alpha_i^r \in [0, 1]$. The shape of this function was chosen to fit to the empirical likelihood ratio distribution, i.e. the paired-unpaired likelihood ratios as function of the SHAPE reactivity values. The scaling parameter β affects all sequence positions equally and can be interpreted as a universal knob to decrease or increase the contribution of SHAPE values in the thermodynamic model for RNA structure prediction.

The goal of the Zarringhalam approach is to minimise the overall difference between the experimentally derived SHAPE data and the predicted RNA structure as measured by the so-called Manhattan distance, i.e. to minimise $\sum_i |\pi_i - \alpha_i^r|$. Unlike the above approach by Deigan, this strategy can be mathematically shown to yield a better fit of the predicted RNA structures to the SHAPE reactivities in terms of Manhattan distance (Zarringhalam et al. 2012).

3.3 Introducing Pseudo-Energy-Like Free Parameters in a Fit to a Thermodynamic Ensemble of RNA Secondary Structures

Both above approaches implicitly assume that all SHAPE reactivity values correspond to a *single RNA secondary structure*. Washietl et al. (2012) stick to the assumption of a naked, already synthesised RNA sequence in thermodynamic equilibrium but interpret the SHAPE reactivity values as ensemble-weighted average values over many identical RNAs with different RNA secondary structures. Many properties of this so-called Boltzmann distribution of RNA secondary structure in thermodynamic equilibrium can be calculated analytically (McCaskill 1990; Miklos et al. 2005).

Their method works as follows. In a first step, SHAPE values for each sequence position i, α_i, are translated into so-called pairing probabilities $p_i(\alpha_i)$ with $p_i(\alpha_i) = 0$ if $\alpha_i > 0.25$ and $p_i(\alpha_i) = 1$ if $\alpha_i \leq 0.25$. Using this simple thresholding procedure, SHAPE reactivity values are thus effectively interpreted as either being paired or unpaired (with 100% probability, i.e. certainty). These position-specific $p_i(\alpha_i)$ values should thus be viewed as pairing status indicators, e.g. denoted by $s_i := p_i(\alpha_i)$, rather than pairing probabilities.

Any discrepancies between the position-specific pairing probabilities $z_i(\theta, \vec{e})$ as they can be explicitly calculated from the Boltzmann ensemble of RNA structures in thermodynamic equilibrium (where θ denotes the set of default parameters of the underlying thermodynamic model and $\vec{e}$ a vector of so-called pseudo-energy corrections e_i introduced for each individual sequence position i) and the position-specific SHAPE-derived pairing status values s_i are assumed to be normally distributed with a position-independent variance σ^2. Every sequence position i in the transcript of L nucleotides length, i.e. $i \in \{1, \ldots L\}$, is assigned a so-called pseudo-energy term e_i. In contrast to the above approaches by Deigan and Zarringhalam, however, these e_i values do not have a link to SHAPE reactivities. Rather, they correspond to position-specific free parameters in a global optimisation problem and have been artificially introduced. Also these e_i terms are assumed to come with a position-independent, overall variance of τ^2. Using a gradient descent method, the method by Washietl et al. then tries to identify the vector of e_i values that minimises the expression:

$$\min_e \frac{1}{\tau^2} \sum_i e_i^2 + \frac{1}{\sigma^2} \sum_i (z_i(\theta, \vec{e}) - s_i)^2$$

This optimisation can be expected to be mathematically challenging as the optimisation procedure is not guaranteed to find the global minimum and can get stuck in local minima. A priori, it is also not clear what the correct interpretation of the resulting e_i values should be. They have no obvious link to SHAPE reactivity values nor to the free parameters of the underlying thermodynamic model (θ). Also, it should be noted that the number of free parameters e_i increases linearly with the length of the input sequence and that the optimisation is done for each input sequence independently.

The current implementation of the Washietl approach into the VIENNAPACKAGE (Lorenz et al. 2016) allows users to explore different ways of converting structure probing data into p_i values and provides several optimisation techniques.

3.4 Using SHAPE Reactivity Values in a Sample and Select Approach Using an Unperturbed Thermodynamic Ensemble of RNA Secondary Structures

All of the above approaches hinge on the validity of the assumption that experimental structure probing data can be interpreted as position-specific pseudo-energy corrections to an underlying thermodynamic model. As the detailed discussion of the above methods shows, even incorporating this assumption into a corresponding strategy for RNA structure prediction is technically and conceptually not entirely straightforward.

Some groups (Ouyang et al. 2013; Quarrier et al. 2010) have decided not to interpret structure probing data as position-specific pseudo-energy corrections at all. Instead, they assume that the in vivo environment introduces unknown changes to the nominal RNA structure of the underlying thermodynamic model (i.e. the MFE-structure as defined earlier) which cannot be modelled by tweaking the underlying parameters of the thermodynamic model. This makes sense as some effects of the in vivo environment, e.g. *trans* interaction partners, can conceptually not be captured by tweaking the free energy parameters of the thermodynamic model for RNA secondary structure prediction. Instead, they propose to address this challenge by sampling RNA secondary structures from the (unperturbed) thermodynamic ensemble of RNA secondary structure (Ding and Lawrence 2003; McCaskill 1990) and re-ranking the sampled RNA structures according to how well they fit the experimentally determined RNA structure probing data. This involves a distance metric such as the Manhattan distance introduced above. For calculating the fit, SHAPE reactivity values are first mapped to discrete paired/unpaired values for each sequence position using a simple thresholding approach before calculating the Manhattan distance to the sampled RNA.

These methods effectively allow for *more than a single RNA secondary structure* to correspond to one set of experimentally determined, position-specific RNA structure probing data, even though these RNA secondary structures conceptually derive from the same Boltzmann ensemble of many identical RNA sequences in thermodynamic equilibrium. By ranking the sampled RNA structures based on fit to the probing data only (rather than the respective probability of the sampled RNA structure in the Boltzmann ensemble), all sampled RNA secondary structures are effectively assumed to have equal prior probability (provided they are sampled at all). The obvious downside of this pragmatic approach is that RNA secondary structure with low probability in the Boltzmann ensemble may never be sampled at all, even if they could provide the best overall fit. Also, this approach only provides limited feedback in terms of insight gained.

3.5 Probabilistic Integration of Experimental RNA Structure Probing Data into Probabilistic Methods for RNA Secondary Structure Prediction

RNA secondary structure prediction does not necessarily need to involve the assumption that any input RNA folds into the minimum-free-energy structure and is in thermodynamic equilibrium. Using probabilistic methods such as stochastic context-free grammars (SCFGs) (Durbin et al. 1998) (or Markov Chain Monte Carlo (MCMC) methods), it is possible to explicitly capture different hypotheses on how RNA secondary structure may arise. This has given rise to a number of RNA secondary structure prediction methods, e.g. PFOLD (Knudsen and Hein 2003), RNA-DECODER (Pedersen et al. 2004a,b), SIMULFOLD (Meyer and Miklos 2007), that yield a high prediction performance for evolutionarily conserved RNA secondary structures. These methods combine a probabilistic model of RNA secondary structures with computationally efficient algorithms to derive the maximum likelihood RNA structure given the underlying RNA structure model. In terms of time-and-memory efficiency, they have the same complexity as thermodynamic methods, e.g. MFOLD (Zuker 2003) and RNAFOLD (Zuker and Stiegler 1981), but offer several conceptual advantages. First, the user can decide the parametrisation of the model. Free parameters can thus be chosen to have a straightforward biological interpretation. Second, given a training set of sufficient size and complexity, the free parameters of the model can be explicitly trained. Third, alternative parametrisations of the same model can be explicitly evaluated and ranked based on likelihood fits to the data. Fourth, the predictive model for RNA secondary structures can be readily extended to take into account additional sources of input information, e.g. evolutionary information in terms of a multiple-sequence alignment (MSA) or experimental RNA structure probing data.

Technically, this can be achieved by replacing the so-called emission probabilities of SCFGs by probabilistic emission models that, for example, read entire alignment columns from an input MSA rather than individual nucleotides from an input sequence. These emission models are probabilistic models that can, for example, explicitly capture how we expect paired and unpaired nucleotides to evolve as function of evolutionary time.

Most importantly, fully probabilistic models allow information of different types (e.g. primary sequence features, RNA structure features, evolution) to be seamlessly merged as the corresponding probabilities for different sources of information can be readily combined in a single predictive framework. This elegantly avoids the need for converting conceptually different sources of information (e.g. chemical RNA structure probing data) into units with a physical interpretation (free energy terms). More importantly, probabilistic models allow us to move beyond the assumption of thermodynamic equilibrium.

3.5.1 Integration into Comparative Methods for RNA Secondary Structure Prediction

PPFOLD 3.0 (Sükösd et al. 2012) (PPFOLD in the following) were the first to integrate external RNA structure probing information into a fully probabilistic model of RNA secondary structure prediction.

The model for RNA structure prediction is identical to PFOLD (Knudsen and Hein 2003), a comparative RNA secondary structure prediction method. It takes as input a multiple-sequence alignment (MSA) and a corresponding evolutionary tree linking the sequences in the MSA and returns as output the maximum-likelihood RNA secondary structure for the input alignment and input tree. PFOLD captures the assumption that RNA secondary structures that have been conserved during evolution are likely to be functional. As far as we know, this is overall a decent assumption to make. In practice, the success of the comparative approach depends on a decent choice of the appropriate evolutionary distances of the sequences in the input alignment. The RNA structure predicted by PFOLD corresponds to the maximum-likelihood RNA secondary structure given the input information and the predictive model and its parameters. The evolutionary relationships of the sequences in the input multiple-sequence alignment are explicitly modelled using two probabilistic models of evolution that capture how unpaired and base-paired nucleotides evolve as function of time, respectively.

The novelty of PPFOLD consists of combining comparative RNA secondary structure prediction with experimental RNA structure probing information. In order to do this, the user needs to specify a probability distribution $P(H|\sigma)$ for a set of experimental probing data H and secondary structures σ. PPFOLD generally assumes that $P(H|D, \sigma) = P(H|\sigma)$, i.e. that there is no dependence on the actual observed nucleotides sequences of the input alignment D. As the discussion of the more recently published method PROBFOLD (Sahoo et al. 2016) below shows, this is probably too simplistic: It can actually be shown that SHAPE-values typically do depend on nucleotide identity. As an alternative to $P(H|\sigma)$, the user can also specify values $P(H_i|i\text{unpaired})$ and $P(H_i|i\text{paired})$, i.e. likelihood values that sequence position i in the input alignment is unpaired or paired given the experimental probing value of H_i for that sequence position. Internally, PPFOLD uses these likelihood values as follows to bias the nominal likelihood values of PFOLD for each paired (i, j) (subscript d for double) and unpaired i (subscript s for single) alignment column, $P_d(i, j)$ and $P_s(i)$:

$$P'_s(i) = P_s(i) \cdot P(H_i|i \text{ unpaired})$$
$$P'_s(i, j) = P_s(i, j) \cdot P(H_i|i \text{ paired}) \cdot P(H_j|j \text{ paired})$$

This assumes that the experimental probing values for the two sequence positions involved in a base-pair are assumed to be independent. The validity of this assumption has since been confirmed by the more recent investigations of PROBFOLD, see below for details.

Similar to PFOLD, PPFOLD naturally reduces to a non-comparative RNA secondary structure prediction method if the input alignment consists of only a single input sequence (although it should be stressed that this is not how PFOLD nor PPFOLD are meant to be used). The authors of PPFOLD deliberately use it with single input sequences in order to make it directly comparable to the non-comparative RNA secondary structure program RNASTRUCTURE (Deigan et al. 2009; Mathews et al. 2004) which also utilises external RNA structure probing data as additional input information. RNASTRUCTURE and PPFOLD (using single sequences) have a similar performance in terms of F-value. The F-value is defined as the harmonic mean of sensitivity and specificity. This is an impressive result given that the RNA secondary structure model of PPFOLD is lightweight compared to the full- fledged thermodynamic model underlying RNASTRUCTURE. PPFOLD thus makes better use of the external RNA structure probing information than RNASTRUCTURE. The performance of PPFOLD w.r.t. RNASTRUCTURE can be further improved in terms of F-value when using PPFOLD with multiple sequence input alignments. As with many comparative RNA secondary structure prediction methods, however, the resulting performance in terms of F-value critically depends on the quality of the input alignment. A poor input alignment (with or without additional probing data) can lower the performance of PPFOLD below the corresponding single-sequence performance with experimental probing data. That is, a poor input alignment can provide more confusion than can be remedied by additional RNA structure probing data.

Note that due to the scarcity of training and testing data, the authors of PPFOLD could not avoid an overlap between their training set (16S and 23S rRNA structures and SHAPE data for *Escherichia coli*) and their test data set (16S rRNA of *E. coli*).

3.5.2 Integration into Non-comparative Methods for RNA Secondary Structure Prediction

Most recently, Sahoo et al. (2016) proposed PROBFOLD, a probabilistic method for non-comparative RNA secondary structure that can integrate information from one or more chemical RNA structure probing experiments. PROBFOLD employs a fully probabilistic stochastic context-free grammar (SCFG) for RNA secondary structure predictions and combines this with probabilistic graphical models (PGMs) (Koller and Friedman 2009) to capture experimental probing data. Compared to PPFOLD PROBFOLD offers a more general modelling approach that is also more readily extendible and more parameter-sparse. The SCFG employed by PROBFOLD is based on the original grammar underlying PFOLD (Knudsen and Hein 2003) with extensions that capture stacking interaction, i.e. correlations between pairs of adjacent base pairs. Overall, the PROBFOLD grammar consists of six production rules in total, three of which emit terminals, i.e. read information from the input sequence. These three production rules require three corresponding emission models called *single*, *pair* and *stack* that model single, pairs and two adjacent pairs of sequence positions, respectively, see Fig. 2 in Sahoo et al. (2016) for a visualisation. The integration of experimental probing data into the RNA secondary structure prediction method happens via three corresponding PGMs that each specify a joint distribution over the RNA primary sequence data and the experimental probing data. Technically, each PGM corresponds to an undirected bipartite graph between so-called factor nodes and so-called variable nodes. The variable nodes represent random variables, whereas the factor nodes correspond to probability distributions between neighbouring random variables. PROBFOLD uses discrete random variables for the efficiency of the calculations. This is technically achieved by discretising the two distributions P^{single} and P^{paired} which model the corresponding distributions of experimental probing data. For this, probing data is first discretised into k bins using normalised histogram models (i.e. multinomials). This implies $k-1$ free parameters specifying the boundaries of these bins. These are chosen to maximise the difference between the probing data distributions of paired and unpaired sequence positions using Kullback Leibler (KL) divergence.

During the development of PROBFOLD, a hierarchy of increasingly complex, fully probabilistic models with an increasing number of free parameters (ranging from 18 to 408, for the final model) was investigated. The final model of PROBFOLD has *only a single user-specified meta-parameter*, corresponding to the number of bins used for discretising the two distributions P^{single} and P^{paired} of the experimental probing data (default is six bins). All other free parameters can be explicitly derived using a dedicated set of training set of known RNA secondary structure with corresponding structure probing data. The final model captures not only stacking interactions between neighbouring base pairs (so-called *stack*-part of the model), but also correlations between the structure probing values of neighbouring positions along the linear sequence (so-called *cor*-part of the model). Due to the scarcity of the training data, the primary sequence and structure probing values are modelled independently in order to keep the number of free parameters low. The trade-off between the sensitivity and the specificity of performance can be explicitly adjusted via a parameter γ. Sahoo et al. carefully evaluate the performance of PROBFOLD, using a dedicated test set which has no overlap with the training set (they are actually the first to do this properly using a cross-evaluation procedure). Reassuringly, they can conclude that over-fitting is not an issue, implying that their method is sensibly parametrised and the number of free parameters in line with the information content provided by their training set.

In terms of performance, they compare PROBFOLD to PPFOLD 3.0 (Sükösd et al. 2012), RNASTRUCTURE V5.6 (Deigan et al. 2009; Mathews et al. 2004), GTFOLD-3.0 (Swenson et al. 2012) and RNAFOLD.ZAR (Lorenz et al. 2011, 2016) (this is how they RNAFOLD in combination with the approach by Zarringhalam for converting the raw SHAPE values) on an independent test data set of 11 RNA structures on which neither of these methods were initially trained. The resulting performance comparison thus allows a fair assessment of the prediction accuracy of several key predictive programs, see Table 3.

The overall performance is measured in terms of F-value, i.e. the harmonic mean of sensitivity and specificity with values of $F \in [0, 1]$ with 1 corresponding to perfect predictions. For PROBFOLD, this is done for a fixed value of γ. PROBFOLD comes second in terms of overall F-value and accuracy across all structures after RNAFOLD.ZAR (F-values 0.77 and 0.71, respectively), but first in terms of performance gain w.r.t. purely sequence-based predictions without any SHAPE input ($\Delta F = 0.29$ (PROBFOLD) compared to $\Delta F = 0.12$ (RNAFOLD.ZAR)). This is impressive given that RNAFOLD employs the state-of-the-art thermodynamic model for predicting RNA secondary structures, whereas PROBFOLD uses a fairly light-weight SCFG with a significantly smaller number of parameters. (In that regard, it is also instructive to compare the baseline performance for single-sequence-only input between PROBFOLD and RNASTRUCTURE, see Table 4.) Of all methods

Table 3 Prediction performance of several computer programs that utilise individual sequences and corresponding SHAPE data as input to make RNA secondary structure predictions (optimal values highlighted in bold)

Performance	PROBFOLD	PPFOLD	RNASTRUCTURE	GTFOLD	RNAFOLD.ZAR
F	0.71	0.55	0.67	0.66	**0.77**
ΔF	**0.29**	0.11	0.02	0.05	0.12

Results and figures from Sahoo et al. (2016). The performance of PROBFOLD, PPFOLD RNASTRUCTURE, GTFOLD and RNAFOLD.ZAR is evaluated on a test set of 11 sequences with corresponding SHAPE data (Cordero et al. 2012b; Rice et al. 2014) and specified in terms of F-value. The F-value corresponds to the harmonic mean of sensitivity and specificity. The ΔF values specify the change in F-value between predictions that are only based on sequence input and predictions that are also based on SHAPE data. The test set consists of 11 small RNA secondary structures comprising SHAPE data for 5S RNA, Adenine riboswitch, cidGMP riboswitch, Glycine riboswitch, P4P6 domain (Tetrahymena ribozyme), Ribonuclease and tRNA phenylalanine (yeast) from Cordero et al. (2012b) and the M-Box riboswitch, Lysine riboswitch, Group I Intron from *T. thermophila* and Group II Intron from *O. iheyensis* from (Rice et al. 2014). Note that this test set contains only rather short sequences (min: 116 nt, max: 425 nt, average: 210 nt)

Table 4 Changes in prediction performance of PROBFOLD and RNASTRUCTURE as different types of RNA structure probing are provided as combined input

	PROBFOLD		RNASTRUCTURE	
Performance	F	ΔF	F	ΔF
seq	0.40	0.00	**0.73**	0.00
seq, CMCT	0.48	0.08	**0.85**	**0.12**
seq, CMCT, DMS	0.54	**0.14**	**0.85**	0.12
seq, CMCT, DMS, SHAPE	0.71	**0.31**	**0.82**	0.09

Results and figures from Sahoo et al. (2016). The performance of PROBFOLD and RNASTRUCTURE for predicting RNA secondary structures is evaluated as function of different kinds of RNA structure probing data supplied as input information (here, *seq* refers to single-sequence-only input). As in Table 3, the performance is specified in terms of F-value with the best performance highlighted in **bold**. The test set here comprises only six sequences for which CMCT, DMS and SHAPE probing data exist, namely 5S RNA, Adenine riboswitch, cidGMP riboswitch, Glycine riboswitch, P4P6 domain (Tetrahymena ribozyme) and tRNA phenylalanine (yeast) from Cordero et al. (2012a,b). Note that this reduced test set is a sub-set of the test set from Table 3 and contains even shorter sequences (min: 116 nt, max: 202 nt, average: 157 nt)

assessed, PROBFOLD is found to be the most robust w.r.t. increasing levels of noise. This is quantitatively assessed using different levels of simulated noise. Based on these results, one can conclude that PROBFOLD makes best use of the external RNA structure probing information. Using a slightly more complex SCFG for modelling RNA secondary structures or employing a comparative approach such as PFOLD should allow PROBFOLD's baseline performance to be further improved in the future.

Apart from the benchmark performance evaluation, the PROBFOLD study offers several important biological insights. First, they find that the SHAPE reactivities for paired and unpaired regions depend significantly on the primary nucleotide sequence. Furthermore, they find that the SHAPE reactivities for neighbouring sequence positions are significantly correlated, both for base-paired and especially for unpaired nucleotides. This is to be expected given that the SHAPE reactivities measure the backbone flexibility of the RNA transcript which is a notion that extends beyond the confines of the single sequence position that ends up being chemically modified. Based on these observations, Sahoo et al. decided to explicitly capture these correlations within the probabilistic models of PROBFOLD. Somewhat surprisingly, they find no evidence that the SHAPE reactivities between two base-pairing nucleotides are correlated. They attribute this to the comparatively high level of noise for low SHAPE reactivities. In PROBFOLD, this finding is captured by modelling the emission models of the left- and right-pairing partner independently using separate distributions.

One of the key advantages of PROBFOLD is that it can seamlessly integrate more than one kind of experimental structure probing data, e.g. DMS and CMCT probing data in addition to SHAPE reactivities. Initial performance results with a model which assumes independence of the different kinds of experimental evidence show that the performance can indeed be significantly improved as more types of experimental evidence are added, see the results in Table 4. Technically, PROBFOLD can also be set up to work with SHAPE-seq data (Lucks et al. 2011).

Conceptually, the theoretical framework underlying PROBFOLD offers a mathematically and conceptually convincing way of integrating experimental RNA structure probing data into models for RNA secondary structure prediction. Unlike most existing methods that are based on thermodynamic models for RNA secondary structure prediction, the number of free parameters in PROBFOLD that are used to integrate experimental RNA structure probing information does not increase with the length of the RNA. Instead, it only depends on the complexity (i.e. parametrisation) of the underlying predictive model. Moreover, these free parameters have a straightforward interpretation in terms of the experimental RNA structure probing data. By employing purely probabilistic concepts, different assumptions about the dependence or independence between probing data and/or between sequence positions can be made explicit and quantitatively assessed, so we can quantitatively test different hypotheses and also learn something about our data from the model. In addition, its free parameters can be readily retrained as more training data or novel types of experimental RNA structure probing data become available. This is a prerequisite for cross-evaluating the performance and for examining if over-fitting is an issue (Table 5).

Table 5 Characteristic features of the computer programs that predict RNA secondary structure by combining sequence data and chemical RNA structure probing data

Features	PROBFOLD	PPFOLD	RNASTRUCTURE	GTFOLD	RNAFOLD.ZAR
Seq input	Single	MSA	Single	Single	Single
Probing input	Multiple	Single	Multiple	Single	Single
Strategy	Prob.	Prob.	Therm.	Therm.	Therm.

All methods (PROBFOLD (Sahoo et al. 2016), RNASTRUCTURE (Deigan et al. 2009; Mathews et al. 2004), GTFOLD (Swenson et al. 2012) and RNAFOLD.ZAR (Lorenz et al. 2011, 2016)) apart from PPFOLD (Sükösd et al. 2012) use single RNAs as sequence input. Only PPFOLD works in a comparative way by using a multiple-sequence alignment (MSA) as input. Technically, it can still be forced to work in single-sequence mode if the input MSA comprises only a single sequence, see the performance evaluation in Table 3, although it is not meant to be used in that way. All methods can utilise SHAPE data as RNA structure probing input. PROBFOLD and RNASTRUCTURE can handle multiple types of RNA structure probing data simultaneously, e.g. SHAPE, DMS and CMCT probing data, see Table 4. Conceptually, all methods can be classified according to the strategy they employ (a) for RNA secondary structure predictions and (b) for integrating RNA structure probing data into the RNA structure predictions. PPFOLD and PROBFOLD are the only programs to work in a fully probabilistic way (prob.). They employ stochastic context-free grammars (SCFGs) as RNA secondary structure models and integrate RNA structure probing information in a fully probabilistic way. RNASTRUCTURE, GTFOLD and RNAFOLD.ZAR employ thermodynamic models for RNA secondary structure prediction (therm.) and aim to predict minimum-free energy structures. They integrate RNA structure probing data into the RNA structure prediction via different types of pseudo-energies

4 Transcriptome-Wide Experimental Methods for Directly Determining RNA Structures and *trans* RNA–RNA Interactions In Vivo

The structural building blocks of RNA secondary structures and of *trans* RNA–RNA interactions are base pairs. Yet, none of the transcriptome-wide methods for chemically probing RNA structures in vivo described above retain direct information on *base pairs*. Rather, information on RNA structure probing is linearised and encoded in *individual sequence positions*. Any direct information on corresponding pairing partners is lost. This is the main reason why major computational efforts are required to covert the raw position-specific experimental data back into actual RNA structures involving base pairs.

This recently changed as three groups simultaneously proposed experimental protocols for directly determining RNA secondary structure features in vivo in a transcriptome-wide fashion: PARIS (Lu et al. 2016), SPLASH (Aw et al. 2016) and LIGR-SEQ (Sharma et al. 2016). PARIS stands for **p**soralen **a**nalysis of **R**NA **i**nteractions and **s**tructures, SPLASH for **s**equencing of **p**soralen cross-linked, **l**igated and **s**elected **h**ybrids and LIGR-SEQ for **lig**ation of interacting **R**NA followed by high-throughput **seq**uencing. In constrast to earlier experimental protocols for probing transcriptome-wide in vivo probing, these three new methods allow to probe RNA structure features in a way which is not specific to any particular RNA-binding protein, see Fig. 1 for an overview.

4.1 Experimental Protocols of PARIS, SPLASH *and* LIGR-SEQ

All three new methods, i.e. PARIS (Lu et al. 2016), SPLASH (Aw et al. 2016) and LIGR-SEQ (Sharma et al. 2016), directly probe so-called duplexes, i.e. stretches of more or less consecutive base pairs. Each duplex can either involve the same or two different RNAs and thus either correspond to an RNA structure feature or a *trans* RNA–RNA interaction. It is important to note that all three experimental protocols process both types of duplexes in an identical manner (and that it is up to their respective, subsequent computational analysis pipelines to detect and distinguish both cases). All three methods are thus methods for both direct RNA structure probing as well as direct probing of *trans* RNA–RNA interactions. Conceptually, all three protocols have common steps but differ in important details. Their overall logical flow is as follows, see also Fig. 1.

4.1.1 Experimental Protocol of PARIS

In the first step of PARIS, duplexes corresponding to RNA structure features or to *trans* RNA–RNA interactions are covalently cross-linked using the psoralen

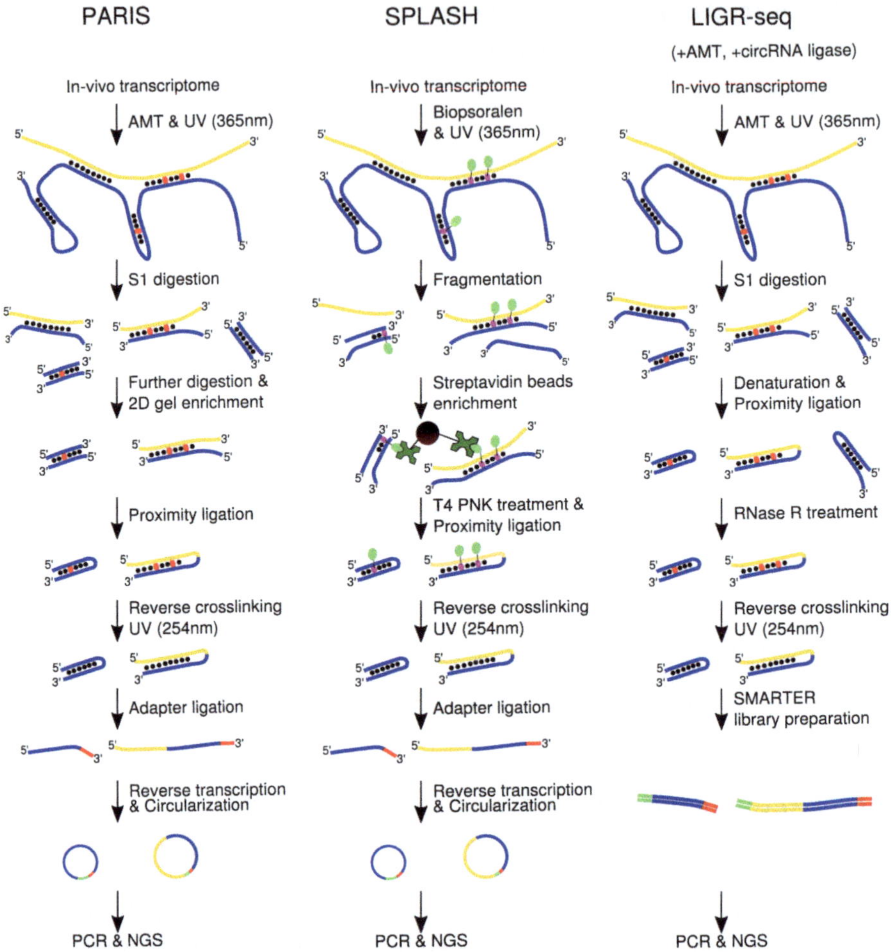

Fig. 1 Overview of the experimental protocols of PARIS, SPLASH and LIGR-SEQ. Lines in yellow and dark blue denote different transcripts. The black dots represent hydrogen bonds between transcripts. A red ellipse denotes the cross-linked psoralen derivative AMT. The complex between psoralen and biotin is shown in pink and light green, see the SPLASH pipeline. In the library preparation step, the red and green regions denote primers and adapters added, during the corresponding preparation protocols. The main difference between the protocols lies in the enrichment strategies for cross-linked duplexes. SPLASH focuses on biotin- dependent enrichment after fragmentation. PARIS utilises 2D-electrophoresis. LIGR-SEQ relies on the fact that AMT-cross-linked duplexes are more resistant to RNase R treatment. LIGR-SEQ requires additional samples to be made, see the text for details. In this figure, we only outline the protocol for making the +AMT+ligase sample

derivative 4'-aminomethyltrioxsalen (AMT) and UV-light at 365 nm. For this, AMT intercalates between base pairs and covalently cross-links preferentially juxtaposed pyrimidines (Calvet and Pederson 1979; Cimino et al. 1985). This effectively staples the two base-pairing arms involved in each duplex together. In the second step, RNase S1 digestion is utilised to remove single-stranded regions of RNA. Subsequently, ShortCut RNase III is used to make duplexes smaller and complete proteinase digestion and RNA purification yield short, directly base-pairing duplexes. In the third step, 2D-electrophoresis is employed for purification and enrichment as cross-linked duplexes appear off-diagonal, corresponding to 0.2%–0.5% of the RNA used as input to the 2D electrophoresis. This step is likely to reduce the overall sensitivity. In the fourth step, the ends of these selected duplexes are proximity-ligated before the cross-linking of the duplexes is reversed using UV-light at 254 nm. The efficiency of the cross-ligation is key for ensuring that information on the base-pairing arms involved in one duplex is not lost. The ligation step concatenates the two arms involved in one duplex into an artificial RNA in which the linear ordering of the two arms is a priori not clear. Finally, pre-adenylated adapters are added to the 3' ends, the resulting RNAs are reverse-transcribed in an adapter-specific way, circularised cDNA are generated and PCR amplification is performed to generate the cDNA libraries for NGS.

PARIS was originally performed in HeLa, HEK293T and mouse embryonic stem (mES) cells. Lu et al. conduct −AMT control experiments and observe no detectable off-diagonal elements in the corresponding 2D electrophoresis.

4.1.2 Experimental Protocol of SPLASH

The overall logical flow of SPLASH is similar to PARIS. Unlike for PARIS, cross-linking of duplexes in the first step is done using a biotinylated version of psoralen (so-called biopsoralen) also using UV-light at 365 nm. The biotin group is key for the subsequent enrichment step. Similar to AMT, biopsoralen also has a preference for cross-linking pyrimidines (Garrett-Wheeler et al. 1984; Hearst 1981). In contrast to AMT, however, biopsoralen typically requires the addition of a mild detergent (e.g. digitonin) to sufficiently increase the cellular uptake. The details of this (i.e. concentrations and duration of treatments with biopsoralen and digitonin) have to be carefully adjusted for each cell type separately. In the second step, cross-linked duplexes are extracted, randomly fragmented using Mg^{2+}-mediated hydrolysis and biotin enriched using streptavidin magnetic beads. Note that due to the random nature of fragmentation procedure a nick can occur in the hybridised region. Therefore, there is a chance that the detected length of the duplex does not correspond to the full length of the original duplex. The enrichment step of SPLASH is thus experimentally more efficient and conceptually more straightforward than the enrichment step of PARIS involving the more loosely-defined off-diagonal in a 2D-electrophoresis. In the third step, the ends of the resulting duplexes are ligated before UV-light at 254 nm is used as in PARIS to reverse the cross-linking. Similarly to PARIS, the fourth step involves the addition of pre-adenylated adapters the 3' ends, the reverse-transcription of the resulting RNAs in an adapter-specific way and the generation of circularised cDNAs. Again, PCR amplification is performed to obtain the final cDNA library for NGS.

The SPLASH protocol was used to examine HeLa cells, human lymphoblastoid cells, human embryonic stem (hES) cells, cells differentiated using retinoic acid and two types of cells from *S. cerevisiae*, namely wild type cells and Prp43 helicase mutant cells. Using between two to four biological replicates for each type of cell, they measure a high correlation (R = 0.75–0.9). Aw et al. (2016) generate several control libraries without cross-linking and without ligation in order to confirm that the duplexes identified by SPLASH are indeed enriched for ligated, cross-linked cases and not due to random background events. Furthermore, they explicitly confirm that cross-linking using biopsoralen is largely independent of solvent accessibility and show that SPLASH can detect RNA structure features with similar precision as the proximity ligation-based approach by Ramani et al. (2015) and has even higher sensitivity regarding *trans* RNA–RNA interactions.

4.1.3 Experimental Protocol of LIGR-SEQ

Conceptually, LIGR-SEQ has the same aims as PARIS and SPLASH, namely the direct detection of duplexes formed via RNA structure features or via *trans* RNA–RNA interactions. Unlike these two protocols, it uses a few features that set it distinctly apart and that have a significant impact on the subsequent computational interpretation of the raw reads.

Similar to PARIS, the first step of LIGR-SEQ consists of in vivo cross-linking of duplexes using AMT and UV-light at 365 nm. In terms of the specificity of the resulting, cross-linked duplexes, LIGR-SEQ is therefore comparable to PARIS (AMT) and SPLASH (biopsoralen). In the second step, RNA is extracted from cells and a limited digest with single-strand S1 endonuclease applied. The third step employs a circRNA ligase to link RNA ends in proximity. The fourth step is an enrichment step which utilises RNase R (a 3'-to'-5' exoribonuclease) to digest linear and structured RNAs whose duplexes have not been cross-linked (Vincent and Deutscher 2006). The pool of surviving RNAs consists of fully circularised RNAs and linear RNAs with cross-linked duplexes (as well as linear RNAs with uncross-linked duplexes whose 3' ends are too short for RNase R to latch on). Some false positives may very well survive the RNase R treatment. The fifth step reverses the cross-linking of duplexes using UV-light at 254 nm. Finally, the resulting RNAs (so-called chimeras in the LIGR-SEQ paper) are used to prepare stranded libraries for NGS. Unlike PARIS and SPLASH, the experimental protocol of LIGR-SEQ includes as default the preparation of an −AMT sample without any AMT-induced cross-linking. All samples are conceptually key for the subsequent computational interpretation of the raw LIGR-SEQ data. Without these, it would be conceptually impossible to define a dedicated probabilistic model which can assign estimated *p*-values to the experimentally detected interactions. Out of the three methods, LIGR-SEQ is currently the only method that is trying to experimentally estimate significance values for its detected interactions. As we will see in the following discussion of the computational analysis pipelines, it is also possible to assign significance values or p-values to proposed RNA structure features based on purely theoretical considerations, but these are conceptually different from the p-values derived by LIGR-SEQ.

4.1.4 Summary of All Three Experimental Protocols

After NGS, the raw data from PARIS, SPLASH and LIGR-SEQ corresponds to reads that each encode the sequence of the two arms involved in a formerly cross-linked duplex. One key difference with respect to chemical RNA structure probing methods is that any duplex can only be probed once as the molecules of the duplex itself end up being examined by the protocol. In contrast to this, methods for chemical RNA structure can probe any individual transcript multiple times and at different time points as they do not consume the investigated molecule itself.

For any given duplex derived by PARIS, SPLASH or LIGR-SEQ, it is unclear if the corresponding duplex derives from an *inter-* or from an *intramolecular* duplex, i.e. from a *trans* RNA–RNA interaction or from RNA structure features. It is also unclear in which linear order the two arms involved in the corresponding duplex appear in the resulting RNA and where their boundary is. These are key challenges to be addressed in the subsequent computational analysis of the raw data.

All three experimental protocols involve a stapler (i.e. AMT (PARIS and LIGR-SEQ) or biopsoralen (SPLASH)) that has a significant bias towards intercalating and cross-linking pyrimidines (Calvet and Pederson 1979; Cimino et al. 1985). Perfectly ordinary duplexes such as those involving G–C base pairs only may thus not be detectable at all using PARIS, SPLASH and LIGR-SEQ. Any absence of detectable duplexes can therefore not necessarily be taken as experimental evidence that the corresponding RNA structure feature of *trans* RNA–RNA interactions does not exist.

In addition, all three experimental protocols involve many steps that each introduce specific errors and biases that add up. As we will see in the following, the overall sensitivity and specificity of the combined step of each experimental protocol is further influenced by the errors and biases introduced by the computational analysis of the raw experimental data. It thus makes sense to consider and, ideally, optimise both in parallel.

4.2 *Computational Protocols of* PARIS, SPLASH *and* LIGR-SEQ

The main tasks of the computational analysis of the raw data from PARIS, SPLASH and LIGR-SEQ are (1) to map the sequenced reads back to the corresponding genome/transcriptome and (2) to figure out, for each read, if it corresponds to an *inter-* or an *intramolecular* duplex. Conceptually, both tasks have to be addressed simultaneously which amounts to the key challenge of the in silico analysis of these experimental data. In contrast to the sequenced reads derived from chemical RNA structure probing experiments, the raw data generated by PARIS, SPLASH and LIGR-SEQ *do not correspond to a consecutive sub-sequence of any single transcript*. Rather, each read either encodes the two separate of a duplex within the same transcript (if the duplex corresponds to an RNA structure feature) or a duplex involving two transcripts (if the duplex corresponds to a *trans* RNA–RNA interaction).

In case of an RNA structure duplex, mapping the corresponding read requires a gapped alignment to a single transcript (with a gap inserted between the two base-paired arms of the duplex encoded in the read) or a chimeric alignment in case of the two parts being non-canonical due to circle formation. This is complicated by the fact that the linear order of the arms in the read need not correspond to the natural linear order of the two arms within the underlying transcript (so-called chiastic reads). In case of a *trans* RNA–RNA duplex, mapping the read involves the identification of a pair of transcripts to which either of the two base-paired arms in the read map. This is conceptually and computationally challenging as the search space of all pairs of transcripts is huge compared to the search space of individual transcripts. Also here, the linear order in which the two arms appear in the read need not correspond to the order in which the respective two transcripts appear (chiastic reads). Furthermore, for both kinds of duplexes, the boundary between the two arms, i.e. where the gap has to be inserted for mapping, is a priori not known. To complicate matters further, it is up to the computational analysis to figure out for each read whether it corresponds to an RNA structure duplex or a *trans* RNA–RNA duplex.

The computational data analyses published in conjunction with the experimental protocols of PARIS, SPLASH and LIGR-SEQ have some main features in common, but differ in key details. As these differences are not exclusively due to the differences in experimental protocols, but partly due to different underlying strategies for interpreting the raw data, we will discuss them here.

4.2.1 Computational Analysis of Raw PARIS Data

Raw PARIS reads are first pre-processed by removing adapters from the 3’ ends and PCR duplicates. The latter is possible due to the insertion of a bar-code (random hexamer) in the middle of the adapter. These reads are then mapped to the corresponding genome using the computer program STAR (Dobin et al. 2013) with a set of input parameters that explicitly allow gapped-reads as well as so-called chiastic reads.

In a chiastic read, the linear order of the mappable parts (in our case, the two arms of a duplex) needs to be inverted. So, a read encoding a 5’-R-L-3’ duplex with a right (R) and left (L) arm of an RNA structure duplex needs to be mapped as 5’-L-3’-gap-5’-R-3’ to the underlying transcript. These chiastic reads naturally arise in all protocols whenever the ligation of a cross-linked, RNA structure-derived duplex happens to fuse the two base-pairing arms of the duplex in the wrong linear order, i.e. 5'-R-L-3' rather than 5'-L-R-3'. Chiastic reads can also arise in duplexes corresponding to *trans* RNA–RNA interactions whenever the mapping of the 5'-R-L-3' read to the (linearly ordered) transcripts of the transcriptome requires the reversal of the linear ordering of the two arms involved in the duplex. The correct mapping of chiastic reads thus always implies the insertion of a gap.

Before the mapping with STAR can actually be performed, a corresponding STAR index needs to be generated. This needs to be done with a carefully adjusted parameter for `genomeSAindexNbases` whenever the index is generated for a so-called mini-genome. The authors of PARIS utilise these mini-genomes in order to artificially reduce the search space for mapping, in particular when searching for specific *trans* RNA–RNA interactions, but also when investigating select genes in terms of RNA structure features (e.g. Xist gene or sub-set of snRNAs only). The parameters of STAR have to be explicitly adjusted whenever mini-genomes are used.

Of all the resulting STAR-mapped PARIS reads, only gapped and chiastic ones are retained. Of the gapped reads, only those are retained whose gap is not due to splicing.

In the next step, the retained mapped reads are grouped into so-called duplex groups (DGs). This is done using a greedy algorithm involving two steps. In the first step, the mapped reads are clustered into initial DGs such that all reads in a DG share at least 5 nt common overlap in both duplex arms (these two regions of overlap define the so-called core regions of the DG). Any mapped read is thereby either merged with an already existing DG or used to start a new DG. In the second step, DGs are merged into single DGs if they are close to each other and "well-defined" for both arms, see supplementary information of PARIS (Lu et al. 2016) for details.

Once the DGs have been established, each duplex group DG is assigned a so-called connection score which is defined as $cs(DG) = N_{\text{span}}(DG)/\sqrt{N_{\text{left}}(DG) \cdot N_{\text{right}}(DG)}$, where $N_{\text{span}}(DG)$ is the number of reads spanning the two duplex arms of DG and $N_{\text{left}}(DG)$ and $N_{\text{right}}(DG)$ are the number of unique reads overlapping the left and the right arm of DG, respectively. Note that $N_{\text{left}}(DG)$ can be different from $N_{\text{right}}(DG)$ as the reads covering each arm of the DG can also be assigned to other duplex groups overlapping DG only in one arm. Any duplex group DG with a connection score $cs(DG) < 0.01$ is then discarded to focus the subsequent analysis on duplexes that are supported by a significant portion of overlapping transcript reads.

The resulting duplexes typically involve two arms of 20–30 nt. The specific base pairs involved in a duplex between these two arms can, however, not be directly inferred from any DG. Rather, they have to be predicted based on the arms of the DG.

Lu et al. (2016) find that known miRNA-mRNA interactions cannot be detected, either because the duplex involved in the seed region is fairly short (around 5 nt length) and/or because binding of the duplex by the Argonaute protein shields the duplex from cross-linking.

Lu et al. try to assign a statistical significance to each detected duplex (whether corresponding to an RNA structure feature or a *trans* RNA–RNA interaction). For this, they compare the free energy of the MFE structure predicted for a multiple-sequence alignment underlying this DG to the corresponding, predicted free energies for 100 randomised versions of this multiple-sequence alignment. They thereby obtain a Z-score (Gesell and von Haeseler 2006). By utilising a procedure which focuses on the multiple-sequence alignment underlying the DG only, however, the Z-score cannot assess the statistical significance of seeing this DG by chance within the same transcript, let alone within the entire transcriptome which is what one would ideally like to know. Lu et al. evaluate the overall performance of PARIS by examining select RNAs (rRNA, snRNA, microRNA, telomerase RNA). This is done by visually comparing corresponding DGs to known features.

4.2.2 Computational Analysis of Raw SPLASH Data

Conceptually, the overall logical flow of the computational analysis of SPLASH is similar to the above for PARIS. Key details, however, differ and these turn out to be important.

To start with, transcriptomes for mapping purposes are generated by downloading the corresponding reference transcriptomes (taking the longest known isoform for each coding or non-coding gene as representative transcript) and by manually adding in select classes of non-coding genes. Any sequence duplicates from the joint set are then removed.

In the first step, the raw SPLASH paired-end reads are pre-processed by removing adapters and merging overlapping paired-end reads into corresponding single reads. In the next step, only these single merged reads are retained and mapped to the respective reference transcriptome using BWA MEM (version 0.7.12) (Li and Durbin 2010) using parameter `-T 20` to lower the minimum length of mapped regions to 20 nt. These mapped reads are then post-processed by sorting them and converting them to BAM-format using SAMTOOLS. Reads are then filtered for potential PCR duplicates by examining sets of reads with identical start coordinates and identical CIGAR strings and by retaining only the first read in each such set (Ramani et al. 2015).

In the original SPLASH analysis, the authors decide to deliberately focus their entire subsequent analysis on long-range features, i.e. RNA structure features and *trans* RNA–RNA interactions where the two arms involved in the corresponding duplex are far apart in terms of the underlying search space. Technically, this is achieved by retaining only split alignments more than 50 nt apart from the BAM-file of mapped reads. The authors of SPLASH then apply several measures to increase the quality of the retained, mapped reads. Reads with a mapping quality below 20 are discarded. In addition, ambiguously mapped reads and mapped reads with similarly scored second best hits are discarded (e.g. pseudo-genes). To lower the number of false positives, any read spanning a known splice-junction is removed using STAR (Dobin et al. 2013) to map splits reads from the transcriptome back to the corresponding genome. For this, reference sets of known splice junctions are assumed to be correct and complete.

The quoted overall sensitivity of SPLASH of 78% is based on its performance for the known RNA structure features of the 80S ribosome. The overall precision is reported to be 75%. In order to estimate the false discovery rate, independently cross-linked total RNAs from human yeast were pooled to prepare and analyse SPLASH libraries for any human-yeast interactions. Based on this strategy, SPLASH is reported to have a false discovery rate $< 3.7\%$.

In order to assign a statistical significance or p-value to the interactions detected by SPLASH, the free energy of the pairwise interaction in the detected duplex is compared to the free energy of many shuffled randomised versions of the sequences underlying the same pairwise interaction. The randomisation procedure keeps the di-nucleotide content preserved. SPLASH thus employs the same strategy as PARIS for estimating p-values to its detected interactions (in PARIS, this is done by shuffling multiple-sequence alignments; in SPLASH this is done by randomising only the sequences involved in the duplex). Both procedures are based on the validity of the assumption that true interactions in vivo have a lower minimum-free energy than interactions between corresponding randomised version of the same sequences. This assumption, however, is generally not justified (Rivas and Eddy 2000). In any case, the resulting p-value could not be interpreted as the probability of observing a corresponding RNA structure duplex or *trans* RNA–RNA interaction feature by chance. For this, entire transcripts (in case of RNA structure features) or pairs of transcripts (in case of *trans* RNA–RNA interactions) would need to be examined.

This could, for example, be achieved using TRANSAT (Wiebe and Meyer 2010), a fully probabilistic method that takes a multiple-sequence alignment and a corresponding evolutionary tree as input and detects evolutionarily conserved duplexes (so-called helices) in the input alignment. Any predicted helices are assigned a log-likelihood score as well as a p-value. This p-value corresponds to the chance of observing the duplex in the same transcript by chance.

4.2.3 Computational Analysis of Raw LIGR-SEQ Data

Raw LIGR-SEQ data consists of stranded, single-end reads. Similar to the above procedures for PARIS and SPLASH, these raw reads first need to be computationally post-processed before their actual interpretation in terms of biological contents can begin.

For this, LIGR-SEQ proposes a dedicated computational analysis pipeline called ALIGATER consisting of several steps. Unlike PARIS and SPLASH, the pipeline comprises a dedicated probabilistic model which is used to estimate p-values for the detected interactions. The first step removes the random bar-codes from the 5' ends. In the second step, these trimmed reads are mapped to the corresponding transcriptome using BOWTIE2 with a set of especially adjusted input parameters that aim to maximise sensitivity while keeping the computational run-time of the analysis reasonable. In the third step, these initial BOWTIE2 alignments in BAM-format are re-analysed such that blocks for each read are recursively chained into longer alignments in order to detect chimeras. This procedure can also handle circular ligation products and identifies the best path through the read. This step assigns a score to each chained alignment and is conceptually key for all of the subsequent analysis. The key corresponding input parameter for this procedure (the so-called chaining penalty) has to be carefully adjusted depending on the library quality as well as the specs of the specific class of transcripts being investigated. Reads with best-scoring chained alignments are then assigned an individual LIGQ score which retains detailed information on the corresponding alignments.

These LIGQ scores are subsequently used to carefully address several potential problems by either discarding or re-classifying chimeras. For example, artifacts due the mis-mapping of spliced transcript and of near-identical sequence duplicates (due to repeats, pseudo-genes or paralogues) are identified via near-identical matches to contiguous stretches of the underlying genome overlapping the ligation site and discarded. Other artifacts that incorrectly identify *intra*-molecular interactions as *inter*-molecular ones are re-classified based on corresponding supporting evidence. Overall, five different post-processing steps are executed, resulting in a strategy that re-classifies events rather than simply discard them and that aims for high sensitivity.

Another significant, conceptual difference of LIGR-SEQ with respect to the two other protocols, i.e. SPLASH and PARIS, is that it proposes an *experimental* strategy for estimating the statistical significance of the detected duplexes. This is achieved via a dedicated probabilistic model that judges the observed versus the expected ratios of chimeric reads. Each observed to expected ratio (i.e. OE_{+AMT} or OE_{-AMT}) corresponds to the corresponding experiments (i.e. +AMT or −AMT) with and without ligase. For this, separate +AMT and −AMT control experiments are performed without the ligation step in order to assess the expected background levels of spurious ligation events. The resulting LIGR-SEQ reads are then computationally processed as described above to detect interaction events (chimeras). Any pair of genes g_x and g_y is assigned a probability for spurious *trans* interactions $P_B(g_x, g_y)$ (using subscript B for background) which is assumed to only be a function of the respective relative whole gene abundance $P(g_x)$ of gene g_x and $P(g_y)$ of gene g_y, respectively. Mathematically, it corresponds to the probability of two independent draws from a multinomial distribution that is proportional to the relative abundance of each gene in the transcriptome. This defines their so-called null model.

The relative whole gene abundance for each gene g is measured in terms of reads per million without length adjustment (the RNase R treatment prevents this normalisation) and denoted RPM(g). So, $P_B(g_x, g_y) \propto P(g_x)P(g_y)$ if $x \neq y$ and if g_x and g_y have experimentally confirmed interactions events. In contrast, $P_B(g_x, g_y) = 0$ if $x = y$ or if $x \neq y$ and no interactions between these two genes are detected. The normalised probability for spurious interactions between gene g_x

and g_y, $p_B(g_x, g_y)$ is then written as (using $P(g_j) = \text{RPM}(g_j)/\sum_i \text{RPM}(g_i)$):

$$p_B(g_x, g_y) = \frac{P_B(g_x, g_y)}{\sum_i \sum_j P_B(g_i, g_j)} = \frac{\text{RPM}(g_x)\text{RPM}(g_y)}{\sum_i \sum_{j \text{ with } j \neq i} \text{RPM}(g_i)\text{RPM}(g_j)}$$

This null model assumes that the probability of a direct, spurious *trans* RNA–RNA interaction between two genes g_x and g_y in the transcriptome is only a function of the abundance of the relative whole gene abundance for each gene in the transcriptome. This model does not capture the primary sequence identity of each gene which is likely to also influence the probability of spurious *trans* RNA–RNA interactions. Assuming the validity of their null model, each experimentally detected interaction between genes g_x and g_y can then be assigned a p-value based on the number of observed reads k that are supporting it. This allows to explicitly filter for significant, AMT-induced interactions. Technically, this is achieved by first defining an enrichment score r_{AMT} which is defined as the ratio between $\text{OE}_{+\text{AMT}}$ and $\text{OE}_{-\text{AMT}}$, i.e. $r_{\text{AMT}} = \text{OE}_{+\text{AMT}}/\text{OE}_{-\text{AMT}}$. For real, AMT-induced interactions, we expect $\text{OE}_{+\text{AMT}} > \text{OE}_{-AMT}$ and require $r_{\text{AMT}} > 1.1$, more than 2 reads ($k > 2$), a p-value $< \alpha$ and an RPM of more than 10 in support. Similarly, interactions with $r_{\text{AMT}} < 0.9$ (and more than 2 reads ($k > 2$), a p-value $< \alpha$ and an RPM of more than 10) are considered false positives and allow to explicitly estimate the false positive rate of the overall protocol. In addition, LIGR-SEQ utilises two biological replicates. These allow to assess the overall technical reproducibility of the protocol (Spearman $Rho = 0.38$, $p < 8 \cdot 10^{-6}$).

Overall, the false discovery rate of LIGR-SEQ is estimated to range between 4.4% for highly expressed transcripts (> 250 RPM) and 25% for sparsely expressed transcripts (> 10 RPM). These numbers can be viewed as worst-case estimates as some known, stable interactions can be detected in both +AMT and -AMT samples. The high sensitivity of LIGR-SEQ can be explicitly confirmed based on known interactions in select groups of genes, e.g. known RNA structure features in the 80S ribosome (Anger et al. 2013) and *trans* RNA–RNA interactions between the 28S and 5S rRNA.

Overall, LIGR-SEQ is the only of the three protocols for measuring RNA structure features and *trans* RNA–RNA interactions in vivo that tries to assign experimentally estimated significance values to the detected features. This is done by proposing an explicit null model and by utilising dedicated, experimentally determined control samples. As mentioned above, TRANSAT (Wiebe and Meyer 2010) could be readily used to assign p-values to any experimentally determined duplexes in order to estimate their statistical significance in terms of the probability of seeing each duplex in the underlying transcript by chance.

5 Outlook

The last few years have seen an explosion of novel experimental and computational methods for determining RNA structures and *trans* RNA–RNA interactions in vivo. All experimental protocols require substantial computational strategies for analysing and for converting the raw experimental data into actual RNA structures or *trans* RNA–RNA interactions. Experimental and computational approaches are closely intertwined and therefore require simultaneous optimisation in order to optimise the overall performance.

Significant future improvements could be made in various ways.

First, we need to fully acknowledge the complexities of transcriptomes in vivo, in particular on the computational side of things. Any transcript in vivo may be long (long in this case meaning longer than 200 nt), may have various, unknown *trans* interaction partners (which may introduce RNA structure changes, e.g. Mazloomian and Meyer (2015)), may assume more than a single functional RNA structure or *trans* RNA–RNA interaction throughput its cellular life (e.g. Zhu and Meyer 2015; Lai et al. 2013) and, in particular, is unlikely to ever experience true thermodynamic equilibrium as a naked RNA. In particular for long RNAs such as coding transcripts, there is no reason to assume that they fold into a minimum-free energy structure spanning the entire transcript.

As advances in the field of ab initio RNA structure prediction showed, we may tackle this challenge best by employing a comparative strategy, i.e. by simply trying to identify RNA structure features or *trans* RNA–RNA interactions that have been conserved during well-chosen evolutionary times. Conceptually, this is currently the only way to detect the overall effects of various complexities in vivo *without having to explicitly model them*. Probabilistic methods are particularly well suited to seamlessly integrating experimental probing data into RNA structure predictions. In order for this line of research to flourish, we require gold-standard data sets of experimental probing data from different experimental probing protocols that examine the same in vivo situation using different methods. This needs, in particular, to include transcripts longer than 200 nt (see the captions of Tables 3 and 4 for the specs of the current data sets) from diverse biological classes of transcripts, not only short and non-coding RNAs that are known to contain global RNA structures spanning the entire transcript. There is, for example, by now ample evidence that short- and long-range RNA structure features are involved in regulating key cellular processes such as alternative splicing (Meyer and Miklos 2005; Raker et al. 2009; Pervouchine et al. 2012; Mazloomian and Meyer 2015). These gold-standard data sets thus have to be large and diverse enough to allow for parameter training as well as cross-evaluation procedures to avoid and evaluate potential issues due to over-fitting. The same applies to methods for predicting *trans* RNA–RNA interaction, where the currently assembled benchmark set (Lai and Meyer 2016) could be significantly increased, diversified and complemented by different kinds of experimental probing data.

On the experimental side of things, it would be beneficial to further reduce the inherent biases and limitation that the current methods have. PARIS, SPLASH and LIGR-SEQ are currently all based on psoralen-derivatives for cross-linking. This makes them blind to duplexes without juxtaposed pyrimidines. It would thus be great to remedy this by identifying intercalators that have complementary chemical specificities. The mapping of raw duplexes could be significantly facilitated by introducing artificial, known linker-sequences during the ligation of duplex-ends. Conceptually, another major step forward could be made by devising experimental protocols that are capable of detecting RNA structure diversity, i.e. cases where different copies of the same transcript engage in different RNA structures or *trans* RNA–RNA interactions in vivo. Right now, any RNA structure variation is misinterpreted as noise when interpreting chemical RNA structure probing data. Using specific variants of SHAPE-MAP (Smola et al. 2015b) may be able to change this conceptually by allowing structure probing information from individual transcripts to be retained throughout the entire protocol. Overall, Smola et al. propose three strategies. The standard Randomer workflow which uses random primers and default fragmentation and library preparation for creating a map of SHAPE-induced mutations, see Fig. 2. Due the fragmentation procedure, probing information on entire transcripts is typically lost. They propose two other strategies for addressing this problem. One is to perform size selection on RNAs with short lengths (< 500 nt) in order to retain full probing information on their entire sequences. This will, however, ignore a large proportion of typical transcriptomes (the average length for human mRNAs is 2.7 kb). To specifically address transcripts longer than 500 nt, i.e. particular isoforms of one gene, the so-called Amplicon workflow can be applied. In that strategy, specific primers, unique to one isoform, can be used to amplify only a region of the transcript. Then, multiple non-overlapping regions can be sequenced similar to the Randomer strategy to produce isoform specific information. This experimental strategy should in particular allow us to gain conceptually novel biological insight into how long coding or non-coding transcripts in eukaryotic genes use RNA structure features as mechanisms of gene regulation at RNA level. In the long run, the most elegant way of retaining RNA structure information on entire individual transcripts would be to combine chemical RNA structure probing with single-molecule sequencing techniques. This, however, will require significant changes of the currently existing protocols.

These are truly exciting times for in vivo transcriptome research, with many significant recent contributions both on the experimental and the computational side. Only by simultaneously optimising both experimental and computational procedures, however, will we be able to combine the best of both worlds. Both

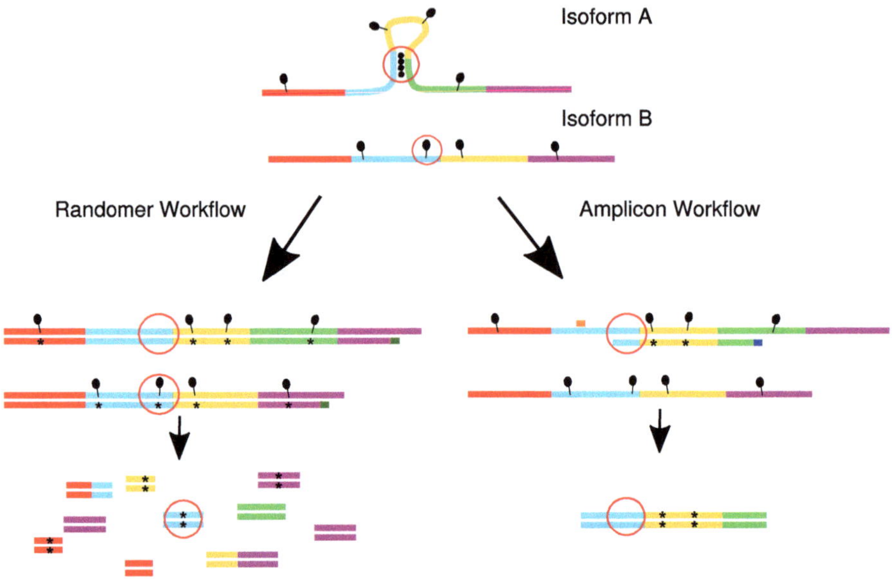

Fig. 2 Overview of the strategies recently proposed by SHAPE-MAP (Smola et al. 2015b). Shown here are two isoforms A and B of the same gene with partially overlapping sequences, where only one isoform assumes an RNA structure. Black ellipses correspond to the adducts produced by the SHAPE reagent. Black stars indicate mutations indicated during reverse transcription. The primer used in the Randomer workflow is shown in dark green. Region-specific primers of the Amplicon workflow are shown in orange and blue. The unpaired region that is paired in isoform A and unpaired in isoform B is highlighted by a red circle. The addition of SHAPE reagents to isoform B in combination with the Randomer workflow will produce a signal confirming that the region is unpaired. To confirm the presence of the RNA structure feature in isoform A, an alternative approach is required. This can be achieved with the Amplicon workflow using primers that are specific for a region in isoform A. This ensures that the adduct that is specific to isoform B is not amplified and thereby ignored

aspects currently come with a range of in-built assumptions and limitations. Questioning and, ideally, further reducing those will be key to discovering truly novel features

References

Anger A, Armache J, Berninghausen O, Habeck M, Subklewe M, Wilson D, Beckmann R (2013) Structures of the human and drosophila 80S ribosome. Nature 497(7447):80–85

Aultman K, Chang S (1982) Partial P1 nuclease digestion as a probe of tRNA structure. Eur J Biochem 124(3):471–476

Aw J, Shen Y, Wilm A, Sun M, Lim X, Boon K, Tapsin S, Chan Y, Tan C, Sim A, Zhang T, Susanto T, Fu Z, Nagarajan N, Wan Y (2016) In vivo mapping of eukaryotic RNA interactomes reveals principles of Higher-Order organization and regulation. Mol Cell 62(4):603–617
Bevilacqua P, Ritchey L, Su Z, Assmann S (2016) Genome-Wide analysis of RNA secondary structure. Annu Rev Genet 50:235–266
Calvet J, Pederson T (1979) Heterogeneous nuclear RNA double-stranded regions probed in living HeLa cells by crosslinking with the psoralen derivative aminomethyltrioxsalen. Proc Natl Acad Sci USA 76(2):755–759
Cheng C, Chou F, Kladwang W, Tian S, Cordero P, Das R (2015) Consistent global structures of complex RNA states through multidimensional chemical mapping. Elife 4:e07600
Cimino G, Gamper H, Isaacs S, Hearst J (1985) Psoralens as photoactive probes of nucleic acid structure and function: organic chemistry, photochemistry, and biochemistry. Annu Rev Biochem 54:1151–1193
Cordero P, Kladwang W, VanLang C, Das R (2012a) Quantitative dimethyl sulfate mapping for automated RNA secondary structure inference. Biochemistry 51(36):7037–7039
Cordero P, Lucks J, Das R (2012b) An RNA mapping DataBase for curating RNA structure mapping experiments. Bioinformatics 28(22):3006–3008
Deigan K, Li T, Mathews D, Weeks K (2009) Accurate SHAPE-directed RNA structure determination. Proc Natl Acad Sci USA 106(1):97–102
Del Campo C, Bartholomäus A, Fedyunin I, Ignatova Z (2015) Secondary structure across the bacterial transcriptome reveals versatile roles in mRNA regulation and function. PLoS Genet 11(10):e1005613
Ding Y, Lawrence C (2003) A statistical sampling algorithm for RNA secondary structure prediction. Nucleic Acids Res 31(24):7280–7301
Ding Y, Tang Y, Kwok C, Zhang Y, Bevilacqua P, Assmann S (2014) In vivo genome-wide profiling of RNA secondary structure reveals novel regulatory features. Nature 505(7485):696–700
Ding Y, Kwok C, Tang Y, Bevilacqua P, Assmann S (2015) Genome-wide profiling of in vivo RNA structure at single-nucleotide resolution using structure-seq. Nat Protoc 10(7):1050–1066
Dobin A, Davis C, Schlesinger F, Drenkow J, Zaleski C, Jha S, Batut P, Chaisson M, Gingeras T (2013) STAR: ultrafast universal RNA-seq aligner. Bioinformatics 29(1):15–21
Durbin R, Eddy S, Krogh A, Mitchison G (1998) Biological sequence analysis: probabilistic models of proteins and nucleic acids. Cambridge University Press, Cambridge
Eddy S (2014) Computational analysis of conserved RNA secondary structure in transcriptomes and genomes. Annu Rev Biophys 43:433–456
Ehresmann C, Baudin F, Mougel M, Romby P, Ebel J, Ehresmann B (1987) Probing the structure of RNAs in solution. Nucleic Acids Res 15(22):9109–9128
Fang R, Moss W, Rutenberg-Schoenberg M, Simon M (2015) Probing Xist RNA structure in cells using targeted Structure-Seq. PLoS Genet 11(12):e1005668
Flynn R, Zhang Q, Spitale R, Lee B, Mumbach M, Chang H (2016) Transcriptome-wide interrogation of RNA secondary structure in living cells with icSHAPE. Nat Protoc 11(2): 273–290
Garrett-Wheeler E, Lockard R, Kumar A (1984) Mapping of psoralen cross-linked nucleotides in RNA. Nucleic Acids Res 12(7):3405–3423
Gesell T, von Haeseler A (2006) In silico sequence evolution with site-specific interactions along phylogenetic trees. Bioinformatics 22(6):716–722
Guerrier-Takada C, Gardiner K, Marsh T, Pace N, Altman S (1983) The RNA moiety of ribonuclease P is the catalytic subunit of the enzyme. Cell 35(3):849–857
Harris K Jr, Crothers D, Ullu E (1995) In vivo structural analysis of spliced leader RNAs in trypanosoma brucei and leptomonas collosoma: a flexible structure that is independent of cap4 methylations. RNA 1(4):351–362

Hearst J (1981) Psoralen photochemistry and nucleic acid structure. J Invest Dermatol 77(1):39–44
Hector R, Burlacu E, Aitken S, Le Bihan T, Tuijtel M, Zaplatina A, Cook A, Granneman S (2014) Snapshots of pre-rRNA structural flexibility reveal eukaryotic 40S assembly dynamics at nucleotide resolution. Nucleic Acids Res 42(19):12138–12154
Higgs PG (2000) RNA secondary structure: physical and computational aspects. Q Rev Biophys 33(3):199–253
Homan P, Favorov O, Lavender C, Kursun O, Ge X, Busan S, Dokholyan N, Weeks K (2014) Single-molecule correlated chemical probing of RNA. Proc Natl Acad Sci USA 111(38):13858–13863
Incarnato D, Neri F, Anselmi F, Oliviero S (2014) Genome-wide profiling of mouse RNA secondary structures reveals key features of the mammalian transcriptome. Genome Biol 15(10):491
Kertesz M, Wan Y, Mazor E, Rinn J, Nutter R, Chang H, Segal E (2010) Genome-wide measurement of RNA secondary structure in yeast. Nature 467(7311):103–107
Kielpinski L, Vinther J (2014) Massive parallel-sequencing-based hydroxyl radical probing of RNA accessibility. Nucleic Acids Res 42(8):e70
Knapp G (1989) Enzymatic approaches to probing of RNA secondary and tertiary structure. Methods Enzymol 180:192–212
Knudsen B, Hein J (2003) Pfold: RNA secondary structure prediction using stochastic context-free grammars. Nucleic Acids Res 31(13):3423–3428
Koller D, Friedman N (2009) Probabilstic graphical models. MIT, Cambridge
Kwok C, Ding Y, Tang Y, Assmann S, Bevilacqua P (2013) Determination of in vivo RNA structure in low-abundance transcripts. Nat Commun 4:2971
Lai D, Meyer IM (2016) A comprehensive comparison of general RNA-RNA interaction prediction methods. Nucleic Acids Res 44(7):e61
Lai D, Proctor JR, Meyer IM (2013) On the importance of cotranscriptional RNA structure formation. RNA 19(11):1461–1473
Latham J, Cech T (1989) Defining the inside and outside of a catalytic RNA molecule. Science 245(4915):276–282
Lavender C, Gorelick R, Weeks K (2015) Structure-based alignment and consensus secondary structures for three HIV-related RNA genomes. PLoS Comput Biol 11(5):e1004230
Lengyel J, Hnath E, Storms M, Wohlfarth T (2014) Towards an integrative structural biology approach: combining Cryo-TEM, X-ray crystallography, and NMR. J Struct Funct Genom 15(3):117–124
Li H, Durbin R (2010) Fast and accurate long-read alignment with Burrows-Wheeler transform. Bioinformatics 26(5):589–595
Li F, Zheng Q, Ryvkin P, Dragomir I, Desai Y, Aiyer S, Valladares O, Yang J, Bambina S, Sabin L, Murray J, Lamitina T, Raj A, Cherry S, Wang L, Gregory B (2012a) Global analysis of RNA secondary structure in two metazoans. Cell Rep 1(1):69–82
Li F, Zheng Q, Vandivier L, Willmann M, Chen Y, Gregory B (2012b) Regulatory impact of RNA secondary structure across the arabidopsis transcriptome. Plant Cell 24(11):4346–4359
Lorenz R, Bernhart S, Höner Zu Siederdissen C, Tafer H, Flamm C, Stadler P, Hofacker I (2011) ViennaRNA package 2.0. Algorithms Mol Biol 6:26
Lorenz R, Luntzer D, Hofacker I, Stadler P, Wolfinger M (2016) SHAPE directed RNA folding. Bioinformatics 32(1):145–147
Loughrey D, Watters K, Settle A, Lucks J (2014) SHAPE-Seq 2.0: systematic optimization and extension of high-throughput chemical probing of RNA secondary structure with next generation sequencing. Nucleic Acids Res 42(21):e165
Low J, Garcia-Miranda P, Mouzakis K, Gorelick R, Butcher S, Weeks K (2014) Structure and dynamics of the HIV-1 frameshift element RNA. Biochemistry 53(26):4282–4291

Lu Z, Zhang Q, Lee B, Flynn R, Smith M, Robinson J, Davidovich C, Gooding A, Goodrich K, Mattick J, Mesirov J, Cech T, Chang H (2016) RNA duplex map in living cells reveals Higher-Order transcriptome structure. Cell 165(5):1267–1279

Lucks J, Mortimer S, Trapnell C, Luo S, Aviran S, Schroth G, Pachter L, Doudna J, Arkin A (2011) Multiplexed RNA structure characterization with selective 2'-hydroxyl acylation analyzed by primer extension sequencing (shape-seq). Proc Natl Acad Sci USA 108(27):11063–11068

Mathews DH, Sabina J, Zuker M, Turner DH (1999) Expanded sequence dependence of thermodynamic parameters improves prediction of RNA secondary structure. J Mol Biol 288(5):911–940

Mathews D, Disney M, Childs J, Schroeder S, Zuker M, Turner D (2004) Incorporating chemical modification constraints into a dynamic programming algorithm for prediction of RNA secondary structure. Proc Natl Acad Sci USA 101(19):7287–7292

Mauger D, Golden M, Yamane D, Williford S, Lemon S, Martin D, Weeks K (2015) Functionally conserved architecture of hepatitis C virus RNA genomes. Proc Natl Acad Sci USA 112(12):3692–3697

Mazloomian A, Meyer IM (2015) Genome-wide identification and characterization of tissue-specific RNA editing events in D. melanogaster and their potential role in regulating alternative splicing. RNA Biol 12(12):1391–1401

McCaskill J (1990) The equilibrium partition function and base pair binding probabilities for RNA secondary structure. Biopolymers 29(6–7):1105–1119

McGinnis J, Dunkle J, Cate J, Weeks K (2012) The mechanisms of RNA SHAPE chemistry. J Am Chem Soc 134(15):6617–6624

Merino E, Wilkinson K, Coughlan J, Weeks K (2005) RNA structure analysis at single nucleotide resolution by selective 2'-hydroxyl acylation and primer extension (shape). J Am Chem Soc 127(12):4223–4231

Meyer I (2017) In silico methods for co-transcriptional RNA secondary structure prediction and for investigating alternative RNA structure expression. Methods 120:3–16

Meyer I, Miklos I (2004) Co-transcriptional folding is encoded within RNA genes. BMC Mol Biol 5:10

Meyer I, Miklos I (2005) Statistical evidence for conserved, local secondary structure in the coding regions of eukaryotic mRNAs and pre-mRNAs. Nucleic Acids Res 33(19):6338–6348

Meyer IM, Miklos I (2007) SimulFold: simultaneously inferring RNA structures including pseudoknots, alignments, and trees using a Bayesian MCMC framework. PLOS Comput Biol 3(8):1441–1454

Miklos I, Meyer I, Nagy B (2005) Moments of the Boltzmann distribution for RNA secondary structures. Bull Math Biol 67(5):1031–1047

Moazed D, Stern S, Noller H (1986) Rapid chemical probing of conformation in 16 S ribosomal RNA and 30 S ribosomal subunits using primer extension. J Mol Biol 187(3):399–416

Morgan S, Higgs PG (1996) Evidence for kinetic effects in the folding of large RNA molecules. Annu Rev Biophys 105:7152–7157

Mortimer S, Weeks K (2007) A fast-acting reagent for accurate analysis of RNA secondary and tertiary structure by SHAPE chemistry. J Am Chem Soc 129(14):4144–4145

Mortimer S, Trapnell C, Aviran S, Pachter L, Lucks J (2012) Shape-seq: high-throughput RNA structure analysis. Curr Protoc Chem Biol 4(4):275–297

Ouyang Z, Snyder M, Chang H (2013) SeqFold: genome-scale reconstruction of RNA secondary structure integrating high-throughput sequencing data. Genome Res 23(2):377–387

Pedersen J, Forsberg R, Meyer I, Hein J (2004a) An evolutionary model for protein-coding regions with conserved RNA structure. Mol Biol Evol 21(10):1913–1922

Pedersen J, Meyer I, Forsberg R, Simmonds P, Hein J (2004b) A comparative method for finding and folding RNA secondary structures within protein-coding regions. Nucleic Acids Res 32(16):4925–4936

Pervouchine DD, Khrameeva EE, Pichugina MY, Nikolaienko OV, Gelfand MS, Rubtsov PM, Mironov AA (2012) Evidence for widespread association of mammalian splicing and conserved long-range RNA structures. RNA 18(1):1–15

Poulsen L, Kielpinski L, Salama S, Krogh A, Vinther J (2015) SHAPE selection (shapes) enrich for RNA structure signal in SHAPE sequencing-based probing data. RNA 21(5):1042–1052

Proctor JR, Meyer IM (2013) CoFold: an RNA secondary structure prediction method that takes co-transcriptional folding into account. Nucleic Acids Res 41(9):e102

Qi L, Lucks J, Liu C, Mutalik V, Arkin A (2012) Engineering naturally occurring trans-acting non-coding RNAs to sense molecular signals. Nucleic Acids Res 40(12):5775–5786

Quarrier S, Martin J, Davis-Neulander L, Beauregard A, Laederach A (2010) Evaluation of the information content of RNA structure mapping data for secondary structure prediction. RNA 16(6):1108–1117

Raker VA, Mironov AA, Gelfand MS, Pervouchine DD (2009) Modulation of alternative splicing by long-range RNA structures in Drosophila. Nucleic Acids Res 37(14):4533–4544

Ramani V, Qiu R, Shendure J (2015) High-throughput determination of RNA structure by proximity ligation. Nat Biotechnol 33(9):980–984

Rice G, Leonard C, Weeks K (2014) RNA secondary structure modeling at consistent high accuracy using differential SHAPE. RNA 20(6):846–854

Righetti F, Nuss A, Twittenhoff C, Beele S, Urban K, Will S, Bernhart S, Stadler P, Dersch P, Narberhaus F (2016) Temperature-responsive in vitro RNA structurome of yersinia pseudotuberculosis. Proc Natl Acad Sci USA 113(26):7237–7242

Rivas E, Eddy S (2000) Secondary structure alone is generally not statistically significant for the detection of noncoding RNAs. Bioinformatics 16(7):583–605

Rouskin S, Zubradt M, Washietl S, Kellis M, Weissman J (2014) Genome-wide probing of RNA structure reveals active unfolding of mRNA structures in vivo. Nature 505(7485):701–705

Sahoo S, Świtnicki M, Pedersen J (2016) ProbFold: a probabilistic method for integration of probing data in RNA secondary structure prediction. Bioinformatics 32(17):2626–2635

Sclavi B, Woodson S, Sullivan M, Chance M, Brenowitz M (1997) Time-resolved synchrotron x-ray "footprinting", a new approach to the study of nucleic acid structure and function: application to protein-DNA interactions and RNA folding. J Mol Biol 266(1):144–159

Seetin M, Kladwang W, Bida J, Das R (2014) Massively parallel RNA chemical mapping with a reduced bias MAP-seq protocol. Methods Mol Biol 1086:95–117

Sharma E, Sterne-Weiler T, O'Hanlon D, Blencowe B (2016) Global mapping of human RNA-RNA interactions. Mol Cell 62(4):618–626

Siegfried N, Busan S, Rice G, Nelson J, Weeks K (2014) RNA motif discovery by SHAPE and mutational profiling (shape-map). Nat Methods 11(9):959–965

Smola M, Calabrese J, Weeks K (2015a) Detection of RNA-protein interactions in living cells with SHAPE. Biochemistry 54(46):6867–6875

Smola M, Rice G, Busan S, Siegfried N, Weeks K (2015b) Selective 2'-hydroxyl acylation analyzed by primer extension and mutational profiling (shape-map) for direct, versatile and accurate RNA structure analysis. Nat Protoc 10(11):1643–1669

Soper SFC, Dator RP, Limbach PA, Woodson SA (2013) In vivo x-ray footprinting of pre-30s ribosomes reveals chaperone-dependent remodeling of late assembly intermediates. Mol Cell 52(4):506–516

Spitale R, Flynn R, Zhang Q, Crisalli P, Lee B, Jung J, Kuchelmeister H, Batista P, Torre E, Kool E, Chang H (2015) Structural imprints in vivo decode RNA regulatory mechanisms. Nature 519(7544):486–490

Steen K, Rice G, Weeks K (2012) Fingerprinting noncanonical and tertiary RNA structures by differential SHAPE reactivity. J Am Chem Soc 134(32):13160–13163

Steif A, Meyer IM (2012) The hok mRNA family. RNA Biol 9(12):1399–1404

Sükösd Z, Knudsen B, Kjems J, Pedersen C (2012) PPfold 3.0: fast RNA secondary structure prediction using phylogeny and auxiliary data. Bioinformatics 28(20):2691–2692

Swenson MS, Anderson J, Ash A, Gaurav P, Sükösd Z, Bader DA, Harvey SC, Heitsch CE (2012) GTfold: enabling parallel RNA secondary structure prediction on multi-core desktops. BMC Res Notes 5:341

Talkish J, May G, Lin Y, Woolford J Jr, McManus C (2014) Mod-seq: high-throughput sequencing for chemical probing of RNA structure. RNA 20(5):713–20

Tyrrell J, McGinnis J, Weeks K, Pielak G (2013) The cellular environment stabilizes adenine riboswitch RNA structure. Biochemistry 52(48):8777–8785

Underwood J, Uzilov A, Katzman S, Onodera C, Mainzer J, Mathews D, Lowe T, Salama S, Haussler D (2010) FragSeq: transcriptome-wide RNA structure probing using high-throughput sequencing. Nat Methods 7(12):995–1001

Vincent H, Deutscher M (2006) Substrate recognition and catalysis by the exoribonuclease RNase R. J Biol Chem 281(40):29769–29775

Wan Y, Qu K, Ouyang Z, Kertesz M, Li J, Tibshirani R, Makino D, Nutter R, Segal E, Chang H (2012) Genome-wide measurement of RNA folding energies. Mol Cell 48(2):169–181

Wan Y, Qu K, Ouyang Z, Chang H (2013) Genome-wide mapping of RNA structure using nuclease digestion and high-throughput sequencing. Nat Protoc 8(5):849–869

Wan Y, Qu K, Zhang Q, Flynn R, Manor O, Ouyang Z, Zhang J, Spitale R, Snyder M, Segal E, Chang H (2014) Landscape and variation of RNA secondary structure across the human transcriptome. Nature 505(7485):706–709. https://doi.org/10.1038/nature12946

Washietl S, Hofacker I, Stadler P, Kellis M (2012) RNA folding with soft constraints: reconciliation of probing data and thermodynamic secondary structure prediction. Nucleic Acids Res 40(10):4261–4272

Watters K, Abbott T, Lucks J (2016a) Simultaneous characterization of cellular RNA structure and function with in-cell SHAPE-Seq. Nucleic Acids Res 44(2):e12

Watters K, Yu A, Strobel E, Settle A, Lucks J (2016b) Characterizing RNA structures in vitro and in vivo with selective 2'-hydroxyl acylation analyzed by primer extension sequencing (shape-seq). Methods 103:34–48

Watts J, Dang K, Gorelick R, Leonard C, Bess J Jr, Swanstrom R, Burch C, Weeks K (2009) Architecture and secondary structure of an entire HIV-1 RNA genome. Nature 460(7256): 711–716

Weeks K (2010) Advances in RNA structure analysis by chemical probing. Curr Opin Struct Biol 20(3):295–304

Wells S, Hughes J, Igel A, Ares M Jr (2000) Use of dimethyl sulfate to probe RNA structure in vivo. Methods Enzymol 318:479–493

Wiebe NJP, Meyer IM (2010) TRANSAT-a method for detecting the conserved helices of functional RNA structures, including transient, pseudo-knotted and alternative structures. PLOS Comput Biol 6(6):e1000823

Wilkinson K, Merino E, Weeks K (2006) Selective 2'-hydroxyl acylation analyzed by primer extension (shape): quantitative RNA structure analysis at single nucleotide resolution. Nat Protoc 1(3):1610–1616

Woese C, Magrum L, Gupta R, Siegel R, Stahl D, Kop J, Crawford N, Brosius J, Gutell R, Hogan J, Noller H (1980) Secondary structure model for bacterial 16S ribosomal RNA: phylogenetic, enzymatic and chemical evidence. Nucleic Acids Res 8(10):2275–2293

Zarringhalam K, Meyer M, Dotu I, Chuang J, Clote P (2012) Integrating chemical footprinting data into RNA secondary structure prediction. PLoS One 7(10):e45160

Zaug A, Cech T (1995) Analysis of the structure of tetrahymena nuclear RNAs in vivo: telomerase RNA, the self-splicing rRNA intron, and U2 snRNA. RNA 1(4):363–374

Zheng Q, Ryvkin P, Li F, Dragomir I, Valladares O, Yang J, Cao K, Wang L, Gregory B (2010) Genome-wide double-stranded RNA sequencing reveals the functional significance of base-paired RNAs in arabidopsis. PLoS Genet 6(9):e1001141

Zhu JYA, Meyer IM (2015) Four RNA families with functional transient structures. RNA Biol 12(1):5–20

Zhu JYA, Steif A, Proctor JR, Meyer IM (2013) Transient RNA structure features are evolutionarily conserved and can be computationally predicted. Nucleic Acids Res 41(12):6273–6285

Zuker M (2003) Mfold web server for nucleic acid folding and hybridization prediction. Nucleic Acids Res 31(13):3406–3415

Zuker M, Stiegler P (1981) Optimal computer folding of large RNA sequences using thermodynamics and auxiliary information. Nucleic Acids Res 9(1):133–148

Is Autogenous Posttranscriptional Gene Regulation Common?

Gary D. Stormo

Contents

Abstract The goal of this chapter is to provide evidence and justification for the hypothesis that autogenous, posttranscriptional regulation of gene expression is common. Several examples are known, mostly from bacteria, bacteriophage, and yeast species. Each was identified either by accident or by a concerted effort to understand the regulation of specific genes. The paucity of examples from higher eukaryotes may be due to the difficulty of identifying them using common approaches for uncovering regulatory interactions. An alternative approach is proposed that can fill the gap.

Keywords Gene expression regulation · Autogenous regulation · Posttranscriptional regulation · Protein–RNA interactions · Feedback regulation

1 Introduction

The expression of proteins is a multistep process and every step can be regulated. Most studies of gene regulation have focused on the initiation of transcription, recruitment by transcription factors (TFs) of the RNA polymerase to the promoter,

G. D. Stormo (✉)
Department of Genetics and Edison Family Center for Genome Sciences and Systems Biology, Washington University School of Medicine, St Louis, MO, USA
e-mail: stormo@wustl.edu

N. Rajewsky et al. (eds.), *Systems Biology*, RNA Technologies,
https://doi.org/10.1007/978-3-319-92967-5_10

and beginning of transcription (Busser et al. 2008; Chasman and Roy 2017; Haraksingh and Snyder 2013; Qu and Fang 2013). In fact, many papers on regulatory elements and networks deal exclusively with transcriptional regulation (Bar-Joseph et al. 2003; Elkon and Agami 2017; Wyrick and Young 2002). There is some justification for the emphasis on the regulation of transcription initiation because posttranscriptional regulation can only occur once an mRNA has been initiated. But the number of posttranscriptional steps, each an opportunity for regulation, suggests that the final protein synthesis levels may be most tightly controlled after transcriptional initiation. In fact, several reports show that the levels of protein expression and the corresponding mRNA are only weakly correlated for many proteins (Liu et al. 2016; Lu et al. 2007; Schwanhausser et al. 2011; Vogel et al. 2010; Vogel and Marcotte 2012), and those studies often use steady-state levels of mRNA, which already includes posttranscriptional regulatory events involved in mRNA degradation.

While the initiation of transcription is essential to gene expression, the multiple steps between that initiation step and the completion of the protein product allow for a myriad of regulatory processes impacting alternative splicing and other processing events, mRNA degradation rates, and translation efficiency, among others. Many studies have examined posttranscriptional gene regulation, with primary emphases on translational regulation (Cline and Bock 1966; Hinnebusch et al. 2016; Larsson et al. 2013; Sonenberg and Hinnebusch 2009; Valencia-Sanchez et al. 2006), regulation by microRNAs (Carthew 2006; Pasquinelli 2012; Valencia-Sanchez et al. 2006), modifications of mRNA sequences (Gilbert et al. 2016), and very large number of RNA-binding proteins (RBPs) that can influence many of those steps (Dassi 2017; Dassi and Quattrone 2012; Matia-Gonzalez et al. 2015; Mitchell et al. 2013) including localization and sequestering of mRNA (Holt and Bullock 2009; Martin and Ephrussi 2009; Yan et al. 2017). Much has been learned, and it is currently a topic of extensive investigation, but most of the work is based on RBPs and miRNAs that regulate many genes. The work often uses methods similar to those applied to TFs that control many genes, such as localization approaches that identify the locations of binding sites for DNA- and RNA-binding proteins on genomic DNA and mRNA. A limitation of those studies is that they can easily miss examples of autoregulation, in which one protein regulates its own expression, and therefore has only one binding site because its primary role is not as a regulatory factor.

Early studies of the regulation of gene expression found examples of autoregulation (Cove 1974; Goldberger 1974; Savageau 1975). Similar to feedback regulation of enzymatic pathways, where the product can limit to the activity of the pathway to control the synthesis rate of the product, protein synthesis pathways were discovered in which the product of the pathway, the protein itself, could regulate some step in the pathway to control the intracellular concentration of the protein. The best studied example was the lambda repressor which controls its own transcription both positively and negatively as part of the genetic switch that determines the fate of the infected cell (Ptashne et al. 1976, 1980, 1982). The repressor first activates its own expression in the competition with the Cro protein to control the

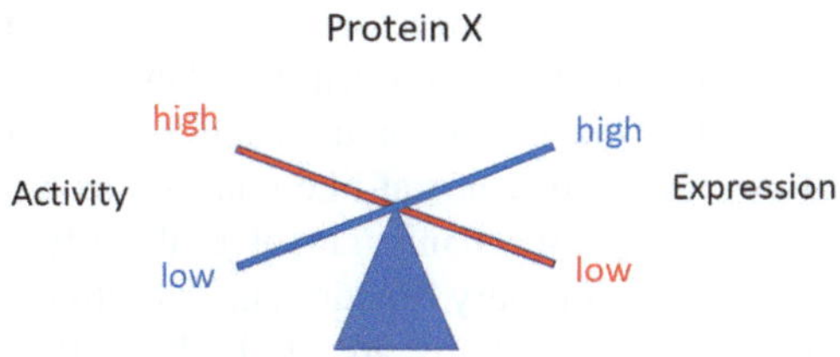

Fig. 1 Autoregulation of gene X requires that its activity and its expression are anticorrelated

fate of the infection, and if it "wins" that competition it then represses its own synthesis to limit its concentration within a narrow range. Since then many other examples of transcriptional autoregulation have been discovered and mathematical analyses demonstrating the advantages of autoregulation have been described (Alon 2007; Bateman 1998; Pinho et al. 2014; Wall et al. 2003). In the simplest sense, autoregulation can be thought of as a direct means of the protein to control its own expression so as to maintain a constant, and appropriate, level of activity in the cell. In this context, activity may refer to an enzymatic function or simply the appropriate concentration of the protein. The diagram in Fig. 1 shows feedback regulation between protein X activity and expression such that they are anticorrelated, but that could be accomplished in several ways. Autoregulation implies the direct interaction between protein X and some step in its synthesis.

In the following section I describe several examples of posttranscriptional autoregulation that have been discovered and studied in detail. I do not discuss examples where the protein's primary role is as a regulator of expression and it simply includes its own synthesis among its targets. I also omit examples from RNA phages and viruses that can only regulate expression from the RNA. I focus on proteins that have a different primary function but exhibit autoregulation as well. General principles can be extracted from the examples that illustrate why I suspect posttranscriptional autoregulation may be common but difficult to detect.

2 Examples of Posttranscriptional Autoregulation

Proteins that are involved in the process of protein synthesis can easily evolve to be autoregulated if their own translation is sensitive to their activity. Two examples from *E. coli* are initiation factor 3 (IF3) and release factor 2 (RF2) (Betney et al. 2010). IF3 is a translation initiation factor whose primary function is to aid in the selection of the start codon for translation, which is most frequently an AUG codon, but other codons are used less commonly. IF3's own start codon is uniquely AUU so when IF3 activity is high its own mRNA is unlikely to be translated, but when its activity is low its expression increases. RF2 is a release factor responsible for releasing the protein chain when a UGA stop codon is encountered. The mRNA for RF2 contains an in-frame UGA codon within the coding region, so when RF2 activity is high translation of the full-length protein is low, but when RF2 activity is low the expression increases due to frame-shifting by a ribosome stalled at the stop codon but inefficiently released. These regulatory events are not on-off decisions, but rather the effect on translation efficiency as a function of protein activity allows for a "thermostat-like control" of protein levels (Betney et al. 2010). The autoregulation mechanism involves the mRNA having evolved to take advantage of the protein's primary function; no evolution of the protein's activity was required.

Ribosomal proteins are all RBPs with their primary binding site within the ribosomal RNAs (rRNAs). In *E. coli*, the synthesis of many ribosomal proteins was found to be regulated at the step of translational initiation (Nomura 2011; Nomura et al. 1984). This was originally discovered because duplication of some ribosomal protein operons led to overexpression of the mRNA, but not the proteins, and could even be observed in vitro, ruling out rapid protein degradation as the mechanism (Fallon et al. 1979; Nomura et al. 1984; Yates et al. 1980). A single protein within an operon is often sufficient to inhibit expression from all of the genes in that operon, because they are "translationally coupled," whereby translation of downstream genes requires translation of the gene preceding it in the operon. So only a subset of the ribosomal proteins is sufficient to control the translation of nearly all of them. This is accomplished by the ribosomal protein binding to the translation initiation site of one gene, typically the first gene in the operon, because that region of the mRNA has evolved to resemble the natural binding site of the protein in the rRNA (Bellur and Woodson 2009; Draper 1989; Schlax et al. 2001; Wu et al. 1994). The binding site on the mRNA must have a lower affinity than the primary rRNA-binding site to ensure that those primary sites are fully occupied before expression of the protein is turned off. Although not as common, similar autoregulatory functions are observed for some ribosomal proteins in yeast, sometimes involving the regulation of splicing (Lu et al. 2015; Warner and McIntosh 2009).

Another example of intrinsic RBPs that also regulate their own synthesis are tRNA synthetase genes (Yao et al. 2014). For example, the threonyl-tRNA synthetase gene of *E. coli* was shown to inhibit its own translation by binding to its mRNA at a structure that mimics the sequence and structure of the threonyl-tRNA (Romby et al. 1990; Romby and Springer 2003; Springer et al. 1985; Torres-Larios et al. 2002). When the activity of the protein is sufficient to charge all of the threonyl-tRNAs in the cell, so that no more of the protein is needed, it binds to its secondary ligand, its own mRNA, to repress its synthesis.

Both of these examples, ribosomal proteins and aminoacyl-tRNA synthetases, have normal functions that require RNA binding, in one case to rRNA and in the other tRNA. When those primary functions are fully satisfied the proteins become autoregulatory repressors to inhibit further synthesis of themselves. As with IF3 and RF2, this requires no evolution of the protein itself, just the evolution of the mRNA sequence to mimic the normal RNA-binding site of the protein. The regulatory site must be a somewhat weaker binder to the protein than the primary site, but better than potential off-target RNAs in the cell. There are many other known examples in bacteria, and a few in yeast, where proteins whose function involves binding to

RNA have obtained a secondary function of controlling their own expression by a simple co-opting of their normal activity through mRNA mimicry.

Three other examples from *E. coli* show specific autoregulation for proteins that either do not bind RNA as their normal function or bind to it nonspecifically. The SuhB gene is an inositol monophosphatase that also controls its own expression by modulating the half-life of its mRNA (Inada and Nakamura 1996). The gene for polynucleotide phosphorylase (Pnp) regulates its own synthesis through both modulating an RNA-cleavage event and repressing translation (Carzaniga et al. 2015). SecA is involved in protein translocation across the plasma membrane and also has domains for DNA and RNA helicase activity. It regulates its own translation through binding to RNA structures in its mRNA, a function that does not require the helicase activity (Kiser and Schmidt 1999; Schmidt et al. 2001).

Two other examples of posttranscriptional autoregulation that do not involve RBPs were discovered in the bacteriophage T4 (Gold 1988; Uzan and Miller 2010). The single-stranded DNA (ssDNA)-binding protein (gene 32) is required for DNA replication, repair, and recombination and binds to ssDNA nonspecifically. Inactive fragments of the protein were found to be much more highly expressed than the wild-type protein, and that was shown to be due to loss of translational repression by the wild-type protein (Russel et al. 1976). Although the protein will bind to RNA, its affinity is about 200-fold lower than for DNA, and the ssDNA binding is not sequence specific, so it was a surprise that it could specifically regulate its own translation. An initial model proposed that its own mRNA is uniquely unstructured around the ribosome-binding site, and due to gene 32's highly cooperative binding it would bind there before binding to any other mRNAs, providing the necessary specificity (von Hippel et al. 1982). Further analysis proposed, and then showed, that a pseudoknotted RNA structure at the 5' end of the mRNA, conserved in other T-even phages even though the sequences differ, provided a nucleation site for the cooperative binding that contributed to the specificity of regulation (McPheeters et al. 1988; Shamoo et al. 1993). The T4 DNA polymerase gene (gene 43) also regulates its own synthesis by binding to a hairpin near the ribosome-binding site of its own mRNA (Andrake et al. 1988; Petrov and Karam 2002; Tuerk et al. 1990). The binding of the protein to its own mRNA occurs when no more DNA polymerase is needed to complete replication of the phage genome.

3 Principles of Posttranscriptional Autoregulation

Evolution of autoregulation can be accomplished easily, only requiring modification of the mRNA, not the protein. Furthermore, any step in the process between the initiation of transcription and the completion of the protein product can be regulated, offering many opportunities for autogenous control of expression. In examples like IF3 and RF2, the mRNA has taken advantage of the normal function of the protein, in one case to effect translation initiation and in the other termination, so that its expression becomes lower when the protein's activity becomes higher, providing a direct, but subtle, feedback to control the synthesis rate. For RBPs, the mRNA simply has to evolve to mimic the normal ligand of the protein in such a way as to alter translation or splicing or some other step in the synthesis process. For other proteins whose function does not involve protein synthesis or RNA binding, the evolution of autoregulation may seem more difficult. But in fact, RNA has an enormous potential for binding with high affinity and specificity to a wide variety of molecules. This was demonstrated with the invention of SELEX, first to identify the sequence requirements for the gene 43 regulatory hairpin (Tuerk and Gold 1990), and then shown to be a general approach that could select RNA aptamers for specific binding to essentially any target (Gold et al. 1995). Nature's use of RNA versatility is evident in the widespread occurrence of riboswitches (McCown et al. 2017). Based on the identification of an RNA structure involved in the regulation of the genes for thiamin synthesis, and the lack of identification of a regulatory protein after extensive search (Miranda-Rios et al. 2001), we speculated that the mRNA may bind thiamin directly leading to an alteration in structure that could control the synthesis of the proteins in the operon, and that such a mechanism may be quite common (Stormo and Ji 2001). This was soon shown to be the case for regulation of the thiamin operon (Winkler et al. 2002) and then discovered to be a widespread mechanism in bacteria to regulate the expression of enzymes for the synthesis of various small molecules (Breaker 2011; Mandal and Breaker 2004; McCown et al. 2017; Mironov et al. 2002; Vitreschak et al. 2004). The versatility of RNA to specifically bind essentially any target means that every protein is, at least potentially, an RNA-binding protein. The extent to which this capability is used by cells, for autoregulation or other functions, is unknown, but could be quite common.

Another principle evident from the known examples of posttranscriptional autoregulation is that they are difficult to find. For the T4 proteins it was the accidental observation that inactive versions of the proteins, either temperature sensitive or premature chain termination mutants, were greatly overexpressed. That could be seen because after infection only the T4 genes are expressed and the inactive fragments are stable. In a complex mixture of proteins from cells, especially when inactive versions are rapidly degraded, such an effect would not be noticed. In the case of IF3 and RF2 the highly unusual features of the genes, the unique AUU start codon for IF3 and the in-frame stop codon in RF2, led to the search for the consequences of those features and the discovery of their use in autoregulation. The ribosomal proteins were under intensive study to understand how the expression of the rRNA and proteins was coordinated. Each of the other cases was also discovered through concerted efforts to understand the regulation of specific genes. Riboswitches are examples of feedback regulation by the end product of an enzymatic pathway, but not by the direct action of a protein to regulate its own synthesis that is the topic of this chapter. But it is informative to consider their discovery. The regulatory mechanisms for those pathways had been under study for several years, but only after the initial example of a riboswitch was identified did it become clear that was a common mechanism and many more examples were identified in rapid succession (Mandal and Breaker 2004; Vitreschak et al. 2004). Only a few examples have been found in eukaryotes but that may be due to the challenges in finding them. Similarly, once miRNAs were discovered and found to be more widespread than just *C. elegans*, there were directed searches for them and many were rapidly identified, shown to be widely expressed and involved in regulating large numbers of genes (Pasquinelli 2012). While some examples of posttranscriptional autoregulation have been known for many years, there have not been concerted efforts to find them. There are certainly efforts to identify proteins that posttranscriptionally regulate gene expression, but they are primarily focused on RBPs that are likely to have many targets (Matia-Gonzalez et al. 2015; Mitchell et al. 2013), and would miss an autoregulatory protein that only binds to, or effects the translation from, its own mRNA. Furthermore, even once a protein is identified as being autoregulatory, determining the mechanism can be challenging. In cases where the protein binds to the mRNA, specificity is often accomplished by complex structural binding sites. And since there is only one example in the cell, typical binding site motif discovery methods aren't useful. One can, sometimes, take advantage of phylogenetically conserved binding sites (see below), but the structural complexity can still leave the problem challenging.

4 Strategies for Discovery

There are two strategies we imagine for identifying posttranscriptional autoregulatory genes, one a mixture of computational and experimental methods and another that is purely experimental and more direct. One common feature of the autoregulatory examples, except for IF3 and RF2 which are special cases that don't involve protein–mRNA binding, is that the binding site is composed of structure and sequence that provide the specificity necessary for regulation of a single gene. By looking for conserved structural motifs, overlapping or near the translation initiation site of the mRNA, across a set of related bacterial species, we discovered several significant examples in *Shewanella* species (Xu et al. 2009). Many of them are almost certainly true cases of regulatory sites because they correspond to known examples from other species, such as riboswitches, attenuator hairpins, or ribosomal protein-binding sites. There were also several novel predictions that have not, to my knowledge, been tested which would be required to verify their existence and function. Their existence also doesn't prove that they represent autoregulatory events, as opposed to alternative functions, and that would have to be shown experimentally. Currently we are working on an experimental approach that takes advantage of resources in yeast. There is a collection of strains in which green fluorescent protein (GFP) has been fused with nearly every yeast gene, and when that gene is expressed the cells are green (Huh et al. 2003). We have designed versions of the genes in which all of the potential *cis*-regulatory elements (CREs) have been removed, which we refer to as CRE-less versions of the genes. This is done by changing the promoter and the 5' and 3' untranslated regions and the introns, which is where regulatory sites typically reside. We have also shuffled the codons of the gene to maintain the same protein sequence but remove potential binding sites within the coding region. These CRE-less gene versions are fused to red fluorescent protein (RFP) behind an inducible promoter. If a gene is autoregulated, turning on the CRE-less version should downregulate the wild-type protein and cells should go from green to red, which can be easily identified with fluorescence sorting. This will allow us to test large libraries of CRE-less genes and identify any that are autoregulated, as well as other examples of regulation. The mechanism of regulation and the exact nature of the CREs will have to be determined by further analysis, such as by replacing some of the potential CREs to identify those required for regulation. We think this will identify new examples of posttranscriptional autoregulation in yeast, and we speculate that there could be many of them. Adapting this method to higher eukaryotes, without the resources available in yeast, will be challenging and we are exploring alternative methods. We think such tools, along with many other approaches to study posttranscriptional regulation in general, will lead to much more comprehensive views of the true regulatory networks in cells.

Acknowledgments I thank Drs. Michael White and Basab Roy for comments and suggestions on the manuscript.

References

Alon U (2007) Network motifs: theory and experimental approaches. Nat Rev Genet 8:450–461
Andrake M, Guild N, Hsu T, Gold L, Tuer C et al (1988) DNA polymerase of bacteriophage T4 is an autogenous translational repressor. Proc Natl Acad Sci USA 85:7942–7946
Bar-Joseph Z, Gerber GK, Lee TI, Rinaldi NJ, Yoo JY et al (2003) Computational discovery of gene modules and regulatory networks. Nat Biotechnol 21:1337–1342
Bateman E (1998) Autoregulation of eukaryotic transcription factors. Prog Nucleic Acid Res Mol Biol 60:133–168
Bellur DL, Woodson SA (2009) A minimized rRNA-binding site for ribosomal protein S4 and its implications for 30S assembly. Nucleic Acids Res 37:1886–1896
Betney R, de Silva E, Krishnan J, Stansfield I (2010) Autoregulatory systems controlling translation factor expression: thermostat-like control of translational accuracy. RNA 16:655–663
Breaker RR (2011) Prospects for riboswitch discovery and analysis. Mol Cell 43:867–879
Busser BW, Buly ML, Michelson AM (2008) Toward a systems-level understanding of developmental regulatory networks. Curr Opin Genet Dev 18:521–529
Carthew RW (2006) Gene regulation by microRNAs. Curr Opin Genet Dev 16:203–208
Carzaniga T, Deho G, Briani F (2015) RNase III-independent autogenous regulation of *Escherichia coli* polynucleotide phosphorylase via translational repression. J Bacteriol 197:1931–1938
Chasman D, Roy S (2017) Inference of cell type specific regulatory networks on mammalian lineages. Curr Opin Syst Biol 2:130–139
Cline AL, Bock RM (1966) Translational control of gene expression. Cold Spring Harb Symp Quant Biol 31:321–333
Cove DJ (1974) Evolutionary significance of autogenous regulation. Nature 251:256
Dassi E (2017) Handshakes and fights: the regulatory interplay of RNA-binding proteins. Front Mol Biosci 4:67
Dassi E, Quattrone A (2012) Tuning the engine: an introduction to resources on post-transcriptional regulation of gene expression. RNA Biol 9:1224–1232

Draper DE (1989) How do proteins recognize specific RNA sites? New clues from autogenously regulated ribosomal proteins. Trends Biochem Sci 14:335–338
Elkon R, Agami R (2017) Characterization of noncoding regulatory DNA in the human genome. Nat Biotechnol 35:732–746
Fallon AM, Jinks CS, Strycharz GD, Nomura M (1979) Regulation of ribosomal protein synthesis in *Escherichia coli* by selective mRNA inactivation. Proc Natl Acad Sci USA 76:3411–3415
Gilbert WV, Bell TA, Schaening C (2016) Messenger RNA modifications: form, distribution, and function. Science 352:1408–1412
Gold L (1988) Posttranscriptional regulatory mechanisms in *Escherichia coli*. Annu Rev Biochem 57:199–233
Gold L, Polisky B, Uhlenbeck O, Yarus M (1995) Diversity of oligonucleotide functions. Annu Rev Biochem 64:763–797
Goldberger RF (1974) Autogenous regulation of gene expression. Science 183:810–816
Haraksingh RR, Snyder MP (2013) Impacts of variation in the human genome on gene regulation. J Mol Biol 425:3970–3977
Hinnebusch AG, Ivanov IP, Sonenberg N (2016) Translational control by 5'-untranslated regions of eukaryotic mRNAs. Science 352:1413–1416
Holt CE, Bullock SL (2009) Subcellular mRNA localization in animal cells and why it matters. Science 326:1212–1216
Huh WK, Falvo JV, Gerke LC, Carroll AS, Howson RW et al (2003) Global analysis of protein localization in budding yeast. Nature 425:686–691
Inada T, Nakamura Y (1996) Autogenous control of the suhB gene expression of *Escherichia coli*. Biochimie 78:209–212
Kiser KB, Schmidt MG (1999) Regulation of the *Escherichia coli* secA gene is mediated by two distinct RNA structural conformations. Curr Microbiol 38:113–121
Larsson O, Tian B, Sonenberg N (2013) Toward a genome-wide landscape of translational control. Cold Spring Harb Perspect Biol 5:a012302
Liu Y, Beyer A, Aebersold R (2016) On the dependency of cellular protein levels on mRNA abundance. Cell 165:535–550
Lu P, Vogel C, Wang R, Yao X, Marcotte EM (2007) Absolute protein expression profiling estimates the relative contributions of transcriptional and translational regulation. Nat Biotechnol 25:117–124
Lu H, Zhu YF, Xiong J, Wang R, Jia Z (2015) Potential extra-ribosomal functions of ribosomal proteins in *Saccharomyces cerevisiae*. Microbiol Res 177:28–33
Mandal M, Breaker RR (2004) Gene regulation by riboswitches. Nat Rev Mol Cell Biol 5:451–463
Martin KC, Ephrussi A (2009) mRNA localization: gene expression in the spatial dimension. Cell 136:719–730
Matia-Gonzalez AM, Laing EE, Gerber AP (2015) Conserved mRNA-binding proteomes in eukaryotic organisms. Nat Struct Mol Biol 22:1027–1033
McCown PJ, Corbino KA, Stav S, Sherlock ME, Breaker RR (2017) Riboswitch diversity and distribution. RNA 23:995–1011
McPheeters DS, Stormo GD, Gold L (1988) Autogenous regulatory site on the bacteriophage T4 gene 32 messenger RNA. J Mol Biol 201:517–535
Miranda-Rios J, Navarro M, Soberon M (2001) A conserved RNA structure (thi box) is involved in regulation of thiamin biosynthetic gene expression in bacteria. Proc Natl Acad Sci USA 98:9736–9741
Mironov AS, Gusarov I, Rafikov R, Lopez LE, Shatalin K et al (2002) Sensing small molecules by nascent RNA: a mechanism to control transcription in bacteria. Cell 111:747–756
Mitchell SF, Jain S, She M, Parker R (2013) Global analysis of yeast mRNPs. Nat Struct Mol Biol 20:127–133
Nomura M (2011) Journey of a molecular biologist. Annu Rev Biochem 80:16–40
Nomura M, Gourse R, Baughman G (1984) Regulation of the synthesis of ribosomes and ribosomal components. Annu Rev Biochem 53:75–117

Pasquinelli AE (2012) MicroRNAs and their targets: recognition, regulation and an emerging reciprocal relationship. Nat Rev Genet 13:271–282
Petrov VM, Karam JD (2002) RNA determinants of translational operator recognition by the DNA polymerases of bacteriophages T4 and RB69. Nucleic Acids Res 30:3341–3348
Pinho R, Garcia V, Irimia M, Feldman MW (2014) Stability depends on positive autoregulation in Boolean gene regulatory networks. PLoS Comput Biol 10:e1003916
Ptashne M, Backman K, Humayun MZ, Jeffrey A, Maurer R et al (1976) Autoregulation and function of a repressor in bacteriophage lambda. Science 194:156–161
Ptashne M, Jeffrey A, Johnson AD, Maurer R, Meyer BJ et al (1980) How the lambda repressor and cro work. Cell 19:1–11
Ptashne M, Johnson AD, Pabo CO (1982) A genetic switch in a bacterial virus. Sci Am 247:128–130, 132, 134–140
Qu H, Fang X (2013) A brief review on the human encyclopedia of DNA elements (ENCODE) project. Genomics Proteomics Bioinformatics 11:135–141
Romby P, Springer M (2003) Bacterial translational control at atomic resolution. Trends Genet 19:155–161
Romby P, Moine H, Lesage P, Graffe M, Dondon J et al (1990) The relation between catalytic activity and gene regulation in the case of *E. coli* threonyl-tRNA synthetase. Biochimie 72:485–494
Russel M, Gold L, Morrissett H, O'Farrell PZ (1976) Translational, autogenous regulation of gene 32 expression during bacteriophage T4 infection. J Biol Chem 251:7263–7270
Savageau MA (1975) Significance of autogenously regulated and constitutive synthesis of regulatory proteins in repressible biosynthetic systems. Nature 258:208–214
Schlax PJ, Xavier KA, Gluick TC, Draper DE (2001) Translational repression of the *Escherichia coli* alpha operon mRNA: importance of an mRNA conformational switch and a ternary entrapment complex. J Biol Chem 276:38494–38501
Schmidt MO, Brosh RM Jr, Oliver DB (2001) *Escherichia coli* SecA helicase activity is not required in vivo for efficient protein translocation or autogenous regulation. J Biol Chem 276:37076–37085
Schwanhausser B, Busse D, Li N, Dittmar G, Schuchhardt J et al (2011) Global quantification of mammalian gene expression control. Nature 473:337–342
Shamoo Y, Tam A, Konigsberg WH, Williams KR (1993) Translational repression by the bacteriophage T4 gene 32 protein involves specific recognition of an RNA pseudoknot structure. J Mol Biol 232:89–104
Sonenberg N, Hinnebusch AG (2009) Regulation of translation initiation in eukaryotes: mechanisms and biological targets. Cell 136:731–745
Springer M, Plumbridge JA, Butler JS, Graffe M, Dondon J et al (1985) Autogenous control of *Escherichia coli* threonyl-tRNA synthetase expression in vivo. J Mol Biol 185:93–104
Stormo GD, Ji Y (2001) Do mRNAs act as direct sensors of small molecules to control their expression? Proc Natl Acad Sci USA 98:9465–9467
Torres-Larios A, Dock-Bregeon AC, Romby P, Rees B, Sankaranarayanan R et al (2002) Structural basis of translational control by *Escherichia coli* threonyl tRNA synthetase. Nat Struct Biol 9:343–347
Tuerk C, Gold L (1990) Systematic evolution of ligands by exponential enrichment: RNA ligands to bacteriophage T4 DNA polymerase. Science 249:505–510
Tuerk C, Eddy S, Parma D, Gold L (1990) Autogenous translational operator recognized by bacteriophage T4 DNA polymerase. J Mol Biol 213:749–761
Uzan M, Miller ES (2010) Post-transcriptional control by bacteriophage T4: mRNA decay and inhibition of translation initiation. Virol J 7:360
Valencia-Sanchez MA, Liu J, Hannon GJ, Parker R (2006) Control of translation and mRNA degradation by miRNAs and siRNAs. Genes Dev 20:515–524
Vitreschak AG, Rodionov DA, Mironov AA, Gelfand MS (2004) Riboswitches: the oldest mechanism for the regulation of gene expression? Trends Genet 20:44–50

Vogel C, Marcotte EM (2012) Insights into the regulation of protein abundance from proteomic and transcriptomic analyses. Nat Rev Genet 13:227–232

Vogel C, Abreu Rde S, Ko D, Le SY, Shapiro BA et al (2010) Sequence signatures and mRNA concentration can explain two-thirds of protein abundance variation in a human cell line. Mol Syst Biol 6:400

von Hippel PH, Kowalczykowski SC, Lonberg N, Newport JW, Paul LS et al (1982) Autoregulation of gene expression: quantitative evaluation of the expression and function of the bacteriophage T4 gene 32 (single-stranded DNA binding) protein system. J Mol Biol 162:795–818

Wall ME, Hlavacek WS, Savageau MA (2003) Design principles for regulator gene expression in a repressible gene circuit. J Mol Biol 332:861–876

Warner JR, McIntosh KB (2009) How common are extraribosomal functions of ribosomal proteins? Mol Cell 34:3–11

Winkler WC, Cohen-Chalamish S, Breaker RR (2002) An mRNA structure that controls gene expression by binding FMN. Proc Natl Acad Sci USA 99:15908–15913

Wu H, Jiang L, Zimmermann RA (1994) The binding site for ribosomal protein S8 in 16S rRNA and spc mRNA from *Escherichia coli*: minimum structural requirements and the effects of single bulged bases on S8-RNA interaction. Nucleic Acids Res 22:1687–1695

Wyrick JJ, Young RA (2002) Deciphering gene expression regulatory networks. Curr Opin Genet Dev 12:130–136

Xu X, Ji Y, Stormo GD (2009) Discovering cis-regulatory RNAs in *Shewanella* genomes by support vector machines. PLoS Comput Biol 5:e1000338

Yan S, Acharya S, Groning S, Grosshans J (2017) Slam protein dictates subcellular localization and translation of its own mRNA. PLoS Biol 15:e2003315

Yao P, Poruri K, Martinis SA, Fox PL (2014) Non-catalytic regulation of gene expression by aminoacyl-tRNA synthetases. Top Curr Chem 344:167–187

Yates JL, Arfsten AE, Nomura M (1980) In vitro expression of *Escherichia coli* ribosomal protein genes: autogenous inhibition of translation. Proc Natl Acad Sci USA 77:1837–1841

The Interplay of Non-coding RNAs and X Chromosome Inactivation in Human Disease

Francesco Russo, Federico De Masi, Søren Brunak, and Kirstine Belling

Contents

Abstract Non-coding RNAs (ncRNAs) represent key molecular players in biological processes and human disease. Several ncRNA types have been discovered including microRNAs (miRNAs) of around 23 nucleotides and long non-coding RNAs (lncRNAs) that are above 200 nucleotides in length. One of the first functional ncRNAs discovered was the lncRNA named *X inactive specific transcript* (*XIST*). *XIST* is the main actor in a fundamental process called X chromosome inactivation (XCI) where, in females, one of the two X chromosomes is silenced to balance the extra gene expression dosage. In this book chapter, we present the emerging evidence for the importance of XCI in diseases such as gastric and bladder cancer and genetic pathologies such as Klinefelter (47,XXY) and Turner (45,X0) syndromes. Furthermore, a new role for the crosstalk between *XIST* and miRNAs is discussed. Finally, new evidence for sex bias of XCI in human tissues and development of cancer is presented.

F. Russo (✉) · S. Brunak · K. Belling
Novo Nordisk Foundation Center for Protein Research, Faculty of Health and Medical Sciences, University of Copenhagen, Copenhagen, Denmark
e-mail: francesco.russo@cpr.ku.dk

F. De Masi
Intomics A/S, Lyngby (Copenhagen), Denmark

N. Rajewsky et al. (eds.), *Systems Biology*, RNA Technologies,
https://doi.org/10.1007/978-3-319-92967-5_11

Keywords Non-coding RNAs · X chromosome inactivation · Human disease · Aneuploidies · Cancer · Comorbidities

1 Introduction

The central dogma of molecular biology describes the flow of information from DNA passing through RNA to protein. Each step of this process is finely regulated by a multitude of finely interconnected mechanisms. Gene expression regulation can be modulated through many molecules including transcription factors and non-coding RNAs (ncRNAs) such as microRNAs (miRNAs) and long non-coding RNAs (lncRNAs). miRNAs are a family of small, ncRNAs (around 21–23 nucleotides) that regulate gene expression in a sequence-specific manner by binding mRNAs (or other ncRNAs), usually leading to their down-regulation (Bartel 2009). Since their first description (Lee et al. 1993), miRNAs changed the way researchers study the central dogma. This new class of functional non-coding sequences in the genome opened a new field in biology and new opportunities emerged for discovering novel genes influencing molecular mechanisms in physiology and disease (Bartel 2004). Several studies showed the link between miRNAs and human pathologies, especially in cancer (Calin et al. 2002, 2004, 2005; Iorio et al. 2005; Song et al. 2013). The first evidence of the involvement of miRNAs in cancer came in 2002 where it was shown that *miR-15* and *miR-16* are located at chromosome 13q14, a region frequently deleted in chronic lymphocytic leukaemia (CLL), and that these miRNAs were deleted or down-regulated in more than 60% of human CLL cases (Calin et al. 2002). The deletion of miRNAs in cancer suggested their potential role as tumour suppressors (Calin et al. 2002, 2004; Iorio et al. 2005; Lagana et al. 2010). Further studies showed that miRNAs can also act as oncogenes (Zhang et al. 2007). A typical example is *miR-21*, up-regulated in most cancer types where it regulates cell proliferation, apoptosis and migration by suppressing the expression of tumour suppressors (Pfeffer et al. 2015). Several other studies revealed the importance of miRNAs in fundamental biological processes and in many diseases and it is estimated that miRNAs regulate approximately 30% of the human protein-coding genome (Filipowicz et al. 2008).

Another emerging class of ncRNAs is lncRNAs (>200 nucleotides in length). They were also discovered in the early 1990s—the same period as miRNAs. The first two examples of functional lncRNAs were *H19* (Brannan et al. 1990) and *XIST* (Brockdorff et al. 1992; Brown et al. 1992), key players in epigenetic regulation. *XIST* is of particular interest for its role in X chromosome inactivation (XCI) and it is located in the region named X inactivation centre (XIC) (Fig. 1).

This multistep process was discovered in 1961 by Mary Lyon (1961), where one X chromosome is randomly silenced in females in order to balance gene expression derived from the presence of two X chromosomes (Avner and Heard 2001). Recent work showed the involvement of XCI in the phenotype of diseases caused by aneuploidies such as Klinefelter syndrome (47,XXY), where an extra

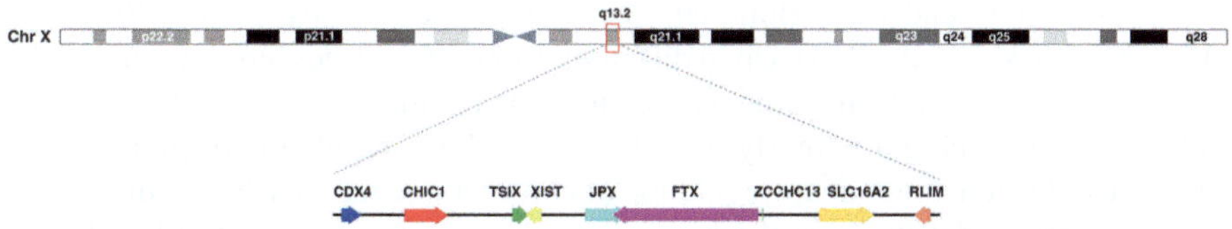

Fig. 1 Location of *XIST* in the X chromosome in the X inactivation centre. Arrows indicate genes and their direction of transcription

X chromosome mediates unbalanced gene expression genome-wide (Belling et al. 2017). Moreover, an emerging role for XCI has also been shown in several cancer types (Chaligne and Heard 2014) and recent works underlined the importance of sex bias of XCI in human tissues and development of cancer (Dunford et al. 2017; Tukiainen et al. 2017).

In this book chapter, we discuss key concepts regarding the involvement of ncRNAs in the regulation of XCI and related diseases with particular focus on genetic diseases and cancer.

2 X Chromosome Inactivation

XCI evolved as a strategy for gene dosage compensation in mammals where, in females, one X chromosome is inactivated. Both X and Y chromosomes originated from autosomes, but X retained more than 95% of the ancestral genes while Y only 2% (Mueller et al. 2013; Soh et al. 2014), generating the gene dosage imbalance between males and females. Two specific regions named pseudoautosomal region 1 (PAR1) and 2 (PAR2) have remained highly homologue (98–100%) between the X and Y. These regions are located at the terminal ends of the chromosomes and escape XCI (Raudsepp et al. 2012).

XCI consists of imprinted inactivation of the paternal X chromosome (Xp) (Kay et al. 1993; Takagi et al. 1978) and then reactivation of Xp during the formation of the blastocyst. At this point, random XCI of Xp or the maternal X chromosome (Xm) is observed (Okamoto et al. 2005) and this status is inherited for the entire process of cell division. During XCI, *XIST* forms a "coat" together with histones on the X chromosome to be inactivated (Avner and Heard 2001). This event represents the starting point of the gradual silencing of the complete chromosome and causes chromatin remodelling (Avner and Heard 2001) and consequently the formation of transcriptionally "silent" Barr bodies (Barr and Bertram 1949).

In the last years, additional knowledge regarding the regulation of XCI came from the use of murine embryonic stem cells (ESCs). The control of XCI during ESC differentiation involves several pluripotency genes. In undifferentiated female ESCs, *Xist* has been shown to be occupied and transcriptionally suppressed by *Oct4*, *Sox2* and *Nanog* whereas *Xite* and *Tsix* genes (negative regulators of XCI (Lee

2005)) are bound and transcriptionally activated by *Oct4*, *Sox2*, *Klf4*, *Zfp42* and *cMyc* (Minkovsky et al. 2013). Upon differentiation, XCI ensues through a multistep process of the initiation, silencing and maintenance of the silenced X. The initiation and onset of silencing are tightly linked with the down-regulation of pluripotency factors and the concomitant up-regulation of chromatin regulators that mediate XCI, such as *Satb1* and *PRC2*. Furthermore, the introduction of *Oct4*, *Sox2*, *Klf4* and *cMyc* into differentiated cells gives rise to induced pluripotent stem cells, which is accompanied by X chromosome reactivation in mouse (Okita and Yamanaka 2011).

XCI is not solely relevant in females; XCI also occur in males, where the single X chromosome is active in all somatic cells, but is inactivated during spermatogenesis and reactivated after fertilisation in female pre-implantation embryos (Epstein et al. 1978; Goto and Monk 1998; Kratzer and Gartler 1978; Monk and Harper 1978).

3 X Chromosome Inactivation in Human Tissues

For many years, it was unclear whether XCI is shared between cells and tissues but a recent study performed by the GTEx consortium opened a new perspective on the landscape of XCI across human tissues (Tukiainen et al. 2017). Tukiainen and colleagues analysed over 5500 transcriptomes from 449 individuals spanning 29 tissues from GTEx and 940 single-cell transcriptomes combined with genomic sequence data. They showed that incomplete XCI affects 23% of X-chromosomal genes, underlining the importance of taking into account these sex biases in gene expression as important variables for phenotypic diversity. This diversity might represent a crucial aspect to consider in the light of personalised medicine, since individuals can vary in their gene dosage due to different incompleteness of XCI and therefore having different treatment outcomes. Overall, Tukiainen and colleagues showed that the sex bias of XCI is specific to escape genes, but some genes had an unexpected trend of gene expression (Tukiainen et al. 2017). In particular eight genes not classified as full escapees followed a similar profile compared to well-established escape genes. One gene, a lncRNA called RP11-706O15.3, without an assigned XCI status and not characterised, showed a similar sex bias pattern to escape genes. RP11-706O15.3 is located between the escape gene *PRKX* and the variable escape gene *NLGN4X* (Tukiainen et al. 2017), and recently it was shown to be up-regulated in Klinefelter syndrome patients compared to healthy individuals (Belling et al. 2017), hypothesising that it potentially escapes XCI and/or might be involved in the XCI process itself. Future analyses of the sequence, structure and functions of this novel lncRNA could be useful in order to understand its role as escape gene.

4 X Chromosome Inactivation in Human Disease

Genomic instability leads to extra or fewer copies of (A) whole chromosomes (i.e. aneuploidies), or (B) chromosome parts (chromosomal abnormalities) (Thompson et al. 2010). This often occurs in cancer and causes increased DNA damage and replication stress (Passerini et al. 2016). However, aneuploidy is not only a hallmark of cancer, but also the basis of some genetic diseases such as Klinefelter (47,XXY) and Turner syndromes (45,X0), which usually occur as a result of incorrect segregation of chromosome pairs during meiosis (Theisen and Shaffer 2010).

Patients affected by aneuploidies have an increased risk of certain other diseases, referred to as comorbidities (Belling et al. 2017) and defined as diseases that co-occur on top of a primary disease of interest in an individual (Hu et al. 2016) (Fig. 2).

The occurrence and severity of comorbidities in patients affected by aneuploidies are potentially influenced by the aneuploidy per se, the inter-individual genetic variation and the presence of mosaicism (Bonomi et al. 2017). Moreover, XCI and varying incompleteness of XCI and genes that escape this process could explain the observed comorbidities and the variability of the phenotype observed in patients affected by X chromosome abnormalities (Belling et al. 2017; Bonomi et al. 2017).

The advent of novel data and computational approaches allows researchers to extract comorbidities population-wide by analysing National Patient Registries (NPRs) (Belling et al. 2017; Hu et al. 2016; Jensen et al. 2014). The study of comorbidities extracted from the Danish NPR in a systems biology framework, integrating several data types such as gene expression and protein–protein interaction networks, led to increased understanding of genes and pathways involved in comorbidities of patients affected by Klinefelter syndrome (Belling et al. 2017). *XIST* is up-regulated

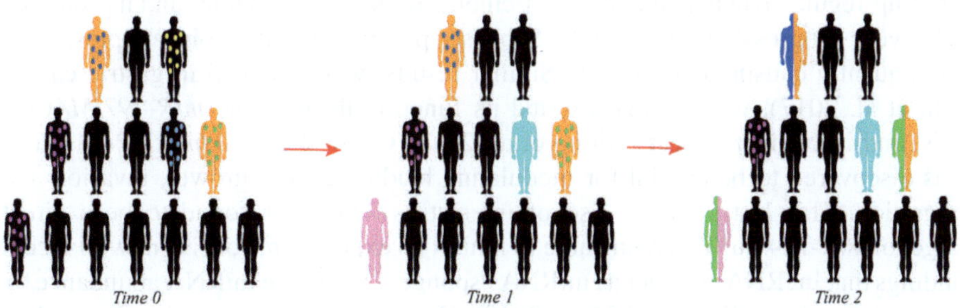

Fig. 2 Graphical representation of comorbidities. Coloured dots in body shapes indicate specific risk factors for diseases, including genetic background and environmental factors. Single-coloured body shape (e.g. pink) indicates one patient with one disease, while multicoloured body shapes (e.g. pink and green) indicate disease comorbidities. At time 0 patients affected by chromosomal abnormality (in orange) can have the same or different risk factors. These patients can develop additional diseases based on these risk factors and these additional diseases can be significantly associated with the initial disease (the chromosomal abnormality) compared to the background population

in Klinefelter patients pointing to the fact that the extra X chromosome is silenced presumably in a similar manner as in females (Belling et al. 2017). Moreover, several genes located in PAR1 that escape the inactivation are aberrantly expressed in Klinefelter patients compared to healthy individuals (Belling et al. 2017).

Genes escaping XCI seem to be crucial to understand the involvement of XCI in human diseases. Surely, the importance of having two X chromosomes for the normal development in females is clear, but less is known about the impact of having only one X chromosome in patients affected by Turner syndrome (Goto and Monk 1998; Zinn et al. 1993). In a recent article, Zhang and colleagues showed the utility of using "X-chromosome inactivation specific differentially methylated CpG sites" (XIDMSs) that are highly methylated in inactive X chromosomes and unmethylated in active X chromosomes for the early detection of Turner syndrome patients (Zhang et al. 2017). This potential marker could help to limit the development of Turner syndrome comorbidities and improve the treatment of these patients.

5 Xist, miRNAs and Cancer

An increasing evidence points to the involvement of *XIST* in diseases including different cancer types.

In the last years *XIST* has been proposed as a biomarker for early detection of gastric cancer (Lu et al. 2017) and non-small cell lung cancer (Tantai et al. 2015) and to predict prognosis in colorectal cancer (Chen et al. 2017). Other studies shed light on mechanisms of action of *XIST* in cancer and of particular interest is the crosstalk between *XIST*, miRNAs and protein-coding genes. Mo and colleagues discovered that *XIST* is involved in hepatocellular carcinoma cell proliferation where it exerts its function through the *miR-139-5p/PDK1* axis (Mo et al. 2017). They found that *XIST* up-regulated in hepatocellular carcinoma tissues and cell lines and it promoted cell cycle progression from G1 to S phase, protecting cells from apoptosis and consequently causing cell growth. Similar results were obtained in gastric cancer (Ma et al. 2017), where *XIST* exerted its function through the *miR-497/MACC1* axis, and in bladder cancer (Xiong et al. 2017) where the *XIST/miR-124/AR* axis was discovered to be crucial for modulating bladder cancer growth, invasion and migration. This last study is also of interest as *XIST* was found to be a direct target of *miR-124* but also being able to inhibit its expression, consistent with recent findings that lncRNAs can act as miRNA "sponges" to reduce miRNA abundance by sequestering them (Ebert and Sharp 2010). However, *XIST* is considered a nuclear lncRNA (Cerase et al. 2015), meaning that probably miRNAs are bound inside the nucleus. Even though miRNAs are usually cytoplasmic, some studies showed that they can cross the nuclear membrane (Hwang et al. 2007). A well-known example is human *miR-29b* that contains additional sequence elements that control its post-transcriptional behaviour (Hwang et al. 2007), and it is predominantly localised in the nucleus, in contrast to many other miRNAs (Hwang et al. 2007).

Future studies are needed in order to elucidate the crosstalk between *XIST* and miRNAs, and eventually other RNA classes.

6 X Chromosome Inactivation and Cancer Sex Bias

In the previous paragraphs, we showed the importance of XCI escape in the potential development of diseases (Belling et al. 2017) and also the variability of this phenomenon in different tissues, which contributes to defining a sex bias in gene expression (Tukiainen et al. 2017). A recent work points out an additional important aspect for escape genes defined as "Escape from X-Inactivation Tumor Suppressor" (EXITS) genes (Dunford et al. 2017). By comparing somatic alterations from >4100 cancers across 21 tumour types, Dunford and colleagues hypothesised that EXITS genes would protect females from developing different cancer types and could explain the male predominance in cancer incidence ($\geq$2:1 male predominance for some individual cancer types based on US data (Edgren et al. 2012)) that was considered largely unexplained (Cook et al. 2009; Dunford et al. 2017). The idea behind this hypothesis is that mutations in tumour-suppressor genes that escape XCI could consist of a significant fraction of excess male cancers because males would require only a single deleterious mutation (since only one X is present in males) while females would require two. Moreover, an alternative situation in males considers the presence of mutations in genes of X and Y chromosomes located in PARs. Tumours with mutations in those genes have an increased probability to occur in males who also have somatic loss of chromosome Y, a phenomenon that can be explained with the age and lifestyle of patients (Dunford et al. 2017).

7 Conclusions

In this book chapter, we presented recent findings in the field of XCI and related molecular mechanisms and diseases. Recently, *XIST* has been mentioned as a candidate prognostic biomarker for cancer patients and a potential target for new therapies. Additional studies are needed to improve our understanding of the role of *XIST* to promote cancer cell growth, invasion, progression and metastasis. Future studies are needed to discover additional molecular mechanisms behind *XIST* and its role in chromosomal abnormalities and associated comorbidities.

Recent advance in manipulation of single genes led to testing the concept that gene imbalance due to an extra chromosome can be corrected by manipulating *XIST* (Jiang et al. 2013). In 2013, Jiang and colleagues (2013) were able to use genome editing with zinc finger nucleases introducing an inducible *XIST* transgene into the *DYRK1A* locus on chromosome 21 in Down syndrome pluripotent stem cells. Surprisingly, *XIST* was able to act like in the traditional XCI process, causing stable heterochromatin modifications, chromosome-wide transcriptional silencing and DNA methylation and forming a chromosome 21 Barr body. This amazing result can lead to future interventions based on the new concept of "chromosome therapy" through the use of RNA technology.

Acknowledgements We would like to thank the Novo Nordisk Foundation for supporting our research (grant agreement NNF14CC0001).

References

Avner P, Heard E (2001) X-chromosome inactivation: counting, choice and initiation. Nat Rev Genet 2:59–67

Barr ML, Bertram EG (1949) A morphological distinction between neurones of the male and female, and the behaviour of the nucleolar satellite during accelerated nucleoprotein synthesis. Nature 163:676

Bartel DP (2004) MicroRNAs: genomics, biogenesis, mechanism, and function. Cell 116:281–297

Bartel DP (2009) MicroRNAs: target recognition and regulatory functions. Cell 136:215–233

Belling K, Russo F, Jensen AB et al (2017) Klinefelter syndrome comorbidities linked to increased X chromosome gene dosage and altered protein interactome activity. Hum Mol Genet 26:1219–1229

Bonomi M, Rochira V, Pasquali D, Balercia G, Jannini EA et al (2017) Klinefelter syndrome (KS): genetics, clinical phenotype and hypogonadism. J Endocrinol Invest 40:123–134

Brannan CI, Dees EC, Ingram RS, Tilghman SM (1990) The product of the H19 gene may function as an RNA. Mol Cell Biol 10:28–36

Brockdorff N, Ashworth A, Kay GF et al (1992) The product of the mouse Xist gene is a 15 kb inactive X-specific transcript containing no conserved ORF and located in the nucleus. Cell 71:515–526

Brown CJ, Hendrich BD, Rupert JL, Lafreniere RG, Xing Y et al (1992) The human XIST gene: analysis of a 17 kb inactive X-specific RNA that contains conserved repeats and is highly localized within the nucleus. Cell 71:527–542

Calin GA, Dumitru CD, Shimizu M et al (2002) Frequent deletions and down-regulation of micro-RNA genes miR15 and miR16 at 13q14 in chronic lymphocytic leukemia. Proc Natl Acad Sci USA 99:15524–15529

Calin GA, Sevignani C, Dumitru CD et al (2004) Human microRNA genes are frequently located at fragile sites and genomic regions involved in cancers. Proc Natl Acad Sci USA 101:2999–3004

Calin GA, Ferracin M, Cimmino A et al (2005) A microRNA signature associated with prognosis and progression in chronic lymphocytic leukemia. N Engl J Med 353:1793–1801

Cerase A, Pintacuda G, Tattermusch A, Avner P (2015) Xist localization and function: new insights from multiple levels. Genome Biol 16:166

Chaligne R, Heard E (2014) X-chromosome inactivation in development and cancer. FEBS Lett 588:2514–2522

Chen DL, Chen LZ, Lu YX et al (2017) Long noncoding RNA XIST expedites metastasis and modulates epithelial-mesenchymal transition in colorectal cancer. Cell Death Dis 8:e3011

Cook MB, Dawsey SM, Freedman ND et al (2009) Sex disparities in cancer incidence by period and age. Cancer Epidemiol Biomark Prev 18:1174–1182

Dunford A, Weinstock DM, Savova V et al (2017) Tumor-suppressor genes that escape from X-inactivation contribute to cancer sex bias. Nat Genet 49:10–16

Ebert MS, Sharp PA (2010) Emerging roles for natural microRNA sponges. Curr Biol 20:R858–R861

Edgren G, Liang L, Adami HO, Chang ET (2012) Enigmatic sex disparities in cancer incidence. Eur J Epidemiol 27:187–196

Epstein CJ, Smith S, Travis B, Tucker G (1978) Both X chromosomes function before visible X-chromosome inactivation in female mouse embryos. Nature 274:500–503

Filipowicz W, Bhattacharyya SN, Sonenberg N (2008) Mechanisms of post-transcriptional regulation by microRNAs: are the answers in sight? Nat Rev Genet 9:102–114

Goto T, Monk M (1998) Regulation of X-chromosome inactivation in development in mice and humans. Microbiol Mol Biol Rev 62:362–378

Hu JX, Thomas CE, Brunak S (2016) Network biology concepts in complex disease comorbidities. Nat Rev Genet 17:615–629

Hwang HW, Wentzel EA, Mendell JT (2007) A hexanucleotide element directs microRNA nuclear import. Science 315:97–100

Iorio MV, Ferracin M, Liu CG et al (2005) MicroRNA gene expression deregulation in human breast cancer. Cancer Res 65:7065–7070

Jensen AB, Moseley PL, Oprea TI et al (2014) Temporal disease trajectories condensed from population-wide registry data covering 6.2 million patients. Nat Commun 5:4022

Jiang J, Jing Y, Cost GJ et al (2013) Translating dosage compensation to trisomy 21. Nature 500:296–300

Kay GF, Penny GD, Patel D, Ashworth A, Brockdorff N et al (1993) Expression of Xist during mouse development suggests a role in the initiation of X chromosome inactivation. Cell 72:171–182

Kratzer PG, Gartler SM (1978) HGPRT activity changes in preimplantation mouse embryos. Nature 274:503–504

Lagana A, Russo F, Sismeiro C, Giugno R, Pulvirenti A et al (2010) Variability in the incidence of miRNAs and genes in fragile sites and the role of repeats and CpG islands in the distribution of genetic material. PLoS One 5:e11166

Lee JT (2005) Regulation of X-chromosome counting by Tsix and Xite sequences. Science 309:768–771

Lee RC, Feinbaum RL, Ambros V (1993) The *C. elegans* heterochronic gene lin-4 encodes small RNAs with antisense complementarity to lin-14. Cell 75:843–854

Lu Q, Yu T, Ou X, Cao D, Xie T, Chen X (2017) Potential lncRNA diagnostic biomarkers for early gastric cancer. Mol Med Rep 16:9545–9552

Lyon MF (1961) Gene action in the X-chromosome of the mouse (*Mus musculus* L.). Nature 190:372–373

Ma L, Zhou Y, Luo X, Gao H, Deng X et al (2017) Long non-coding RNA XIST promotes cell growth and invasion through regulating miR-497/MACC1 axis in gastric cancer. Oncotarget 8:4125–4135

Minkovsky A, Barakat TS, Sellami N, Chin MH, Gunhanlar N et al (2013) The pluripotency factor-bound intron 1 of Xist is dispensable for X chromosome inactivation and reactivation in vitro and in vivo. Cell Rep 3:905–918

Mo Y, Lu Y, Wang P et al (2017) Long non-coding RNA XIST promotes cell growth by regulating miR-139-5p/PDK1/AKT axis in hepatocellular carcinoma. Tumour Biol 39:1010428317690999

Monk M, Harper M (1978) X-chromosome activity in preimplantation mouse embryos from XX and XO mothers. J Embryol Exp Morphol 46:53–64

Mueller JL, Skaletsky H, Brown LG et al (2013) Independent specialization of the human and mouse X chromosomes for the male germ line. Nat Genet 45:1083–1087

Okamoto I, Arnaud D, Le Baccon P, Otte AP, Disteche CM et al (2005) Evidence for de novo imprinted X-chromosome inactivation independent of meiotic inactivation in mice. Nature 438:369–373

Okita K, Yamanaka S (2011) Induced pluripotent stem cells: opportunities and challenges. Philos Trans R Soc Lond B Biol Sci 366:2198–2207

Passerini V, Ozeri-Galai E, de Pagter MS et al (2016) The presence of extra chromosomes leads to genomic instability. Nat Commun 7:10754

Pfeffer SR, Yang CH, Pfeffer LM (2015) The role of miR-21 in cancer. Drug Dev Res 76:270–277

Raudsepp T, Das PJ, Avila F, Chowdhary BP (2012) The pseudoautosomal region and sex chromosome aneuploidies in domestic species. Sex Dev 6:72–83

Soh YQ, Alföldi J, Pyntikova T et al (2014) Sequencing the mouse Y chromosome reveals convergent gene acquisition and amplification on both sex chromosomes. Cell 159:800–813

Song SJ, Poliseno L, Song MS et al (2013) MicroRNA-antagonism regulates breast cancer stemness and metastasis via TET-family-dependent chromatin remodeling. Cell 154:311–324

Takagi N, Wake N, Sasaki M (1978) Cytologic evidence for preferential inactivation of the paternally derived X chromosome in XX mouse blastocysts. Cytogenet Cell Genet 20:240–248

Tantai J, Hu D, Yang Y, Geng J (2015) Combined identification of long non-coding RNA XIST and HIF1A-AS1 in serum as an effective screening for non-small cell lung cancer. Int J Clin Exp Pathol 8:7887–7895

Theisen A, Shaffer LG (2010) Disorders caused by chromosome abnormalities. Appl Clin Genet 3:159–174

Thompson SL, Bakhoum SF, Compton DA (2010) Mechanisms of chromosomal instability. Curr Biol 20:R285–R295

Tukiainen T, Villani AC, Yen A et al (2017) Landscape of X chromosome inactivation across human tissues. Nature 550:244–248

Xiong Y, Wang L, Li Y, Chen M, He W et al (2017) The long non-coding RNA XIST interacted with MiR-124 to modulate bladder cancer growth, invasion and migration by targeting androgen receptor (AR). Cell Physiol Biochem 43:405–418

Zhang B, Pan X, Cobb GP, Anderson TA (2007) MicroRNAs as oncogenes and tumor suppressors. Dev Biol 302:1–12

Zhang Q, Guo X, Tian T et al (2017) Detection of turner syndrome using X-chromosome inactivation specific differentially methylated CpG sites: a pilot study. Clin Chim Acta 468:174–179

Zinn AR, Page DC, Fisher EM (1993) Turner syndrome: the case of the missing sex chromosome. Trends Genet 9:90–93

Novel Insights of the Gene Translational Dynamic and Complex Revealed by Ribosome Profiling

Zhe Wang and Zhenglong Gu

Contents

Z. Wang (✉)
Division of Infectious Diseases, Weill Medical College of Cornell University, New York, NY, USA
e-mail: zhw2001@med.cornell.edu

Z. Gu
Division of Nutritional Sciences, Cornell University, Ithaca, NY, USA

N. Rajewsky et al. (eds.), *Systems Biology*, RNA Technologies,
https://doi.org/10.1007/978-3-319-92967-5_12

Abstract Biology research has entered into the big data era. Systems biology approaches, therefore, have become essential tools to elucidate the whole landscape of how cells separate, grow, and resist different stresses. In 2009, a novel RNA technology, termed ribosome profiling, was invented by Dr. Jonathan Weissman Lab from UCSF. Ribosome profiling (Ribo-Seq) is a powerful tool which can provide the most direct readout of the intracellular translation state of a protein including information on the location of translation start/stop sites, ribosome distribution pattern, and even the moving rate of the translating ribosome, at the whole-genome scale and single-nucleotide resolution.

To date, many researchers including our lab have successfully applied ribosome profiling method for diverse purposes. We thus review in this chapter the underlying mechanism and recent advances as regards this fantastic tool. Firstly, we introduce the working mechanism, advantages, and study history of ribosome profiling. Secondly, we discuss the data analysis pipeline, also compare different statistical algorithms and data visualization software. Finally, we review the extensive applications of Ribo-seq, for example, identification of uORF, computation of global translation efficiency (TE), the study of the posttranscriptional regulatory role of RNA binding protein and others. We hope this chapter would be useful for interested systems biology researchers as well as RNA biologists.

Keywords Ribosome profiling · Ribo-Seq · RNA-Seq · Translation · Deep sequencing · mRNA · Ribosome footprint

1 Introduction

In the classic "central dogma of molecular biology," ribosome-operated mRNA translation is the final step. This process is so crucial and thus it is tightly regulated by the cell, by controlling the complex translational machinery which comprises hundreds of ribosomal subunits and a series of regulatory factors (RNA binding proteins and small RNAs). Previously, researchers have shown that dysregulation of this process will trigger specific cell defects, and sometimes even cause severe human disease. The recent advances in structural biology have significantly improved the understanding of the 3D structure of the ribosome. However, the regulatory mechanism, especially in a systematic way, remains largely unclear.

For a long while, genome-wide mRNA expression profiling has been utilized as a tool for examining the cellular physiological states and for investigating the regulatory mechanism of gene expression. Most of the abovesaid profiling was conducted by expression microarray or RNA-seq approaches. The abovesaid two assays can be used to quantify the steady abundances of the gene transcripts in cells. Until now, mRNA expression profiling has been extensively applied in almost every branch of biology and has already become the routine method for understanding the complex gene expression pattern. However, many researchers have indicated that endogenous protein synthesis rate is hard to be predicted by relying on mRNA abundance. In other words, transcript abundance is often poorly correlated with protein abundance in vivo, suggesting the critical role of posttranscriptional regulation.

The advance of mass spectrometry-based proteomics provides a straightforward approach for systematically measuring the protein level in vivo. However, it can only capture the steady protein abundance. Also, it is hard to reach the depth and breadth of coverage provided by the deep sequencing-based method. A novel deep sequencing-based technique, termed ribosomal profiling (Ribo-seq), was invented to provide a way for quantitating translatome at both genome-wide scale and single-nucleotide resolution (Ingolia et al. 2009). The biochemical mechanism of Ribo-seq initially stems from the long-term observation that the moving ribosome on a transcript can efficiently cover about 30 nucleotide-long mRNA fragments, thus avoiding digestion by nuclease. The abovesaid ribosome-protected mRNA fragments, also called ribosome footprints (RFPs), can be then sequenced, to provide a landscape comprising the accurate record of the position of the ribosome at the moment when the translatome is quenched. The quantitation of RFPs on a specific mRNA strand, therefore, indicates the proxy for the efficiency of protein synthesis. At the same time, the localization of RFPs allows us to globally map the initiation and termination sites of each translation product and thus Ribo-seq also provides a unique way to explore the whole ORF spectrum in a living cell.

In this chapter, we discuss the biochemical mechanism of Ribo-seq method, its unique properties and strengths, and the varied versions of it. Secondly, since Ribo-seq is a big data-based systems biology approach, we also review the relative data analysis pipelines, available software, and public databases. We hope this chapter would be useful for interested systems biology researchers as well as RNA biologists.

2 The Experimental Design, Readout Type, and Data Analysis Pipeline

For instantaneously snapshotting the translatome, cycloheximide (CHX, for eukaryotes) or chloramphenicol (CHL, for prokaryotes) are typically used to treat ribosomes before sample collection (Ingolia et al. 2009; Oh et al. 2011). Both CHX and CHL can bind the large subunit of ribosome and then cause the inhibition of translational elongation. Afterward, the "frozen" ribosome–mRNA complexes are separated by ultracentrifugation based on the molecular weights. Nuclease treatment is then applied to remove all of the unprotected mRNA regions, so eventually, around 30 nucleotide-long shelled RFPs can be obtained and purified. They are then reverse-transcribed, barcoded, and a standard DNA library constructed for deep sequencing. With the recent revolution in sequencing power, researchers now can deeply profile all translating ribosomes. The budding yeast genome, for example, encodes about 6000 proteins with an average mRNA coding region of approximately

300 nucleotides in length. Nuclease digestion of ribosome–mRNA complexes will yield more than 10 million RFPs per sample, which will be decoded accurately in a state-of-the-art deep sequencing platform, such as the HiSeq system of Illumina.

The similarities between ribosome profiling and RNA-seq for analyzing gene expression allow for the use of current bioinformatics tools with ribosome profiling data. For example, we published a study which aimed to quantitate the regulatory differences between a laboratory *S. cerevisiae* strain BY4742 and a pathogenic *S. cerevisiae* strain YJM789 (Sun et al. 2016). After obtaining the Ribo-seq and RNA-seq data sets, the raw reads from RFP and mRNA are cleaned by removing the adaptors/barcodes and filtered by quality using custom Perl scripts. The cleaned mRNA reads are mapped to the corresponding genome using SOAP, with no more than two mismatches allowed. The uniquely aligned reads from both mRNA and RFP are assigned to genomic features, such as coding sequence (CDS), intron, 5′-UTR, or 3′-UTR, based on the position of the 5′-most nucleotide. To quantify the mRNA and RFP reads, we first obtain the base level read coverage of all ortholog genes from both species. A minimum of 50 reads mapping to the CDS region is required to retain the gene for further transcriptional and translational analysis. For testing the statistical significance of mRNA and RFP divergence, we sought to generate a sequence-specific null distribution accounting for the influence of sequencing bias and gene length for both mRNA and RFP. We let L_B and L_Y denote, respectively, the mappable lengths of BY4742 and YJM789 orthologs, and let $\pi_B = [\pi_B(A), \pi_B(T), \pi_B(G), \pi_B(C)]$ and $\pi_Y = [\pi_Y(A), \pi_Y(T), \pi_Y(G), \pi_Y(C)]$ denote the corresponding marginal nucleotide frequencies of each ortholog. We began to form the null ortholog pairs by resampling, with replacement, the base level counts from either BY4742 or YJM789 ortholog using the same length and nucleotide frequency from the orthologs (L_B, π_B or L_Y, π_Y). The resampling was repeated 10,000 times and the null distribution of the log2-transformed ratio of BY4742 to YJM789 was derived from two resamplings of both replicates. We compared the observed base level counts log2-ratio of BY4742 to YJM789 with the underlying null distributions to obtain two-sided *P* values.

3 Optimization and Varied Versions of Ribosome Profiling

3.1 Use Harringtonine to Define Translation Start Sites and Monitor Translational Kinetics

In a standard ribosome profiling protocol, cycloheximide (for eukaryotes) or chloramphenicol (for prokaryotes) are normally employed as elongation inhibitors. Weissman and coworkers pioneered the use of drug harringtonine for ribosome profiling (Ingolia et al. 2011). Harringtonine is a cephalotaxine alkaloid that inhibits protein synthesis at low micromolar concentrations. They found that harringtonine treatment causes a profound accumulation of ribosomes at the sites of translation initiation in mouse embryonic stem cells. They proposed that this pattern occurs because harringtonine binds to free 60S subunits but not those that are joined into an 80S ribosome. Therefore, the elongating ribosomes are resistant to harringtonine, but a 60S subunit-bound drug will form an 80S at the initiation site without moving forward. They then used a support vector machine (SVM)-based machine learning strategy to systematically analyze harringtonine-treated ribosome profiling and eventually identified 13,454 candidate translation initiation sites within ~5000 transcripts that actively expressed in the mESCs. In the same study, they also developed a ribosome profiling-based pulse-chase strategy for determining rates of translation elongation, by combined application of harringtonine and cycloheximide. The design is as follows: first treat the cell with harringtonine to stop the translation initiation, then leave a short time for run-off elongation before applying cycloheximide to halt translation by all active ribosomes, which will eventually produce a series of snapshots that could be assembled into a moving picture of translation in vivo. Interestingly, they found that the kinetics of elongation are significantly consistent and are independent of mRNA length, protein abundance, and codon usage. The last observation is especially surprising since it is usually assumed that codons corresponding to low-abundance tRNAs are decoded more slowly than those read by richly abundant tRNAs.

3.2 *Optimization of Nuclease Digestion Process*

Gene translation is a quite fast process (~6 codons/s) and sensitive to a series of biochemical factors, such as temperature, pH, and ionic strength. Therefore, ribosome profiling has to be conducted carefully to avoid potential bias. A critical issue in ribosome profiling is nuclease treatment of ribosome–mRNA complexes since there is a dual requirement for maintaining the stability of ribosomal particles and converting polysomes to monosomes entirely. Gladyshev and coworkers systematically compared the efficacy of ribonucleases I, A, S7, and T1, in various species including *E. coli*, *S. cerevisiae*, *C. elegans*, *D. melanogaster*, and *M. musculus* (Gerashchenko and Gladyshev 2017). Their result shows that although RNase I is the nuclease of choice in the several published cases of yeast, it is not necessarily the case with other species. S7 and T1 are less aggressive to the ribosome and thus provide a valuable substitution. In the same year, Buskirk and coworkers published their research on the endonuclease RelE (Hwang and Buskirk 2017). The in vivo data proves that RelE predominantly cleaves mRNA after the second nucleotide in empty A sites. They then explored the possibility of using RelE to replace MNase in the ribosome profiling preparation procedure. The result is promising since RelE can yield precise 3′-ends and reveals a clear reading frame which is hard to acquire in bacteria.

3.3 Development of Simplified Protocols for Ribosome Profiling

Until now, ribosome profiling is a time-consuming and expensive method. To combat this challenge, several groups have been engaged in developing simplified ribosome profiling protocols. Palsson and coworkers published a streamlined Ribo-seq protocol for characterization of microorganisms (Latif et al. 2015). Without using sucrose gradient fraction separation process, they directly treat with MNase after repeated freeze–thaw lysis to *E. coli* cells. Monosomes are then recovered using a size exclusion spin column analogous to those used for ARTseq (Epicenter, Madison, WI). Next, the recovered ribosomes are treated with Qiazol (Qiagen, Germantown, MD) to recover RNA footprints. Footprints are subsequently isolated using kit-based purification assays and exclude the gel purification steps in the previous protocol. Therefore, they reduced the total experimental time from 7–8 days to 3–4 days. From one single simple, they successfully obtained ~1.3 million reads, and the processed data sets show a strong linear correlation between biological replicates analogous to those produced by the established protocol. With a similar aim, Nicchitta's group also published a simple and inexpensive pipeline for preparing human lymphocyte samples (Reid et al. 2015). In this protocol, they skip the ribosome purification step as well and directly use MNase to digest polysomes. They estimate the cost of this simplified method from cell lysis to library completion at about $50 per sample, and the protocol can reasonably be finished within 2 days.

3.4 The Proximity-Specific Ribosome Profiling

The localization of protein synthesis plays a key role in multiple processes, including development, cellular motility, and synaptic plasticity. However, there are rare molecular approaches suitable to accurately quantitate this spatial information. In 2014, Weissman's group developed a novel tool termed proximity-specific ribosome profiling which allows for precise characterization of localized protein synthesis (Jan et al. 2014). The underlying mechanism of this superior method is to biotinylate ribosomes in vivo in the subcellular location dependent way. They performed this experiment in five steps: (1) inserting a non-perturbing ribosome tag (RPL16 and RPS2 in yeast) composing of a tobacco etch virus (TEV) protease-cleavable AviTag; (2) genetic targeting of a biotin ligase (BirA) to a subcellular location of interest, such as endoplasmic reticulum (ER) membrane; (3) controlling the intracellular ribosome biotinylation by giving brief biotin pulses; (4) preventing the post-lysis biotinylation; and (5) separating the biotinylated ribosomes and purifying through TEV cleavage. Weissman's group first used this strategy to discover the principles of ER cotranslational translocation in budding yeast *S. cerevisiae* and human embryonic kidney-293 (HEK-293) cell line as well. They discovered that the vast majority of secretory proteins conduct cotranslational targeting in vivo, independent of their dependence on signal recognition particle (SRP), and they then proposed a model wherein a pioneering round of translation is in charge of recruiting the ribosome-nascent chain (RNC) to the ER surface, after which the message remains tethered to the ER by ongoing translation by downstream ribosomes. Their result also reveals that the ER-associated ribosomes are highly dynamic since the bimodal enrichment distribution is rapidly collapsed into a single population on the order of just minutes. In another study, they used proximity-specific ribosome profiling to study the targeting and plasticity of mitochondrial proteins in *S. cerevisiae* (Williams et al. 2014). To mark ribosomes selectively on the mitochondrial surface, they inserted BirA to the C terminus of *OM45*, a major component of the outer membrane (OM). After a 2-min pulse of biotin, of the enriched genes, 87% were annotated as mitochondrial. A significant subgroup of expressed mitop2 genes was cotranslationally targeted. After comparing the mitochondrial and ER localization, they demonstrated that the majority of proteins target to a specific organelle. Except for fumarate reductase Osm1, they are known to localize into mitochondria, but a conserved ER isoform of Osm1 was found in this study. This dual localization mechanism is enabled by alternative translation initiation sites encoding distinct targeting signals.

4 Specific Software and Databases for Ribosome Profiling

Along with the extensive application of ribosome profiling in different fronts, much specific software for analyzing and visualizing Ribo-seq results have been developed worldwide. Ohler's group developed RiboTaper (https://ohlerlab.mdc-berlin.de/software/) which majorly aims to identify translated regions by the characteristic three-nucleotide periodicity of Ribo-seq data (Calviello et al. 2016). From the dataset of HEK293 cells, they found the active translation that covered ORF annotation for over 11,000 genes. They also reported several hundreds of uORFs and ORFs in annotated noncoding genes (ncORFs). In 2016, Milles's group published a software named SPECtre (https://github.com/mills-lab/spectre) (Chun et al. 2016). SPECtre is a spectral coherence-based classifier and shows a marked improvement in accuracy for identifying active translation and exhibits overall high accuracy at a low false discovery rate (FDR). Barbry's group developed RiboProfiling (https://www.bioconductor.org/packages/release/bioc/html/RiboProfiling.htm) (Popa et al. 2016), which is a Bioconductor package and provides a full pipeline to cover all key steps for Ribo-seq analysis. Brierley's group developed the riboSeqR R package (http://www.bioconductor.org/packages/riboSeqR) (Chung et al. 2015). This package parses data aligned to a transcriptome, providing frame-calling and plotting functions. Baranov's group developed GWIPS-viz browser (http://gwips.ucc.ie) (Michel et al. 2014), which provides access to the genomic alignments of Ribo-seq data and corresponding RNA-seq data along with relevant annotation tracks— allowing for the cross-species comparison of orthologous genes. Based on GWIPS-viz, the same group then published the RiboGalaxy (http://ribogalaxy.ucc.ie) which is a freely available Galaxy-based web server (Michel et al. 2016). RiboGalaxy is a user-friendly and also powerful tool for Ribo-seq analysis. Relying on it, people can compare their ribosome profiles to existing ribo-seq tracks from published studies. Also, people can evaluate the quality of their ribo-seq data, determine the strength of the three-periodicity signal, and generate meta-gene ribosome profiles. Another Galaxy toolbox is RiboTools (https://testtoolshed.g2.bx.psu.edu/view/rlegendre/ribo_tools) (Legendre et al. 2015) which is designed for the accurate analysis of k-mer length distribution, translation ambiguities, and translation read-through events. In 2017, Rätsch's group published a statistical framework and an analysis tool, RiboDiff (http://bioweb.me/ribodiff), to detect genes with changes in translation efficiency across experimental treatments (Zhong et al. 2017).

The more the number of ribosome profiling datasets decoded, the more the realization of importance of sORFs. Many follow-up studies have shown that some micropeptides could be translated from these sORFs, exhibiting critical functional roles in cells. Therefore, comprehensive collection and analysis of these sORFs have become a need in this field. In 2015, Menschaert's group published a database termed sORFs.org (http://www.sorfs.org) for sORFs identified using ribosome profiling (Olexiouk et al. 2016). Currently, sORFs.org harbors 263,354 sORFs, originating from three cells lines: HCT116 (human), E14_Mesc (mouse), and S2 (fruit fly). RPFdb (www.rpfdb.org) is another valuable database, developed by Xie's group (Xie et al. 2016). RPFdb is designed to comprehensively host, analyze, and visualize RPF data, and presently it contains 777 samples from 82 studies in eight species. RPFdb can be queried by keywords of studies or by genes. Meanwhile, it also provides a genome browser to query and visualize context-specific translated mRNAs.

5 Representative Applications of Ribosome Profiling in Different Fields

5.1 Application of Ribosome Profiling for Decoding the Functional Roles of Small RNA

MicroRNA (miRNA) is a small noncoding RNA molecule (containing about 22 nucleotides) found in plants, animals, and some viruses that functions in RNA silencing and posttranscriptional regulation of gene expression. miRNAs play the regulatory role by pairing to the target mRNAs to direct their repression—which could happen at translational level or mRNA level. In 2010, Bartel and Weissman's group combined Ribo-seq and RNA-seq to understand the relative contributions of the abovesaid two outcomes (Guo et al. 2010). They examined the impact of introducing miR-1 or miR-155 in HeLa cells, and the impact of knocking out *mir-223* in mouse neutrophil. The abovesaid miRNAs were selected because they have been reported to regulate thousands of proteins. The result shows that for both ectopic and endogenous regulatory interactions, only a small fraction of repression observed by Ribo-seq (11%–16%) was attributable to reduced translational efficiency. In other words, mRNA reduction consistently mirrored RPF reduction.

5.2 Application of Ribosome Profiling for Analyzing the Regulatory Roles of RNA Binding Protein

The concept of RNA regulons was proposed to specifically describe the observation that mRNA-binding proteins (RBPs) usually bind and orchestrate the fate of target mRNAs encoding functionally related proteins. It is increasingly clear that RNA regulons play an essential role in determining the stability, subcellular localization, and translation of their targets, and thus are vital for phenotypic outcomes and even disease states in various organisms including humans.

A study for understanding the functional roles of the CCHC-type zinc finger nucleic acid-binding protein (CNBP/ZNF9) has been published to describe its essential role in embryonic development of mammals (Benhalevy et al. 2017) by first using photoactivatable ribonucleoside-enhanced cross-linking and immunoprecipitation (PAR-CLIP) to identify ~8000 CNBP binding sites on over 4000 mRNAs. Furthermore, the G-rich elements are enriched among the CNBP's binding sites. By using Ribo-seq, they found that CNBP did not affect target mRNA abundance but rather increased their translational efficiency, which could be explained by the fact that CNBP improves translation by preventing G-quadruplex structure.

In the same year, Großhans's group reported the molecular mechanism of LIN41, a well-known RBP that regulates animal development (Aeschimann et al. 2017). Combing the RNA binding experiment and Ribo-seq assay, they found that LIN41 binds to the 3′-UTRs and regulate their degradation of *mab-10*, *mab-3*, and *dmd-3*. Meanwhile, LIN41 binds to the 5′-UTR and repress the translation of the A isoform of *lin-29*. This research thus presents an unusual case of an RBP exists position-dependent dual activity-which could emerge as a common feature shared by other RBPs.

5.3 Application of Ribosome Profiling for Exploring the Molecular Mechanism of Cancer

The mammalian target of rapamycin (mTOR) kinase is a crucial regulator of protein synthesis that links nutrient sensing to cancer. Earlier studies have shown that 4EBP1 and p70S6K1/2 are two major regulators of protein synthesis downstream of mTOR1, a subunit of mTOR complex. However, the detailed regulatory mechanism of mTOR remains incomplete. Ruggero's group first used Ribo-seq to discover the translational landscape mTOR signaling during the procedure of prostate cancer metastasis (Hsieh et al. 2012). In human prostate cells, mTOR is constitutively hyperactivated. By treating cells with PP242-anmTOR ATP site inhibitor, three primary downstream mTOR effectors 4EBP1, p70S6K1/2, and AKT are obviously inhibited. Ribo-seq result then shows that 144 target mRNAs selectively decreased at the translational level, with limited changes in transcription. Furthermore, they also developed a clinically relevant ATP site inhibitor of mTOR, INK128. Its selective decrease in the expression of YB1, MTA1, vimentin, and CD44 at the translation level, therefore, provides a therapeutic way for cancer initiation and progression.

Tumor growth is highly related to the availability of certain amino acids in the environment for protein synthesis. Relying on the hypothesis that cytosolic global occupancy of ribosome positions could be used as an index of alterations in the availability of amino acid for protein synthesis, a Ribo-seq-based protocol termed "diricore" (differential-ribosome-measurements-of-codon-reading) was developed. "Diricore" analyzes the subsequence and 5′-end of the RPFs. The results show diricore signals at asparagine codons and high levels of asparagine synthetase (ASNS) (Loayza-Puch et al. 2016). When they applied diricore to kidney cancer, it showed a clear signal representing the limitation of proline pool, which is then proven to be biochemically linked to PYCR1, a critical enzyme that catalyzes the last step in proline synthesis. In another study from the same group, they applied diricore approach and successfully found that epithelial breast cells respond rapidly to TGFβ1 and the specific restriction of leucine pool (Loayza-Puch et al. 2017).

Ribo-seq was also applied to an epidermis-specific in vivo model—a transgenic mouse of inducible SOX2, which is extensively expressed in oncogenic RAS-associated cancers (Sendoel et al. 2017). Surprisingly, when searched the 5′-UTR of the top 10% of efficiently translated uORFs in $SOX2^{+}$ epidermis, they discovered an obvious correlation between improved translation, increased length, and decreased minimum free energy of 5′-UTRs. Also, many of these SOX2-induced uORFs start from CUG codon rather than the conventional AUG codon of canonical ORFs; this unique pattern is consistent with another result that depletion of conventional eIF2 complexes has adverse effects on normal but not oncogenic growth.

5.4 Application of Ribosome Profiling for Studying Evolution

Understanding how gene regulation evolves is a crucial area in the current evolutionary field. Gene regulation occurs at various levels. However, despite the higher biofunctional importance of protein levels, our understanding of gene regulation is still primarily based on studies of mRNA levels. In contrast, our knowledge of how translational regulation evolves has lagged far behind.

Excitingly, Ribo-seq currently provides an excellent tool for tracing the global gene translation during evolution, thus allowing for the parallel comparison with the universal changed pattern in mRNA level. Recently, several groups have applied Ribo-seq and RNA-seq assays to closed yeast species and their interspecific hybrids, for quantitating the relative contribution of changes in mRNA level and gene translation to regulatory evolution (McManus et al. 2014; Artieri and Fraser 2014; Wang et al. 2015). Surprisingly, we consistently found that translation is much more conserved than transcription, mostly due to the buffering effect of translational regulation for the transcriptional divergence. In this study, we also found that the *trans* effects are more responsible than the *cis* effects for the discrepancy at both levels of gene regulation. For genes whose transcription or translation are affected by both the *trans* and *cis* factors (the CT genes), these factors usually function in the opposite directions (the compensating CT effect), indicating the stability of regulation for these genes. The results from our three groups have shown the underappreciated complexity of posttranscriptional regulatory divergence. These results thus demonstrate that surveys of various levels of gene regulation from different genetic backgrounds can enable a more comprehensive understanding of the gene regulation evolutionary modes in nature.

5.5 Application of Ribosome Profiling for Understanding the Physiology of Pathogenic Bacteria

Mycobacterium tuberculosis (Mtb), the causative agent of tuberculosis (TB), kills more humans than any other bacterium, yet humans remain its only major natural reservoir. However, until now functions of almost half of the Mtb genes are still unclear. The first ribosome profiling study for understanding this ancient pathogenic bacterium has been published in 2015 (Shell et al. 2015). Surprisingly, one-quarter of the mycobacterial transcripts are leaderless, lacking a 5′-UTR and Shine-Dalgarno ribosome-binding site. This result suggests that leaderless translation is a major feature of mycobacterial genomes, which is not common in many other bacteria, such as *E. coli*, but more like the way in some archaeal species and mitochondria. Leaderless transcripts are likely recognized by 70S ribosomes, rather than 30S subunits. These 70S ribosomes will be modified by stress-induced endoribonuclease, which cleaves the 3′ end of the 16S rRNA, thus removing the anti-Shine-Dalgarno sequence. Ribo-seq was also used to decode the genome of *Mycobacterium abscessus* (Mab), another deadly respiratory pathogenic non-tuberculous mycobacterial species (Miranda-CasoLuengo et al. 2016). They identified 126 new ribosomally protected ORFs and 80% of which are ≤ 50 amino acids in length. To test if these sORFs are likely to be coding, the fragment length organization similarity score (FLOSS) of the abovesaid sORFs is calculated to be compared with those of known coding regions within the genome. The analysis indicates that some Mab sORFs are potential to be translated. With a similar aim, a study was published to identify the small unannotated genes in *Salmonella*, an important foodborne pathogen (Baek et al. 2017). Based on Ribo-seq data they uncovered 130 uORFs. Of them, 98% are sORFs putatively encoding peptides/proteins ≤ 100 amino acids and some of them are uniquely expressed in the infection-relevant low $Mg^{2+}$and/or low pH condition.

5.6 Application of Ribosome Profiling for Investigating the Pathogen–Host Interaction

Many viruses take a common strategy called host shutoff to repress cellular mRNA translation but allow the efficient translation of own viral mRNA at the same time. Ribo-seq was used to discover the mechanisms that are being utilized by the Influenza A virus (IAV) to induce host shutoff (Bercovich-Kinori et al. 2016). The analysis shows that IAV genes transcripts are not preferentially translated during its infection, indicated by the fact that viral genes translation efficiencies mainly fall into the normal range of host gene translation. Instead, the host shutoff is driven by the viral dominance over the mRNA pool. For example, at 8hpi IAV mRNAs take over 53.8% of the translation activity in vivo as 57.3% of the mRNAs in the cells are viral. Through Go enrichment analysis they then found that many genes responsive to eIF2α phosphorylation are translationally induced after IAV infection. It is not entirely unexpected because that eIF2α phosphorylation is a stress response that will limit the overall protein synthesis rates but enhancing the translation of specific genes that take part in the adaptive stress response.

Unlike IAV and many other viruses that suppress cellular protein synthesis, host mRNA translation and polyribosome formation are stimulated by human cytomegalovirus (HCMV). Furthermore, these key protein synthesis factors, including eIF4A, eIF4E, eIF4F, eIF4G, and PABP1, increase in response to HCMV infection. Ian Mohr's group used polysome profiling to find that the translationally activated cellular mRNAs of HCMV encode proteins critical for DNA damage response, proliferation, ribosome biogenesis, chromatin organization, and so on (McKinney et al. 2014). At the same time, the host mRNAs repressed by HCMV include those involved in differentiation and acquired immune response. In an earlier study, Weissman and coworkers used Ribo-seq to identify the range of HCMV-translated ORFs and quantitate their temporal expression during its infecting human foreskin fibroblasts (HFFs). They eventually identified a total of 751 translated ORFs, corresponding to annotated 165–252 ORFs from previous genome sequencing data (Stern-Ginossar et al. 2012; Fields et al. 2015)

Bacteriophage lambda is one of the most extensively studied organisms and provides a valuable model to understand the interaction of a virus with its host. In 2013, by collaborating with Jeffrey Robert's group, we published a high-resolution view of bacteriophage lambda gene expression in the process of its lytic growth, by Ribo-seq (Liu et al. 2013a, b). Being consistent with several other ribosome profiling types of research, we determined numerous translated small proteins which might be relevant to lysogen and phage growth, acting as intercell signaling factors, toxins, and membrane components.

Legionella pneumophila is a gram-negative bacterium which is the causative agent of Legionnaires' disease. Previous studies have shown the *Legionella pneumophila* infection causes the global pathogen-induced block of host translation. However, a seemingly contradictory fact is that the host still can provoke the production of specific pro-inflammatory cytokines during the infection stage. A study was published to explain the abovesaid mystery (Barry et al. 2017). By using Ribo-seq, RNA-seq, and ribosome run-off assays, they found that mRNA superinduction, rather than selective mRNA translation, is the strategy by which host cells produce inflammatory cytokines when facing the global translation inhibition. To be successful, the magnitude of mRNA superinduction (1000-fold) has to exceed the magnitude of the block in protein synthesis (20-fold).

5.7 *Application of Ribosome Profiling for Studying the Molecular Mechanism of the Cell Cycle*

The cell cycle or cell-division cycle is the series of events that take place in a cell leading to its division and duplication of its DNA to produce two daughter cells. However, it has to be precisely programmed. Many fundamental research of cell cycle and cell division have been performed in budding and fission yeasts because of a number of their advantages: unicellularity, homologous to human, ease of genetic manipulation, and others. Pioneering microarray and RNA-seq studies provided a transcriptional landscape during yeast sexual reproduction progression but failed to catch many informative modulations, especially in extensive posttranscriptional regulation. Relying on Ribo-seq, Weissman's group first measured the global protein production through the budding yeast *S. cerevisiae* meiotic sporulation program (Brar et al. 2012). By using the traditional synchronization procedures and an estrogen-activatable derivative of the Ndt80 transcription factor guided assay, they successfully separated yeast cells to 25 meiotic time points and found pervasive translational controlling events in meiosis; thus, this work provided the molecular landscapes of the broad restructuring of meiotic cells.

In a research work that has related aims, McAdams's group monitored the translational dynamic in *Caulobacter* cell cycle control (Schrader et al. 2016). *Caulobacter* requires tightly temporal and spatial control of gene expression to finish an asymmetric cell division, yielding distinct daughter cells. The authors selected six key time points in a single cell cycle and performed Ribo-seq and RNA-seq to find the global characteristics of *Caulobacter* transcription and translation. The results show that the highly expressed proteins during cell cycle are typically coordinately controlled between translation and transcription, suggesting that the scheduling of translational regulation is organized by the same cyclical regulatory circuit.

5.8 Application of Ribosome Profiling for Engineering Bioethanol Producers

Needless to say, microbially produced ethanol or other aliphatic alcohols have been recognized as one of our critical sustainable energy sources. However, these alcohols usually are toxic which limits their mass production in bacteria or fungi. Dissecting the molecular mechanism of alcohol-imposed toxic effects and understanding how microbes evolve resistance, therefore, become essential for both fundamental research and application in bioenergy production.

Several vital mutations which determine the ethanol tolerance in *E. coli* (Haft et al. 2014) have been successfully identified. Interestingly, many of the abovesaid mutations are biochemically related to gene transcription and translation; for example, RpsQ is a ribosomal protein involved in decoding, MetJ is a feedback repressor of methionine biosynthesis, and Rho is a well-known transcription factor. With the hypothesis that ethanol may induce compensatory changes in the decoding center of the ribosome, they investigated the mechanism by multiple Omics techniques including Ribo-seq assay (Haft et al. 2014). The result first suggested that ethanol can induce toxic translational misreading, an effect which can be further strengthened by adding streptomycin, an antibiotic that can cause translational misreading. They then designed a Ribo-seq experiment to compare three typical physiological conditions. The result indicated that ribosomes were widely distributed across mRNAs before stress. During acute stress, the relative ribosomal occupancy near the $3'$ ends of genes decreased from $\sim$0.95 to $\sim$0.75, suggesting a widespread aberrant termination of translation. Further analysis shows that ethanol has weak effects on ribosome occupancy at most codons but significantly takes effect on the occupancy at nonstart AUG codons. Nonstart AUG occupancy dramatically increased during acute toxicity and only partially recovered during chronic toxicity, which may be ameliorated by the adaptive inactivation of the MetJ repressor of methionine biosynthesis genes. Together, relying on Ribo-seq and other Omics assays, the authors elucidated that ethanol-induced inhibition and uncoupling of mRNA and protein synthesis through direct effects on ribosomes and RNA polymerase conformations are significant contributors to ethanol.

5.9 Application of Ribosome Profiling for Investigating Plant Molecular Biology

To understand the molecular mechanism of translational regulation in photomorphogenic *Arabidopsis thaliana*, Ribo-seq was adopted to map the translatome in *Arabidopsis* etiolated seedlings in the dark and after light exposure (Liu et al. 2013a, b). The results show that genes involved in the organization and function of chloroplasts can be translationally enhanced by light. The uORG initiated by ATG but not CTG mediated translational repression of the downstream main open reading frame. Later on, the translational regulation under hypoxia stress in seedlings of *Arabidopsis thaliana* was investigated (Juntawong et al. 2014). When the seedlings lacked enough oxygen, the frequency of ribosomes at the start codon decreased, consistent with a widespread decline in translational initiation. Interestingly, the abundance of mRNA of hypoxia-upregulated genes increased in polysome complexes during the stress. However, the number of ribosomes per transcripts was not enhanced compared to normoxic conditions. Ribo-seq was also used to discover the phytohormone signal transduction pathway. A study unveiled a new translation-based branch of ethylene response in *Arabidopsis* (Merchante et al. 2015). By using Ribo-seq, they found that the signaling molecule *EIN2* and the nonsense-mediated decay proteins UPFs play a central role in an ethylene-induced translational regulatory mechanism, which eventually targets to another ethylene signaling component *EBF2*. These findings represent a typical example of gene-specific regulation of translation responding to a critical growth regulator. Most recently, a study optimized the buffer of RNase used in Ribo-seq (Hsu et al. 2016). This improvement offers a significantly improved footprint precision in *Arabidopsis thaliana*. This superresolution method can map over 90% of the footprints to the main reading frame, and therefore help to uncover many small ORFs in annotated noncoding RNAs and pseudogenes.

Except for *Arabidopsis*, a study reported the dynamic of chloroplast translation during chloroplast differentiation in maize (Zoschke et al. 2013; Chotewutmontri and Barkan 2016). The rate of protein production of most genes increases early in development and declines when the photosynthetic apparatus is mature. The differential gene expression in bundle sheath and mesophyll chloroplasts results primarily from differences in mRNA abundance, but the divergence in translational efficiency can keep amplifying the mRNA-level effects in some instances. They also demonstrated that ACG does not serve as a start codon in maize chloroplasts since editing of ACG to AUG at the *rpl2* start codon is essential for translation initiation.

6 Concluding Remarks and Perspective

As discussed above, ribosome profiling has proved to be a fantastic innovation with awesome power in systematically decoding translatome at single nucleotide resolution. Depending on the depth of current sequencing capacity, ribosome profiling has become a very sensitive method that allows for the measurement of relatively rates of translation events. The highly parallel sequencing readout of all ribosome positions yields more quantitative and detailed information than alternative methods, such as the pulsed label-based approach. Ribosome profiling also provides precise footprints of positional information, thus allowing for identifying a number of novel translational events: sORF, uORF, ncORF, ribosome pausing, translational initiation at non-AUG codons. The abovesaid information will provide the experimental evidence for revising the current in silico-based genome annotations. Ribosome profiling can also offer instantaneous translation efficiency measurements which describe the real-time cell decision-making process, and therefore has distinct biological significance at the steady-state translation level (Brar and Weissman 2015).

Much remains to be learned from the complex and dynamic ribosome profiles. For example, why do ribosomes pause at specific regions, and is there some unique sequence/motif in charge of this? What are the in vivo functional roles of individual translational initiation/elongation/termination factors, and can ribosome profiling tell us about this? Why transcription and translation are not often changed in a coordinated way, and what is the biochemical mechanism and functional significance behind it? These interesting questions remain to be explored, and ribosome profiling will be a critical tool in helping us understand these concepts.

Acknowledgements We apologize for not being able to cite many works owing to lack of space.

References

Aeschimann F, Kumari P, Bartake H et al (2017) LIN41 post-transcriptionally silences mRNAs by two distinct and position-dependent mechanisms. Mol Cell 65:476–489

Artieri CG, Fraser HB (2014) Evolution at two levels of gene expression in yeast. Genome Res 24:411–421

Baek J, Lee J, Yoon K et al (2017) Identification of unannotated small genes in *Salmonella*. G3 (Bethesda) 7:983–989

Barry KC, Ingolia NT, Vance RE (2017) Global analysis of gene expression reveals mRNA superinduction is required for the inducible immune response to a bacterial pathogen. eLife 6:e22707

Benhalevy D, Gupta SK, Danan CH et al (2017) The human CCHC-type zinc finger nucleic acid-binding protein binds G-rich elements in target mRNA coding sequences and promotes translation. Cell Rep 18:2979–2990

Bercovich-Kinori A, Tai J, Gelbart IA et al (2016) A systematic view on influenza induced host shutoff. eLife 5:e18311

Brar GA, Weissman JS (2015) Ribosome profiling reveals the what, when, where and how of protein synthesis. Nat Rev Mol Cell Biol 16:651–664

Brar GA, Yassour M, Friedman N et al (2012) High-resolution view of the yeast meiotic program revealed by ribosome profiling. Science 335:552–557

Calviello L, Mukherjee N, Wyler E et al (2016) Detecting actively translated open reading frames in ribosome profiling data. Nat Methods 13:165–170

Chotewutmontri P, Barkan A (2016) Dynamics of chloroplast translation during chloroplast differentiation in maize. PLoS Genet 12:e1006106

Chun SY, Rodriguez CM, Todd PK et al (2016) SPECtre: a spectral coherence-based classifier of actively translated transcripts from ribosome profiling sequence data. BMC Bioinformatics 17:482

Chung BY, Hardcastle TJ, Jones JD et al (2015) The use of duplex-specific nuclease in ribosome profiling and a user-friendly software package for Ribo-seq data analysis. RNA 21:1731–1745

Fields AP, Rodriguez EH, Jovanovic M et al (2015) A regression-based analysis of ribosome-profiling data reveals a conserved complexity to mammalian translation. Mol Cell 60:816–827

Gerashchenko MV, Gladyshev VN (2017) Ribonuclease selection for ribosome profiling. Nucleic Acids Res 45:e6

Guo H, Ingolia NT, Weissman JS et al (2010) Mammalian microRNAs predominantly act to decrease target mRNA levels. Nature 466:835–840
Haft RJ, Keating DH, Schwaegler T (2014) Correcting direct effects of ethanol on translation and transcription machinery confers ethanol tolerance in bacteria. Proc Natl Acad Sci USA 111:E2576–E2585
Hsieh AC, Liu Y, Edlind MP et al (2012) The translational landscape of mTOR signalling steers cancer initiation and metastasis. Nature 485:55–61
Hsu PY, Calviello L, Wu HL et al (2016) Super-resolution ribosome profiling reveals unannotated translation events in Arabidopsis. Proc Natl Acad Sci USA 113:E7126–E7135
Hwang JY, Buskirk AR (2017) A ribosome profiling study of mRNA cleavage by the endonuclease RelE. Nucleic Acids Res 45:327–336
Ingolia NT, Ghaemmaghami S, Newman JR et al (2009) Genome-wide analysis in vivo of translation with nucleotide resolution using ribosome profiling. Science 324:218–223
Ingolia NT, Lareau LF, Weissman JS (2011) Ribosome profiling of mouse embryonic stem cells reveals the complexity and dynamics of mammalian proteomes. Cell 147:789–802
Jan CH, Williams CC, Weissman JS (2014) Principles of ER cotranslational translocation revealed by proximity-specific ribosome profiling. Science 346:1257521
Juntawong P, Girke T, Bazin J et al (2014) Translational dynamics revealed by genome-wide profiling of ribosome footprints in Arabidopsis. Proc Natl Acad Sci USA 111:E203–E212
Latif H, Szubin R, Tan J et al (2015) A streamlined ribosome profiling protocol for the characterization of microorganisms. Biotechniques 58:329–332
Legendre R, Baudin-Baillieu A, Hatin I et al (2015) RiboTools: a Galaxy toolbox for qualitative ribosome profiling analysis. Bioinformatics 31:2586–2588
Liu X, Jiang H, Gu Z et al (2013a) High-resolution view of bacteriophage lambda gene expression by ribosome profiling. Proc Natl Acad Sci USA 110:11928–11933
Liu MJ, Wu SH, Wu JF et al (2013b) Translational landscape of photomorphogenic Arabidopsis. Plant Cell 25:3699–3710
Loayza-Puch F, Rooijers K, Buil LC et al (2016) Tumour-specific proline vulnerability uncovered by differential ribosome codon reading. Nature 530:490–494
Loayza-Puch F, Rooijers K, Zijlstra J et al (2017) TGFβ1-induced leucine limitation uncovered by differential ribosome codon reading. EMBO Rep 18:549–557
McKinney C, Zavadil J, Bianco C et al (2014) Global reprogramming of the cellular translational landscape facilitates cytomegalovirus replication. Cell Rep 6:9–17
McManus CJ, May GE, Spealman P et al (2014) Ribosome profiling reveals post-transcriptional buffering of divergent gene expression in yeast. Genome Res 24:422–430
Merchante C, Brumos J, Yun J et al (2015) Gene-specific translation regulation mediated by the hormone-signaling molecule EIN2. Cell 163:684–697
Michel AM, Fox G, M Kiran A et al (2014) GWIPS-viz: development of a ribo-seq genome browser. Nucleic Acids Res 42:D859–D864
Michel AM, Mullan JP, Velayudhan V et al (2016) RiboGalaxy: a browser based platform for the alignment, analysis and visualization of ribosome profiling data. RNA Biol 13:316–319
Miranda-CasoLuengo AA, Staunton PM, Dinan AM et al (2016) Functional characterization of the *Mycobacterium abscessus* genome coupled with condition specific transcriptomics reveals conserved molecular strategies for host adaptation and persistence. BMC Genomics 17:553
Oh E, Becker AH, Sandikci A et al (2011) Selective ribosome profiling reveals the cotranslational chaperone action of trigger factor in vivo. Cell 147:1295–1308
Olexiouk V, Crappé J, Verbruggen S et al (2016) sORFs.org: a repository of small ORFs identified by ribosome profiling. Nucleic Acids Res 44:D324–D329
Popa A, Lebrigand K, Paquet A et al (2016) RiboProfiling: a bioconductor package for standard Ribo-seq pipeline processing. F1000Res 5:1309
Reid DW, Shenolikar S, Nicchitta CV (2015) Simple and inexpensive ribosome profiling analysis of mRNA translation. Methods 91:69–74
Schrader JM, Li GW, Childers WS et al (2016) Dynamic translation regulation in Caulobacter cell cycle control. Proc Natl Acad Sci USA 113:E6859–E6867

Sendoel A, Dunn JG, Rodriguez EH et al (2017) Translation from unconventional 5′ start sites drives tumour initiation. Nature 541:494–499
Shell SS, Wang J, Lapierre P et al (2015) Leaderless transcripts and small proteins are common features of the mycobacterial translational landscape. PLoS Genet 11:e1005641
Stern-Ginossar N, Weisburd B, Michalski A et al (2012) Decoding human cytomegalovirus. Science 338:1088–1093
Sun X, Wang Z, Guo X et al (2016) Coordinated evolution of transcriptional and post-transcriptional regulation for mitochondrial functions in yeast strains. PLoS One 11:e0153523
Wang Z, Sun X, Zhao Y et al (2015) Evolution of gene regulation during transcription and translation. Genome Biol Evol 7:1155–1167
Williams CC, Jan CH, Weissman JS (2014) Targeting and plasticity of mitochondrial proteins revealed by proximity-specific ribosome profiling. Science 346:748–751
Xie SQ, Nie P, Wang Y et al (2016) RPFdb: a database for genome wide information of translated mRNA generated from ribosome profiling. Nucleic Acids Res 44:D254–D258
Zhong Y, Karaletsos T, Drewe P et al (2017) RiboDiff: detecting changes of mRNA translation efficiency from ribosome footprints. Bioinformatics 33:139–141
Zoschke R, Watkins KP, Barkan A (2013) A rapid ribosome profiling method elucidates chloroplast ribosome behavior in vivo. Plant Cell 25:2265–2275

Biophysical Analysis of miRNA-Dependent Gene Regulation

Andrea Riba, Matteo Osella, Michele Caselle, and Mihaela Zavolan

Contents

Abstract microRNAs (miRNAs) are short (~22 nucleotides long) RNAs that are encoded in the genome of species ranging from viruses to man. Together with proteins of the Argonaute family, they form RNA-induced silencing complexes, which bind target mRNAs, reducing their stability and translation rate. A miRNA typically has hundreds of evolutionarily conserved binding sites across the transcriptome, and frequently, a given mRNA carries binding sites for multiple miRNAs. In this chapter we discuss behaviors that miRNA-containing regulatory networks can exhibit, with specific examples from various experimental systems.

Keywords microRNA · Ultrasensitivity · Noise · Competing endogenous RNA · Binding hierarchy

A. Riba (✉)
Institut de Génétique et Biologie Moléculaire et Cellulaire, Illkirch, France
e-mail: andrea.riba@igbmc.fr

M. Osella · M. Caselle
Dipartimento di Fisica, Torino, Italy

M. Zavolan (✉)
Biozentrum, University of Basel, Basel, Switzerland
e-mail: mihaela.zavolan@unibas.ch

N. Rajewsky et al. (eds.), *Systems Biology*, RNA Technologies,
https://doi.org/10.1007/978-3-319-92967-5_13

1 Introduction

Since the discovery of the first microRNA (miRNA) more than 20 years ago (Lee et al. 1993; Wightman et al. 1993), these molecules have been implicated in virtually every aspect of multicellular organism biology, including development (Reinhart et al. 2000), metabolism (Krützfeldt et al. 2005; Lynn 2009), and immune defense (Xiao and Rajewsky 2009). Through base-pairing interactions, miRNAs direct ribonucleoprotein effector complexes to mRNA targets. Computational analyses were instrumental in unraveling determinants of miRNA-target interactions (Lewis et al. 2005; Gaidatzis et al. 2007; Khorshid et al. 2013; Agarwal et al. 2015), revealing their structural basis (Chandradoss et al. 2015), their effect on target gene expression (Hausser et al. 2013; Eichhorn et al. 2014), and ultimately providing genome-wide predictions of miRNA targets (Gumienny and Zavolan 2015; Agarwal et al. 2015).

The posttranscriptional regulation of mRNAs by miRNAs is eminently combinatorial: a miRNA species interacts with hundreds of targets, and a given mRNA carries binding sites for many miRNAs (Lewis et al. 2005). Although most genes appear to be targeted by miRNAs, highly conserved and presumably optimized miRNA binding sites are in transcripts encoding epigenetic regulators and transcriptional factors (Gruber and Zavolan 2013). As these regulators have themselves many targets, a perturbation in miRNA expression can be propagated through these complex networks of primary and secondary targets. Thus, the changes in gene expression and the phenotypic consequences of perturbations in miRNA expression are difficult to predict. At the same time, the circuits that are composed of epigenetic, transcriptional, and posttranscriptional regulators can exhibit complex behaviors in which the ability of miRNAs and mRNAs to reciprocally titrate each other plays as essential role (Hausser and Zavolan 2014). Attempts have been made to link specific circuit topologies to specific types of cellular responses such as transitions between cell states that occur in development or during disease-related transformations (Shenoy and Blelloch 2014; Cora' et al. 2017).

2 miRNA-Target Interaction

The first miRNA-target sites to be discovered, such as those of the let-7 miRNA in the lin-14 and lin-41 genes (Reinhart et al. 2000), exhibit extensive complementarity to the miRNA. Thus, the first generation of miRNA-target prediction programs was based on the premise that target sites would have a high degree of complementarity to the miRNA and would be predictable by programs akin to those used to predict RNA secondary structure. However, it rapidly became clear that conserved sequence elements, located in 3′ untranslated regions of mRNAs and implicated in translation regulation, are perfectly complementary to 5′ halves of miRNAs only (Lai 2002). This led to a new generation of miRNA-target prediction programs that strongly emphasized the miRNA "seed" region, which corresponds to nucleotides 2–7 of the miRNA (Lewis et al. 2005). The structural basis of miRNA seed interaction appears to be its pre-structuring in the Argonaute protein, which holds the 5′ end of the miRNA in a relatively rigid and accessible conformation, ready for the interaction with the target (Wang et al. 2008). Indeed, a study employing fluorescence resonance energy transfer demonstrated that Argonaute initially scans mRNAs for regions complementary to the nucleotides 2–4 of the miRNA, the interaction being then propagated to nucleotides 2–8 (Chandradoss et al. 2015). The conformational changes induced by the initial interaction also exposed nucleotides 13–16 that can further establish bonds with the miRNA (Schirle et al. 2014). These dynamical rearrangements in the interacting RNAs cannot be readily modeled to predict functionally relevant miRNA-target interactions. However, by measuring the impact of single point mutations on the dissociation constant, the relative importance of individual nucleotides of the miRNA for the interaction with target sites can be evaluated (Wee et al. 2012).

High-throughput approaches have been also developed to isolate the RNA fragments to which miRNAs bind within cells. This has been done initially by cross-linking RNAs to proteins with ultraviolet light, followed by the immunoprecipitation of the Argonaute protein and the sequencing of RNA fragments that are complexed with Argonaute (Chi et al. 2009; Hafner et al. 2010). These large collections of miRNA binding sites can be used within probabilistic models to learn parameters of miRNA-target interaction (Khorshid et al. 2013). These parameters recapitulate remarkably well the relative importance of different regions of the miRNA in target recognition. Modifications of this cross-linking immunoprecipitation (CLIP) approach have been used to isolate not only miRNA-target sites but also the interacting miRNA (Helwak et al. 2013). The resulting data surprisingly revealed an abundance of sites that do not see to carry extensive complementarity to the miRNA seed, and it remains unclear whether these are functional or rather reflect transient interactions. Computational analyses of data obtained with the more standard CLIP approaches showed that chimeric sequences, composed of a miRNA and an interacting target site, can be readily identified, albeit with low frequency, and they seem to largely reflect miRNA seed type of interactions (Grosswendt et al. 2014). These data too can be used to infer even miRNA-specific models of miRNA-target interaction (Fig. 1), and these models are surprisingly accurate in predicting the effect of mutations on the dissociation constants measured in vitro (Breda et al. 2015).

The energy of miRNA-target site interaction, however difficult to obtain, is still insufficient for identifying mRNAs that respond to miRNAs within cells. This is partly due to the fact that miRNAs seem to impact various aspects of target dynamics, as will be described in more detail below. In addition, other aspects that have been found relevant such as the structural accessibility of putative target sites and their nucleotide composition need to be taken into consideration. That much remains to be understood is underscored by the fact that the feature that remains most predictive for the response of a miRNA target to miRNA perturbation is the degree of evolutionary conservation of the miRNA seed-matching site. This

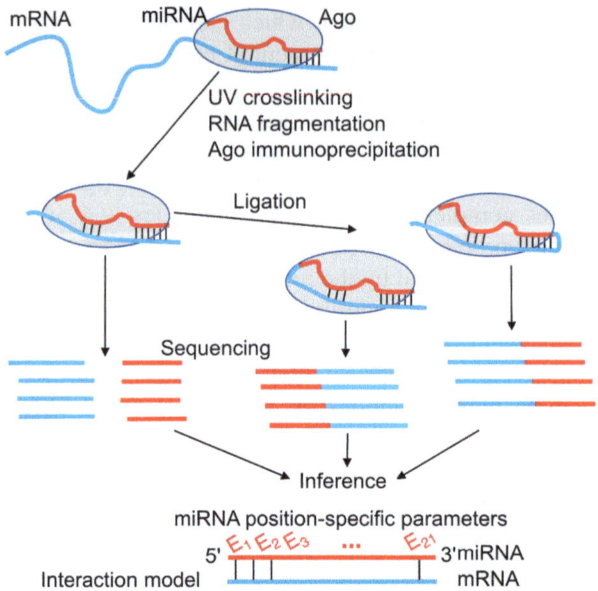

Fig. 1 High-throughput identification and modeling of miRNA-target interactions. Shown are cross-linking and immunoprecipitation-based approaches to isolate RNAs that bind miRNA-containing Argonaute complexes. In the "standard" approach (shown on the left), miRNAs and mRNA fragments are isolated and sequenced, while in more recent approaches, ligation of the miRNA and mRNA fragments is carried out prior to sequencing. This results in the sequencing of "chimeric" reads, composed of a miRNA part (in red) and a target part (in blue). These allow identification of direct interactions. Computational approaches can be used to infer parameters of miRNA-target interactions from these high-throughput data sets

effective parameter likely captures a host of factors that have not been explicitly modeled so far.

3 Modeling the Effect of miRNA-Target Interactions on Gene Expression

Mathematical models have been widely used to develop insights into the properties that miRNAs confer to gene expression regulatory networks. The simplest model of constitutive gene expression views the dynamics of an mRNA species as a *birth-death* process, where transcription and degradation occur at fixed rates, k_m and γ_m, respectively. Each mRNA molecule serves as template for proteins that are synthesized at rate k_p and degraded with rate γ_p. This two-step model of gene expression is sufficient to explain the observed gamma distribution of protein molecule numbers per cell, which contrasts with the Poisson distribution, which would be expected from a single-step birth-death process (Friedman et al. 2006;

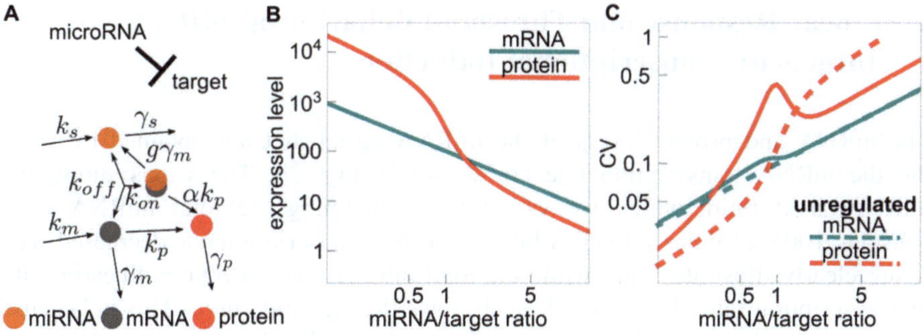

Fig. 2 Basic model of a miRNA regulating a single mRNA target. (**a**) Reaction scheme: the small RNA is produced at rate k_s and decays at rate γ_s, while its target mRNA is produced at rate k_m and decays at rate γ_m. The two molecules associate at rate k_{on} and dissociate at rate k_{off}. Proteins are synthesized at rate k_p per mRNA when the mRNA is free and rate $\alpha k_p$when the mRNA is bound by a miRNA-containing silencing complex. In the latter case, the mRNA decays at rate $g\gamma_m$. Finally, proteins decay at rate γ_p. (**b**) Threshold-linear behavior in protein expression: protein molecules only start accumulating once the mRNA level has reached a level that is sufficient for saturating miRNA-silencing complexes. (**c**) Coefficient of variation of mRNA and protein levels in single cells at steady state (mRNA, green; protein, red; miRNA-regulated targets, solid lines; unregulated genes, dashed lines). Parameters: k_m = variable, $\gamma_m = 0.1$/h, $g = 1.55$, $k_{on} = 0.24$/h, $k_{off} = 0.16$/h, $\gamma_p = 0.02$/h, $k_p = 0.5$/h, $\alpha = 0$ (full translation inhibition), $k_s = 2$/h, $\gamma_s = 0.02$/h (Hausser and Zavolan 2014). For the unregulated gene parameters were adjusted to yield the same average mRNA and protein as for the regulated target, at each specific miRNA:target ratio

Taniguchi et al. 2010). Variants of this constitutive gene expression model have been used to study miRNA-dependent gene regulation, which can be accounted for by allowing the mRNA to bind a free miRNA [or, more precisely, a free, miRNA-containing RNA-induced silencing complex (RISC)] and to decay and translate at different rates than the unbound mRNA. In the model depicted in Fig. 2 (see also Hausser and Zavolan 2014), binding of the miRNA enhances mRNA degradation by a factor g and reduces translation by a factor α.

The change in the probability of observing specific numbers of various molecular species (mRNA, protein, small RNA, small RNA-mRNA complex) per cell, as a function of time, is described by the master equation (1). With a linear noise approximation (LNA; Van Kampen 1992), the number of molecules of different types at steady state can be obtained. Here the εstep operator (Van Kampen 1992) gives the probabilities of states that can yield the current state, through the small change (shown in the superscript) in the species indicated by the subscript.

$$\frac{\partial P}{\partial t} = \begin{matrix} \left[k_s\left(\varepsilon_s^{-1}-1\right)+\gamma_s\left(\varepsilon_s^{+1}-1\right)s+\right. \\ k_m\left(\varepsilon_m^{-1}-1\right)+\gamma_m\left(\varepsilon_m^{+1}-1\right)m+k_p\left(m+\alpha c\right)\left(\varepsilon_p^{-1}-1\right)+\gamma_p\left(\varepsilon_p^{+1}-1\right)p+ \\ \left. k_{on}\left(\varepsilon_c^{-1}\varepsilon_m^{+1}\varepsilon_s^{+1}-1\right)ms+k_{off}\left(\varepsilon_m^{-1}\varepsilon_s^{-1}\varepsilon_c^{+1}-1\right)c+g\gamma_m\left(\varepsilon_c^{+1}\varepsilon_s^{-1}-1\right)c\right] \\ P\left(m,p,s,c,t\right) \end{matrix} \tag{1}$$

4 Linear Response and Threshold Behavior of miRNA Targets to Transcriptional Induction

The mRNA and protein levels of the miRNA target that are obtained by varying the mRNA transcription rate are shown in Fig. 2b. The *x*-axis shows the miRNA:target ratio, which we varied by maintaining the total miRNA level constant, 100 molecules per cell, while the mRNA transcription rate decreased. The figure clearly illustrates that, while the total mRNA level decreases linearly with its transcription rate, the protein level follows the sigmoid curve. This is because protein molecules can only accumulate when the mRNA level is at a sufficiently high level to saturate the miRNA-containing silencing complexes. If the mRNA transcription rate is lower than this specific threshold, the synthesized mRNAs are bound (titrated) by the miRNA and degraded/silenced. Theoretical work showed that the threshold is strictly dependent on the stoichiometry of miRNAs and their targets (Bosia et al. 2013; Riba et al. 2014). As a cell typically expresses multiple targets of a miRNA, each with its specific affinity, the context specificity of target responses is immediately apparent.

That small RNAs enable a distinct model of gene regulation through their ability to sequester targets, has been initially shown in the bacterium *Escherichia coli* (Levine et al. 2007). Subsequent work proposed that molecular titration is a pervasive mechanism underlying "ultrasensitive" ("all or none") responses in biological systems (Buchler and Louis 2008). Studies in whole organisms such as the worm *Caenorhabditis elegans* (Reinhart et al. 2000) and the zebra fish *Danio rerio* (Shkumatava et al. 2009) indicated that, just as the small RNAs in bacteria, miRNAs also generate thresholds in the expression pattern of the targets, rendering developmental processes more robust (Hornstein and Shomron 2006). Work using fluorescent miRNA-target reporters indicated that molecular titration enables the emergence of genetic circuits that are resistant to noise (Mukherji et al. 2011). The slope of the transition toward the regime of protein accumulation depends on target affinity, which can be modulated by changing the number of binding sites for a specific family of miRNAs (Mukherji et al. 2011).

5 miRNA Effects on Target Noise

Analysis of a constitutive gene expression model (Ozbudak et al. 2002) revealed that if the dynamics of the process is stochastic, as expected within cells, the variance in the number of protein molecules within a cell (a measure of "noise" in protein levels) is proportional to the number of protein molecules generated from a mRNA during the mRNA's lifetime, which is known as the "burst size," $b = \frac{k_p}{\gamma_m}$. Strikingly, the two rates that determine the burst size are precisely those that miRNAs have been reported to influence and that also, coherently, in the direction of decreasing the burst size. The situation is complicated by the expression of the miRNA regulator itself being stochastic. Thus, a miRNA target will experience not only its own "intrinsic noise" but also the "extrinsic noise" caused by fluctuations in the expression of the regulator. How the noise of miRNA-regulated proteins compares to that of unregulated ones depends on the balance of these two effects. This behavior is illustrated in Fig. 2c: at low rates of target transcription (high miRNA:target ratio), the miRNA decreases the burst size thus reducing the variability in its target expression compared to unregulated genes with the same average mRNA/protein level expression. However, above the threshold, where the miRNA is not effective anymore in repressing its target (low miRNA:target ratio), the noise in miRNA expression propagates to the target, increasing its variability relative to an unregulated target. This effect has also been observed experimentally, with target reporters (Schmiedel et al. 2015).

A circuit especially suited for noise reduction is the so-called "incoherent" feed-forward loop. It contains a transcription factor that induces both the expression of a target and of a miRNA, the miRNA then repressing the target. It is called "incoherent" because the transcription factor and the miRNA exert opposing effects on the common target. Extensive theoretical work showed that this circuit reduces target expression noise even in the presence of fluctuations of the regulator (Osella et al. 2011) and is surprisingly abundant among mammalian gene regulatory networks (Re et al. 2009).

The impact of miRNAs on cellular decision-making under noisy gene expression has also been observed in a few in vivo systems. In particular, miR-7 has been reported to buffer intrinsic noise in target expression during development of the sensory organ precursor of the fly (Li et al. 2009). In contrast, the removal of miRNAs in pluripotent stem cells drives cells into a low-noise ground state with increased self-renewal (Kumar et al. 2014), which is consistent with an increase in the noise of highly expressed targets in the presence of miRNAs.

6 Competing Endogenous RNAs and the Importance of miRNA-Target Interaction Affinity

The combinatorial nature of miRNA-target interactions prompted investigations into possible crosstalks induced by miRNAs between their targets, whereby specific miRNA targets that carry high affinity sites are highly susceptible to fluctuations in the level of the pool of other targets (which have been called competing endogenous RNAs = ceRNAs). The model shown in Fig. 3 and described by the master equation (2) allows one to study these behaviors. Compared to the model from Eq. (1), here we have additional terms corresponding to a "pool" of targets that compete for the

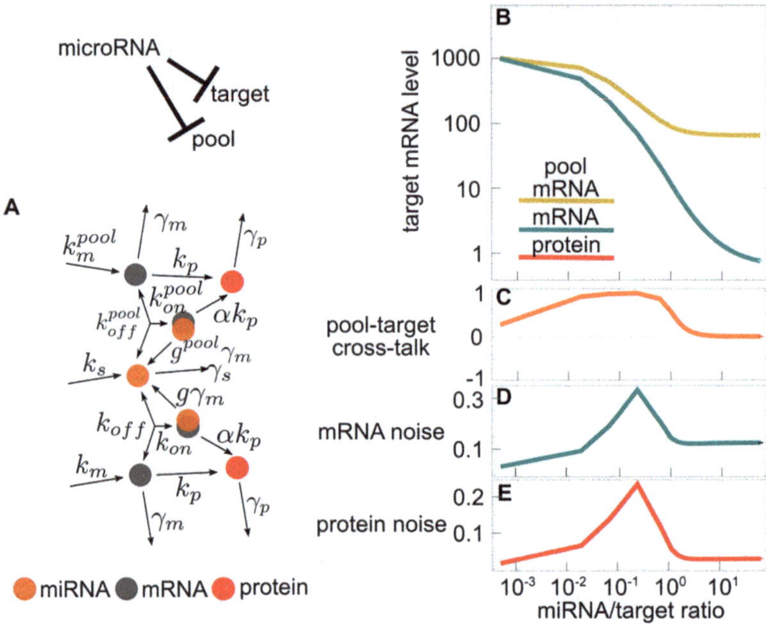

Fig. 3 Model of a miRNA interacting with and regulating multiple targets at the same time. (**a**) Reaction scheme showing the molecular species of this system. (**b**) Threshold behavior of the targets in response to the induction in miRNA expression. (**c**) Pearson correlation coefficient between "pool" and selected target mRNA levels as a measure of the crosstalk between miRNA targets. (**d**, **e**) The fluctuations in mRNA (**d**) and protein levels (**e**) are maximal at the threshold. Parameters: $k_m = k_m^{pool} = 100/\text{h}$, $\gamma_m = 0.1/\text{h}$, $k_{on} = k_{on}^{pool} = 0.24/\text{h}$, $k_{off} = 0.0016/\text{h}$, $k_{off}^{pool} = 0.16/\text{h}$, $g = 1.55 \times 10^3$, $g^{pool} = 1.55 \times 10$, $\gamma_p = 0.02/\text{h}$, $k_p = 0.5/\text{h}$, $\alpha = 1$, $k_s = $ variable, $\gamma_s = 0.02/\text{h}$

miRNAs with a selected target:

$$
\begin{aligned}
\frac{\partial P}{\partial t} = \Big[& k_s\left(\mathcal{E}_s^{-1}-1\right) + \gamma_s\left(\mathcal{E}_s^{+1}-1\right)s + \\
& k_m\left(\mathcal{E}_m^{-1}-1\right) + \gamma_m\left(\mathcal{E}_m^{+1}-1\right)m + k_p\left(m+\alpha c\right)\left(\mathcal{E}_p^{-1}-1\right) + \gamma_p\left(\mathcal{E}_p^{+1}-1\right)p + \\
& k_m^{pool}\left(\mathcal{E}_{m_{pool}}^{-1}-1\right) + \gamma_m\left(\mathcal{E}_{m_{pool}}^{+1}-1\right)m_{pool} + k_p\left(m_{pool}+\alpha c_{pool}\right)\left(\mathcal{E}_{p_{pool}}^{-1}-1\right) + \\
& \gamma_p\left(\mathcal{E}_{p_{pool}}^{+1}-1\right)p_{pool} + k_{on}\left(\mathcal{E}_c^{-1}\mathcal{E}_m^{+1}\mathcal{E}_s^{+1}-1\right)ms + k_{off}\left(\mathcal{E}_m^{-1}\mathcal{E}_s^{-1}\mathcal{E}_c^{+1}-1\right)c + \\
& g\gamma_m\left(\mathcal{E}_c^{+1}\mathcal{E}_s^{-1}-1\right)c + k_{on}^{pool}\left(\mathcal{E}_{c_{pool}}^{-1}\mathcal{E}_{m_{pool}}^{+1}\mathcal{E}_s^{+1}-1\right)m_{pool}s + \\
& k_{off}^{pool}\left(\mathcal{E}_{m_{pool}}^{-1}\mathcal{E}_s^{-1}\mathcal{E}_{c_{pool}}^{+1}-1\right)c_{pool} + g^{pool}\gamma_m\left(\mathcal{E}_{c_{pool}}^{+1}\mathcal{E}_s^{-1}-1\right)c_{pool}\Big] \\
& P\left(m, m_{pool}, p, p_{pool}, s, c, c_{pool}, t\right)
\end{aligned}
\tag{2}
$$

Theoretical work showed that if the miRNA:target ratio is close to 1 (in the regime of matched stoichiometries), the mRNAs and proteins of the different targets fluctuate together at the cost of increased noise (Figliuzzi et al. 2013; Bosia et al. 2013; Riba et al. 2014). A target whose level increases through a stochastic fluctuation will draw to itself free miRNA molecules, decreasing their availability to other targets and thereby leading to a temporarily increased level of the other targets. A negative fluctuation will have the opposite effect, and thus, targets of a specific miRNA would fluctuate together in the same direction. If the translation rates of individual mRNAs remain relatively unchanged, this mechanism may enable the cell to maintain the relative proportions of target proteins in spite of stochastic fluctuations in transcription (Riba et al. 2014; Martirosyan et al. 2017). However, as fluctuations in the regulator's level will similarly be propagated across the network of targets, the noise in both mRNA and protein levels of the targets will increase (Fig. 3d,e). For the figure, we have increased the values of the g and g^{pool} parameters, denoting the strength of miRNA downregulation, relative to Fig. 2, to magnify the effect of crosstalk.

Demonstrating these behaviors in in vivo systems remains challenging, because the interaction of individual targets with the miRNA depends on their relative affinities (Jens and Rajewsky 2015). Binding sites are generally classified by the degree of complementarity to the 5′ end of the miRNA (miRNA "seed") (Bartel 2009), but seed complementarity alone is not sufficient to explain how miRNA targets respond. Many other features such as the distance between the binding sites and the stop codon, their structural accessibility and evolutionary conservation (Hausser et al. 2009), as well as the abundance of other target sites (Arvey et al. 2010; Garcia et al. 2011) determine how strongly targets are downregulated by the miRNA. Thus, the threshold, noise reduction, and crosstalk for individual targets are difficult to observe beyond reporter constructs.

Although many cases of ceRNAs have been reported in the literature (e.g., Poliseno et al. 2010; Cesana et al. 2011; Wang et al. 2013; Liang et al. 2015; Laneve et al. 2017), the evidence is strongly debated (Denzler et al. 2014, 2016). More extensive measurements of mRNA, protein and miRNA copy numbers per cell, as well as affinities of miRNA-target interactions will be needed to accurately interpret the available data. The PTEN pseudogene, which has been reported to titrate miRNAs from its PTEN paralog to significantly affect its expression (Poliseno et al. 2010), appears to be a case where the two ceRNAs share extensive regulatory inputs, as they both carry many binding sites for miRNAs of multiple families (miR-20a, miR-19b, miR-21, miR-26a, miR-214). This is consistent with the idea that ceRNAs are high affinity targets close to their specific miRNA-imposed threshold of expression.

7 Analysis of miRNA-Containing Regulatory Circuits in In Vivo Systems

Having developed an intuition of the expected behaviors of miRNA-containing regulatory networks, we review some of the best experimentally characterized examples pertaining to different modes of regulation (Fig. 4). As transcription factors and epigenetic regulators are frequently targeted by miRNAs (Gruber and Zavolan 2013), it is perhaps expected that some of the best characterized miRNA-containing circuits contain mixtures of these regulators.

In rare cases, the miRNA-containing regulatory circuits are very short negative feedback loops, the miRNA being encoded in an intron of a target, with which it is co-expressed, to then repress the target or a specific isoform (Bosia et al. 2012). Examples are provided by miR-128b, encoded in the intron of a specific isoform of the cyclic AMP-regulated phosphoprotein of 21 kD (ARPP-21) (Megraw et al. 2010) and important for target downregulation in the brain, and miR-26b, which is encoded in the intron of the cytidine small phosphatase 2 (CTDSP2) (Dill et al. 2012). CTDSP2 regulates the RNA polymerase II-dependent transcription by modulating the phosphorylation status of RNAPII carboxy-terminal domain (CTD) and is recruited by the REST complex to genes that are silenced in neurons. Five potential binding sites are predicted in the 3′ UTR of CTDSP2, which is rather uncommon but likely contributes to the robustness of target downregulation. The reinforcement of inhibition from miR-124, which is expressed when the REST

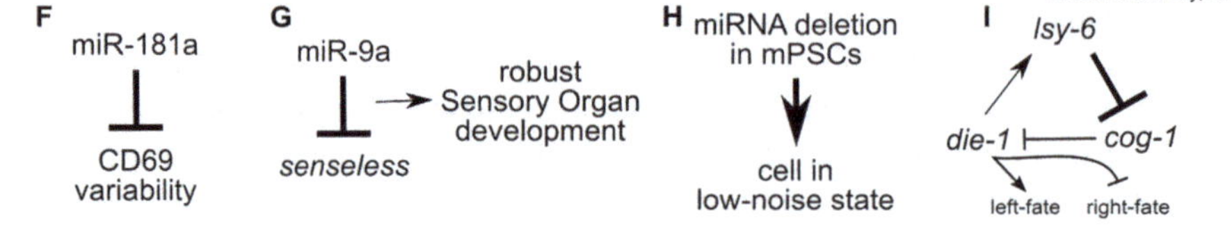

Fig. 4 Examples of experimentally characterized miRNA-containing regulatory circuits. (**a**) The intron-encoded miR-26b regulates its host gene CTDSP2. (**b**) Lin-41 is an essential let-7 target in worm development. (**c**) A regulatory network involving let-7 leads to a stable switch in cell state upon inflammation. (**d**) miR-200 family of miRNAs and ZEB1/2 transcription factors form a toggle switch for the conversion between epithelial and mesenchymal cell types. (**e**) miR-139 and miR-24 increase the variability of vascular development of zebrafish. (**f**) miR-181a reduces the population of cells with high CD69 expression, and thereby the cell-to-cell variability in CD69 protein. (**g**) miR-9a establishes a robust development of the sensory organ in fly. (**h**) Depletion of miRNAs through deletion of DGCR8 leads to a low-noise state of pluripotent stem cells. (**i**) Epigenetic regulation of lsy-6 miRNAs by the die-1 transcription factor is at the core of left-right asymmetric development of chemosensory neurons in worm

inhibition is released (Dill et al. 2012), further argues for the importance of robust CTFSP2 downregulation.

A miRNA of very high biological (as well as historical) relevance is let-7, the second discovered member of the miRNA class of posttranscriptional regulators (Reinhart et al. 2000). Although computational analysis revealed from early on that let-7 has many putative targets (Lewis et al. 2003; Johnson et al. 2005), very recent data showed that the phenotype of the let-7 mutation in the worm can be fully rescued by a complementary mutation in a single binding site, located in the 3′ UTR of the E3 ubiquitin protein ligase lin-41/TRIM71 mRNA (Ecsedi et al. 2015). This is puzzling, because other predicted let-7 targets have also been validated experimentally and found to influence developmental processes: let-60/RAS has been found to partially rescue the developmental phenotype in the worm (Johnson et al. 2005); hbl-1 affects the division of seam cells and their fusion during vulval development of the worm (Abrahante et al. 2003); HMGA2 (worm homolog—smg-3) promotes oncogenic transformation in mammalian cells (Mayr et al. 2007). Thus, lin-41/TRIM71 may be the most limiting target in worm development rather than the unique let-7 target. Another possibility, which needs to be explored with single-cell analyses of various developmental stages, is that the pattern of miRNA and target expression is such that different targets interact with the miRNA at different developmental stages. The let-7 binding sites in lin-41 are unusual in that they do not form perfect Watson-Crick base pairs with the miRNA seed region, but have extensive complementarity to the region beyond the seed (Vella et al. 2004).

Another regulatory loop that involves the let-7 miRNA has been uncovered in human breast cells: Src activation triggers NF-kB, which activates LIN28 transcription. As LIN28 inhibits the processing of the let-7 miRNA (Heo et al. 2008), the consequence of Src activation is a reduction in the repressive activity of the let-7 miRNA on its interleukin 6 (IL6) target. IL6 activates NF-kB, thus completing a positive feedback loop that is responsible for maintaining the cell state triggered by Src activation even in the absence of the initial stimulus (Iliopoulos et al. 2009). Thus, the transformation is inherited though not genetically, but rather through a change in gene expression driven by a positive feedback loop. The relationship between let-7 and LIN28 is likely more complex as well, because one of the two LIN28 paralogs in human, LIN28B, is itself likely regulated by let-7. This is because the LIN28B 3′ UTR has four predicted let-7 binding sites, which are predicted to base pair not only with the let-7 seed region but also with the 3′ end of the miRNA. Thus, LIN28B and let-7 likely form a double-negative feedback loop (also known as toggle switch), LIN28 inhibiting the processing of the let-7 miRNA and let-7 posttranscriptionally repressing LIN28 expression.

The toggle switch is a common type of genetic circuit, being in fact a minimal system that allows a system to make a binary decision. Such a decision is made by cells that switch between epithelial and mesenchymal cell states, either during normal organ development, or during cancer progression (Thiery et al. 2009). The transitions between these two states are known as mesenchyme-to-epithelium transition or MET and, correspondingly, EMT. The two stable states are enforced by a double-negative feedback loop involving the ZEB1/2 transcription factors, which stabilize the mesenchymal state, and miRNAs of the miR-141/200 family, which promote the epithelial phenotype (Burk et al. 2008). Similar to other situations when a strong miRNA-dependent phenotype can be linked to a specific gene, the ZEB1 3′ UTR has five binding sites for miR-141/200 miRNAs, two with complementarity not only to the seed but also to the miRNA 3′ end. Theoretical analyses suggest that the frequent involvement of miRNAs in multistable regulatory loops (such as the toggle switch) which underlie cell fate decisions (Ivey and Srivastava 2010) could be related to the specific noise properties of miRNA-mediated regulation. Specifically, miRNAs can increase the stability of the alternative phenotypic states that are regulated by the toggle switch to stochastic fluctuations (Osella et al. 2014). The robustness of cell state transitions to stochastic fluctuations in gene expression is presumably of very high importance during developmental processes.

A very recently described in vivo system where a miRNA influences phenotypic variability is the vasculature development in zebra fish (Kasper et al. 2017). In the wild-type zebra fish, miR-139 and miR-24 are expressed in the endothelium. Loss of these individual miRNAs leads to increased variability in the structure of the vascular tree and increased sensitivity to stress. Specifically, loss of miR-139 leads to an increased variance in the number of filopodia, as endothelial cells form abnormal numbers (either more or less) of filopodia. In contrast, stepwise depletion of miR-24 reduces the size of filopodia, a variation in a single direction. The dual behavior of the miRNAs can be rationalized by the above-presented models. Although the response to miR-24 depletion is still compatible with the derepression of one or more targets, the response to miR-139 depletion suggests a more complex response, whereby the wild-type miRNA expression corresponds to reduced target expression noise.

The miRNA regulation of phenotypic heterogeneity has not only been described in development but also in signaling. For example, miR-181a, whose expression is downregulated as lymphocytes mature, regulates T-cell receptor (TCR)-induced signal transduction (Li et al. 2007) and affects the noise of its target, CD69 (Blevins et al. 2015). Deletion of miR-181a leads to an increased fraction of cells with high CD69 expression, while the genomic deletion of the miRNA-processing enzyme Dicer resulted in increased CD69 expression by 20–50% at the protein level. Two other miRNAs, miR-17 and miR-20a, also regulate CD69, forming an incoherent feed-forward loop with the transcriptional activator MYC. As discussed above, incoherent feed-forward loops are typical noise reduction circuits.

miR-9a may also provide an example of noise regulation in development, specifically during the sensory organ development in fly (Nolo et al. 2000). With genomically integrated reporter constructs containing miR-9a sites, as well as strains with different copy numbers of the miRNA, Cassidy et al. (2013) showed that the interaction between miR-9a and its target, *senseless*, perturbs a switch in cell fate. This experimental system, as presumable many others involving early developmental transitions, is relatively difficult to test, because the number of cells assuming a specific fate is very small (differences in cell numbers were generally less than 1 on average, but nevertheless $\sim$20% of the maximum and statistically significant).

Finally, single-cell analyses of pluripotent stem cells showed that removal of miRNAs through the knockout of the pri-miRNA processing factor DGCR8 drives these cells into a low-noise state (Kumar et al. 2014). The mechanism is poorly understood, although it seems to involve key miRNA regulators such as MYC and LIN28, as well as key indirect targets of embryonically expressed miRNAs, such as the de novo DNA methyltransferase DNMT3b.

A complex regulatory network controlling the left-right asymmetric development of chemosensory neurons in the worm is controlled by lsy-6 and miR-273 (Chang et al. 2004). The regulatory loop involving these miRNAs as well as two transcription factors, *die-1* and *cog-1*, is triggered by Delta/Notch signaling, which breaks the symmetry by remodeling the chromatin at lsy-6 locus, which results in the expression of lsy-6 in ASEL cells (Cochella and Hobert 2012). The two transcription factors act antagonistically, through the miRNAs whose expression they regulate: *die-1* promotes the expression of lsy-6, which inhibits *cog-1* in ASEL cells, while *cog*-1 promotes the production of miR-273, which represses *die-1*. Both transcription factor mRNAs have two noncanonical binding sites with extensive 3′ complementarity, mostly for miR-273.

8 Conclusions

As other regulators, be they epigenetic, transcriptional, and posttranscriptional, miRNAs typically have many targets. Thus, predicting the effect of perturbations in miRNA expression is challenging, particularly during processes that unfold over some time, such as development. Nevertheless, simplified computational models can provide important insights. In this chapter, we have started from a basic, two-step model, of constitutive gene expression, and explored the expected dynamics first of a single miRNA target and then of a population of targets, as the relative levels of the miRNA and of the targets vary.

Owing to their ability to titrate their targets, miRNAs enrich their target dynamics and modulate specific aspects of it. In particular, substantial computational as well as experimental work demonstrates that miRNAs modulate the cell-to-cell variability in their targets' expression. The sign of this modulatory effect depends on the specific concentration regime of miRNAs and targets. Targets that are expressed spuriously, at low level, can be rapidly degraded in the presence of the miRNA, which thereby reduces the "intrinsic" noise in target expression. This mechanism may be important for enforcing specific target expression levels, in spite of the stochasticity inherent in gene expression. However, for targets to be strongly repressed, they need to have high affinity for the miRNA. Many of the examples that we discussed indeed involve genes that carry multiple binding sites, with extensive complementarity to a specific miRNA. Furthermore, for a target to undergo strong repression, the miRNA should be expressed at a sufficiently high level and not be entirely scavenged by other targets. In spite of these many constraints, numerous examples of miRNAs reducing the noise in their targets' expression have been uncovered.

The miRNA regulator also undergoes stochastic fluctuations in gene expression. These fluctuations propagate to its targets having the effect of increased fluctuations in the highly expressed compared to unregulated targets. This behavior may be relevant for transitions between cellular states, and in this context, it is intriguing that impairing miRNA biogenesis seems to lead to a low-noise, rather than a high-noise state. One of the main challenges currently is to unravel the in vivo hierarchy of miRNA targets, defined by their affinity as well as the decay rate in the presence of the miRNA. miRNA-target responses have been mostly studied with reporter constructs, which are generally expressed at high levels and are probably insensitive to the presence of other targets. The progress of single-cell analyses may eventually allow one to measure the expression of miRNAs as well as of their targets in individual cells. This will enable the inference of dynamic parameters of individual targets and ultimately more accurate predictions of gene expression dynamics.

References

Abrahante JE, Daul AL, Li M et al (2003) The *Caenorhabditis elegans* hunchback-like gene lin-57/hbl-1 controls developmental time and is regulated by microRNAs. Dev Cell 4:625–637

Agarwal V, Bell GW, Nam J-W, Bartel DP (2015) Predicting effective microRNA target sites in mammalian mRNAs. Elife 4. https://doi.org/10.7554/eLife.05005

Arvey A, Larsson E, Sander C et al (2010) Target mRNA abundance dilutes microRNA and siRNA activity. Mol Syst Biol 6:363

Bartel DP (2009) MicroRNAs: target recognition and regulatory functions. Cell 136:215–233

Blevins R, Bruno L, Carroll T et al (2015) microRNAs regulate cell-to-cell variability of endogenous target gene expression in developing mouse thymocytes. PLoS Genet 11:e1005020

Bosia C, Osella M, Baroudi ME et al (2012) Gene autoregulation via intronic microRNAs and its functions. BMC Syst Biol 6:131

Bosia C, Pagnani A, Zecchina R (2013) Modelling competing endogenous RNA networks. PLoS One 8:e66609

Breda J, Rzepiela AJ, Gumienny R et al (2015) Quantifying the strength of miRNA-target interactions. Methods 85:90–99

Buchler NE, Louis M (2008) Molecular titration and ultrasensitivity in regulatory networks. J Mol Biol 384:1106–1119

Burk U, Schubert J, Wellner U et al (2008) A reciprocal repression between ZEB1 and members of the miR-200 family promotes EMT and invasion in cancer cells. EMBO Rep 9:582–589

Cassidy JJ, Jha AR, Posadas DM et al (2013) miR-9a minimizes the phenotypic impact of genomic diversity by buffering a transcription factor. Cell 155:1556–1567

Cesana M, Cacchiarelli D, Legnini I et al (2011) A long noncoding RNA controls muscle differentiation by functioning as a competing endogenous RNA. Cell 147:358–369

Chandradoss SD, Schirle NT, Szczepaniak M et al (2015) A dynamic search process underlies microRNA targeting. Cell 162:96–107

Chang S, Johnston RJ, Frøkjær-Jensen C et al (2004) MicroRNAs act sequentially and asymmetrically to control chemosensory laterality in the nematode. Nature 430:785–789

Chi SW, Zang JB, Mele A, Darnell RB (2009) Argonaute HITS-CLIP decodes microRNA–mRNA interaction maps. Nature 460:479–486

Cochella L, Hobert O (2012) Embryonic priming of a miRNA locus predetermines postmitotic neuronal left/right asymmetry in *C. elegans*. Cell 151:1229–1242
Cora' D, Re A, Caselle M, Bussolino F (2017) MicroRNA-mediated regulatory circuits: outlook and perspectives. Phys Biol 14:045001
Denzler R, Agarwal V, Stefano J et al (2014) Assessing the ceRNA hypothesis with quantitative measurements of miRNA and target abundance. Mol Cell 54:766–776
Denzler R, McGeary SE, Title AC et al (2016) Impact of microRNA levels, target-site complementarity, and cooperativity on competing endogenous RNA-regulated gene expression. Mol Cell 64:565–579
Dill H, Linder B, Fehr A, Fischer U (2012) Intronic miR-26b controls neuronal differentiation by repressing its host transcript, ctdsp2. Genes Dev 26:25–30
Ecsedi M, Rausch M, Großhans H (2015) The let-7 microRNA directs vulval development through a single target. Dev Cell 32:335–344
Eichhorn SW, Guo H, McGeary SE et al (2014) mRNA destabilization is the dominant effect of mammalian microRNAs by the time substantial repression ensues. Mol Cell 56:104–115
Figliuzzi M, Marinari E, De Martino A (2013) MicroRNAs as a selective channel of communication between competing RNAs: a steady-state theory. Biophys J 104:1203–1213
Friedman N, Cai L, Xie XS (2006) Linking stochastic dynamics to population distribution: an analytical framework of gene expression. Phys Rev Lett 97:168302
Gaidatzis D, van Nimwegen E, Hausser J, Zavolan M (2007) Inference of miRNA targets using evolutionary conservation and pathway analysis. BMC Bioinformatics 8:69
Garcia DM, Baek D, Shin C et al (2011) Weak seed-pairing stability and high target-site abundance decrease the proficiency of lsy-6 and other microRNAs. Nat Struct Mol Biol 18:1139–1146
Grosswendt S, Filipchyk A, Manzano M et al (2014) Unambiguous identification of miRNA: target site interactions by different types of ligation reactions. Mol Cell 54:1042–1054
Gruber AJ, Zavolan M (2013) Modulation of epigenetic regulators and cell fate decisions by miRNAs. Epigenomics 5:671–683
Gumienny R, Zavolan M (2015) Accurate transcriptome-wide prediction of microRNA targets and small interfering RNA off-targets with MIRZA-G. Nucleic Acids Res 43:1380–1391
Hafner M, Landthaler M, Burger L et al (2010) Transcriptome-wide identification of RNA-binding protein and microRNA target sites by PAR-CLIP. Cell 141:129–141
Hausser J, Zavolan M (2014) Identification and consequences of miRNA-target interactions – beyond repression of gene expression. Nat Rev Genet 15:599–612
Hausser J, Landthaler M, Jaskiewicz L et al (2009) Relative contribution of sequence and structure features to the mRNA binding of Argonaute/EIF2C–miRNA complexes and the degradation of miRNA targets. Genome Res 19:2009–2020
Hausser J, Syed AP, Selevsek N et al (2013) Timescales and bottlenecks in miRNA-dependent gene regulation. Mol Syst Biol 9:711
Helwak A, Kudla G, Dudnakova T, Tollervey D (2013) Mapping the human miRNA interactome by CLASH reveals frequent noncanonical binding. Cell 153:654–665
Heo I, Joo C, Cho J et al (2008) Lin28 mediates the terminal uridylation of let-7 precursor MicroRNA. Mol Cell 32:276–284
Hornstein E, Shomron N (2006) Canalization of development by microRNAs. Nat Genet 38(Suppl):S20–S24
Iliopoulos D, Hirsch HA, Struhl K (2009) An epigenetic switch involving NF-kappaB, Lin28, let-7 microRNA, and IL6 links inflammation to cell transformation. Cell 139:693–706
Ivey KN, Srivastava D (2010) MicroRNAs as regulators of differentiation and cell fate decisions. Cell Stem Cell 7:36–41
Jens M, Rajewsky N (2015) Competition between target sites of regulators shapes post-transcriptional gene regulation. Nat Rev Genet 16:113–126
Johnson SM, Grosshans H, Shingara J et al (2005) RAS is regulated by the let-7 microRNA family. Cell 120:635–647
Kasper DM, Moro A, Ristori E et al (2017) MicroRNAs establish uniform traits during the architecture of vertebrate embryos. Dev Cell 40:552–565.e5

Khorshid M, Hausser J, Zavolan M, van Nimwegen E (2013) A biophysical miRNA-mRNA interaction model infers canonical and noncanonical targets. Nat Methods 10:253–255
Krützfeldt J, Rajewsky N, Braich R et al (2005) Silencing of microRNAs in vivo with "antagomirs". Nature 438:685–689
Kumar RM, Cahan P, Shalek AK et al (2014) Deconstructing transcriptional heterogeneity in pluripotent stem cells. Nature 516:56–61
Lai EC (2002) Micro RNAs are complementary to 3' UTR sequence motifs that mediate negative post-transcriptional regulation. Nat Genet 30:363–364
Laneve P, Po A, Favia A et al (2017) The long noncoding RNA linc-NeD125 controls the expression of medulloblastoma driver genes by microRNA sponge activity. Oncotarget 8:31003–31015
Lee RC, Feinbaum RL, Ambros V (1993) The *C. elegans* heterochronic gene lin-4 encodes small RNAs with antisense complementarity to lin-14. Cell 75:843–854
Levine E, Zhang Z, Kuhlman T, Hwa T (2007) Quantitative characteristics of gene regulation by small RNA. PLoS Biol 5:e229
Lewis BP, Shih I-H, Jones-Rhoades MW et al (2003) Prediction of mammalian microRNA targets. Cell 115:787–798
Lewis BP, Burge CB, Bartel DP (2005) Conserved seed pairing, often flanked by adenosines, indicates that thousands of human genes are microRNA targets. Cell 120:15–20
Li Q-J, Chau J, Ebert PJR et al (2007) miR-181a is an intrinsic modulator of T cell sensitivity and selection. Cell 129:147–161
Li X, Cassidy JJ, Reinke CA et al (2009) A microRNA imparts robustness against environmental fluctuation during development. Cell 137:273–282
Liang W-C, Fu W-M, Wong C-W et al (2015) The lncRNA H19 promotes epithelial to mesenchymal transition by functioning as miRNA sponges in colorectal cancer. Oncotarget 6:22513–22525
Lynn FC (2009) Meta-regulation: microRNA regulation of glucose and lipid metabolism. Trends Endocrinol Metab 20:452–459
Martirosyan A, De Martino A, Pagnani A, Marinari E (2017) ceRNA crosstalk stabilizes protein expression and affects the correlation pattern of interacting proteins. Sci Rep 7:43673
Mayr C, Hemann MT, Bartel DP (2007) Disrupting the pairing between let-7 and Hmga2 enhances oncogenic transformation. Science 315:1576–1579
Megraw M, Sethupathy P, Gumireddy K et al (2010) Isoform specific gene auto-regulation via miRNAs: a case study on miR-128b and ARPP-21. Theor Chem Acc 125:593–598
Mukherji S, Ebert MS, Zheng GXY et al (2011) MicroRNAs can generate thresholds in target gene expression. Nat Genet 43:854–859
Nolo R, Abbott LA, Bellen HJ (2000) Senseless, a Zn finger transcription factor, is necessary and sufficient for sensory organ development in *Drosophila*. Cell 102:349–362
Osella M, Bosia C, Corá D, Caselle M (2011) The role of incoherent microRNA-mediated feedforward loops in noise buffering. PLoS Comput Biol 7:e1001101
Osella M, Riba A, Testori A et al (2014) Interplay of microRNA and epigenetic regulation in the human regulatory network. Front Genet 5:345
Ozbudak EM, Thattai M, Kurtser I et al (2002) Regulation of noise in the expression of a single gene. Nat Genet 31:69–73
Poliseno L, Salmena L, Zhang J et al (2010) A coding-independent function of gene and pseudogene mRNAs regulates tumour biology. Nature 465:1033–1038
Re A, Corá D, Taverna D, Caselle M (2009) Genome-wide survey of microRNA–transcription factor feed-forward regulatory circuits in human. Mol Biosyst 5:854–867
Reinhart BJ, Slack FJ, Basson M et al (2000) The 21-nucleotide let-7 RNA regulates developmental timing in *Caenorhabditis elegans*. Nature 403:901–906
Riba A, Bosia C, El Baroudi M et al (2014) A combination of transcriptional and microRNA regulation improves the stability of the relative concentrations of target genes. PLoS Comput Biol 10:e1003490

Schirle NT, Sheu-Gruttadauria J, MacRae IJ (2014) Structural basis for microRNA targeting. Science 346:608–613
Schmiedel JM, Klemm SL, Zheng Y et al (2015) Gene expression. MicroRNA control of protein expression noise. Science 348:128–132
Shenoy A, Blelloch RH (2014) Regulation of microRNA function in somatic stem cell proliferation and differentiation. Nat Rev Mol Cell Biol 15:565–576
Shkumatava A, Stark A, Sive H, Bartel DP (2009) Coherent but overlapping expression of microRNAs and their targets during vertebrate development. Genes Dev 23:466–481
Taniguchi Y, Choi PJ, Li G-W et al (2010) Quantifying *E. coli* proteome and transcriptome with single-molecule sensitivity in single cells. Science 329:533–538
Thiery JP, Acloque H, Huang RYJ, Nieto MA (2009) Epithelial-mesenchymal transitions in development and disease. Cell 139:871–890
Van Kampen NG (1992) Stochastic processes in physics and chemistry. Elsevier, New York
Vella MC, Choi E-Y, Lin S-Y et al (2004) The *C. elegans* microRNA let-7 binds to imperfect let-7 complementary sites from the lin-41 3'UTR. Genes Dev 18:132–137
Wang Y, Sheng G, Juranek S et al (2008) Structure of the guide-strand-containing argonaute silencing complex. Nature 456:209–213
Wang Y, Xu Z, Jiang J et al (2013) Endogenous miRNA sponge lincRNA-RoR regulates Oct4, Nanog, and Sox2 in human embryonic stem cell self-renewal. Dev Cell 25:69–80
Wee LM, Flores-Jasso CF, Salomon WE, Zamore PD (2012) Argonaute divides its RNA guide into domains with distinct functions and RNA-binding properties. Cell 151:1055–1067
Wightman B, Ha I, Ruvkun G (1993) Posttranscriptional regulation of the heterochronic gene lin-14 by lin-4 mediates temporal pattern formation in *C. elegans*. Cell 75:855–862
Xiao C, Rajewsky K (2009) MicroRNA control in the immune system: basic principles. Cell 136:26–36

Modeling and Analyzing the Flow of Molecular Machines in Gene Expression

Yoram Zarai, Michael Margaliot, and Tamir Tuller

Contents

Abstract Gene expression is a fundamental cellular process by which proteins are synthesized based on the information encoded in the genetic material. During this process, macromolecules such as ribosomes or RNA polymerases scan the genetic material in a sequential manner. We review several deterministic, continuous-time

Y. Zarai · M. Margaliot · T. Tuller (✉)
Faculty of Electrical Engineering, Tel Aviv University, Tel Aviv, Israel
e-mail: tamirtul@post.tau.ac.il

N. Rajewsky et al. (eds.), *Systems Biology*, RNA Technologies,
https://doi.org/10.1007/978-3-319-92967-5_14

models for the flow of such macromolecules. These models are both easy to simulate and amenable to rigorous mathematical analysis. We demonstrate how these models can be used to predict the expression levels of genes and to study important biological phenomena such as competition for finite resources, sensitivity of gene expression to various biophysical factors, and optimization of the protein production rate.

Keywords Ribosome flow model · Stability · Entrainment · Biotechnology · Gene expression · Synthetic biology

1 Introduction

Gene expression is the process that transforms the information encoded in the genes into functional proteins. This is a tightly regulated process that is closely related to cell activation and proliferation. Gene expression consists of several stages. During *transcription*, instructions encoded in regions of the DNA called genes are copied into molecules called messenger-RNA (mRNA), and during the *translation* process, the information inscribed in the mRNA is translated into a chain of amino acids that is folded co- and post-translationally to yield a functional protein (Alberts et al. 2007; Chandar and Viselli 2012). All cells, from bacteria to human, express their genetic information in this way—a principle so fundamental that Francis Crick called it the *central dogma of molecular biology* (Crick 1970).

The genetic information encoded in the DNA is composed of a sequence of nucleotides. Four types of nucleotides, differing in their nitrogen-containing bases, are used in the DNA: adenine (A), thymine (T), guanine (G), and cytosine (C). When transcribed into mRNA, thymine is replaced by uracil (U). Each sequence of three consecutive nucleotides in the mRNA, called a *codon*, corresponds to a specific amino acid or to a control signal. Out of the $4^3 = 64$ possible codons, one (usually AUG, referred to as the "start-codon") determines where protein synthesis begins (i.e., it indicates the first amino acid in the protein), three codons (UAA, UAG, and UGA, referred to as "stop codons") signal the completion of protein synthesis, and the remaining 61 codons are used to encode the standard 20 amino acids (Alberts et al. 2007). This redundant genetic code is universal across all present-day organisms.

During transcription [translation] complex molecular machines called RNA polymerases (RNAPs) [ribosomes] scan the DNA [mRNA] and "read" the genetic information. The flow of these molecular machines plays an important role in gene expression. For example, in order to increase the protein production rate several ribosomes may read the same mRNA molecule simultaneously. A ribosome that is stalled may then lead to the formation of "traffic jams" behind it, and consequently to depletion of the pool of free ribosomes. To prevent this, the cell operates a surveillance and rescue system for stalled ribosomes (Mills and Green 2017). Translation undergoes selection for optimization, as it is known to be one of the most energy-consuming processes in the cell (Tuller et al. 2010; Alberts et al. 2007). Another testimony of the importance of ribosome flow is the fact that about half of the currently existing antibiotics target the bacterial ribosome by interfering with translation initiation, elongation, termination, and other regulatory mechanisms (Myasnikov et al. 2016; Johansson et al. 2014).

Mathematical models of the flow of biological machines are becoming more and more important as new experimental techniques provide more and more data on the location of such machines inside the cell (Ingolia 2014; Newhart and Janicki 2014; Mayer and Churchman 2016), sometimes in real time (Iwasaki and Ingolia 2016). Such models are particularly important in the context of synthetic biology and biotechnology, as they can be used to obtain qualitative and quantitative predictions of the effects of manipulating the genetic machinery.

Here, we review the *ribosome flow model* (RFM) that is a deterministic, continuous-time mathematical model for the flow of interacting particles. This model is highly amenable to analysis using tools from systems and control theory. The RFM is a dynamic mean-field approximation of an important stochastic model called the asymmetric simple exclusion process that is briefly reviewed in the next section. Section 3 describes the RFM, and Sect. 4 analyzes its dynamical properties. Extensions of the RFM that can be used to model more sophisticated features of gene expression are reviewed in Sect. 5. Section 6 describes some of the biological implications of the analysis results, and Sect. 7 describes how the RFM can be used to integrate large-scale genomic data. The final section concludes and describes several directions for further research.

2 Asymmetric Simple Exclusion Process

The standard model for the flow of molecular machines like RNAPs, ribosomes, and biological motors is the *asymmetric simple exclusion process* (ASEP) (MacDonald et al. 1968; MacDonald and Gibbs 1969; Spitzer 1970; Zia et al. 2011; Shaw et al. 2003). ASEP is a general stochastic model describing particles that can hop from a site to a neighboring site along an ordered (usually 1D) lattice. Each site may be either free or occupied by a single particle, and hops may take place only to a free target site. This represents the fact that the particles have volume and cannot overtake one another. Effectively, this generates interactions between the moving particles. The motion is assumed to be directionally asymmetric, i.e., there is some preferred direction of motion. In the *totally asymmetric simple exclusion process* (TASEP), the motion is unidirectional.

TASEP has two main boundary configurations. In TASEP with *open boundary conditions*, the two sides of the chain are connected to two particle reservoirs, and particles can hop into the chain (if the first site is empty) and out of the chain (if the last site is full). In TASEP with *periodic boundary conditions* the chain is closed, so that a particle that hops from the last site returns to the first site. Thus, here the particles hop around a ring, and the total number of particles is conserved.

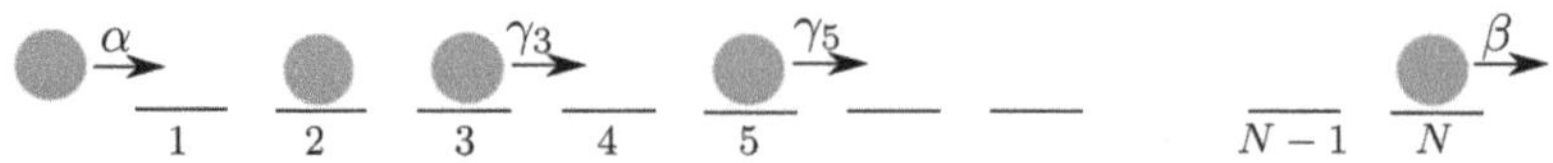

Fig. 1 TASEP with open boundary conditions

In the context of using TASEP to model mRNA transcription [translation], the lattice models the DNA [mRNA] molecule that is coarse-grained to a set of consecutive sites. The particles model the moving RNAPs [ribosomes]. Initiation time and the time that an RNAP [ribosome] spends transcribing [translating] each site are considered random and nucleotide [codon] dependent.

Figure 1 depicts TASEP with open boundary conditions. The input [exit] rate into [from] the N-site chain is denoted by α [β], and γ_i is the hopping (or elongation) rate from site i to site $i+1$. During a short time interval $[t, t+\Delta T]$, a particle can enter the chain with probability $\alpha \Delta T$ (but only if the first site is empty), exit the chain with probability $\beta \Delta T$ (but only if site N is occupied), and hop from site i to site $i+1$ with a probability $\gamma_i \Delta T$ (but only if site $i+1$ is empty).

TASEP with open boundary conditions exhibits nontrivial phenomena such as boundary-induced phase transitions and shock fronts (Blythe and Evans 2007), and has become a paradigmatic model for nonequilibrium statistical mechanics (Chou et al. 2011; Blythe and Evans 2007; Evans and Blythe 2002; Derrida 1998). The phase transitions represent situations where small changes in the rates yield a large quantitative and qualitative change in the behavior. For example, a small increase in the entry rate α may lead to a sharp increase in the density of particles along the chain, and thus to a sharp decrease in the flow rate.

TASEP and its different variants have been used to model and analyze numerous natural and artificial processes including traffic flow, pedestrian dynamics, molecular motor traffic, genome evolution, gene expression, the movement of ants along a trail, and more (Schadschneider et al. 2011; Pinkoviezky and Gov 2013; Zur and Tuller 2016). However, due to the indirect interactions between the particles, analysis of TASEP is difficult, and closed-form results exist only for the case of the *homogeneous* TASEP, i.e. when all the internal elongation rates γ_i are assumed to be equal (Derrida et al. 1992, 1993). In the nonhomogeneous case, one must resort to ad-hoc arguments and extensive and time-consuming Monte Carlo simulations. This holds even in cases when only one or two rates differ from all the others (Kolomeisky 1998; Chou and Lakatos 2004; Dong et al. 2007a; Tripathy and Barma 1998). A model that is more amenable to analysis, and also easier to simulate, is the RFM.

3 Ribosome Flow Model (RFM)

The RFM (Reuveni et al. 2011) is a *deterministic*, nonlinear, continuous-time compartmental model for interacting particles flow that can be derived via a dynamic mean-field approximation of TASEP with open boundary conditions

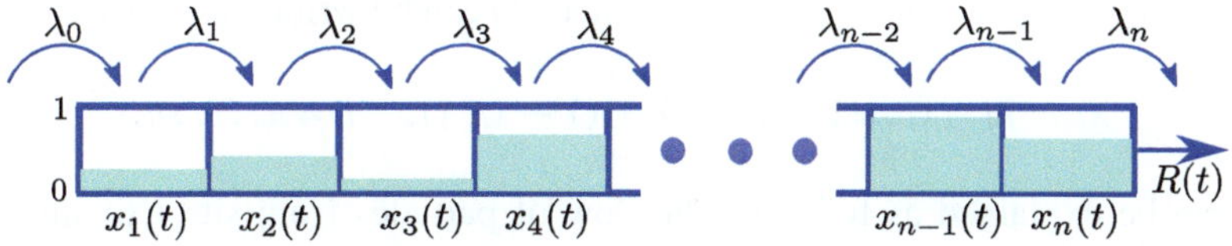

Fig. 2 The RFM models unidirectional flow along a chain of n sites. The state variable $x_i(t) \in [0, 1]$ represents the density at site i at time t. The parameter $\lambda_i > 0$ controls the transition rate from site i to site $i + 1$, with λ_0 [λ_n] controlling the initiation [exit] rate. The output rate at time t is $R(t) := \lambda_n x_n(t)$

(Zarai et al. 2017e). The model includes n consecutive sites. The normalized occupancy level (or density) of site i at time t is described by a state variable $x_i(t) : \mathbb{R}_+ \to [0, 1]$, $i = 1, \ldots, n$, where $x_i(t) = 0$ [$x_i(t) = 1$] means that site i is completely free [full] at time t.

The transition between sites i and site $i + 1$ is regulated by a parameter $\lambda_i > 0$. In particular, λ_0 [λ_n] controls the initiation [exit] rate into [from] the chain. The rate at which particles exit the chain at time t is referred to as the production rate, and is denoted by $R(t)$ (see Fig. 2). Note that x_i is dimensionless, and every rate λ_i has units of 1/time.

When modeling the flow of biological machines like RNAPs [ribosomes] the chain models a DNA [mRNA] molecule coarse-grained into n sites of nucleotides [codons], and $R(t)$ is the rate at which RNAPs [ribosomes] detach from the molecule, i.e., the rate at which mRNAs [proteins] are produced. The values of the λ_is can be determined based on biophysical properties. For example, in the context of translation, these properties include the number of available free ribosomes and nucleotide context surrounding initiation codons, the codon compositions in each site, the tRNA pool of the organism, folding of the mRNA molecule, and the interaction between the nascent peptide and the ribosomal exit tunnel (Reuveni et al. 2011; Tuller et al. 2011; Dana and Tuller 2012).

The dynamics of the RFM is given by n nonlinear first-order ordinary differential equations:

$$
\begin{aligned}
\dot{x}_1 &= \lambda_0(1 - x_1) - \lambda_1 x_1(1 - x_2), \\
\dot{x}_2 &= \lambda_1 x_1(1 - x_2) - \lambda_2 x_2(1 - x_3), \\
\dot{x}_3 &= \lambda_2 x_2(1 - x_3) - \lambda_3 x_3(1 - x_4), \\
&\vdots \\
\dot{x}_{n-1} &= \lambda_{n-2} x_{n-2}(1 - x_{n-1}) - \lambda_{n-1} x_{n-1}(1 - x_n), \\
\dot{x}_n &= \lambda_{n-1} x_{n-1}(1 - x_n) - \lambda_n x_n.
\end{aligned} \tag{1}
$$

If we define $x_0(t) := 1$ and $x_{n+1}(t) := 0$ then (1) can be written more succinctly as

$$\dot{x}_i = \lambda_{i-1}x_{i-1}(1 - x_i) - \lambda_i x_i(1 - x_{i+1}), \quad i = 1, \ldots, n. \tag{2}$$

This can be explained as follows. The flow of particles from site i to site $i + 1$ is $\lambda_i x_i(t)(1 - x_{i+1}(t))$. This flow is proportional to $x_i(t)$, i.e., it increases with the occupancy level at site i, and to $(1 - x_{i+1}(t))$, i.e., it decreases as site $i + 1$ becomes fuller. This corresponds to a "soft" version of the simple exclusion principle in ASEP. Note that the maximal possible flow from site i to site $i + 1$ is the transition rate λ_i. Equation (2) thus states that the change in the state variable x_i as a function of time equals the flow entering site i from site $i - 1$, minus the flow exiting site i to site $i + 1$.

A system where each state variable describes the amount of "material" in some compartment, and the dynamics describes the flow of material between the compartments and also with the surrounding environment, is called a *compartmental system* (Jacquez and Simon 1993). Compartmental systems proved to be useful models in various biological domains including physiology, pharmacokinetics, population dynamics, and epidemiology (Brauer 2008; Holza and Fahrb 2001; Jacquez 1996). The RFM is clearly a nonlinear compartmental system, with x_i denoting the normalized amount of "material" in compartment i, and the flow satisfying a "soft" simple exclusion principle.

It turns out that the RFM enjoys a rich set of important dynamical properties. Some of these properties are reviewed in the next section. Section 6 describes several biological implications of these properties.

4 Dynamical Properties of the RFM

Let $x(t, a)$ denote the solution of (1) at time $t \geq 0$ for the initial condition $x(0) = a$. Since the state variables correspond to normalized occupancy levels, we always assume that a belongs to the closed n-dimensional unit cube: $C^n := \{x \in \mathbb{R}^n : x_i \in [0, 1], i = 1, \ldots, n\}$. Let $\text{int}(C^n)$ denote the interior of C^n, and let ∂C^n denote the boundary of C^n.

4.1 Repelling Boundaries and Persistence

The next result shows in particular that both C^n and $\text{int}(C^n)$ are invariant sets of the RFM dynamics. In other words, if the initial condition at $t = 0$ corresponds to a state with all densities in [0, 1] ((0, 1)) then the densities remain in [0, 1] ((0, 1)) for all time $t \geq 0$.

Proposition 1 (Margaliot and Tuller 2012; Margaliot et al. 2014) *For any* $a \in C^n$ *the solution of* (1) *satisfies* $x(t, a) \in \text{int}(C^n)$ *for all* $t > 0$.

4.2 Strong Monotonicity

For two vectors $a, b \in \mathbb{R}^n$, we write $a \leq b$ if $a_i \leq b_i$ for all i, and $a \ll b$ if $a_i < b_i$ for all i. A dynamical system is called *cooperative* if for any two initial conditions $a \leq b$ the solutions $x(t, a)$ and $x(t, b)$ emanating from a and b satisfy $x(t, a) \leq x(t, b)$ for all $t \geq 0$ (Smith 1995). In other words, the flow preserves the ordering between the initial conditions. The next result shows that the RFM is a cooperative system.

Proposition 2 (Margaliot and Tuller 2012) *For any* $a, b \in C^n$, *with* $a \leq b$, *the solutions of the RFM satisfy*

$$x(t, a) \leq x(t, b), \quad \textit{for all } t \geq 0. \tag{3}$$

Furthermore, if $a \leq b$ *and* $a \neq b$ *then*

$$x(t, a) \ll x(t, b), \quad \textit{for all } t > 0. \tag{4}$$

This has the following interpretation. We say that a density profile b is "more occupied" than a if $b_i \geq a_i$ for all i, that is, the density at each site in profile b is large than or equal to the density in the corresponding site in profile a. If this is the case at time zero then the dynamics of the RFM guarantees that this relation between the corresponding density profiles remains true for all time $t \geq 0$.

4.3 Contractivity

A dynamical system is called *contractive* if there exists a vector norm $|\cdot|$ and $\gamma > 0$ such that for any two initial conditions a, b

$$|x(t, a) - x(t, b)| \leq \exp(-\gamma t)|a - b|, \quad \text{for all } t \geq 0.$$

In other words, the distance between any two trajectories contracts to zero at an exponential rate. This also means that the initial condition is "quickly forgotten". Differential analysis and, in particular, contraction theory proved to be a powerful tool for analyzing nonlinear dynamical systems (Lohmiller and Slotine 1998; Russo et al. 2010; Aminzare and Sontag 2014).

The RFM satisfies a slightly weaker, but still quite useful, property. Let $|\cdot|_1 : \mathbb{R}^n \to \mathbb{R}_+$ denote the L_1 norm, i.e., for $z \in \mathbb{R}^n$, $|z|_1 = |z_1| + \cdots + |z_n|$.

Proposition 3 (Margaliot et al. 2014) *For any $\varepsilon > 0$ there exists $\ell = \ell(\varepsilon) > 0$ such that the solutions of the RFM satisfy*

$$|x(t,a) - x(t,b)|_1 \leq (1+\varepsilon)\exp(-\ell t)|a-b|_1, \tag{5}$$

for all $a, b \in C^n$ and all $t \geq 0$.

This is contractivity up to an arbitrarily small overshoot $(1+\varepsilon)$ (Margaliot et al. 2016).

4.4 Global Asymptotic Stability

Since the compact and convex set C^n is an invariant set of the dynamics, it contains a steady-state point e. By Proposition 1, $e \in \text{int}(C^n)$. Applying (5) with $b = e$ yields the following result.

Corollary 1 *The RFM admits a unique steady-state point $e \in \text{int}(C^n)$ that is globally asymptotically stable, i.e.*

$$\lim_{t\to\infty} x(t,a) = e, \quad \text{for all } a \in C^n.$$

Thus, any set of rate values λ_i, $i = 0, 1 \ldots, n$, induces a *unique* steady-state density and any solution of the RFM converges to this density, regardless of the initial density. In particular, the production rate $R(t) = \lambda_n x_n(t)$ converges to the *steady-state production rate*:

$$R := \lambda_n e_n. \tag{6}$$

At steady state, i.e., for $x = e$, the left-hand side of all equations in (1) is zero, so $R = \lambda_i e_i(1 - e_{i+1})$, $i = 0, \ldots, n$, where $e_0 := 1$ and $e_{n+1} := 0$. It follows that for any $c > 0$

$$R(c\lambda_0, \ldots, c\lambda_n) = cR(\lambda_0, \ldots, \lambda_n),$$

so R is *positively homogeneous of order one* with respect to (w.r.t.) the rates $\lambda_0, \ldots, \lambda_n$. This means that if we multiply all the rates by a factor $c > 0$ then the steady-state production rate will also increase by the same factor c. Similarly, e_i, $i = 1, \ldots, n$, is positively homogeneous of order zero w.r.t. the rates, i.e., $e_i(c\lambda_0, \ldots, c\lambda_n) = e_i(\lambda_0, \ldots, \lambda_n)$, for all i.

Solving the set of nonlinear equations that define the steady state is not trivial. It turns out that there exists a better representation of the mapping from the rates to the steady state. Let $\mathbb{R}^k_{++}$ denote the set of k-dimensional vectors with positive entries.

4.5 Steady-State Spectral Representation

Consider the RFM with dimension n and rates $\lambda_0, \dots, \lambda_n$. Define the $(n+2)\times(n+2)$ Jacobi matrix

$$A := \begin{bmatrix} 0 & \lambda_0^{-1/2} & 0 & 0 & \dots & 0 & 0 \\ \lambda_0^{-1/2} & 0 & \lambda_1^{-1/2} & 0 & \dots & 0 & 0 \\ 0 & \lambda_1^{-1/2} & 0 & \lambda_2^{-1/2} & \dots & 0 & 0 \\ & & & \vdots & & & \\ 0 & 0 & 0 & \dots & \lambda_{n-1}^{-1/2} & 0 & \lambda_n^{-1/2} \\ 0 & 0 & 0 & \dots & 0 & \lambda_n^{-1/2} & 0 \end{bmatrix}. \tag{7}$$

This is a symmetric matrix, so all its eigenvalues are real. Since A is (componentwise) nonnegative and irreducible, it admits a unique maximal eigenvalue $\sigma > 0$ (called the Perron eigenvalue or Perron root), and a corresponding eigenvector $\zeta \in \mathbb{R}^{n+2}_{++}$ (the Perron eigenvector) (Horn and Johnson 2013).

Theorem 1 (Poker et al. 2014) *Consider an RFM with dimension n and rates $\lambda_0, \dots, \lambda_n$. Let A be the matrix defined in* (7). *Then:*

$$R = \sigma^{-2} \text{ and } e_i = \lambda_i^{-1/2}\sigma^{-1}\frac{\zeta_{i+2}}{\zeta_{i+1}}, \quad i = 1, \dots, n. \tag{8}$$

This means that the steady-state density and production rate in the RFM can be obtained from the spectral properties of A. In particular, this makes it possible to determine R and e even for very large chains using efficient and numerically stable algorithms for computing the eigenvalues and eigenvectors of a Jacobi matrix.

Theorem 1 has several more important implications. For example, it implies that $R = R(\lambda_0, \dots, \lambda_n)$ is a *strictly concave function* on $\mathbb{R}^{n+1}_{++}$ (Poker et al. 2014). Also, it implies that the sensitivity of the steady state w.r.t. a perturbation in the rates becomes an eigenvalue sensitivity problem. Indeed, Theorem 1 implies that

$$\frac{\partial}{\partial \lambda_i} R = \frac{2}{\sigma^3 \lambda_i^{3/2} \zeta' \zeta} \zeta_{i+1}\zeta_{i+2}, \quad i = 0, \dots, n. \tag{9}$$

This provides a spectral expression for the change in the steady-state production rate caused by a small change in one of the rates. Note that (9) implies in particular that $\frac{\partial}{\partial \lambda_i} R > 0$ for all i. In other words, an increase in any of the rates yields an increase in R.

4.6 Entrainment

Many biological systems are excited by periodic signals, for example the 24 h solar day or the periodic cell cycle division process. An important question is whether the system *entrains* (or phase-locks or synchronizes) to the excitation, that is, whether its behavior converges to a periodic pattern with the same period as the excitation. It is well-known that stable linear time-invariant systems entrain (Zadeh and Desoer 1963). Nonlinear systems, even seemingly simple, may not entrain. For example, their trajectories may display a chaotic pattern rather than converge to a periodic pattern (Nikolaev et al. 2017). There are however two important classes of nonlinear systems that do entrain: contractive systems (Russo et al. 2010), and cooperative systems that admit a first integral (see, e.g., Margaliot et al. 2017).

A function $f : \mathbb{R} \to \mathbb{R}$ is called T-periodic if $f(t) = f(t+T)$ for all t. For example, $\sin(t)$ is $2\pi k$ periodic, for all integers k. Consider the RFM under the following assumptions:

(1) The initiation rate $\lambda_0(t)$ and transition rate $\lambda_i(t)$, $i = 1, \ldots, n$, are continuous, strictly positive and uniformly bounded *functions of time*, i.e., there exist $\delta_1, \delta_2 \in \mathbb{R}$ such that for all $i \in \{0, \ldots, n\}$

$$0 < \delta_1 \leq \lambda_i(t) \leq \delta_2, \quad \text{for all } t \geq 0. \tag{10}$$

(2) There exists a minimal $T > 0$ such that all the $\lambda_i(t)$s are T-periodic.

We refer to this case as the *periodic ribosome flow model* (PRFM). Note that this includes in particular the case where some of the rates are constant, as a constant function is T-periodic for every T. However, item (2) above implies that the case where *all* the rates are constant is ruled out, as then the minimal T is zero. Indeed, this case is just the RFM. The periodicity of the rate/s may be the result for example of periodicity in the abundance of certain tRNA molecules.

Theorem 2 (Margaliot et al. 2014) *The PRFM admits a* unique *periodic solution $\gamma : \mathbb{R}_+ \to \text{int}(C^n)$, with period T, and for any $a \in C^n$ the trajectory emanating from a at time $t = 0$ converges to γ as $t \to \infty$.*

This means that the RFM entrains to periodic excitations in its rates. In particular, the production rate $R(t) = \lambda_n(t)x_n(t)$ converges to the T-periodic function $\lambda_n(t)\gamma_n(t)$.

Example 1 Figure 3 depicts $x_i(t)$, $i = 1, 2, 3$, as a function of t, for a PRFM with $n = 3$, $\lambda_0(t) \equiv 3/5$, $\lambda_1(t) = 1 + \frac{7}{20}\sin(\frac{\pi t}{5})$, $\lambda_2(t) = \frac{4}{5} + \frac{3}{5}\cos(\frac{\pi t}{5} + \frac{1}{3})$, $\lambda_3(t) \equiv 4/5$, and initial condition $x_i(0) = 1/2$, $i = 1, 2, 3$. Note that all the rates here are periodic, with a minimal common period $T = 10$. It may be seen that each state variable converges to a periodic function with period $T = 10$. □

Some features of the flow of molecular machines are not captured by the RFM. For example, the RFM, like ASEP, is based on the assumption that the particle size is equal to the site size, yet it is known that every ribosome covers several codons. Several extensions of the RFM have been introduced. These include (1) the *ribosome flow model on a ring* (RFMR) (Raveh et al. 2015; Zarai et al. 2017f) that is a mean-field approximation of TASEP with periodic boundary conditions; (2) a model for *bidirectional* flow that can describe, for example, the motion of

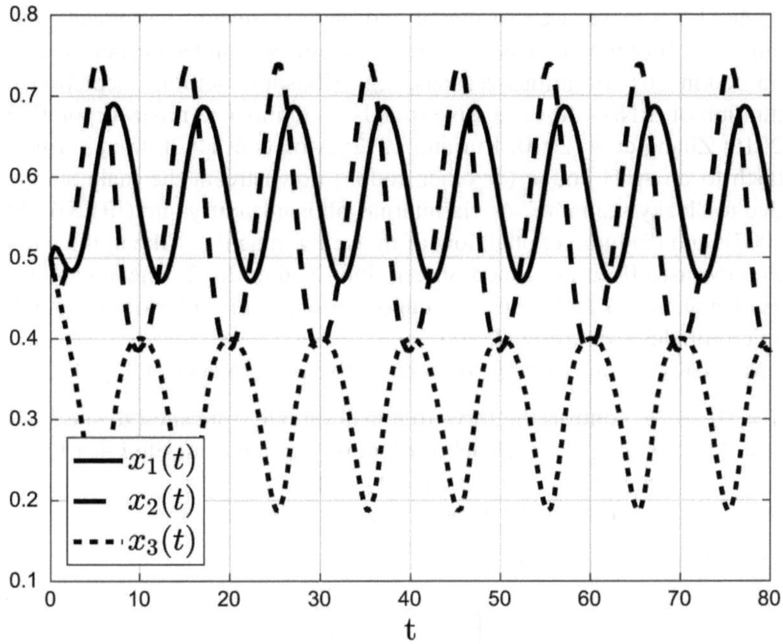

Fig. 3 Densities $x_i(t)$, $i = 1, 2, 3$, as a function of t, for the PRFM in Example 1

RNA Polymerases II during transcription (Edri et al. 2014); (3) an RFM with positive feedback from the production rate to the initiation rate that can describe mRNA circularization and ribosome cycling (Margaliot and Tuller 2013); and (4) a generalization of the RFM that includes nearest-neighbor interactions between the particles (Zarai et al. 2017a).

5 Extensions of the RFM

In this section, we review two models that are motivated by biological phenomena that are not encapsulated by the RFM.

5.1 *Bidirectional Flow with Langmuir Kinetics*

The *mean-field approximation of ASEP with Langmuir kinetics* (MFALK) is a deterministic flow model that encapsulates bidirectional flow along the chain and also the possibility of particles to attach/detach from any site along the chain.

The MFALK can be used to model and analyze several biological processes including (1) ribosomes that detach from the mRNA molecule before reaching the stop codon due to various reasons, e.g., ribosome stalling, depletion in the concentration of tRNAs, or successive rounds of amino acid misincorporation (Sin et al. 2016; Zhang et al. 2010; Kurland 1992; Alberts et al. 2007); (2) ribosomes that attach to an ATG codon (or other codon) downstream the main start codon (e.g., due to "leaky scanning" or via internal ribosome entry site (IRES)) (Alberts et al. 2007); and (3) bidirectional flow of molecular machines like in the motion of RNA Polymerases II during gene transcription (Nudler 2012; Cheung and Cramer 2011; Edri et al. 2014), and the movement of motor proteins like kinesin and dynein along microtubules (Alberts et al. 2007).

The MFALK contains four sets of nonnegative parameters (see Fig. 4):

- $\lambda_i, i = 0, \ldots, n$, controls the forward transition rate from site i to site $i+1$,
- $\gamma_i, i = 0, \ldots, n$, controls the backward transition rate from site $i+1$ to site i,
- $\beta_i, i = 1, \ldots, n$, controls the attachment rate to site i,
- $\alpha_i, i = 1, \ldots, n$, controls the detachment rate from site i.

Let $x_0(t) \equiv 1$, $x_{n+1}(t) \equiv 0$, and

$$z_i(t) := \begin{cases} 0, & i = 0, \\ x_i(t), & i \in \{1, \ldots, n\}, \\ 1, & i = n+1. \end{cases}$$

Then, the dynamical equations describing the MFALK with n sites are given by

$$\begin{aligned} \dot{x}_i &= \lambda_{i-1} x_{i-1}(1 - x_i) - \lambda_i x_i (1 - x_{i+1}) \\ &\quad + \gamma_i z_{i+1}(1 - z_i) - \gamma_{i-1} z_i (1 - z_{i-1}) + \beta_i (1 - x_i) - \alpha_i x_i, \end{aligned} \tag{11}$$

for all $i \in \{1, \ldots, n\}$. The first two terms on the right-hand side of (11) are just like in the RFM. The term $\gamma_i z_{i+1}(1 - z_i) - \gamma_{i-1} z_i (1 - z_{i-1})$ represents backward flow

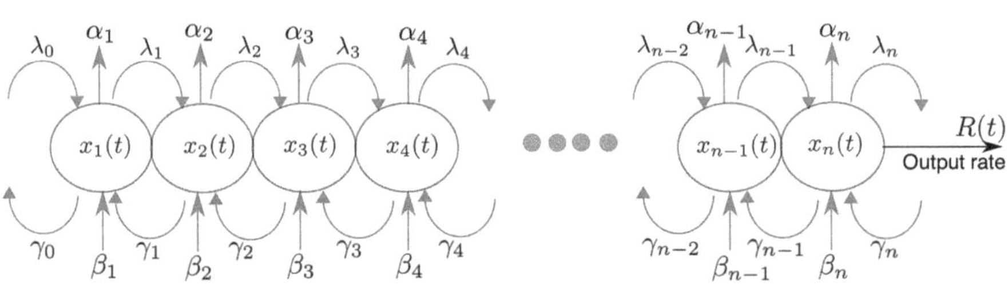

Fig. 4 Topology of the MFALK. The state variable $x_i(t) \in [0, 1]$ describes the density of site i at time t. The parameter λ_i [γ_i] controls the transition rate from site i [$i+1$] to site $i+1$ [i]. The parameter α_i [β_i] controls the detachment [attachment] rate from [to] site i. $R(t)$ denotes the output rate at time t

with soft simple exclusion. The term $\beta_i(1 - x_i)$ represents attachment of particles from the environment to site i, and $\alpha_i x_i$ represents detachment from site i.

The *output rate* from site n at time t is the total flow from this site to the environment:

$$R(t) := (\lambda_n + \alpha_n)x_n(t) - (\gamma_n + \beta_n)(1 - x_n(t)). \tag{12}$$

Note that $R(t)$ may be positive, zero, or negative.

The RFM is a special case of the MFALK with $\alpha_i = \beta_i = \gamma_i = 0$ for all i. Zarai et al. (2017c) have shown that some of the dynamical properties of the RFM described in Sect. 4 hold also for the MFALK.

5.2 *Ribosome Flow Model with Extended Objects*

In the RFM and ASEP, the particle size is equal to the site size. This assumption is not always satisfied in real biological flow. For example, each ribosome typically covers between 9 and 11 codons, and the geometry (e.g., length of the exit tunnel) can be longer than 30 codons (Alberts et al. 2007). Each RNAP typically covers between 42 and 51 nucleotides (Rice et al. 1993). It is interesting to note that the pioneering paper by MacDonald et al. (1968) already considered a version of TASEP with extended objects as a model for ribosome flow during translation (see also Lodish 1974).

The *ribosome flow model with extended objects* (RFMEO) describes the unidirectional flow of particles that cover ℓ site units, with $1 \le \ell \le n$. The RFMEO is a mean-field approximation of TASEP with extended objects (Zarai et al. 2017e).

Assume, without loss of generality, that the particle is "processing" (i.e., transcribing or translating) the left-most site it is covering, and refer to this part of the particle as the *reader*. A similar assumption is used in TASEP with extended objects (Dong et al. 2007b; Shaw et al. 2004a,b; Lakatos and Chou 2003; Shaw et al. 2003). Thus, the statement "the particle is at site i" means that the reader is located at site i; the particle is processing site i; its corresponding transition rate is λ_i; and sites $i, \ldots, i + \ell - 1$ are covered by this particle.

Let $x_i(t)$ denote the normalized *reader* occupancy level at site i at time t, and let $y_i(t)$ denote the normalized *coverage* occupancy level at site i at time t, that is,

$$y_i(t) := \sum_{j=\max\{1, i-\ell+1\}}^{i} x_j(t), \quad i = 1, \ldots, n.$$

The dynamics of the RFMEO with n sites is given by n nonlinear first-order ordinary differential equations:

$$\dot{x}_i = q_{i-1}(x) - q_i(x), \quad i = 1, \ldots, n, \tag{13}$$

where q_{i-1} $[q_i]$ is the flow into [out of] site i. This flow is given by

$$q_i(x) := \lambda_i x_i(1 - y_{i+\ell}), \quad i = 0, \dots, n, \tag{14}$$

with $x_0(t) \equiv 1$, and $y_j(t) \equiv 0$ for all $j > n$.

To explain (13), consider for example the equation for the change in the density at site 1, namely,

$$\dot{x}_1 = q_0(x) - q_1(x) = \lambda_0(1 - y_\ell) - \lambda_1 x_1(1 - y_{\ell+1}).$$

The term $\lambda_0(1 - y_\ell)$ represents the entry rate into site 1. Indeed, since the entering particle will cover sites $1, 2, \dots, \ell$, this entry rate decreases with the coverage density $y_\ell = x_1 + \cdots + x_\ell$. (In the literature on TASEP with extended objects this is referred to as the "complete-entry" flow (Dong et al. 2007b)). The term $\lambda_1 x_1(1 - y_{\ell+1})$ is the flow from site 1 to site 2. This increases with the occupancy at site 1 and, similarly, decreases with the coverage occupancy $y_{\ell+1}$. We note that the dynamical equations describing the RFMEO are the same for any chosen reader location (e.g., choosing the reader at location $\ell/2$ results in exactly the same RFMEO equations).

The output rate of particles from the chain is denoted by $R(t) := \lambda_n x_n$. Note that in the special case $\ell = 1$ we have $y_i = x_i$ for all $i = 1, \dots, n$, and then (13) reduces to the RFM.

Some of the dynamical properties of the RFM described in Sect. 4 hold also for the RFMEO (Zarai et al. 2017e).

The next section describes several biological implications of the dynamical properties of the RFM.

6 Implications of the Analysis of the RFM to Gene Expression Modeling and Engineering

Analysis of the RFM yields results that are relevant to modeling and engineering gene expression. Some of these are reviewed in this section.

6.1 Constrained Maximization of the Steady-State Production Rate

As mentioned above, translation is known to be one of the most energy-consuming processes in the cell (Alberts et al. 2007; Tuller et al. 2010), and it is natural to assume that evolution shaped this process to *maximize* protein production subject to the limited biocellular budget. If this is indeed so then one can estimate various parameters of the translation machinery by solving an appropriate mathematical optimization problem. The same problem also arises in the context of synthetic biology, namely, re-engineering heterologous genes in order to maximize their production rate in a host organism (Romanos et al. 1992; Moks et al. 1987; Binnie et al. 1997).

To study this using the RFM, consider the problem of finding the rates $\lambda_0, \ldots, \lambda_n$ that maximize the steady-state production rate R subject to the constraint: $\sum_{i=0}^{n} w_i \lambda_i \leq b$. Here the positive w_is allow a different weighting of each rate, and $b > 0$ is related to the limited biomolecular budget in the cell. The strict concavity of R implies that this problem is a *convex optimization problem* (Boyd and Vandenberghe 2004). It thus admits a unique solution λ_i^*, $i = 0, \ldots, n$, that can be determined using highly-efficient numerical algorithms that scale well with n (Poker et al. 2014).

Example 2 Figure 5 depicts the optimal rates λ_i^* for an RFM with $n = 8$, and with a homogeneous constraint, i.e., $w_0 = \cdots = w_8 = b = 1$. The optimal values were found numerically using a simple search algorithm that is guaranteed to converge for convex optimization problems. It may be noticed that the optimal rates are symmetric w.r.t. the center of the chain, and increase toward the center of the chain. This implies that when considering a homogeneous constraint (i.e., when assigning equal weighting to all the rates), the most important rates are those near the center of the chain. □

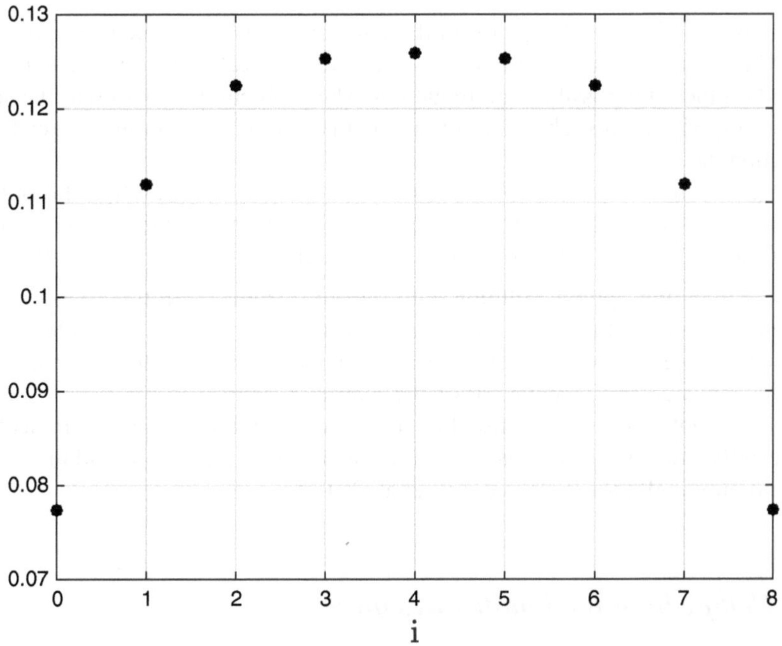

Fig. 5 Optimal rates λ_i^*, as a function of i, for a RFM with $n = 8$ and the constraint $\sum_{i=0}^{8} \lambda_i \leq 1$

6.2 Optimal Down-Regulation of Translation

Down-regulation of translation is important in cell biology, medicine, and biotechnology. For example, in many organisms small RNA genes, such as microRNAs, hybridize to the mRNA in specific locations (Ghildiyal and Zamore 2009; Inui et al. 2010) in order to down-regulate translation initiation or elongation (Fabian et al. 2010; Filipowicz et al. 2008) and/or promote mRNA degradation. Cancer cells are often targeted via generating tumor-specific RNA interference (RNAi) genes that down-regulate the oncogenes (Tavazoie et al. 2008; Zhang et al. 2003; Devi 2006). Furthermore, many viral therapeutic treatments and viral vaccines are based on the attenuation of mRNA translation in the viral genes (Ben-Yehezkel et al. 2015; Goz and Tuller 2015; Wang et al. 2005; Coleman et al. 2008; Perez et al. 2009).

In order to study down-regulation of translation using the RFM, consider the following problem.

Problem 1 *Given an RFM with n sites, rates* $\bar{\lambda}_0, \ldots, \bar{\lambda}_n$*, and a total reduction budget* $b \in [0, \min_i\{\bar{\lambda}_i\})$*, define the set*

$$\Omega^{n+1}(\bar{\lambda}, b) := \left\{ \left[\bar{\lambda}_0 - \varepsilon_0, \ldots, \bar{\lambda}_n - \varepsilon_n\right] : \varepsilon_i \geq 0, \sum_{i=0}^{n} \varepsilon_i = b \right\}.$$

Find a set of rates $\lambda^* \in \Omega^{n+1}$ *such that* $R(\lambda^*) = \min_{\lambda \in \Omega^{n+1}} R(\lambda)$.

In other words, Ω^{n+1} is the set of all the rates that can be obtained by applying a total reduction b to the given rates $\bar{\lambda}_i$. From a mathematical point of view, b provides a bound on the total possible rate reduction. The problem then is to find the set of rates corresponding to such a total reduction that provides the minimal stead-state production rate.

Using the strict concavity of R, Zarai et al. (2017d) showed that the optimal solution always corresponds to reducing *all* the reduction budget b from a single rate. If there exists an index $j \in \{0, \ldots, n\}$, such that $\frac{\partial}{\partial \lambda_j} R(\bar{\lambda}) > \frac{\partial}{\partial \lambda_i} R(\bar{\lambda})$, for all $i \neq j$, then the optimal solution is to reduce b from $\bar{\lambda}_j$. In this case $\bar{\lambda}_j$ is the "bottleneck" rate, as the sensitivity of R w.r.t. this rate is maximal. Note that the sensitivities here can be determined using efficient algorithms for computing the Perron eigenvalue and eigenvector of the matrix A defined in Eq. (7). If such an index j does not exist, one may use the spectral representation described in Sect. 4.5 to efficiently evaluate R after reducing b from each rate separately, and then select the minimum of all these values (Zarai et al. 2017d).

6.3 Competition for Limited Resources

Biological evidence suggests that the competition for RNAPs and ribosomes, as well as other various transcription and translation factors, plays a key role in the cellular economy of gene expression. The limited availability of these resources is one of the reasons why gene, mRNA, and protein levels in the cell do not necessarily correlate (Tuller et al. 2010; Sharp et al. 2010; Tuller and Zur 2015; Richter and Smith 1981; Vind et al. 1993; Jens and Rajewsky 2015; Ceroni et al. 2015; Brackley et al. 2011; Greulich et al. 2012), and should be taken into account when designing gene expression circuits that involve heterologous gene expression (Ceroni et al. 2015; Gyorgy et al. 2015; Dana and Tuller 2012). The competition for these resources leads to an indirect coupling between the concurrent gene expression processes in the cell. This is particularly relevant when many identical intracellular processes, all using the same resources, take place in parallel (Brackley et al. 2011).

In order to analyze the effect of competition for the limited resources in the context of translation, Raveh et al. (2016) introduced a network of RFMs interconnected via a dynamic pool of free ribosomes. The pool feeds the initiation sites in all the RFMs, and the ribosomes exiting every RFM are fed back into the pool. The total number of ribosomes in this network is conserved, so if more ribosomes bind to some RFM the pool is depleted, and consequently the initiation rates in all the RFMs decrease. This network can also be applied to study competition for resources in other scenarios, e.g., competition for RNAPs during transcription.

It was shown in Raveh et al. (2016) that the network always converges to a steady state. If we fix all the rates and the total initial density of ribosomes in the network then this steady state is unique. This allows to address questions such as how do these steady-state profiles change when one of the rates in one of the RFMs is modified? Suppose, without loss of generality, that a rate λ_i in the first RFM in the network is increased to a value $\bar{\lambda}_i > \lambda_i$. It was shown that in the modified network the steady-state values change as follows. The production rate in the first RFM increases, and the other production rates and the pool occupancy either *all increase or all decrease*. This result has an intuitive biophysical interpretation: If the modified transition rate is a bottleneck rate in the mRNA chain, then increasing it leads to a faster flow of ribosomes through this mRNA molecule. This increases the number of free ribosomes in the pool and, therefore, the production rates in all other mRNAs as well. On the other-hand, if the modified transition rate is located upstream of the bottleneck rate then increasing it worsens the "traffic jam" of ribosomes along this mRNA, the pool is depleted, and the production rates in all the other mRNAs decrease. These results highlight the importance of modeling the translation of not only a single mRNA molecule, but rather a set of molecules that are indirectly coupled by the competition for shared resources.

As noted above, new experimental procedures provide more and more data on gene expression. Mathematical or computational models are needed to integrate and explain this data. The next section reviews several applications of the RFM in this context.

7 Applications of the RFM to Studying Gene Expression Based on Large-Scale Genomic Data

We describe several applications of the RFM to the process of gene expression based on genomic data analysis.

7.1 Parameter Estimation for the RFM and Predictions of Experimental Data

In order to analyze genomic data using the RFM, the various parameters of the RFM, i.e., the λ_is, must be inferred and set to their intracellular values. The codon decoding rates can be inferred from Ribo-seq experiments (Ingolia et al. 2009; Dana and Tuller 2014a) or from computational algorithms that use transcript features (e.g., adaptation to the tRNA pool, local mRNA folding, and the interaction of the nascent peptide with the ribosome exit tunnel (Tuller et al. 2011)) in order to predict the RFM rates. Similarly, the elongation rates of the RNAP can be estimated based on NGS experiments such as NET-seq (Edri et al. 2014; Churchman and Weissman 2011; Cohen et al. 2017).

Another fundamental parameter that must be estimated is the initiation rate λ_0. It is not trivial to experimentally measure initiation; however, there exist complex models that enable the prediction of the translation initiation rate based on, for example, the nucleotide composition of the transcript near the start codon or via estimation from experimental data of ribosome densities (Salis et al. 2009; Zur and Tuller 2013; Ciandrini et al. 2013). For example, the model of Salis et al. (2009) for prokaryote is based on summing five free energy terms: (1) the folding energy of the mRNA subsequence prior to binding with the 30S complex; (2) the energy released when the last nine nucleotides of the 16S rRNA cofold and hybridize with the mRNA subsequence at the 16S rRNA-binding site; (3) the energy released when the tRNA(fMet) anticodon hybridizes to the start codon; (4) the energy released when the standby site (near the 16S rRNA-binding site) folds; and (5) an energetic penalty for a nonoptimal distance between the 16S rRNA-binding site and the start codon. An eukaryotic translation initiation model should consider, among others, the nucleotide composition surrounding the start codon, the additional AUGs surrounding the start codon, and the strength of the mRNA folding surrounding the start codon (Zur and Tuller 2013; Kozak 1986; Ben-Yehezkel et al. 2015).

Several papers compared predictions of the RFM with biological measurements. For example, protein levels and ribosome densities in translation (Reuveni et al. 2011), and RNAP densities in transcription (Edri et al. 2014). The results demonstrate high correlation between gene expression measurements and the RFM predictions. A publicly available application enables biologists to easily use the RFM to derive such predictions (Zur and Tuller 2012).

7.2 *Effect of Ribosomal Drop-Off on Production Rate*

Translation is the most energetically consuming process in the cell, and ribosome drop-off before reaching the stop codon results in truncated, nonfunctional and possibly deleterious proteins. Nevertheless, there seems to be a certain minimal abortion rate even in non-stressed conditions (Sin et al. 2016; Kurland and Mikkola 1993). In recent years ribosome drop-off has been modeled and studied in several interesting papers (Bonnin et al. 2017; Sin et al. 2016; Keiler et al. 1996; Keiler 2015; Zaher and Green 2011; Chadani et al. 2010; Subramaniam et al. 2014; Gilchrist and Wagner 2006). It was suggested that in some cases ribosome drop-off is important for proof reading (Zaher and Green 2009), and that ribosome stalling and abortion play a role in ribosome homeostasis and thus translational regulation (Shoemaker et al. 2010; Zupanic et al. 2014; Mills and Green 2017).

To analyze the effect of ribosomal drop-off on the protein production rate, Zarai et al. (2017c) used the MFALK with backward and attachment rates set to zero (i.e., $\beta_i = \gamma_i = 0$, for all i). The detachment rates α_i were estimated based on biological data for *S. cerevisiae* using values from (Sin et al. 2016; Kurland and Mikkola 1993). The transition rates λ_i were estimated using Ribo-seq data for the codon decoding rates (Dana and Tuller 2014b).

The results show, for example, that as the drop-off rate per site increases from 10^{-4} to 10^{-2}, the average steady-state density decreases by about 30% and the steady-state production rate decreases by about 50%. This demonstrates the significant ramifications that ribosomal drop-off is expected to have on translation.

7.3 *Coupling by Sharing Finite Intracellular Resources and Entrainment*

As noted above, oscillations in one or more of the RFM rates induce a periodic solution γ of the RFM and every trajectory converges to γ. A more general question is what will be the effect of oscillations in one or more rates in an RFM that is connected, via the competition for shared resources, to other RFMs?

Zarai and Tuller (2017) studied this question using an RFM-based whole cell simulation of translation in *S. cerevisiae* that includes competition for ribosomes (see Fig. 6). It was shown that fluctuations in mRNA levels of a single gene or a group of genes induce a global periodic behavior in the network: The ribosomal densities and mRNA production rates of all *S. cerevisiae* mRNAs oscillate. By numerically measuring the oscillation amplitudes, it was demonstrated that fluctuations of endogenous and heterologous genes can cause a significant fluctuation of up to 50% in the steady-state production rates of the rest of the genes. Furthermore, oscillating mRNAs that experience high ribosomal occupancy (e.g., ribosomal traffic jams) have the largest impact on the translation of the *S. cerevisiae* genome. Synonymously mutating these oscillating mRNAs can alleviate the fluctuations in all *S. cerevisiae* genes.

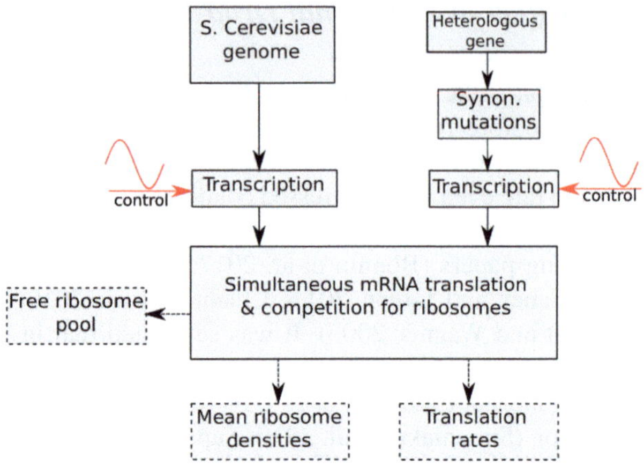

Fig. 6 Flow diagram of the translation network analyzed in Zarai and Tuller (2017). Dashed blocks specify numerical measurements

8 Discussion

In his 1948 essay "Intelligent Machinery", Turing described a machine that consisted of: "... an unlimited memory capacity obtained in the form of an infinite tape marked out into squares, on each of which a symbol could be printed. At any moment there is one symbol in the machine; it is called the scanned symbol." (Turing 2004).

Evolution came up with a related idea a long time ago. Genetic information is coded as an ordered list of symbols, and machines like RNAPs and ribosomes "read" and process this information symbol by symbol. The relation between genetics and the Turing machine is well-known. For example, Adleman (1994) states that "One can imagine the eventual emergence of a general purpose computer consisting of nothing more than a single macromolecule conjugated to a ribosomelike collection of enzymes that act on it". For another possible biomolecular embodiment of a Turing machine, see Shapiro (2012) and the references therein. Shapiro (2012) also points out that "Molecular machines such as DNA polymerase, RNA polymerase and the ribosome are most naturally understood as simple finite-state transducers, a special case of the Turing machine."

The flow of these molecular machines along the genetic material is of great importance in gene expression. We reviewed several deterministic continuous-time models for the flow of biological machines based on the RFM. The RFM is both easy to simulate and amenable to rigorous analysis using tools from systems and control theory, which are not traditionally employed in the context of gene expression modeling and analysis. This enables efficient modeling and analysis of large-scale genomic data (Zur and Tuller 2016; Zarai and Tuller 2017). We furthermore showed how the RFM can be used for rigorous analysis of processes such as translation and transcription. This can decipher novel mechanisms in gene expression dynamics and evolution. Recent large-scale experimental approaches enable fitting the parameters of the RFM to the intracellular conditions; the fitted RFM provides not only basic predictions (such as production rate and ribosome densities, which can be measured directly), but also more advanced predictions that currently cannot be fully measured at a genomic level like optimality and sensitivity of translation, and the effect of ribosome drop-off on the production rate.

There are many open problems related to the RFM dynamics and gene expression regulation and evolution. For example, the proofs of entrainment in contractive systems are based on implicit arguments and as such provide no information on the periodic solution of the RFM (or network of RFMs), except for its period. An important problem is obtaining more information on the geometry of the periodic solution, its amplitude, average, and dependence on various parameters ((see Margaliot and Coogan 2017) for some related ideas). Another important problem is to better understand the level of optimality of the translation process in living cells and the evolution of this process.

Directions for further research can benefit from combining biological experiments with analysis tools from systems and control theory. On the one hand, important biological assertions on gene expression can be naturally addressed in the framework of the RFM. One example from a recent review paper (Mills and Green 2017) is the statement: "Moreover, the translation of mRNAs with low initiation rates is likely to be most negatively affected by changes in ribosome concentration." As another example, a "comparative genomics" analysis of the RFMs fitted to various organisms may teach us about the evolution of translation.

On the other hand, many ideas and tools from systems and control theory applied to the RFM may immediately lead to important biological ramifications. Examples include applying systematic approaches for parameter estimation to deduce the transition rates, as well as using tools from control theory to understand what density profiles can be obtained by manipulating one or more of the RFM rates (Zarai et al. 2017b). As another example, let us show how a classical topic from random matrix theory is related to an important biological question. In practice, many identical mRNA molecules undergo simultaneous translation in the cell and the ribosomes translating these may experience different transition rates due to, say, the variability of tRNA abundance in different locations in the cell. This can be modeled by assuming that the RFM rates are not constant, but rather they are random variables (RVs) with some known distribution supported over $\mathbb{R}_{++}$. A natural question is what will be the average protein production rate? In the context of the matrix A given in (7) this amounts to the following question: given the distributions of the RVs λ_i what is the average value of the maximal eigenvalue of A?

In addition, most of the research so far focused on using models like ASEP and RFM to study translation [transcription] on a single, isolated mRNA [DNA] molecule. We think that tools from the theory of networked systems can be applied successfully to study networks of interconnected RFMs modeling the large-scale concurrent cellular processes of gene expression.

Finally, one bottleneck in the field is related to our ability to directly and accurately measure variables related to ribosome movement, e.g., initiation rate, ribosome/RNAP abortions, ribosome/RNAP collisions over a single mRNA/DNA, etc. Such measurements are essential to validate analytical and computational predictions. Current experimental procedures are very noisy, include various biases, cannot directly measure some important translation variables, and are based on average measurements over a pool of cells/mRNAs. In addition, developing experimental approaches for high-resolution analysis of additional types of intracellular machines (e.g., molecular motors movement on the cytoskeletal filaments) is required in order to better understand the universality of the theories and computational models in this field.

References

Adleman LM (1994) Molecular computation of solutions to combinatorial problems. Science 266:1021–1024

Alberts B, Johnson A, Lewis J, Raff M, Roberts K, Walter P (2007) Molecular biology of the cell, 5th edn. Garland Science, New York

Aminzare Z, Sontag ED (2014) Contraction methods for nonlinear systems: a brief introduction and some open problems. In: Proceedings of 53rd IEEE conference on decision and control. Los Angeles, CA, pp 3835–3847

Ben-Yehezkel T, Atar S, Zur H, Diament A, Goz E, Marx T, Cohen R, Dana A, Feldman A, Shapiro E, Tuller T (2015) Rationally designed, heterologous S. cerevisiae transcripts expose novel expression determinants. RNA Biol 12:972–984

Binnie C, Cossar J, Stewart D (1997) Heterologous biopharmaceutical protein expression in streptomyces. Trends Biotechnol 15(8):315–320

Blythe RA, Evans MR (2007) Nonequilibrium steady states of matrix-product form: a solver's guide. J Phys A Math Theor 40(46):R333–R441

Bonnin P, Kern N, Young NT, Stansfield I, Romano MC (2017) Novel mRNA-specific effects of ribosome drop-off on translation rate and polysome profile. PLoS Comput Biol 13(5):e1005555

Boyd S, Vandenberghe L (2004) Convex optimization. Cambridge University Press, Cambridge

Brackley CA, Romano MC, Thiel M (2011) The dynamics of supply and demand in mRNA translation. PLoS Comput Biol 7(10):e1002203

Brauer F (2008) Compartmental models in epidemiology. In: Brauer F, van den Driessche P, Wu J (eds) Mathematical epidemiology. Lecture notes in mathematics, vol 1945. Springer, Berlin, pp 19–79

Ceroni F, Algar R, Stan GB, Ellis T (2015) Quantifying cellular capacity identifies gene expression designs with reduced burden. Nat Methods 12:415–418

Chadani Y, Ono K, Ozawa S, Takahashy Y, Takay K, Nanamiya H, Tozawa Y, Kutsukake K, Abo T (2010) Ribosome rescue by Escherichia coli ArfA (YhdL) in the absence of trans-translation systems. Mol Microbiol 78:796–808

Chandar N, Viselli S (2012) Cell and molecular biology. Wolters Kluwer Health, Philadelphia

Cheung ACM, Cramer P (2011) Structural basis of RNA polymerase II backtracking, arrest and reactivation. Nature 471(7337):249–253

Chou T, Lakatos G (2004) Clustered bottlenecks in mRNA translation and protein synthesis. Phys Rev Lett 93(19):198101

Chou T, Mallick K, Zia RKP (2011) Non-equilibrium statistical mechanics: from a paradigmatic model to biological transport. Rep Prog Phys 74:116601

Churchman LS, Weissman JS (2011) Nascent transcript sequencing visualizes transcription at nucleotide resolution. Nature 469(7330):368–373

Ciandrini L, Stansfield I, Romano M (2013) Ribosome traffic on mRNAs maps to gene ontology: genome-wide quantification of translation initiation rates and polysome size regulation. PLoS Comput Biol 9(1):e1002866

Cohen E, Zafrir Z, Tuller T (2017) A code for transcription elongation speed. RNA Biol 1–14. https://doi.org/10.1080/15476286.2017.1384118

Coleman J, Papamichail D, Skiena S, Futcher B, Wimmer E, Mueller S (2008) Virus attenuation by genome-scale changes in codon pair bias. Science 320:1784–1787

Crick F (1970) Central dogma of molecular biology. Nature 227(5258):561–563

Dana A, Tuller T (2012) Efficient manipulations of synonymous mutations for controlling translation rate–an analytical approach. J Comput Biol 19:200–231

Dana A, Tuller T (2014a) The effect of tRNA levels on decoding times of mRNA codons. Nucleic Acids Res 42(14):9171–9181

Dana A, Tuller T (2014b) Mean of the typical decoding rates: a new translation efficiency index based on the analysis of ribosome profiling data. G3 5(1):73–80

Derrida B (1998) An exactly soluble non-equilibrium system: the asymmetric simple exclusion process. Phys Rep 301(1):65–83

Derrida B, Domany E, Mukamel D (1992) An exact solution of a one-dimensional asymmetric exclusion model with open boundaries. J Stat Phys 69(3–4):667–687

Derrida B, Evans MR, Hakim V, Pasquier V (1993) Exact solution of a 1D asymmetric exclusion model using a matrix formulation. J Phys A Math Gen 26(7):1493

Devi G (2006) siRNA-based approaches in cancer therapy. Cancer Gene Ther 13(9):819–829

Dong J, Schmittmann B, Zia RK (2007a) Towards a model for protein production rates. J Stat Phys 128(1–2):21–34

Dong JJ, Schmittmann B, Zia RKP (2007b) Inhomogeneous exclusion processes with extended objects: the effect of defect locations. Phys Rev E 76:051113

Edri S, Gazit E, Cohen E, Tuller T (2014) The RNA polymerase flow model of gene transcription. IEEE Trans Biomed Circuits Syst 8(1):54–64

Evans M, Blythe R (2002) Nonequilibrium dynamics in low-dimensional systems. Physica A 313(1):110–152

Fabian M, Sonenberg N, Filipowicz W (2010) Regulation of mRNA translation and stability by microRNAs. Annu Rev Biochem 79:351–379

Filipowicz W, Bhattacharyya S, Sonenberg N (2008) Mechanisms of post-transcriptional regulation by microRNAs: are the answers in sight? Nat Rev Genet 9(2):102–114

Ghildiyal M, Zamore P (2009) Small silencing RNAs: an expanding universe. Nat Rev Genet 10:94–108

Gilchrist MA, Wagner A (2006) A model of protein translation including codon bias, nonsense errors, and ribosome recycling. J Theor Biol 239(4):417–434

Goz E, Tuller T (2015) Widespread signatures of local mRNA folding structure selection in four Dengue virus serotypes. BMC Genomics 16(10):S4

Greulich P, Ciandrini L, Allen RJ, Romano MC (2012) Mixed population of competing totally asymmetric simple exclusion processes with a shared reservoir of particles. Phys Rev E 85:011142

Gyorgy A, Jimenez JI, Yazbek J, Huang H, Chung H, Weiss R, Del Vecchio D (2015) Isocost lines describe the cellular economy of genetic circuits. Biophys J 109:639–46

Holza M, Fahrb A (2001) Compartment modeling. Adv Drug Deliv Rev 48:249–264

Horn RA, Johnson CR (2013) Matrix analysis. Cambridge University Press, Cambridge

Ingolia NT (2014) Ribosome profiling: new views of translation, from single codons to genome scale. Nat Rev Genet 15(3):205–213

Ingolia NT, Ghaemmaghami S, Newman JR, Weissman JS (2009) Genome-wide analysis in vivo of translation with nucleotide resolution using ribosome profiling. Science 324(5924):218–223

Inui M, Martello G, Piccolo S (2010) MicroRNA control of signal transduction. Nat Rev Mol Cell Biol 11(4):252–263

Iwasaki S, Ingolia NT (2016) Seeing translation. Science 352(6292):1391–1392

Jacquez JA (1996) Compartmental analysis in biology and medicine, 3rd edn. BioMedware, Ann Arbor, MI

Jacquez JA, Simon CP (1993) Qualitative theory of compartmental systems. SIAM Rev 35(1):43–79

Jens M, Rajewsky N (2015) Competition between target sites of regulators shapes post-transcriptional gene regulation. Nat Rev Genet 16(2):113–126

Johansson M, Chen J, Tsai A, Kornberg G, Puglisi J (2014) Sequence-dependent elongation dynamics on macrolide-bound ribosomes. Cell Rep 7:1534–1546

Keiler K (2015) Mechanisms of ribosome rescue in bacteria. Nat Rev Microbiol 13:285–297

Keiler K, Waller P, Sauer R (1996) Role of a peptide tagging system in degradation of proteins synthesized from damaged messenger RNA. Science 271:990–993

Kolomeisky AB (1998) Asymmetric simple exclusion model with local inhomogeneity. J Phys A Math Gen 31(4):1153

Kozak M (1986) Point mutations define a sequence flanking the aug initiator codon that modulates translation by eukaryotic ribosomes. Cell 44(2):283–92

Kurland C (1992) Translational accuracy and the fitness of bacteria. Ann Rev Genet 26:29–50

Kurland C, Mikkola R (1993) The impact of nutritional state on the microevolution of ribosomes. In: Kjelleberg S (ed) Starvation in bacteria. Plenum Press, New York, NY, pp 225–238

Lakatos G, Chou T (2003) Totally asymmetric exclusion processes with particles of arbitrary size. J Phys A Math Gen 36:20272041

Lodish HF (1974) Model for the regulation of mRNA translation applied to haemoglobin synthesis. Nature 251:385–388

Lohmiller W, Slotine JJE (1998) On contraction analysis for non-linear systems. Automatica 34:683–696

MacDonald CT, Gibbs JH (1969) Concerning the kinetics of polypeptide synthesis on polyribosomes. Biopolymers 7(5):707–725

MacDonald CT, Gibbs JH, Pipkin AC (1968) Kinetics of biopolymerization on nucleic acid templates. Biopolymers 6:1–25

Margaliot M, Coogan S (2017) Approximating the frequency response of contractive systems. CoRR abs/1702.06576. http://arxiv.org/abs/1702.06576

Margaliot M, Tuller T (2012) Stability analysis of the ribosome flow model. IEEE/ACM Trans Comput Biol Bioinform 9:1545–1552

Margaliot M, Tuller T (2013) Ribosome flow model with positive feedback. J R Soc Interface 10:20130267

Margaliot M, Sontag ED, Tuller T (2014) Entrainment to periodic initiation and transition rates in a computational model for gene translation. PLoS ONE 9(5):e96039

Margaliot M, Sontag ED, Tuller T (2016) Contraction after small transients. Automatica 67:178–184

Margaliot M, Grüne L, Kriecherbauer T (2018) Entrainment in the master equation. Roy Soc Open Sci 5(4). https://doi.org/10.1098/rsos.172157

Mayer A, Churchman L (2016) Genome-wide profiling of rna polymerase transcription at nucleotide resolution in human cells with native elongating transcript sequencing. Nat Protoc 11:813–833

Mills EW, Green R (2017) Ribosomopathies: there's strength in numbers. Science 358(6363). https://doi.org/10.1126/science.aan2755

Moks T, Abrahmsen L, Holmgren E, Bilich M, Olsson A, Pohl G, Sterky C, Hultberg H, Josephson SA (1987) Expression of human insulin-like growth factor I in bacteria: use of optimized gene fusion vectors to facilitate protein purification. Biochemistry 26(17):5239–5244

Myasnikov AG, Kundhavai Natchiar S, Nebout M, Hazemann I, Imbert V, Khatter H, Peyron JF, Klaholz BP (2016) Structure-function insights reveal the human ribosome as a cancer target for antibiotics. Nat Commun 7:12856

Newhart A, Janicki SM (2014) Seeing is believing: Visualizing transcriptional dynamics in single cells. J Cell Physiol 229(3):259–265

Nikolaev EV, Rahi SJ, Sontag E (2017) Subharmonics and chaos in simple periodically-forced biomolecular models. bioRxiv p 145201

Nudler E (2012) RNA polymerase backtracking in gene regulation and genome instability. Cell 149(7):1438–1445

Perez JT, Pham AM, Lorini MH, Chua MA, Steel J, tenOever BR (2009) MicroRNA-mediated species-specific attenuation of influenza A virus. Nat Biotechnol 27(6):572–576

Pinkoviezky I, Gov N (2013) Transport dynamics of molecular motors that switch between an active and inactive state. Phys Rev E 88(2):022714

Poker G, Zarai Y, Margaliot M, Tuller T (2014) Maximizing protein translation rate in the nonhomogeneous ribosome flow model: a convex optimization approach. J R Soc Interface 11(100):20140713

Raveh A, Zarai Y, Margaliot M, Tuller T (2015) Ribosome flow model on a ring. IEEE/ACM Trans Comput Biol Bioinform 12(6):1429–1439

Raveh A, Margaliot M, Sontag E, Tuller T (2016) A model for competition for ribosomes in the cell. J R Soc Interface 13(116):20151062

Reuveni S, Meilijson I, Kupiec M, Ruppin E, Tuller T (2011) Genome-scale analysis of translation elongation with a ribosome flow model. PLoS Comput Biol 7(9):e1002127

Rice GA, Chamberlin MJ, Kane CM (1993) Contacts between mammalian RNA polymerase II and the template DNA in a ternary elongation complex. Nucleic Acids Res 21(1):113–118

Richter JD, Smith LD (1981) Differential capacity for translation and lack of competition between mRNAs that segregate to free and membrane-bound polysomes. Cell 27:183–191

Romanos M, Scorer C, Clare J (1992) Foreign gene expression in yeast: a review. Yeast 8(6):423–488

Russo G, di Bernardo M, Sontag ED (2010) Global entrainment of transcriptional systems to periodic inputs. PLoS Comput Biol 6:e1000739

Salis H, Mirsky E, Voigt C (2009) Automated design of synthetic ribosome binding sites to control protein expression. Nat Biotechnol 27(10):946–950

Schadschneider A, Chowdhury D, Nishinari K (2011) Stochastic transport in complex systems: from molecules to vehicles. Elsevier, Amsterdam

Shapiro E (2012) A mechanical turing machine: blueprint for a biomolecular computer. Interface Focus 2(4):497–503

Sharp PM, Emery LR, Zeng K (2010) Forces that influence the evolution of codon bias. Philos Trans R Soc Lond B 365(1544):1203–1212

Shaw LB, Zia RK, Lee KH (2003) Totally asymmetric exclusion process with extended objects: a model for protein synthesis. Phys Rev E Stat Nonlin Soft Matter Phys 68:021910

Shaw LB, Kolomeisky AB, Lee KH (2004a) Local inhomogeneity in asymmetric simple exclusion processes with extended objects. J Phys A Math Gen 37(6):2105

Shaw LB, Sethna JP, Lee KH (2004b) Mean-field approaches to the totally asymmetric exclusion process with quenched disorder and large particles. Phys Rev E 70(2):021901

Shoemaker C, Eyler D, Green R (2010) Dom34:Hbs1 promotes subunit dissociation and peptidyl-tRNA drop-off to initiate no-go decay. Science 330(6002):369–372

Sin C, Chiarugi D, Valleriani A (2016) Quantitative assessment of ribosome drop-off in E. coli. Nucleic Acids Res 44(6):2528–2537

Smith HL (1995) Monotone Dynamical systems: an introduction to the theory of competitive and cooperative systems. Mathematical surveys and monographs, vol 41. American Mathematical Society, Providence, RI

Spitzer F (1970) Interaction of Markov processes. Adv Math 5:246–290

Subramaniam A, Zid B, O'Shea E (2014) An integrated approach reveals regulatory controls on bacterial translation elongation. Cell 159(5):1200–1211

Tavazoie SF, Alarcón C, Oskarsson T, Padua D, Wang Q, Bos PD, Gerald WL, Massagué J (2008) Endogenous human microRNAs that suppress breast cancer metastasis. Nature 451(7175):147–152

Tripathy G, Barma M (1998) Driven lattice gases with quenched disorder: Exact results and different macroscopic regimes. Phys Rev E 58:1911–1926

Tuller T, Zur H (2015) Multiple roles of the coding sequence 5' end in gene expression regulation. Nucleic Acids Res 43(1):13–28

Tuller T, Carmi A, Vestsigian K, Navon S, Dorfan Y, Zaborske J, Pan T, Dahan O, Furman I, Pilpel Y (2010) An evolutionarily conserved mechanism for controlling the efficiency of protein translation. Cell 141(2):344–354

Tuller T, Veksler I, Gazit N, Kupiec M, Ruppin E, Ziv M (2011) Composite effects of gene determinants on the translation speed and density of ribosomes. Genome Biol 12(11):R110

Turing A (2004) Intelligent machinery. In: Copeland BJ (ed) The essential turing. Clarendon Press, Oxford, pp 411–432

Vind J, Sorensen MA, Rasmussen MD, Pedersen S (1993) Synthesis of proteins in Escherichia coli is limited by the concentration of free ribosomes: expression from reporter genes does not always reflect functional mRNA levels. J Mol Biol 231:678–688

Wang Q, Contag C, Ilves H, Johnston B, Kaspar R (2005) Small hairpin RNAs efficiently inhibit hepatitis C IRES-mediated gene expression in human tissue culture cells and a mouse model. Mol Ther 12(3):562–568

Zadeh LA, Desoer CA (1963) Linear system theory. McGraw-Hill, New York

Zaher S, Green R (2009) Quality control by the ribosome following peptide bond formation. Nature 457:161–166

Zaher H, Green R (2011) A primary role for elastase factor 3 in quality control during translation elongation in Escherichia coli. Cell 147:396–408

Zarai Y, Tuller T (2018) Oscillatory behavior at the translation level induced by mRNA levels oscillations due to finite intracellular resources. PLoS Comput Biol 14(4):e1006055

Zarai Y, Margaliot M, Kolomeisky AB (2017a) A deterministic model for one-dimensional excluded flow with local interactions. PLoS ONE 12(8):1–23

Zarai Y, Margaliot M, Sontag ED, Tuller T (2017b) Controllability analysis and control synthesis for the ribosome flow model. IEEE/ACM Trans Comput Biol Bioinform (to appear)

Zarai Y, Margaliot M, Tuller T (2017c) A deterministic mathematical model for bidirectional excluded flow with langmuir kinetics. PLoS ONE 12(8):e0182178

Zarai Y, Margaliot M, Tuller T (2017d) Optimal down regulation of mRNA translation. Sci Rep 7:41243

Zarai Y, Margaliot M, Tuller T (2017e) Ribosome flow model with extended objects. J R Soc Interface 14(135)

Zarai Y, Ovseevich A, Margaliot M (2017f) Optimal translation along a circular mRNA. Sci Rep 7:9464

Zhang L, Yang N, Mohamed-Hadley A, Rubin S, Coukos G (2003) Vector-based RNAi, a novel tool for isoform-specific knock-down of VEGF and anti-angiogenesis gene therapy of cancer. Biochem Biophys Res Commun 303(4):1169–1178

Zhang G, Fedyunin I, Miekley O, Valleriani A, Moura A, Ignatova Z (2010) Global and local depletion of ternary complex limits translational elongation. Nucleic Acids Res 38(14):4778–4787

Zia R, Dong J, Schmittmann B (2011) Modeling translation in protein synthesis with TASEP: a tutorial and recent developments. J Stat Phys 144:405–428

Zupanic A, Meplan C, Grellscheid SM, Mathers JC, Kirkwood TB, Hesketh JE, Shanley DP (2014) Detecting translational regulation by change point analysis of ribosome profiling data sets. RNA 20(10):1507–1518

Zur H, Tuller T (2012) RFMapp: ribosome flow model application. Bioinformatics 28(12):1663–1664

Zur H, Tuller T (2013) New universal rules of eukaryotic translation initiation fidelity. PLoS Comput Biol 9(7):e1003136

Zur H, Tuller T (2016) Predictive biophysical modeling and understanding of the dynamics of mRNA translation and its evolution. Nucleic Acids Res 44(19):9031–9049

Robust Approaches to Generating Reliable Predictive Models in Systems Biology

Kiri Choi

Contents

Abstract A computational technique is described to reduce the model search space and construct an ensemble of models for systems biology using perturbation data. While doing so, an effective way of representing a network model is developed for computing purposes using adjacency matrix-like data structures. This allows models to include Uni-Uni to Bi-Bi reactions in addition to enzymatic activation and inhibition. It is demonstrated that the technique is effective, fast, and suggests it can be used as an initial filtering step in conjunction with other computational techniques. Finally, other potential methods to construct a set of reliable network models using time-course data are explored.

Keywords Systems biology · Biochemical networks · Network reduction · Machine learning · Ensemble modeling

1 Introduction

Models in systems biology aim to simulate the dynamics of biochemical networks such as signal transduction pathways, metabolic pathways, and gene regulatory networks. These models are typically encoded with a set of ordinary differential

K. Choi (✉)
Department of Bioengineering, University of Washington, Seattle, WA, USA
e-mail: kirichoi@uw.edu

N. Rajewsky et al. (eds.), *Systems Biology*, RNA Technologies,
https://doi.org/10.1007/978-3-319-92967-5_15

equations (ODE) and visualized as network diagrams. Models can be solved and simulated to make predictions under various conditions, furthering our understanding of the network and making them useful for the processes such as drug discovery (Butcher et al. 2004; Kitano 2002a,b).

Systems biology today is heading towards multi-scale modeling, constructing larger and more complex models (Natale et al. 2017). Examples include the whole-cell model of *Mycoplasma genitalium* (Karr et al. 2012) and the central metabolism model of *E. coli* (Millard et al. 2017). However, as the size and complexity of a model grow, validation becomes more and more difficult. A large portion of these models is composed of multiple submodels where each submodel should be validated against the data.

One of the core goals of systems biology, or computational modeling in general, is to construct reliable models. A reliable model is one having precision and accuracy against observations as well as being able to reliably predict new behavior. Thus the reliability of a model is dictated by how close the predicted outcome is compared with measurements under similar conditions. Generally, the reliability of a model can be increased by improving the model itself, collecting additional measurements used for constructing the model (increase the size of the dataset) and by implementing better algorithms (for example, parameter estimation).

There have been significant advances in both experimental and computational techniques that might contribute to improving model reliability. On the experimental side, we now have high-throughput data acquisition techniques for various types of experimental data. Some of the examples include those involving CRISPR-Cas9 (Qi et al. 2013; Cheng et al. 2013; Gilbert et al. 2014; Chavez et al. 2015), proteomics (Shi et al. 2016, 2012), metabolomics (Sévin et al. 2017), genomics (Davey et al. 2011; Van Dijk et al. 2014), and etc., all of which now have multiple ways to acquire large-scale experimental data.

One striking innovation comes from the advancement of CRISPR through the introduction of CRISPRa/i (activation/inhibition) technique. CRISPRa/i screening allows highly selective activation and inhibition of specific target genes (Qi et al. 2013; Cheng et al. 2013; Gilbert et al. 2014; Chavez et al. 2015). Proteomics is another area where there has been significant progress in terms of experimental techniques. These advances allow targeted proteomics to be ultra-sensitive and quantitative, allowing the measurement of low levels of protein abundance (Shi et al. 2012, 2016).

Computational biology, in general, had experienced significant progress as well. There have been numerous attempts to integrate various advanced and effective computational approaches to solve biological problems, in the forms of ensemble modeling (Henriques et al. 2017; Bonneau et al. 2006), information theory (Henriques et al. 2017), machine learning (Bonneau et al. 2006; Yan et al. 2017; Fisher and Woodhouse 2017), inference techniques (Oates et al. 2014; Daniels and Nemenman 2015; McGoff et al. 2016), and others (Pan et al. 2016; Li et al. 2013).

In this chapter, we discuss how we can take advantage of these new ideas and techniques to construct reliable models for systems biology. In particular, we talk about a potentially powerful method to reduce the model search space using perturbation data.

2 Reducing the Search Space

One of the issues in systems biology is that the model search space can be extraordinarily large especially when dealing with multi-scale models, although small models can exhibit the same problem as well. Consider a metabolic network model, for example. One must make sure that the stoichiometry of the network is correct, that the various rate laws and associated regulatory loops for the enzymatic reactions are accurate, and that the numerous parameters involved in rate laws are reasonably accurate. When we consider a large scale model which could have more than 60 metabolites and reactions (Millard et al. 2017), it is evident that modeling efforts with limited prior knowledge of the system can be quite challenging. Thus, for the scope of the problem we are dealing with, it would be immensely useful if we could reduce the search space of models and potentially generate an ensemble of likely models.

One way to achieve rapid, coarse-grained reduction of the model search space is to use perturbation data, similar to what was proposed by Mangan et al. (2016). With the advent of CRISPR-Cas9, we now have unprecedented control over selective activation/inhibition of specific genes for perturbation analysis. Specifically, utilizing CRISPRa/i, one can selectively (and combinatorially) perturb the total amount of species or individual reaction kinetics.

In this section, we present an algorithm that can be implemented to automatically generate an ensemble of reliable models by reducing the model search space using this type of perturbation data. But before doing so, we start by discussing a suitable data structure for a network model, which is a necessary step for any kind of computations. All computations presented in this chapter were done using Python. In particular, the Tellurium (Choi et al. 2016) environment is used in conjunction with the libRoadRunner solver (Somogyi et al. 2015) for model simulations and Antimony language (Smith et al. 2009) for the model description.

2.1 *Representation of Network Models*

It is important to define how to represent a model for computing purposes, especially when we have limited knowledge of the reaction steps present in a network. In systems biology, pathway networks are usually described using a set of chemical reactions from which a set of ordinary differential equations is derived. The same model will be visualized through network diagram with arrows as reactions. A model can have diverse types of motifs ranging from linear chains to dense overlapping regulons (DOR). Reactions can range from simple Uni-Uni[1] reactions to enzyme kinetics.

[1] Uni-Uni refers to reactions of the type $A \to B$.

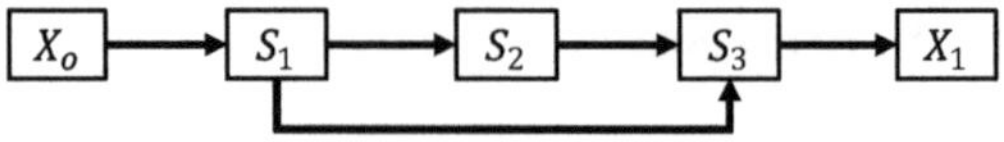

Fig. 1 Coherent type 1 feed-forward loop (C1-FFL). X_o and X_1 represent the boundary species in the model and are fixed during a simulation

The problem we would like to solve is how to reduce the model search space and generate an ensemble of network models based on limited knowledge of the network that is consistent with the experimental data. For our solution to be general, it is essential to define a data structure that can describe a network with enough flexibility to account for potentially diverse types of interactions. One of the easiest ways to do this is using matrices where rows represent participants as reactants and columns represent participants as products; a reaction exists between the species specified on the row and column if the value is 1. There are no reactions between the species specified on the row and column if the value is 0. This description is akin to the adjacency matrix used in computer science and connectivity matrix used in computational neuroscience, except in our case we need to also take directionality into account (a directed graph). For a simple Coherent type 1 feed-forward loop (C1-FFL) (Alon 2007) with three floating species and a boundary input/output (Fig. 1), our description will result in the matrix shown in (1) (m_{c1ffl}).

$$m_{c1ffl} = \begin{array}{c} \\ X_o \\ X_1 \\ S_1 \\ S_2 \\ S_3 \end{array} \begin{array}{c} \begin{array}{ccccc} X_o & X_1 & S_1 & S_2 & S_3 \end{array} \\ \begin{pmatrix} 0 & 0 & 1 & 0 & 0 \\ 0 & 0 & 0 & 0 & 0 \\ 0 & 0 & 0 & 1 & 1 \\ 0 & 0 & 0 & 0 & 1 \\ 0 & 1 & 0 & 0 & 0 \end{pmatrix} \end{array} \tag{1}$$

The above description is sufficient to describe the majority of network motifs with reversible/irreversible reactions, but there are certain types of reactions that may not be easily expressed in this manner. We can expand the proposed notation farther to incorporate more complex dynamics one might see in, e.g., nonlinear kinetics found in enzyme-catalyzed reactions. One way to approach this is to append the matrix with combinations of individual species. This does increase the computational complexity of the problem. However, if we limit the scope of searches to Bi-Bi[2] reactions at maximum, which is a reasonable restriction to impose in many systems, the increase in dimensionality might be acceptable, as we only need to add a combination of selecting two species out of n total species, $C(n, 2)$, to the number of rows and columns. The total number of combinations of r samples

[2]Bi-Bi refers to reactions of the type $A + B \rightarrow C + D$.

out of n objects is given by:

$$C(n,r) = \frac{n!}{r!(n-r)!} \tag{2}$$

From the equation, with $r = 2$, it is evident that when the total number of species increases by one, only n rows and columns are added.

For example, consider a moderate-sized network with ten species in total. In this case, our matrix will be a square matrix with 55 rows and columns (ten species and 45 combinations). It is possible to add an additional row/column to consider production/degradation (one to the number of rows/columns) of species as well. Something to consider when defining the model is that the scope of a model is arbitrary. If there are well-defined inputs and outputs, or the system has a steady-state solution, it is entirely possible to break down a large model into small closed systems to apply computational techniques.

The proposed notation is advantageous because it can represent enzymatic reactions with only minor changes. Consider the simplest examples of enzymatic activation and repression shown in Fig. 2. These reactions can be expressed as a variation of the Bi-Bi reaction shown below.

$$S_1 + X \rightarrow S_2 + X \tag{3}$$

However, in order to account for both enzymatic activation and repression, we cannot rely on the binary system (composed of 0s and 1s) where 0 will denote no reaction and 1 will denote activation. Thus, it is inevitable to introduce an additional state to our network representation. The additional state, with the value of -1, will represent enzymatic repression. Then, we can represent enzymatic activation and repression shown in Fig. 2 as the following set of matrices (4).

$$\begin{array}{c|cccccc} & A & S_1 & S_2 & A+S_1 & A+S_2 & S_1+S_2 \\ \hline A & 0 & 0 & 0 & 0 & 0 & 0 \\ S_1 & 0 & 0 & 0 & 0 & 0 & 0 \\ S_2 & 0 & 0 & 0 & 0 & 0 & 0 \\ A+S_1 & 0 & 0 & 0 & 0 & \mathbf{1} & 0 \\ A+S_2 & 0 & 0 & 0 & \mathbf{0} & 0 & \mathbf{0} \\ S_1+S_2 & 0 & 0 & 0 & 0 & \mathbf{0} & 0 \end{array} \qquad \begin{array}{c|cccccc} & R & S_1 & S_2 & R+S_1 & R+S_2 & S_1+S_2 \\ \hline R & 0 & 0 & 0 & 0 & 0 & 0 \\ S_1 & 0 & 0 & 0 & 0 & 0 & 0 \\ S_2 & 0 & 0 & 0 & 0 & 0 & 0 \\ R+S_1 & 0 & 0 & 0 & 0 & \mathbf{-1} & 0 \\ R+S_2 & 0 & 0 & 0 & \mathbf{0} & 0 & \mathbf{0} \\ S_1+S_2 & 0 & 0 & 0 & 0 & \mathbf{0} & 0 \end{array} \tag{4}$$

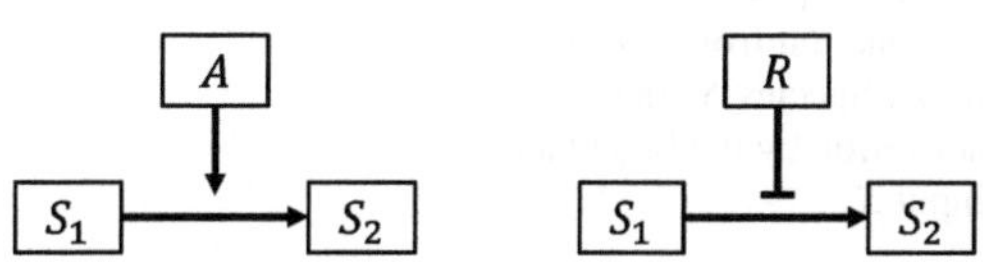

Fig. 2 Simplest cases of enzymatic activation by activator *A* and repression by repressor *R*

Note that a reaction defined in the matrix will be interpreted as an enzymatic activation or repression only when one of the reactants is also present in the products. In the example above, both the reactant and the product contain either an activator or a repressor. This means that only when a reaction with states $+1$ or -1 defined in non-diagonal and non-antidiagonal (similar to a diagonal but runs from top right to bottom left) of Bi-Bi specific quadrant of the network matrix (bottom right) will be interpreted as an enzymatic activation or repression. In the matrix above (4), bold entries indicate which reactions are treated as enzymatic activation or repression. This also means that where the additional state of -1 can be introduced is limited. -1 state can only be present in Uni-Uni reaction (as a type of auto-regulation) or in Bi-Bi reaction (as an enzymatic repression). This way, we can express diverse types of motifs and reactions through the network matrix.

This is very useful in representing complex signaling cascades. For example, consider a simple example shown in Fig. 3. This model can be represented by a 10 by 10 square matrix as shown in Matrix (5) (m_{cascade}) when using mass action kinetics with boundary signal input S_o integrated into the reaction between species S_1 and S_2 (which is possible since boundary input S_o is fixed) for the purpose of simplification.

Now that we have a concrete structure to represent a network model, we introduce a quick and simple algorithm to reduce the model search space and collect an ensemble of network models that are always consistent with perturbation data.

$$
m_{\text{cascade}} = \begin{array}{r} \\ S_1 \\ S_2 \\ S_3 \\ S_4 \\ S_1+S_2 \\ S_1+S_3 \\ S_1+S_4 \\ S_2+S_3 \\ S_2+S_4 \\ S_3+S_4 \end{array}
\begin{array}{c}
\begin{array}{cccccccccc} S_1 & S_2 & S_3 & S_4 & S_1+S_2 & S_1+S_3 & S_1+S_4 & S_2+S_3 & S_2+S_4 & S_3+S_4 \end{array} \\
\begin{pmatrix}
0 & 0 & 0 & 0 & 0 & 0 & 0 & 0 & 0 & 0 \\
1 & 0 & 0 & 0 & 0 & 0 & 0 & 0 & 0 & 0 \\
0 & 0 & 0 & 0 & 0 & 0 & 0 & 0 & 0 & 0 \\
0 & 0 & 1 & 0 & 0 & 0 & 0 & 0 & 0 & 0 \\
0 & 0 & 0 & 0 & 0 & 0 & 0 & 0 & 0 & 0 \\
0 & 0 & 0 & 0 & 0 & 0 & 0 & 0 & 0 & 0 \\
0 & 0 & 0 & 0 & 0 & 0 & 0 & 0 & -1 & 0 \\
0 & 0 & 0 & 0 & 0 & 0 & 0 & 0 & 1 & 0 \\
0 & 0 & 0 & 0 & 0 & 0 & 0 & 0 & 0 & 0 \\
0 & 0 & 0 & 0 & 0 & 0 & 0 & 0 & 0 & 0
\end{pmatrix}
\end{array} \tag{5}
$$

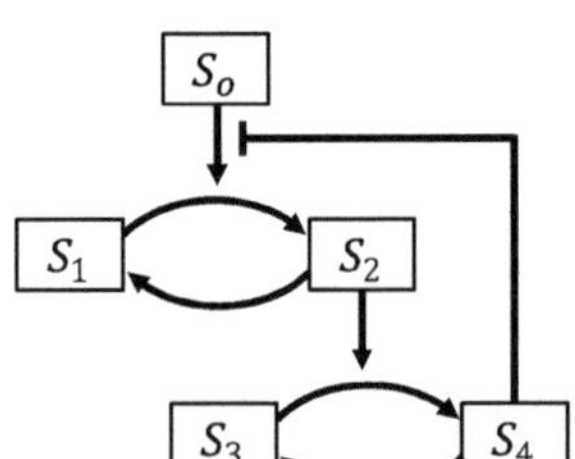

Fig. 3 A simple cascade model involving four floating species. Species S_2 activates the reaction from species S_3 to S_4. Species S_4 inhibits activation by the boundary input S_o

2.2 Network Reduction Technique

Now that we have a way to represent a model, we describe a method to reduce the search space using perturbation data. Perturbation data contain steady-state solutions of the network in question with and without a specified perturbation on certain reactions or species. We can then compare the steady-state solution in the perturbed state against the unperturbed state for each and every floating species. One can create an array of three-valued logic (trilean), i.e. $+1$ when the steady-state solution with perturbation is higher than that without perturbation, -1 when the steady-state solution with perturbation is lower than that without perturbation, and 0 when the difference in steady-state solutions with and without perturbation is smaller than a predefined threshold. One thing to keep in mind is that techniques such as CRISPRa/i allow one to perturb combinations of reactions extracting extra information about the network if there are two or more targets that can be perturbed.

Once arrays of trileans are obtained, random synthetic networks can be generated through the use of the proposed network representation method above. An important feature that is necessary for this step is preserving known information, i.e. keeping reactions that are experimentally perturbed and any other reactions known to exist in the network of interest in each and every randomly generated synthetic network. That way, it is possible to compare the results from the real network and the synthetic network. Known information includes any reactions that are known to exist, but it also includes any reactions that are known to be non-existent, as well as information on rate constants and initial species amount. Knowledge of non-existent reactions is very helpful in removing unwanted reactions which will reduce the search space significantly. After preserving known reactions (and non-reactions) in the network matrix, other reactions are explored, randomly assigning various states to the empty spots in the network matrix. There are also other rules that are enforced when generating random synthetic networks in order to remove meaningless and nonsensical solutions. First of all, we make sure all species in the network are involved in at least one reaction, removing incomplete networks. Also, direct reactions between input and output boundary species are not allowed. Finally, input boundary species remain as inputs and output boundary species remain as outputs.

Once a random synthetic network is generated under these rules, we calculate the steady-state solutions with and without perturbations at known reactions to create another set of arrays of trileans. Combinations of perturbations may be applied to the network if the same was done experimentally. Finally, trileans obtained via experimental measurement can be compared against trileans obtained from the synthetic network. The synthetic network will be accepted if and only if the arrays of trileans are identical and discarded otherwise. At this point, a single iteration is completed and the process starting from the random network generation is repeated. Figure 4 illustrates the general workflow of the network reduction technique.

So how well does this technique work? Consider a simple C1-FFL model shown in Fig. 1. Suppose that we know that a single reaction between species S_1 and S_3 exists, which is the minimal information necessary to use this technique. 10,000

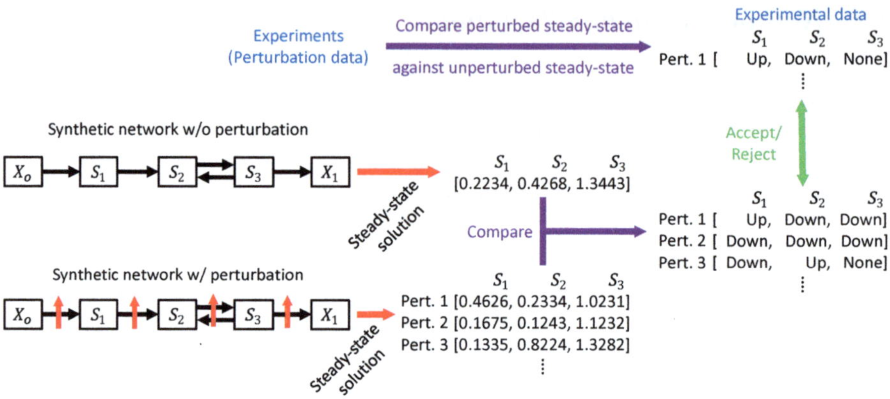

Fig. 4 Illustration of the network reduction technique. Perturbation data are compared with each other to create an array of trileans. Similar steps are taken for synthetic networks, calculating steady-state solutions with and without perturbations present. A synthetic network will be accepted if and only if the array of trileans match with that of experimental results. Set of trileans for combinations of perturbations may be compared if the same was done experimentally

iterations comparing unique and randomly generated networks result in less than 100 accepted networks, indicating more than 99% reduction in the potential network space. The set of accepted networks contained the original network.

What of the cascade model shown in Fig. 3? In this case, let us assume that we know a single reaction between two floating species but nothing else. After running 10,000 iterations, less than 20 networks are accepted, indicating more than 99% reduction in the potential network space as before. The set of accepted networks contained the original network. Figure 5 shows some of the other networks that survived the selection process. In all of the cases, only the reaction between species S_1 and S_2, complete with the repression from species S_4, was given.

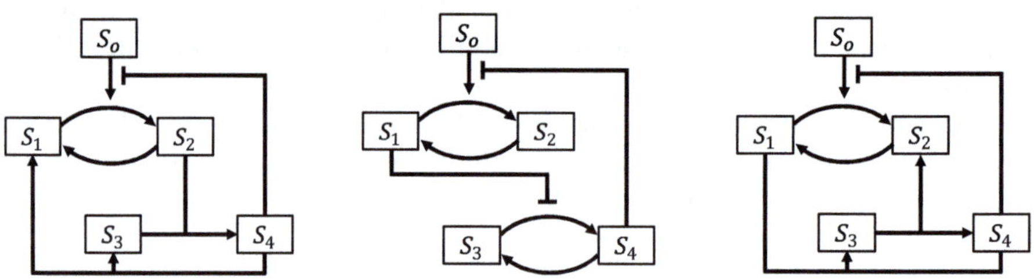

Fig. 5 Examples of various networks that survived the selection process. Only the reaction between species S_1 and S_2, complete with repression from species S_4, was given. In all of the cases, perturbing the reaction between species S_1 and S_2 results in similar qualitative steady-state floating species responses. Only mass action kinetics were used

After testing on various types of networks, we suggest that the technique on average can reduce the search space by more than 95% when an adequate amount of information about the network is given. This technique is appealing for several reasons. The technique makes qualitative comparisons of features. Therefore, it is a coarse-grained way of reducing the search space but faster in the sense that parameter optimization is unnecessary in any of the steps. The technique is also continuous. In a highly modularized workflow, this kind of reduction technique can be used as an initial screening step, continuously collecting and passing accepted models down to its downstream processes. The performance is also reasonable. Based on our experience, once parallelized, a million iterations can be carried out in several hours on a typical modern CPU.

The main benefit of such an algorithm is that the resulting search space might become small enough to run a regression analysis on accepted models to choose the most conceivable model. Furthermore, this algorithm might be useful in conjunction with other advanced computational techniques which might benefit from the reduced model search space. Since we are collecting a number of acceptable models, we can consider the output of the algorithm as an ensemble which will be true for most of the cases where the amount of information given is limited. The size of the ensemble will be determined by the amount of information given and every single model inside the ensemble will be equally acceptable. This approach can help with the robustness of the model when we consider an ensemble as a whole (Gosink et al. 2014). The ensemble resulting from the network reduction algorithm will certainly be interesting to examine. It might be possible to extract a common set of motifs from the resulting ensemble to gain insight on what makes a network acceptable. Understanding the similarities between the accepted networks will help additional endeavors to design the network model. Using synthetic toy models, it should also be possible to quantify the prior knowledge as in which part of the network has the most impact in terms of reducing the search space. Based on our experiences, knowledge on upstream reactions tend to have the most impact on the outcome, but further analysis is necessary for more complex networks. In the final section, we will discuss other potential ways to generate an ensemble of reliable models.

3 Towards the Application of Machine Learning

The network reduction technique presented in the previous section can be a powerful way to filter out a large portion of the model search space. However, there are other types of experimental data we can use to generate an ensemble of networks and to reduce the size of an ensemble further. For example, time-course data, in particular, provide valuable information on the dynamics of the network over specific time periods. As suggested in the previous section, one straightforward application of using time-course data in conjunction with the network reduction technique would be the use of time-course data for parameter estimation to systematically rank models according to the output of some objective function.

However, this type of analysis will not be ideal for multi-scale modeling, where the number of parameters is large and can easily result in over-fitting. Then what options do we have? Are there other ways to generate an ensemble of reliable network models using perturbation and time-course data? One suggestion is to apply machine learning. In particular, a multilayer neural network shows a lot of promise since it can extract various features out of the input data. Extracted features such as rates of change, curvature, etc. could be analyzed further to infer certain motifs based on unique response characteristics. Another important feature will be the relative changes between various species which will be important for identifying activation and repression type of reactions.

When trying to implement machine learning, probably the easiest way for this type of problem might be to use supervised machine learning. One can treat a network as a class, training against the training and validation sets generated from synthetic data to start with. Here, synthetically generated networks are necessary because the workflow must be validated as well. Nevertheless, the result of a supervised machine learning approach will be a single network model. This is not ideal since our model is technically not a classification problem and there can be multiple similar network models that can explain given data equally well. There are types of machine learning processes that introduce stochastic elements and thus obtain an ensemble of outputs. These techniques are typically designed for computer vision problem where the input data is a matrix representation of an image. For example, in imaging experiments, due to limitations in optics or in resolution, systematic distortion in the input data can be introduced, such as the Airy disk. This type of setup might be applicable to our problem if we decide to introduce error in the measurement data. Expanding farther, convolutional neural networks and reinforcement learning might provide even better performance in terms of predictions. These are two techniques that are relatively under-utilized in the field when attempting to apply machine learning to the experimental data we obtain but have huge potential if properly implemented.

Due to the wide range of applications of machine learning, there is a multitude of tools and libraries available, including Tensorflow (Abadi et al. 2015), Keras (Chollet et al. 2015), scikit-learn (Pedregosa et al. 2011), and so on. These libraries are well-supported and easy to use, making it ideal for biologists who are not proficient in computational work. Hopefully, we see continued interest and effort to design workflow utilizing these powerful techniques to aid systems biology in constructing reliable network models.

Acknowledgements KC is supported by NIH grants GM123032-01A1. The content is solely the responsibility of the author and does not necessarily represent the views of the National Institutes of Health. KC wishes to thank Herbert Sauro and Joseph Hellerstein for their help and guidance in completing this chapter.

References

Abadi M, Agarwal A, Barham P et al (2015) Tensorflow: large-scale machine learning on heterogeneous distributed systems. http://download.tensorflow.org/paper/whitepaper2015.pdf
Alon U (2007) Network motifs: theory and experimental approaches. Nat Rev Genet 8(6):450–461
Bonneau R, Reiss DJ, Shannon P et al (2006) The inferelator: an algorithm for learning parsimonious regulatory networks from systems-biology data sets de novo. Genome Biol 7(5):R36
Butcher EC, Berg EL, Kunkel EJ (2004) Systems biology in drug discovery. Nat Biotechnol 22(10):1253–1259
Chavez A, Scheiman J, Vora S et al (2015) Highly efficient cas9-mediated transcriptional programming. Nat Methods 12(4):326–328
Cheng AW, Wang H, Yang H et al (2013) Multiplexed activation of endogenous genes by CRISPR-on, an RNA-guided transcriptional activator system. Cell Res 23(10):1163–1171
Choi K, Medley JK, Cannistra C et al (2016) Tellurium: a python based modeling and reproducibility platform for systems biology. bioRxiv p 054601. https://doi.org/10.1101/054601
Chollet F et al (2015) Keras. https://github.com/fchollet/keras
Daniels BC, Nemenman I (2015) Efficient inference of parsimonious phenomenological models of cellular dynamics using s-systems and alternating regression. PLoS ONE 10(3):e0119821
Davey JW, Hohenlohe PA, Etter PD et al (2011) Genome-wide genetic marker discovery and genotyping using next-generation sequencing. Nat Rev Genet 12(7):499–510
Fisher J, Woodhouse S (2017) Program synthesis meets deep learning for decoding regulatory networks. Curr Opin Syst Biol 4:64–70
Gilbert LA, Horlbeck MA, Adamson B et al (2014) Genome-scale CRISPR-mediated control of gene repression and activation. Cell 159(3):647–661
Gosink LJ, Hogan EA, Pulsipher TC et al (2014) Bayesian model aggregation for ensemble-based estimates of protein pKa values. Proteins Struct Funct Bioinf 82(3):354–363
Henriques D, Villaverde AF, Rocha M et al (2017) Data-driven reverse engineering of signaling pathways using ensembles of dynamic models. PLoS Comput Biol 13(2):e1005379
Karr JR, Sanghvi JC, Macklin DN et al (2012) A whole-cell computational model predicts phenotype from genotype. Cell 150(2):389–401
Kitano H (2002a) Computational systems biology. Nature 420(6912):206–210
Kitano H (2002b) Systems biology: a brief overview. Science 295(5560):1662–1664
Li S, Park Y, Duraisingham S et al (2013) Predicting network activity from high throughput metabolomics. PLoS Comput Biol 9(7):e1003123
Mangan NM, Brunton SL, Proctor JL et al (2016) Inferring biological networks by sparse identification of nonlinear dynamics. IEEE Trans Mol Biol Multi-Scale Commun 2(1):52–63
McGoff KA, Guo X, Deckard A et al (2016) The local edge machine: inference of dynamic models of gene regulation. Genome Biol 17(1):214
Millard P, Smallbone K, Mendes P (2017) Metabolic regulation is sufficient for global and robust coordination of glucose uptake, catabolism, energy production and growth in escherichia coli. PLoS Comput Biol 13(2):e1005396
Natale JL, Hofmann D, Hernández DG et al (2017) Reverse-engineering biological networks from large data sets. arXiv preprint arXiv:170506370
Oates CJ, Dondelinger F, Bayani N et al (2014) Causal network inference using biochemical kinetics. Bioinformatics 30(17):i468–i474
Pan W, Yuan Y, Gonçalves J et al (2016) A sparse bayesian approach to the identification of nonlinear state-space systems. IEEE Trans Automat Control 61(1):182–187
Pedregosa F, Varoquaux G, Gramfort A et al (2011) Scikit-learn: machine learning in python. J Mach Learn Res 12:2825–2830
Qi LS, Larson MH, Gilbert LA et al (2013) Repurposing crispr as an rna-guided platform for sequence-specific control of gene expression. Cell 152(5):1173–1183

Sévin DC, Fuhrer T, Zamboni N et al (2017) Nontargeted in vitro metabolomics for high-throughput identification of novel enzymes in escherichia coli. Nat Methods 14(2):187–194
Shi T, Fillmore TL, Sun X et al (2012) Antibody-free, targeted mass-spectrometric approach for quantification of proteins at low picogram per milliliter levels in human plasma/serum. Proc Natl Acad Sci 109(38):15395–15400
Shi T, Niepel M, McDermott JE et al (2016) Conservation of protein abundance patterns reveals the regulatory architecture of the EGFR-MAPK pathway. Sci Signal 9(436):rs6
Smith LP, Bergmann FT, Chandran D et al (2009) Antimony: a modular model definition language. Bioinformatics 25(18):2452–2454
Somogyi ET, Bouteiller JM, Glazier JA et al (2015) libroadrunner: a high performance SBML simulation and analysis library. Bioinformatics 31(20):3315–3321
Van Dijk EL, Auger H, Jaszczyszyn Y et al (2014) Ten years of next-generation sequencing technology. Trends Genet 30(9):418–426
Yan J, Deforet M, Boyle KE et al (2017) Bow-tie signaling in c-di-GMP: machine learning in a simple biochemical network. PLoS Comput Biol 13(8):e1005677

Hints from Information Theory for Analyzing Dynamic and High-Dimensional Biological Data

Kumar Selvarajoo, Vincent Piras, and Alessandro Giuliani

Contents

Abstract Advances in biological sciences resulted in a data deluge, especially as for gene, protein, and metabolite expression. The issue of computational power needed to analyze such massive datasets is much less critical than the result interpretation task. This work deals with the latter, proposing a soft, data-driven approach, based on simple information theory concepts, as applied to classical multidimensional statistical methods. The proposed approach allows for a strong

K. Selvarajoo (✉)
Biotransformation Innovation Platform (BioTrans), Agency for Science, Technology and Research A*STAR, Singapore, Singapore
e-mail: kumar_selvarajoo@biotrans.a-star.edu.sg

V. Piras
Institute for Integrative Biology of the Cell (I2BC), CEA, CNRS, Université Paris-Sud, Université Paris-Saclay, Gif-sur-Yvette Cedex, France
e-mail: vincent.piras@i2bc.paris-saclay.fr

A. Giuliani
Environment and Health Dept. Istituto Superiore di Sanità, Roma, Italy
e-mail: alessandro.giuliani@iss.it

N. Rajewsky et al. (eds.), *Systems Biology*, RNA Technologies,
https://doi.org/10.1007/978-3-319-92967-5_16

interaction between the interpretative and computational aspects of the problem fostering interdisciplinarity. The application of these methods on transcriptome data relative to immune response and cellular development reveals insightful regulations, not only on the key instructive local processes but also on the subtle, yet robust, global-scale behavior. Furthermore, these techniques are swift in utility, as no detailed a priori knowledge of the biological system in study is required, and avoid "biased" expression cutoffs that are usually required for traditional/reductionist approaches.

Keywords Transcriptome analysis · Information theory · Biostatistical approaches · Single cell · Cellular variability · Physical laws

1 Introduction

The beginning of this millennium has witnessed significant progress made in experimental biology, moving away from the single molecule approaches toward monitoring "whole system" (genome, transcriptome, proteome, and metabolome) response. The large-scale measurement techniques allowed researchers to view the complexity of intracellular molecular concentration readouts and required bioinformatics techniques to make sense of it. However, till today, due to the challenge of machine or operator reliability issues, the low-abundance molecules are often marred by technical or background noise, and consequently, a large chunk of data is discarded (Sultan et al. 2008; Bottomly et al. 2011; Rau et al. 2013).

Initially, this was not a major problem as there has been a general belief that lowly expressed genes may not have significant roles in cellular processes. For example, in a high-throughput microarray study (monitoring gene expression levels for over 22,000 Affymetrix probe IDs) on the toll-like receptor (TLR)4-stimulated macrophages (Hirotani et al. 2005), only 148 genes were analyzed based on an arbitrary, thus "biased," threefold expression change cutoff. This was considered sensible since TLRs are known to induce the expression of proinflammatory cytokines, which numbers about a hundred or so (Dinarello 2007).

However, another study using similar conditions showed almost 3000 genes, belonging to diverse cellular processes, such as cell proliferation, differentiation, and DNA replication, which were induced by TLR4 (Nilsson et al. 2006). This work challenged traditional immunology where only a few major proinflammatory mediators were considered as relevant. Still more important, while a microscopic explanation based on around 100 molecular players of the kind "gene A and gene B are expressed then gene D . . . " is still feasible, a microscopic description based on 3000 interactors is totally out of scope and asks for a total rethinking of what we consider as "an explanation" in biology.

From simple information theory considerations, it is evident that the expression value of a single gene does not carry any relevant information on the system at hand: it needs a context (Thomas and Cover 2006; Smith and MacArthur 2017). For example, when we are thirsty on a very hot day, it does not matter how many cafes there are serving good quality coffee, it is usually water that one resorts to. In this case, the abundance of different coffee brands, or even the theoretical abundance of water in the region, is totally irrelevant; only the availability of water is important, and this behavior is monotonic to most humans.

Mathematician Claude Elwood Shannon postulated that information (or entropy) analysis requires three characteristics: monotonicity, independence, and branching (Shannon 1948). Monotonicity has to do with the universality (context of independence) of a given piece of information. This property is crucial because it reassures us of the invariance of the "meaning" of a given information atom. Independence holds when the information gain from two independent experiments is the sum of the information gain from their combination. Branching breaks a question into parts and bridges them in a tree structure, where following along the path, the information gain should be additive (Bialek 2012; Bonchev and Trinajstić 1977). Thus, analyzing high-throughput experimental data should not overly focus on the expression values alone (i.e., analyzing only highly expressed data), but look out for the principles used in information theory.

It is worth noting the strict analogy of these principles with spectral data analysis techniques like principal component analysis (or analogously singular value decomposition) whose aim is to represent a dataset in terms of mutually independent information atoms (in the form of eigenvectors of the correlation/covariance matrix of the studied variables). This analogy was aptly commented by Soofi (1994), while the quantitative link between complexity and the information entropy of the eigenvalue distribution across principal components of a dataset is demonstrated by Giuliani et al. (2001).

In the following of the chapter, we will try to show how these information theory principles and their statistical counterparts can shed light into transcriptome-wide response of living cells to distinct environmental perturbation and allow generating biological hypotheses on the observed phenomena.

Section 2 deals with the transcriptome-wide innate immune response to TLR4 in macrophages. Section 3 describes $CD4^{+}$ T-cell differentiation. Section 4 presents the gene expression variability between single development cells and across their stages, and Sect. 5 shows the reduction in gene expression noise across diverse cell types following the law of large numbers. Notably, even with the looming technical challenges, in all the presented works, simple information theoretic approaches are able to detect novel global patterns across all gene expression levels.

2 Local Immune and Global Diverse Processes in TLR4-Induced Macrophages

TLRs, currently 13 characterized, are transmembrane microbial pattern-recognition receptors (O'Neill et al. 2013). The TLR4 recognizes Gram-negative bacteria, through lipopolysaccharide (LPS), and triggers two branching (MyD88- and TRIF-dependent) signaling pathways (Fig. 1a). Collectively, the pathways activate key

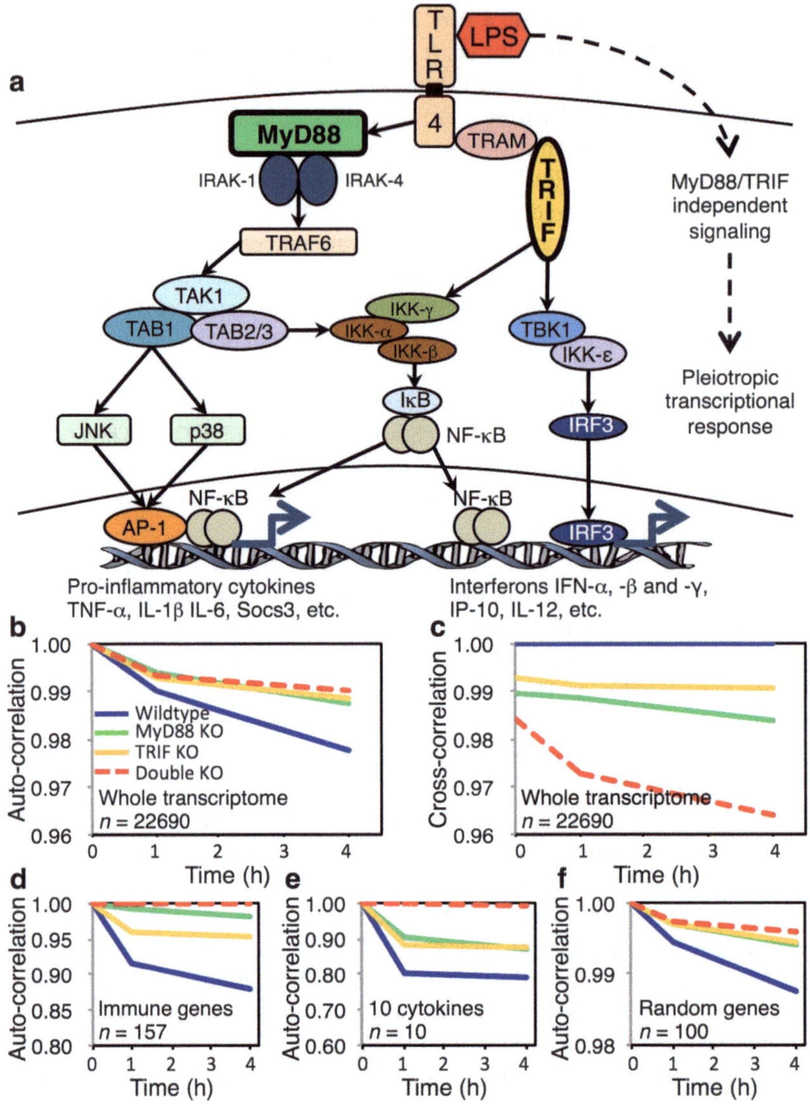

Fig. 1 (**a**) Schematic topology of the toll-like receptor (TLR)4 signaling. *Dotted line* indicates hypothetic pathways activating TLR4 signaling independent response. (**b,c**) Pearson *auto-* (**b**) and *cross*-correlations (**c**) for whole transcriptome. (**d–f**) Pearson *auto*-correlations for 157 important immune genes (**d**), a group of ten major cytokine genes (*tnf*, *il1b*, *il12*, *il6*, *il8*, *ccl3*, *ccl4*, *socs3*, *socs1*, and *cxc10*) (**e**), and random selection of 100 genes (**f**)

transcription factors such as activator protein (AP)-1, nuclear factor-κB (NF-κB), and interferon regulatory factors (IRF)-3, resulting in the induction of proinflammatory cytokines and interferons.

We analyzed LPS-stimulated murine macrophage transcriptome data using Affymetrix microarray chips with 22,690 probe IDs in four experimental conditions (wildtype (WT), MyD88 knockout (KO), TRIF KO, and MyD88/TRIF double KO or DKO) at three time points (0, 1, 4 h) (Tsuchiya et al. 2009a). After performing intensity background adjustment and normalization using Robust Multi-array Average (Irizarry et al. 2003), we applied temporal Pearson correlation metrics (Appendix A) on the transcriptome profiles. The Pearson correlations are computed between the vectors having as components the 22,690 gene expression levels at different times with the vector correspondent to the initial state ($t = 0$ h, *auto*-correlations, Fig. 1b). In the same way, correlations can be computed between gene expression profile of the wildtype (WT) and other conditions (e.g., MyD88 KO) at the same time point (*cross*-correlations, Fig. 1c). These correlations were calculated at different scales: whole transcriptome ($n = 22{,}690$), immune-related genes ($n = 157$), selected cytokine-coding genes ($n = 10$), and randomly extracted genes ($n = 100$) (Fig. 1d–f).

In principle, when two samples containing high-dimensional data (such as microarray, RNA-Seq) are compared, the correlation analyses provide a measure of deviation from unity as a source of difference between the samples. The Pearson correlation coefficient R shows the average (compressed to single value) information of the transcriptome-wide response. Briefly, two samples with identical and completely nonidentical information will show unit ($R = 1$) and null ($R = 0$) correlation, respectively. Perfect correlation ($R = 1$) is an idealized situation that is far from reality, as technical or experimental noise alone interferes and reduces correlation (and clearly the same holds true for $R = 0$ given the inescapable presence of noise-induced apparent correlations). In the present situation, we obtained very high correlation coefficient due to the presence of an ideal (and shared across different samples) expression profile relative to the macrophage cell type.

For the whole transcriptome TLR4 response, in all four conditions, we observed a monotonic deviation of R moving away from unity, with the most pronounced response for cytokine genes. The most notable and worthy of the results is for DKO. Prior to this study, DKO was expected to produce no noticeable transcriptome response (Hirotani et al. 2005). However, we observed very similar global response of DKO with MyD88 KO or TRIF KO (Fig. 1b). This is an indication that LPS is able to invoke a gradual intracellular response independent of the key adaptor molecules MyD88 and TRIF. The *cross*-correlations, on the other hand, showed that DKO response, compared with WT, is the least similar, pointing to different source of mechanisms for activation (Fig. 1c).

Next, comparing correlation coefficients of 157 immune-related and 10 cytokine genes, the *auto*-correlation for DKO was almost unchanged with time, indicating lack of response in DKO (Fig. 1d,e). However, when we analyzed random selection of 100 genes, the DKO response was observed, similarly to the whole transcriptome (Fig. 1f). Together, these data suggest that the source of DKO transcriptome-wide response, although lacking the common immune-related processes, is most distinct from WT compared to MyD88 KO or TRIF KO. Gene Ontology database analysis indicated that DKO induced diverse cellular processes such as nucleotides metabolism, focal adhesion, mRNA transcriptional mechanism, etc. In other words, the overall results suggested the presence of unknown pathways, independent of MyD88 and TRIF, to activate novel genes and biological processes in DKO (Fig. 1a, *dotted arrow*).

Notably, subsequent works from other laboratories provided evidence for the predicted MyD88- and TRIF-independent response: Hagar et al. (2013) and Kayagaki et al. (2011) showed that caspase-11, which plays a pivotal role in shaping inflammasome (a proinflammatory regulatory mechanism), is activated intracellularly without the need for TLR4 for promoting interleukin-1 family of cytokines.

Thus, it is clear from this work that the simple statistical metric R can be used successfully to observe major shifts in transcriptome-wide response to a given stimulus under control and mutant conditions without entering in the microscopic (and impossible to manage due to the huge number of genes) mechanistic details.

The major advantages for this methodology are that (i) it does not require traditional biased threshold (e.g. two- or threefold expression change) cutoff, thereby eliminating data that otherwise still show monotonic response, and (ii) it detects novel global response arising from lowly expressed genes.

3 Adaptive T-Cell Differentiation Response

$CD4^+$ T cells (also known as T helper cells) are adaptive immune cells that receive antigens from the innate immune cells, such as macrophages or dendritic cells, and differentiate into distinct effector subtypes: Th_1, Th_2, Th_9, Th_{17}, Th_{22}, T_{reg}, and T_{FH}. Depending on the co-stimulatory molecules, such as interleukin (IL)-4, IL-6, IL-12, and IFN-γ, the diverse differentiation lineages are achieved (Smith-Garvin et al. 2009; MacLeod et al. 2010; Zhu et al. 2010; Magombedze et al. 2013; Kared et al. 2014). Previous studies largely monitored a limited subset of genes or proteins to analyze the differential roles of the subtypes. However, more recent ones are focusing on a larger spectrum of molecules using high-throughput technologies (Ciofani et al. 2012; Tuomela et al. 2012; Bhairavabhotla et al. 2016).

Here, we analyzed RNA-Seq data based 10,307 non-zero gene expressions of Th_{17} differentiation at 1, 3, 6, 9, 12, 16, 24, and 48 h and compared with Th_1, Th_2, and T_{reg} at end time point (48 h) (Simeoni et al. 2015). Unstimulated $CD4^+$ T cells (T_{naive} at $t = 0$, or Th_0) were used as control cells for the experiments, and it was measured at 1, 3, 6, 16, and 48 h. The Th_0 subtype, with only T-cell receptor stimulation with anti-CD3/CD28, is widely believed to be non-polarized or undifferentiated and in some cases treated as naïve T cells (Calabresi et al. 2002; Negishi et al. 2005; Verma-Gandhu et al. 2007; Newcomb et al. 2009; Swain et al. 2012; Ciofani et al. 2012; Touzot et al. 2014).

Firstly, we used power-law relationship to analyze the expression data since gene expression distribution has been observed to follow this law (Furusawa and Kaneko 2003; Ueda et al. 2004), possibly due to their scale-free organization (Albert 2005).

That is, any deviation from the power law indicates gene expressions that are biased by machine/operator-induced variations or technical noise. Figure 2a shows the fraction of genes versus their expression values in log scale for the different subtypes and time points. It can be seen that the gene expression value scatter largely follows the power law with exponent approximately equal to 1. Only below expression (FPKM) value lower than 10 that the scatter deviates from the power law for all cases (see Fig. S1 of Simeoni et al. 2015). This is conceivable as the lowly expressed genes are more susceptible to noise, technical or biological in nature (Conesa et al. 2016); see also Sect. 5.

It is important to note that the exponent $k \sim 1$ for all investigated T-cell types is consistent with another T-cell differentiation dataset (see Fig. S2 of Simeoni et al. 2015) and several other mammalian cell types we analyzed (Sect. 5). Thus, in contrast to a majority of studies that used arbitrary threshold values for expression cutoff, we used the fundamental statistical power law to remove unreliable expression data and, consequently, retained 6404 genes for analyses.

We next used the Pearson (*auto-* and *cross-*) correlation metrics, discussed in the TLR4 study above, to track the global response of the T-cell differentiation process. The *auto*-correlation analyses showed similar monotonic deviation from unity at T_{naive} state ($t = 0$ h) to stabilize at $R \sim 0.7$ from 16 h onward for both Th_0 and Th_{17} (Fig. 2b). Th_1, Th_2, and T_{reg}, measured only at 48 h, also showed similar $R \sim 0.7$. This result is surprising if Th_0 is considered as undifferentiated or non-polarized, as we would then expect the global gene expressions not to vary significantly from T_{naive} and the *auto*-correlations to remain close to 1. Although it is well known that Th_1, Th_2, and T_{reg} express a number of unique cytokines crucial for their different phenotypes and functions, notably, their global averaged gene expression responses are similar to Th_0. In other words, on a local scale (referring to a few crucial cytokines), all the subtypes are distinct, but on a global scale (majority of genome), they show similar response (Tsuchiya et al. 2009b).

In checking for nonlinear monotonic response, we adopted Spearman rank correlations (Appendix B) and observed similar decreasing profiles (Fig. 2c). Similar temporal correlation values of Th_1, Th_2, Th_{17}, and T_{reg} differentiation were also observed when analyzing another similar dataset (Hu et al. 2013). Thus, these data reaffirm our observations for Th_0 response. *Cross*-correlation revealed that Th_0 global response is more similar to Th_1, Th_2, and T_{reg} ($R > 0.95$) than Th_{17} ($R \sim 0.85$) (Fig. 2d,e).

Evaluating Shannon entropy (Shannon 1948), which measures the disorder of a high-dimensional system, where higher values indicate increasing disconnection between variables and zero value indicates order (Appendix E), we observed relatively low values for T_{naive}, but entropy gradually increases for all subtypes and stabilizes between 16 h and 48 h (Fig. 2f). Th_{17} showed the highest value among the subtypes, while values for Th_0, Th_1, Th_2, and T_{reg} were very similar at 48 h. These results not only show increasing diversity of transcriptome-wide expressions during T-cell differentiation, but they also reveal global similarity between Th_0, Th_1, Th_2, and T_{reg}, raising the question whether Th_0 is control or another subtype.

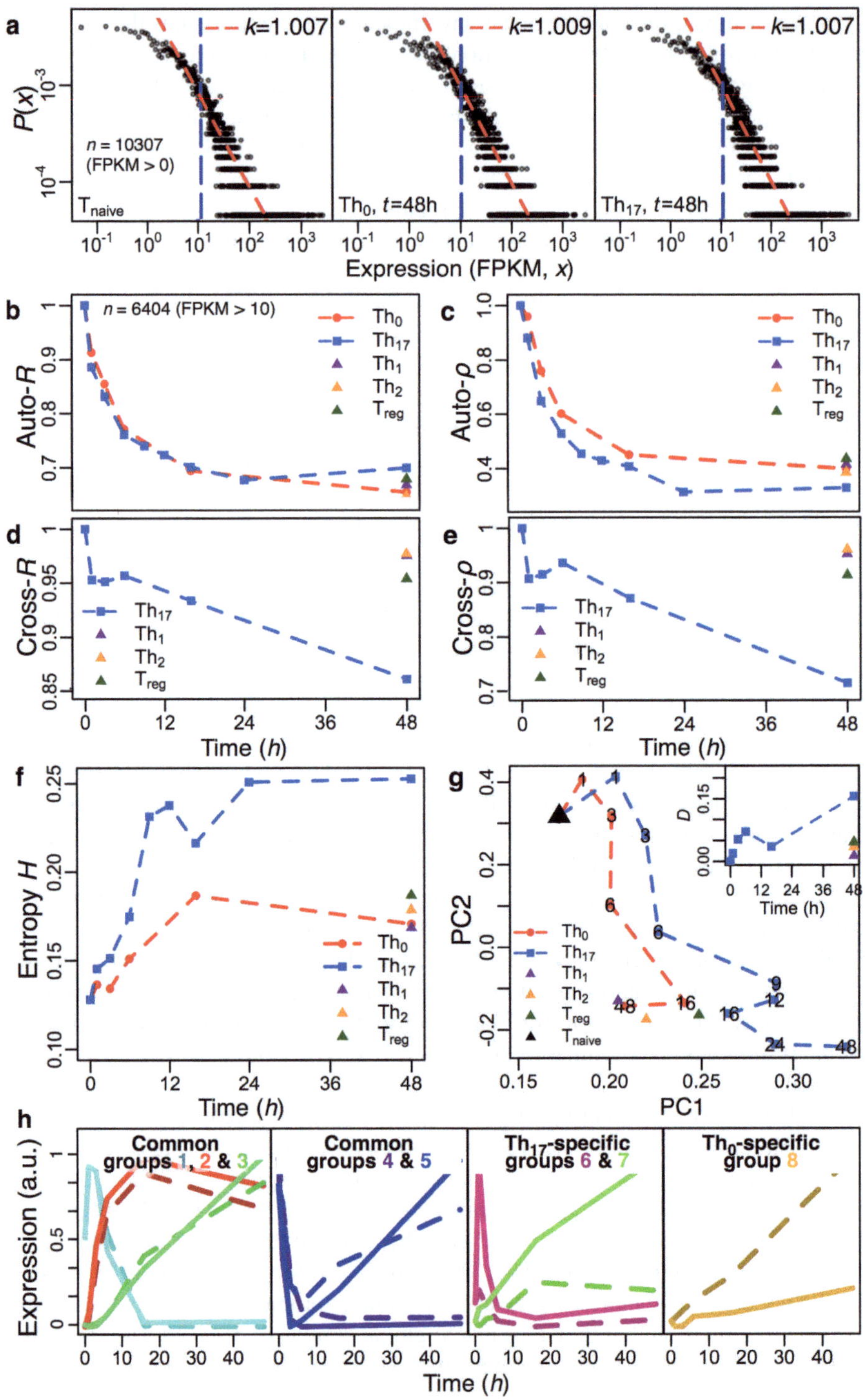
a
k=1.007
k=1.009
k=1.007
P(x)
n = 10307
(FPKM > 0)
T naive
Th0, t=48h
Th17, t=48h
Expression (FPKM, x)
b
n = 6404 (FPKM > 10)
Auto-R
c
Auto-ρ
Th0
Th17
Th1
Th2
Treg
d
Cross-R
e
Cross-ρ
Time (h)
f
Entropy H
g
PC2
PC1
Tnaive
D
Time (h)
h
Expression (a.u.)
Common groups 1, 2 & 3
Common groups 4 & 5
Th17-specific groups 6 & 7
Th0-specific group 8
Time (h)

We next investigated the first two major principal components (PCs) accounting for about 84% and 8% of the expression variance, respectively, and plotted them against each other for all the subtypes at available time points (Fig. 2g). Notably, the PC trajectories of Th_0 and Th_{17} diverged most profoundly at 48 h (Fig. 2g, *insert*), consistent with the observation from the *cross*-correlations. Moreover, at 48 h, the PC values of Th_0, Th_1, Th_2, and T_{reg} are closely clustered and almost equidistant from the T_{naive} state. Thus, correlations, entropy, and principal component analyses reinforce the fact that the global temporal response of Th_0 is far from being that of T_{naive} or Th_{17} but is very close to that of Th_1, Th_2, and T_{reg}.

To further probe the dynamic global response of Th_0 and Th_{17}, we performed hierarchical clustering of the gene expression values (Appendix C). By setting at least twofold change in gene expressions between any time points ($t = 0, 1, 3, 6, 16$, and 48 h), we obtained 5704 genes with 4379 (or 77%) common between Th_0 and Th_{17}. Hierarchical clustering analysis revealed eight major temporal groups of genes for Th_0 and Th_{17}, of which three groups were distinctly different between them (Fig. 2h). Two of the three distinct groups were Th_{17}-specific (largely involving immune response, differentiation, and metabolic processes) and the remaining Th_0-specific (immune response, differentiation and cell cycle, replication). Investigating closer into the function of the more pronouncedly expressed genes (threefold change with at least 100 units), we picked out noncoding small rDNA-derived RNA (*Mir715*), lymphocyte-specific protein 1 (*Lsp1*), heat shock protein 8 (*Hspa8*), cystatin C (*Cst3*), inhibitor of DNA binding 3 *(Id3)*, and adaptor-related protein complex 3, beta 1 subunit (*AP3B1*), significantly upregulated in Th_0, and interleukins (*Il17f, Il17a, Il21*), basic leucine zipper transcription factor ATF-like (*BATF*), cytochrome c oxidase (*Cox4i1*), and cofilin 1 (*CFL1*) were expressed, including novel genes such as microRNA *Mir686* and ribosomal protein-coding genes (e.g., *Rpl24, Rpl28, Rpl29, Rpl36, Rpl41*) for Th_{17}.

Overall, our simple yet robust statistical metrics have identified novel global response for Th_0, which is commonly treated as a control, undifferentiated, or non-polarized. Our data also indicate that Th_0 response is globally similar to that of Th_1, Th_2, and T_{reg}. Taking a closer look into the local Th_0 response revealed 260 Th_0-specific genes largely involved in immune response and differentiation, cell cycle and replication, stress and damage response, and apoptosis.

Fig. 2 (**a**) Gene expression units (x, FPKM) vs. fraction of genes ($P(x)$) for representative datasets. *Dotted red line*: fitted power law. *Dotted blue lines*: expression threshold $x \geq 10$ for power-law fitting. (**b,c**) Pearson (**b**) and Spearman (**c**) *auto*-correlations between unstimulated naïve $CD4^+$ T cell (T_{naive}, $t = 0$ h) and other subtypes at each time point. (**d,e**) Pearson (**d**) and Spearman (**e**) *cross*-correlations between Th_0 and other subtypes. (**f**) Temporal Shannon entropy, H, of the different T-cell subtype transcriptomes. (**g**) Transcriptome-wide expression principal component trajectories for Th_0 and Th_{17} (x-axis: PC1, 84% of total variance, y-axis: PC2, 8%). Th_1, Th_2, and T_{reg} at 48 h, and T_{naive} are also shown at $t = 0$ h. *Insert:* Euclidean distance, D (a.u.) between Th_0 and Th_{17} PC trajectories and between Th_0 and Th_1, Th_2, T_{reg}, PC coordinates at 48 h. (**h**) Eight major average temporal expression profiles (dotted lines, Th_0; solid lines, Th_{17}) of common, Th_{17}-specific, and Th_0-specific groups

It is worth noting that we did not impose any preconceived structure (or biological hypothesis) to the data, allowing the internal correlation structure to emerge and generate biological hypotheses (Giuliani 2017).

4 Single Cell Transcriptional Variability During Embryonic Development

In contrast to cell population or bulk studies presented so far, single cells have shown that individual molecules (transcripts, proteins, or metabolites), within a homogenous cell population, display variable expression/abundance levels (Elowitz et al. 2002; Chang et al. 2008). This variability has been linked to the stochastic nature of molecular network regulations or biological noise (Elowitz et al. 2002; Eldar and Elowitz 2010; Selvarajoo 2012, 2013) and has shown to play pivotal roles for the survival of species to diverse environmental conditions or for cell fate decisions (Maamar et al. 2007; Raj et al. 2010; Eldar and Elowitz 2010; Selvarajoo 2012, 2013).

Although the recent literature catalogues ample works in single cell studies that embody the importance of single molecular variability, there is a general lack of investigation for global regulatory properties at an omics-wide scale. Studying global properties has been instrumental in interpreting collective mechanisms of living organisms, for example, the innate immune response to invading pathogens or the attractor states of cell differentiation process. Here, to understand the global noise patterns of single developmental cells, RNA-Seq transcriptome-wide expressions of oocytes to blastocysts in human and mouse were investigated (Piras et al. 2014).

To better understand the variability and the effects of technical and biological noises, we compared pairwise transcriptome-wide expressions of any two cells taken from the same cell origin. Notably, despite large expression scatter, especially for late developmental stages (Fig. 3a), the Pearson correlation coefficient R between single cells of the same development stage is generally high (Fig. 3b, *dotted lines*). However, the R between cells of distinct origins is significantly lower (Fig. 3b, *solid lines*).

We further investigated Spearman correlation, distance correlation (dependence, Appendix D), and maximum information coefficient (association (Reshef et al. 2011)). Remarkably, all metrics revealed similar trends compared with Pearson correlations (Fig 3b; see *right panels, ρ, dCor* and *MIC*). These results indicate that the global transcriptional program of developmental cells clearly deviates along the stages in time, with faster rate of deviation occurring for mouse when compared with human.

Next, we assessed the diversity of single cell transcriptome using Shannon entropy (Appendix E). For both human and mouse, Shannon entropy values remained low in early stages but gradually increased from two-cell (human) or

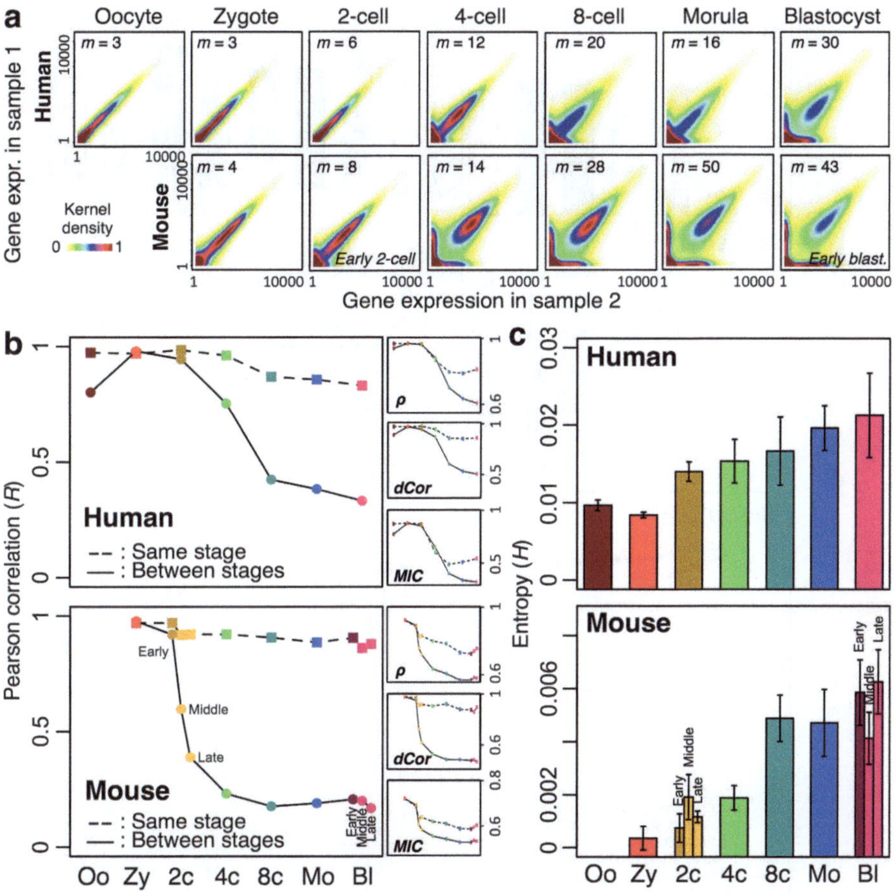

Fig. 3 (**a**) Gene expression values distributions (kernel density estimation) between pairs of single cells in human and mouse, from oocytes to blastocysts. *m*, number of cells; *n*, number of genes. (**b**) Pearson correlation between cells of the same stage (i.e., *cross*-correlations, *dotted lines*) and or between zygote and other stages (i.e., *auto*-correlations, *solid lines*). *Right panels*: corresponding spearman correlation (ρ), distance correlation (*dCor*) and maximum information coefficient (*MIC*). (**c**) Shannon entropy, *H*, of single cell transcriptomes for each stage

four-cell (mouse) stage, to reach high values for morula and blastocyst (Fig. 3c). This result, therefore, shows the disconnection or diversity of transcriptome-wide expressions increases during mammalian development.

To further understand the effects of increasing entropy and diversity in single cell transcriptomes during embryogenesis, we quantified single cells' expression scatter by computing transcriptome-wide average noise (a.k.a. total noise), η^2_{tot}, i.e., summing the squared coefficient of variation, defined as the variance (σ^2) of expression divided by the square mean expression (μ^2), for all genes between all possible pairs of single cells (Appendix F). We observed that η^2_{tot} is low during

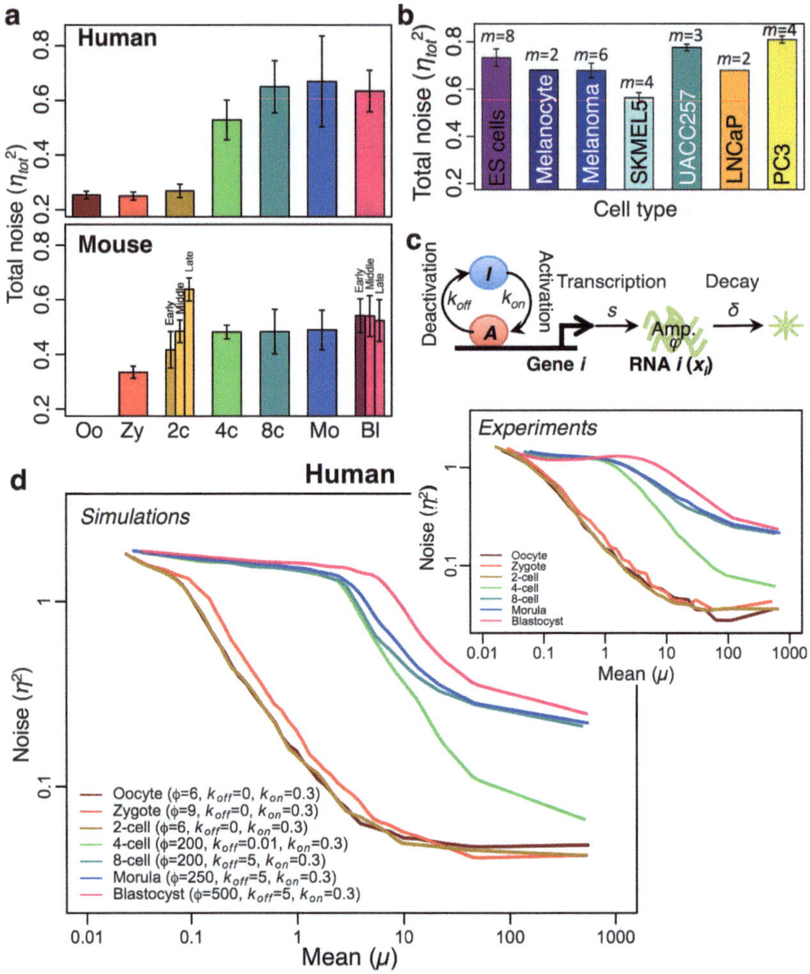

Fig. 4 (**a,b**) Total transcriptome noise, $\eta^2{}_{tot}$, for each embryogenesis stage (**a**) and for single stem, somatic, and cancer cells (**b**). Average for *m* cells, error bars: 1 s.d. (**c**) Single cell transcriptional model. k_{on} and k_{off}: promoter dynamics, *s*: transcription rate, δ: RNA decay rate, φ: amplification factor. (**d**) Simulated noise (η^2) versus mean (μ) expression patterns for each stage using the single cell transcriptional model. Noise is high for low expression genes ($\mu \sim 0.1$) and decreases for high expression genes ($\mu > 100$). *Insert*: η^2 versus μ patterns from experimental data. Patterns for mouse were also analyzed (Piras et al. 2014)

initial embryonic cell differentiation but increases at later stages with significant increase from two- to four-cell stage onward (Fig. 4a). We also compared total noise for embryonic stem, normal somatic, and cancer cells and found similar values as obtained for later stages development cells (Fig. 4b). These data show total noise stabilizes at ~0.7 and may not increase further.

In summary, by studying the distribution of gene expressions between single cells, we observed that the expression scatter increased from two-cell to four-cell stage onward in both human and mouse. Next, we examined the Pearson correlation and Shannon entropy for each developmental stage. Again, we observed that expressions become more variable from the two-cell stage. Subsequently, the global noise character of single cells was investigated by quantifying the squared coefficient of expression variations over mean expression values. Here, we observed clear transition of noise patterns occurring between two-cell and eight-cell stage.

To understand the underlying mechanisms for noise transition along the development, we built a stochastic transcriptional model (Fig. 4c). By estimating the parameter values to match each developmental cell pattern (noise η^2 vs. mean expression μ, see Fig. 4d, compare *panel*—simulations vs. *insert*—experiments), the model indicated that early developmental stages are mainly dominated by low transcriptional activity. That is, the number of transcripts produced per activation event is low (model parameter $\phi \sim 6$–9 for early stages, Fig. 4d). Note that the lower overall transcription in oocytes and early zygote is consistent with (i) transcriptional silencing and (ii) stochastic degradation of maternal RNA that has been observed from oocytes to a four-cell stage in humans (Braude et al. 1988; Tadros and Lipshitz 2009). Transcriptional silencing is likely due to chromatin condensation state that prevents transcriptional machinery from reaching gene promoters (Braude et al. 1988; Debey et al. 1993; De La Fuente 2006).

For the later-stage developmental cells, the model suggests that on top of high transcriptional process ($\phi \sim 200$–500), the cells possess quantal (i.e., binary) activation of most transcription factors (model parameters k_{on} and $k_{off} \neq 0$ for later stages) or are subject to more extrinsic variability such as phenotypic diversity among individual cells. These factors increase the general expression scatter and noise levels. However, investigating expression-independent random noise in our single cell transcriptional model simulations suggests that the levels of extrinsic and/or technical noise in our RNA-Seq data for all cells are relatively low. That is, the relatively high levels of noise for later stages stem from quantal activation rather than technical biases, or in certain cases, such as blastocyst cells, may result from phenotypic variability, as blastocysts consist of different cellular subtypes. Conversely, since phenotypic variability among more homogenous eight-cell stage is similar to blastocyst, we believe that quantal promoter activation is crucial for the increase of noise scatter along development stages. Notably, such quantal promoter activation has been noted to occur for single cell organisms such as *E. coli* (Eldar and Elowitz 2010) and has been shown to be important for the cell fate decision of *B. subtilis* (Maamar et al. 2007).

5 Single Cells to Population Study Reveals Statistical Laws

From recent single cell studies, cell-to-cell transcriptome-scale molecular variability is now established (Islam et al. 2011; Ramsköld et al. 2012; Sasagawa et al. 2013; Shalek et al. 2013; Picelli et al. 2014; Marinov et al. 2014; Zheng et al. 2017). It is, therefore, intriguing how heterogeneous single cells are able to execute well-defined and coordinated biological processes such as cellular growth or immune response. To probe into the question, we first approached a theoretical situation and next compared the analysis with actual experimental data.

We theoretically generated the expression values of 20,000 genes, over a realistic expression range, using a Poisson process (Piras et al. 2012; Raj and Oudenaarden 2009) for two single cells and obtained a high Pearson $R \sim 0.98$ (Fig. 5a). Next, to include extrinsic noise or variability, we added random fluctuations from a gamma

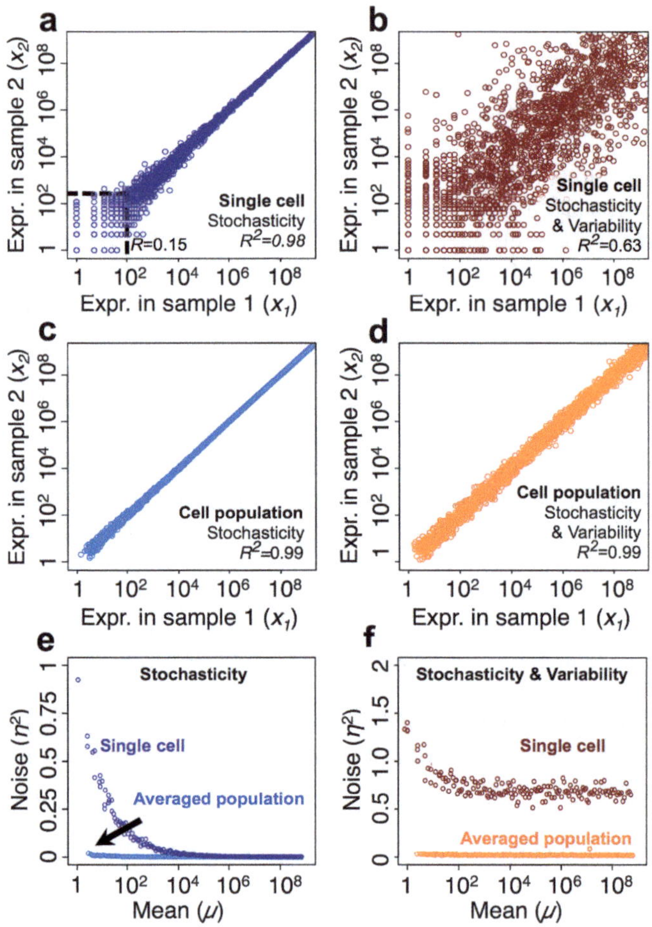

Fig. 5 (**a,b**) Simulated expression values of two single cells ((**a**) stochastic fluctuations only (intrinsic noise), (**b**) stochastic fluctuations and variable (extrinsic) noise with gamma distribution). (**c,d**) Simulated expression values of two-cell populations (average of 30 single cells; (**c**) stochastic noise only, (**d**) stochastic and variable noise). (**e,f**) Noise (η^2) versus mean expression (μ) for 100 pairs of simulated samples (2×100 single cells, or 2×100 populations of 30 cells; (**e**) stochastic noise only, (**f**) stochastic and variable noise)

distribution (Taniguchi et al. 2010) to the single cell data, which reduced $R \sim 0.63$ (Fig. 5b). To simulate population-level gene expressions, we repeated the data generation for 100 single cell pairs and took their averages for the two conditions (Fig. 5c, stochastic only, and Fig. 5d, stochastic and variability). The averaging of the samples resulted in an increase in R values (>0.99) for both conditions.

Next, we computed noise, (η^2) (Appendix F), against mean expression values (μ) for single cell and averaged samples and observed noise is initially high at low expressions and progressively reduced when expression levels increased (Fig. 5e,f). These theoretical data suggest that (i) noise (stochastic only or stochastic and variability) for single cells are higher compared to population average, (ii) stochastic noise becomes negligible at higher expression levels (especially obvious for single cells, i.e., $\eta^2 \approx 0$ for $\mu > 10^4$, Fig. 5e), and (iii) noise from variability reaches finite asymptotic values for single cells ($\eta^2 \approx 0.7$ for $\mu > 10^4$, Fig. 5f), whereas it is negligible for population averages ($\eta^2 \approx 0$, for $\mu > 10^4$, Fig. 5f). In general, these data indicate that single cells possess more noise compared to cell population, and this is especially obvious for the lowly expressed genes, whether the variability in gene expressions is low or high.

To test our statistical predictions, we investigated RNA-Seq datasets of single cells and cell populations in six cell types (Piras and Selvarajoo 2015) (prostate cancer LNCaP, embryonic kidney HEK293T, lymphoblastoid GM12878 cell lines in human, and embryonic stem (ES), primary endoderm (PE), bone marrow-derived dendritic cells (BMDC) in mouse) with consistent experimental protocols (Ramsköld et al. 2012; Sasagawa et al. 2013; Shalek et al. 2013; Picelli et al. 2014; Marinov et al. 2014).

The pairwise gene expression scatter plots, similar to the development cell analysis above, showed a general decrease in variability for the middle and low expression genes as we move from single cells to increasing cell population size for all cell types (Fig. 6a). Notably, the highly expressed genes did not vary pairwise across all cell population sizes. Globally, the Pearson R increases gradually as the population size is increased (Table 1). Spearman's rank (ρ) correlation coefficients also showed a comparable increase when the population size is increased (Table 2). These data reveal the emergence of correlated structure of transcriptomes for all investigated cell types when single cells form into populations.

As described in Section 4, transcriptome-wide expression distributions are expected to follow power law. Here, it can be seen for all samples, above 1 expression unit (RPKM, FPKM, or TPM), power-law distributions with exponent, $k \sim 1$ (Fig. 6b). In other words, removing genes with expression units below 1 (which shows no power-law structure) most likely reduces overall transcriptome-wide noise. Next, we computed transcriptome-wide average noise (Appendix F) for all available samples and observed noise is initially high for single cells and progressively reduced when population size is increased (Fig. 6c, *crosses*). Notably, reduction of noise matched the predictions from the law of large numbers (LLN)

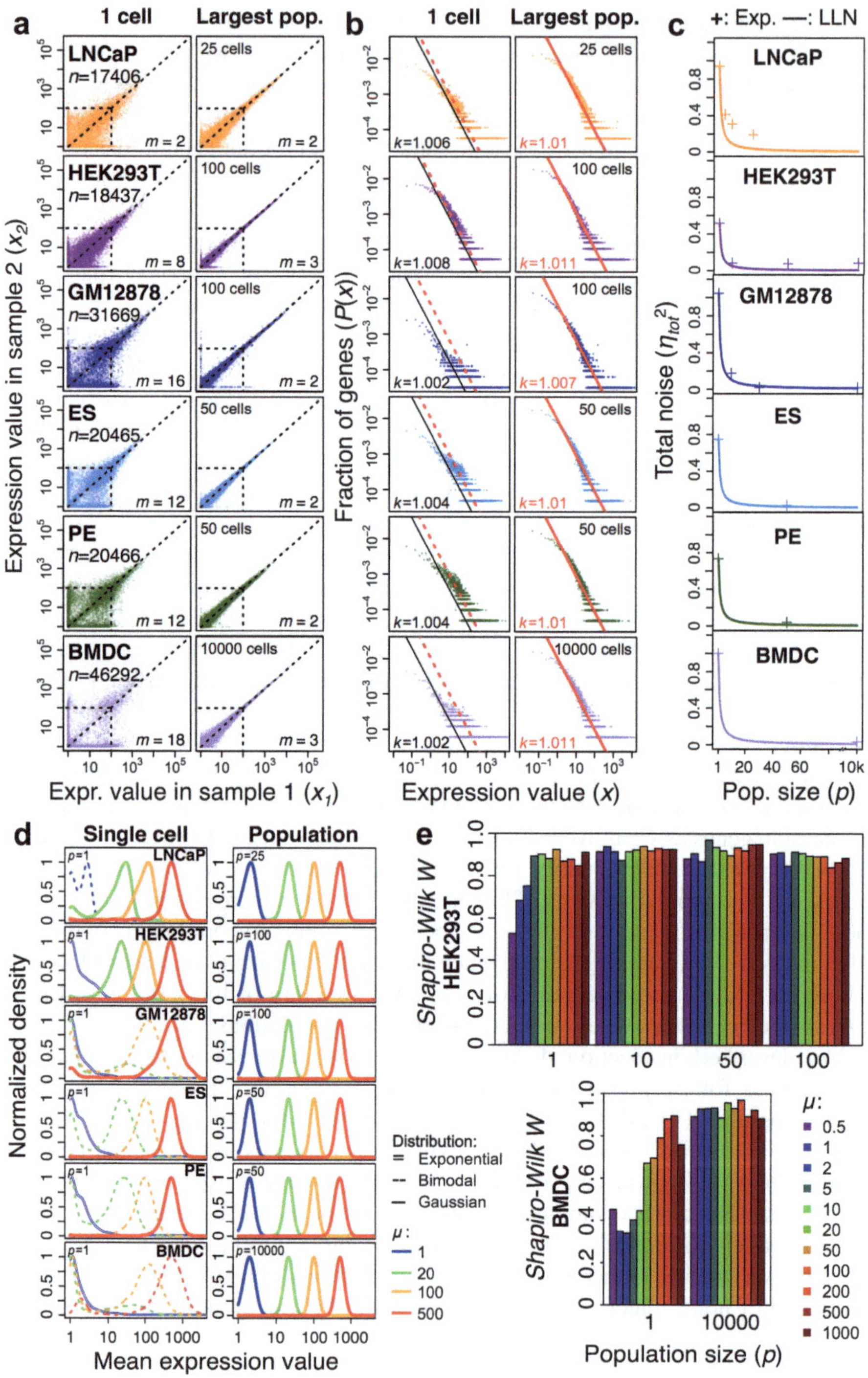
a
1 cell
Largest pop.
LNCaP
n=17406
HEK293T
n=18437
GM12878
n=31669
ES
n=20465
PE
n=20466
BMDC
n=46292
25 cells
100 cells
50 cells
10000 cells
Expression value in sample 2 (x_2)
Expr. value in sample 1 (x_1)
b
Fraction of genes ($P(x)$)
Expression value (x)
c
+: Exp. —: LLN
Total noise (η_{tot}^2)
Pop. size (p)
d
Single cell
Population
Normalized density
Mean expression value
Distribution:
Exponential
Bimodal
Gaussian
e
Shapiro-Wilk W
HEK293T
BMDC
Population size (p)
μ:

Table 1 Pearson correlation values of transcriptomes. Values are obtained by averaging the correlation obtained for all combinations or pairwise samples

Cell type	Population size (p)				
	1	5	10	25–50	≥100
LNCaP	0.701	0.964	0.958	0.986	
HEK293T	0.791		0.994	0.989	0.988
GM12878	0.831		0.977	0.997	0.997
ES	0.944			0.997	
PE	0.924			0.989	
BMDC	0.583				0.998

Table 2 Spearman correlation values of transcriptomes

Cell type	Population size (p)				
	1	5	10	25–50	≥100
LNCaP	0.707	0.859	0.885	0.916	
HEK293T	0.824		0.954	0.960	0.962
GM12878	0.590		0.824	0.873	0.882
ES	0.738			0.952	
PE	0.780			0.951	
BMDC	0.599				0.951

(Fig. 6c, *lines*), indicating that the reduction of transcriptome-wide noise is actually due to the averaging of expression values of each gene in the single cells forming populations.

To further understand the origins of noise reduction when cell populations are formed, we compared the average distributions of gene expressions obtained in single cells and populations at different expression ranges, i.e., for genes whose average expression values, μ, are order of 1 unit (low expression), 20 units (middle-low), 100 units (middle-high), and 500 units (high).

Fig. 6 (**a**) Gene expression values between two representative samples in single cells and the largest available populations. n, number of genes; m, number of available samples. *Dotted squares*: low expression values $x < 100$. (**b**) Gene expression values versus fraction of genes in a representative sample in single cells and largest available populations. *Solid black lines*: fitted theoretical power law, $P(x) = x^{-k}(k-1)/{x_{min}}^{1-k}$, where k is the exponent and x_{min} (=1) is the threshold below which power law fails. *Thick red lines*: fitted power law of the largest available cell population, repeated in single cells plots by a *red dotted line.* (**c**) Transcriptome-wide average (total) noise ($\eta^2{}_{tot}$) versus population size (p) after removing genes with median expression values lower than 1 unit. *Crosses*, experimental values; *lines*, values expected from the law of large numbers; n, number of genes. (**d**) Average distributions (normalized kernel density) of transcriptomes in single cells and populations, for four representative groups of 100 genes with mean expression values ranging from $\mu \sim 1$ to $\mu \sim 500$. (**e**) Shapiro–Wilk test (W) for normality of gene expression distributions. Average W is reported for eight representative groups of 100 genes with mean expression values ranging from $\mu \sim 1$ to $\mu \sim 500$ in single cells and populations. $W \sim 1$ indicates normally (Gaussian) distributed values. As the test requires at least three data samples to be performed ($m \geq 3$), we could only compare single cell versus cell population statistic for HEK293T and BMDC

For single cells, we observed three major patterns (Fig. 6d): (i) exponential-like distributions for low expression genes with many off-cells (where genes are not expressed) and few on-cells (where genes are expressed), (ii) bimodal distributions for middle-low and middle-high expression genes with both off- and on-cells (note that LNCaP and HEK293T displayed bimodality at lower range of expressions), and (iii) gamma/Gaussian-like distributions for high expression genes with a majority of on-cells (note that gamma distributions approximate Gaussians when the mean increases (Peizer and Pratt 1963)). These observations are in consistency with previous studies, which showed that individual genes transcript levels followed gamma distributions (Bengtsson et al. 2005; Taniguchi et al. 2010; Wills et al. 2013), or bimodal distributions (Shalek et al. 2013) in single cells.

In contrast to single cells, a unique unimodal and Gaussian-like distribution pattern was observed for all cell populations at all ranges of expressions (Fig. 6d). This behavior reflects the central limit theorem (CLT), which states that the distribution of sample mean converges to a Gaussian distribution when sample size increases (Grinstead and Snell 2006). To assess the normality (Gaussian) of expression distributions, we used the Shapiro–Wilk test (Royston 1983), which provides a statistic, W, that equals to 1 when the data is normally distributed.

We performed the analysis for HEK293T and BMDC cells where at least three data samples were available and observed that W increased close to 1.0 for single cells with mean expressions above 5 for HEK293T and above 100 for BMDC. However, for cell populations with $p \geq 10$ cells, both cell types displayed W close to 1.0 across all expression ranges (Fig. 6e). That is, high expression genes are quasi-normally distributed in both single cells and populations, while genes with exponential and bimodal expression distributions also attained quasi-normal distributions when their population size increased. In summary, distributions of gene expressions among samples approximate Gaussian at transcriptome-wide scale when cell population size increases, following the statistical law of CLT.

To sum up, it is shown, for each investigated cell type, that the reduction of expression variability from single cells to populations follows both statistical laws of LLN and CLT. The result suggests that gene expression values observed in cell populations are the average of the expression values of individual cells that form the populations. Explicitly, the deterministic population average transcriptome-wide structure of mammalian cells gradually emerges from single cell noisy expressions. Thus, the study of individual cells may not necessarily be a direct representation of population behavior where variability, especially, is concerned. On another perspective, this result tells us of the presence of "phenotypic attractors" imposing top-down rules to single cell expression allowing for an emergent global phenotype behavior. The presence of microscopic (single cell-level) noise is instrumental to cope with environment variability as well as for the onset of state transitions (e.g., epithelial–mesenchymal transition (EMT), linked to carcinogenesis (Lamouille et al. 2014)). Again, these state transitions can be followed in terms of Pearson correlation dynamics at both single cell and tissue levels (Mojtahedi et al. 2016).

6 Final Remarks

In a fundamental paper that appeared in 1948 entitled "Science and Complexity," Warren Weaver (1948), one of the fathers of modern information science, working together with Claude Shannon, proposed a tripartition of science styles. Scientific themes can be subdivided into (1) problems of simplicity, (2) problems of disorganized complexity, and (3) problems of organized complexity.

The first class (simplicity) collected all those problems faced in terms of differential equations and thus well suited for deriving "general laws of nature." These "simple problems" were the ones solved by most "sophisticated" mathematics because they were amenable to a high degree of abstraction (e.g., a planet could be considered an abstract dimensionless "material point").

Problems of disorganized complexity (class 2) allow for superior precision (and, most importantly, for a higher degree of generalization) than class 1 problems. These problems imply a somewhat opposite style of reasoning with respect to the "problems of simplicity." In this case, the efficiency does not stem from the possibility to get an efficient abstract description of the involved players but from totally discarding such "atomic" knowledge in favor of very coarse grain macroscopic descriptors corresponding to gross averages on a transfinite number of atomic elements. This is the case of thermodynamic parameters (e.g., pressure, volume, temperature, etc.).

The two abovementioned approaches have drastic limitations of their applicability range. Class 1 needs the presence of very few involved players interacting in a stable way with a practically null effect of boundary conditions, whereas class 2 needs very large numbers of particles with only negligible interactions among them.

Problems of organized complexity (Weaver's class 3) arise in all those situations in which many (even if not so many as in class 2) are involved with non-negligible interactions. This is the *middle kingdom* of complexity, where biological systems live. The concept of network (or graph) is the archetypal form of organized complexity: a set of nodes (e.g., genes) that are each other connected by mutual correlations (edges). The wiring architecture of these graphs can vary in both space and time; this is why focusing on quantitative statistical descriptors from information theory concepts (like the one we referred in this chapter) is particularly suited to get rid of biological regulation.

Appendix A

The Pearson correlation coefficient *R*(X,Y), or simply *R* for short, between two transcriptome datasets, ***X*** and ***Y***, containing *n* gene expression values, is obtained by (for large *n*), $R\left(X,Y\right)=\sum_{i=1}^{n}\left(x_i-\mu_X\right)\left(y_i-\mu_Y\right)/\sigma_X\sigma_Y$, where x_i and y_i are the expression value of the ith gene in the vectors $\boldsymbol{X}$ and $\boldsymbol{Y}$, respectively; μ_X and μ_Y, the average expression values of each transcriptome dataset; and σ_X and σ_Y, the corresponding standard deviations. Pearson correlation measures linear monotonic relationship between two vectors, where $R = 1$ if the two vectors are identical and $R = 0$ if there is no linear or monotonic relationships between the vectors.

Appendix B

Spearman rank correlation, which measures nonlinear monotonic response, is defined by $\rho(X, Y) = 1 - 6\sum_{i=1}^{n}\left(r_{x_i} - r_{y_i}\right)^2/n\left(n^2 - 1\right)$, where $r_{x,i}$ and $r_{y,i}$ are the ranks of the expression value of the ith gene x_i and y_i, in vectors $\boldsymbol{X}$ and $\boldsymbol{Y}$, respectively. In the context of cellular temporal responses, the decrease in correlation, when comparing two expression vectors (transcriptomes), is a measure of difference between the two vectors. Values close to 1 indicate identical vectors, negative values show anticorrelated vectors, and null values denote absence of nonlinear monotonic relationship between variables (Tsuchiya et al. 2009a, 2009b; Felli et al. 2010; Giuliani et al. 2014).

Appendix C

Hierarchical clustering builds a hierarchy of clusters using two methods: agglomerative and divisive algorithms. We used the former (Ward's) where each observation starts in its own cluster, and pairs of clusters are merged moving up the hierarchy. The Ward's method (Ward 1963) starts with n clusters of size 1 and continues until all the observations are included into one cluster. It begins with the "leaves," looks for groups of leaves to form "branches," and works its way to reach the "trunk."

Here, 5704 genes in Th_0 and Th_{17} differentiation at time $t = 0, 1, 3, 6, 16$, and 48 h (with at least twofold difference between minimum and maximum values and ten expression units) were used to form clusters using the normalized gene expression standard scores, $Z_{i,t} = \left(x_{i,t} - \overline{x}_i\right)/\sigma_{x_i}$, where $x_{i,t}$ is the raw expression value of the ith gene at time t in Th_0 and Th_{17}, $\overline{x}_i$ is the mean value in all samples, and σ_{x_i} is the standard deviation.

In the first step, $n-1$ clusters are formed, one of size two and the remaining of size 1. The pair of the ith gene in Th_0 and Th_{17} that yield the smallest $Z_{i,t}$ forms the first cluster and iterated until the algorithm stops when all sample units are combined into a single large cluster of size n.

Appendix D

Distance correlation (Szekely et al. 2007) measures the statistical dependence between variables ***X*** and ***Y***, which is $dCor(X, Y) = dCov(X, Y) / \sqrt{dVar(X)dVar(Y)}$, where distance covariance of ***X*** and ***Y*** is divided by the product of their distance standard deviations. Note that $dCor(\boldsymbol{X},\boldsymbol{Y}) = 0$ if and only if the two vectors are statistically independent.

Appendix E

Shannon entropy (Shannon 1948) measures the disorder of a high-dimensional system, where higher values indicate increasing disorder. Entropy of each single cell transcriptome, ***X***, is defined as, $H(X) = -\sum_{i=1}^{n} p(x_i) \log_2 p(x_i)$, where $p(x_i)$ represents the probability of gene expression value $x = x_i$. Entropy values were obtained through binning approach, and the number of bins, $b = 26$, was determined from the data using Doane's rule (Doane 1976), such as $b(X) = 1 + log_2 n + log_2(1 + |g_X|/\sigma_g)$, where g_X is the skewness of the expression distribution of each sample and $\sigma_g = \sqrt{6(n-2)/(n+1)(n+3)}$.

Appendix F

The most widely adopted methodology to compute gene expression noise is the squared coefficient of variation, η^2, i.e., the variance in expressions among the total number of cells (σ^2) divided by the squared mean expression (μ^2) (Paulsson 2004; Pedraza and Paulsson 2008). This noise evaluation provides a dimensionless and normalized measurement of the variation of a given variable among multiple observations.

Transcriptome-wide average noise, a.k.a. total noise, for each cell type, is defined as $\eta_{tot}^2 = \frac{1}{n}\sum_{i=1}^{n}\eta_i^2$, where n is the number of genes and η_i^2 is the pairwise noise of the ith gene (variability between any two cells), defined as $\eta_i^2 = \frac{2}{m(m-1)}\sum_{j=1}^{m-1}\sum_{k=j+1}^{m}\eta_{ijk}^2$, where m is the number of cells and η_{ijk}^2 is the expression noise of the ith gene, defined by the variance divided by the squared mean expression (Rosner 2011) in the pair of cells (j,k), such as $\eta_{ijk}^2 = \sigma_{ijk}^2/\mu_{ijk}^2$, where $\mu_{ijk} = (x_{ij} + x_{ik})/2$ is the average value of the ith gene in the pair of single cells (j,k) and $\sigma_{ijk}^2 = (x_{ij} - x_{ik})^2/2$ is the corresponding variance.

References

Albert R (2005) Scale-free networks in cell biology. J Cell Sci 118:4947–4957
Bengtsson M, Ståhlberg A, Rorsman P et al (2005) Gene expression profiling in single cells from the pancreatic islets of Langerhans reveals lognormal distribution of mRNA levels. Genome Res 15:1388–1392
Bhairavabhotla R, Kim YC, Glass DD et al (2016) Transcriptome profiling of human FoxP3+ regulatory T cells. Hum Immunol 77:201–213
Bialek W (2012) Biophysics: searching for principles. Princeton University Press, Princeton
Bonchev D, Trinajstić N (1977) Information theory, distance matrix, and molecular branching. J Chem Phys 67:4517
Bottomly D, Walter NA, Hunter JE et al (2011) Evaluating gene expression in C57BL/6J and DBA/2J mouse striatum using RNA-Seq and microarrays. PLoS One 6:e17820
Braude P, Bolton V, Moore S (1988) Human gene expression first occurs between the four- and eight-cell stages of preimplantation development. Nature 332:459–461
Calabresi PA, Yun SH, Allie R et al (2002) Chemokine receptor expression on MBP-reactive T cells: CXCR6 is a marker of IFN gamma-producing effector cells. J Neuroimmunol 127:96–105
Chang HH, Hemberg M, Barahona M et al (2008) Transcriptome-wide noise controls lineage choice in mammalian progenitor cells. Nature 453:544–547
Ciofani M, Madar A, Galan C et al (2012) A validated regulatory network for Th17 cell specification. Cell 151:289–303
Conesa A, Madrigal P, Tarazona S et al (2016) A survey of best practices for RNA-seq data analysis. Genome Biol 17:13
De La Fuente R (2006) Chromatin modifications in the germinal vesicle (GV) of mammalian oocytes. Dev Biol 292:1–12
Debey P, Szöllösi MS, Szöllösi D et al (1993) Competent mouse oocytes isolated from antral follicles exhibit different chromatin organization and follow different maturation dynamics. Mol Reprod Dev 36:59–74
Dinarello CA (2007) Historical insights into cytokines. Eur J Immunol 37:S34–S45
Doane DP (1976) Aesthetic frequency classification. Am Stat 30:181–183
Eldar A, Elowitz MB (2010) Functional roles for noise in genetic circuits. Nature 467:167–173
Elowitz MB, Levine AJ, Siggia ED et al (2002) Stochastic gene expression in a single cell. Science 297:1183–1186
Felli N, Cianetti L, Pelosi E et al (2010) Hematopoietic differentiation: a coordinated dynamical process towards attractor stable states. BMC Syst Biol 4:85
Furusawa C, Kaneko K (2003) Zipf's law in gene expression. Phys Rev Lett 90:088102
Giuliani A (2017) The application of principal component analysis to drug discovery and biomedical data. Drug Discov Today 22:1069–1076
Giuliani A, Colafranceschi M, Webber CL et al (2001) A complexity score derived from principal components analysis of nonlinear order measures. Physica A 301:567–588
Giuliani A, Filippi S, Bertolaso M (2014) Why network approach can promote a new way of thinking in biology. Front Genet 5:83
Grinstead CM, Snell JL (2006) Introduction to probability, 2nd edn. American Mathematical Society, Providence
Hagar JA, Powell DA, Aachoui Y et al (2013) Cytoplasmic LPS activates caspase-11: implications in TLR4-independent endotoxic shock. Science 341:1250–1253
Hirotani T, Yamamoto M, Kumagai Y et al (2005) Regulation of lipopolysaccharide-inducible genes by MyD88 and Toll/IL-1 domain containing adaptor inducing IFN-beta. Biochem Biophys Res Commun 328:383–392
Hu G, Tang Q, Sharma S et al (2013) Expression and regulation of intergenic long noncoding RNAs during T cell development and differentiation. Nat Immunol 14:1190–1198

Irizarry RA, Hobbs B, Collin F et al (2003) Exploration, normalization, and summaries of high density oligonucleotide array probe level data. Biostatistics 4:249–264

Islam S, Kjällquist U, Moliner A et al (2011) Characterization of the single-cell transcriptional landscape by highly multiplex RNA-seq. Genome Res 21:1160–1167

Kared H, Camous X, Larbi A (2014) T cells and their cytokines in persistent stimulation of the immune system. Curr Opin Immunol 29:79–85

Kayagaki N, Warming S, Lamkanfi M et al (2011) Non-canonical inflammasome activation targets caspase-11. Nature 479:117–121

Lamouille S, Xu J, Derynck R (2014) Molecular mechanisms of epithelial–mesenchymal transition. Nat Rev Mol Cell Biol 15:178–196

Maamar H, Raj A, Dubnau D (2007) Noise in gene expression determines cell fate in *Bacillus subtilis*. Science 317:526–529

MacLeod MK, Kappler JW, Marrack P (2010) Memory CD4 T cells: generation, reactivation and re-assignment. Immunology 130:10–15

Magombedze G, Reddy PB, Eda S et al (2013) Cellular and population plasticity of helper CD4(+) T cell responses. Front Physiol 4:206

Marinov GK, Williams BA, McCue K et al (2014) From single-cell to cell-pool transcriptomes: stochasticity in gene expression and RNA splicing. Genome Res 24:496–510

Mojtahedi M, Skupin A, Zhou J et al (2016) Cell fate decision as high-dimensional critical state transition. PLoS Biol 14:e2000640

Negishi T, Kato Y, Ooneda O et al (2005) Effects of aryl hydrocarbon receptor signaling on the modulation of TH1/TH2 balance. J Immunol 175:7348–7356

Newcomb DC, Zhou W, Moore ML et al (2009) A functional IL-13 receptor is expressed on polarized murine CD4+ Th17 cells and IL-13 signaling attenuates Th17 cytokine production. J Immunol 182:5317–5321

Nilsson R, Bajic VB, Suzuki H et al (2006) Transcriptional network dynamics in macrophage activation. Genomics 88:133–142

O'Neill LA, Golenbock D, Bowie AG (2013) The history of Toll-like receptors – redefining innate immunity. Nat Rev Immunol 13:453–460

Paulsson J (2004) Summing up the noise in gene networks. Nature 427:415–418

Pedraza JM, Paulsson J (2008) Effects of molecular memory and bursting on fluctuations in gene expression. Science 319:339–343

Peizer DB, Pratt JW (1963) A normal approximation for binomial, F, beta, and other common, related tail probabilities, I. J Am Stat Assoc 63:1416–1456

Picelli S, Faridani OR, Björklund AK et al (2014) Full-length RNA-seq from single cells using Smart-seq2. Nat Protoc 9:171–181

Piras V, Selvarajoo K (2015) The reduction of gene expression variability from single cells to populations follows simple statistical laws. Genomics 105:137–144

Piras V, Tomita M, Selvarajoo K (2012) Is central dogma a global property of cellular information flow? Front Physiol 3:439

Piras V, Tomita M, Selvarajoo K (2014) Transcriptome-wide variability in single embryonic development cells. Sci Rep 4:7137

Raj A, van Oudenaarden A (2009) Single-molecule approaches to stochastic gene expression. Annu Rev Biophys 38:255–270

Raj A, Rifkin SA, Andersen E et al (2010) Variability in gene expression underlies incomplete penetrance. Nature 463:913–918

Ramsköld D, Luo S, Wang YC et al (2012) Full-length mRNA-Seq from single-cell levels of RNA and individual circulating tumor cells. Nat Biotechnol 30:777–782

Rau A, Gallopin M, Celeux G et al (2013) Data-based filtering for replicated high-throughput transcriptome sequencing experiments. Bioinformatics 29:2146–2152

Reshef DN, Reshef YA, Finucane HK et al (2011) Detecting novel associations in large data sets. Science 334:1518–1524

Rosner B (2011) Fundamentals of Biostatistics, 7th edn. Duxbury Press, Boston

Royston JP (1983) Some techniques for assessing multivarate normality based on the Shapiro–Wilk W. Appl Stat 32:121–133
Sasagawa Y, Nikaido I, Hayashi T et al (2013) Quartz-Seq: a highly reproducible and sensitive single-cell RNA sequencing method, reveals non-genetic gene-expression heterogeneity. Genome Biol 14:R31
Selvarajoo K (2012) Understanding multimodal biological decisions from single cell and population dynamics. Wiley Interdiscip Rev Syst Biol Med 4:385–399
Selvarajoo K (2013) Uncertainty and certainty in cellular dynamics. Front Genet 4:68
Shalek AK, Satija R, Adiconis X et al (2013) Single-cell transcriptomics reveals bimodality in expression and splicing in immune cells. Nature 498:236–240
Shannon CE (1948) A mathematical theory of communication. Bell Syst Tech J 27:623–666
Simeoni O, Piras V, Tomita M et al (2015) Tracking global gene expression responses in T cell differentiation. Gene 569:259–266
Smith CG, MacArthur BD (2017) Information-theoretic approaches to understanding stem cell variability. Curr Stem Cell Rep 3:225–331
Smith-Garvin JE, Koretzky GA, Jordan MS (2009) T cell activation. Annu Rev Immunol 27:591–619
Soofi ES (1994) Capturing the intangible concept of information. J Am Stat Assoc 89:1243–1254
Sultan M, Schulz MH, Richard H et al (2008) A global view of gene activity and alternative splicing by deep sequencing of the human transcriptome. Science 321:956–960
Swain SL, McKinstry KK, Strutt TM (2012) Expanding roles for CD4+ T cells in immunity to viruses. Nat Rev Immunol 12:136–148
Szekely GJ, Rizzo ML, Bakirov NK (2007) Measuring and testing independence by correlation of distances. Ann Stat 35:2769–2794
Tadros W, Lipshitz HD (2009) The maternal-to-zygotic transition: a play in two acts. Development 136:3033–3042
Taniguchi Y, Choi PJ, Li GW et al (2010) Quantifying *E. coli* proteome and transcriptome with single-molecule sensitivity in single cells. Science 329:533–538
Thomas JA, Cover TM (2006) Elements of information theory, 2nd edn. Wiley, Hoboken
Touzot M, Grandclaudon M, Cappuccio A et al (2014) Combinatorial flexibility of cytokine function during human T helper cell differentiation. Nat Commun 5:3987
Tsuchiya M, Piras V, Choi S et al (2009a) Emergent genome-wide control in wildtype and genetically mutated lipopolysaccarides-stimulated macrophages. PLoS One 4:e4905
Tsuchiya M, Selvarajoo K, Piras V et al (2009b) Local and global responses in complex gene regulation networks. Physica A 388:1738–1746
Tuomela S, Salo V, Tripathi SK et al (2012) Identification of early gene expression changes during human Th17 cell differentiation. Blood 119:e151–e160
Ueda HR, Hayashi S, Matsuyama S et al (2004) Universality and flexibility in gene expression from bacteria to human. Proc Natl Acad Sci U S A 101:3765–3769
Verma-Gandhu M, Verdu EF, Cohen-Lyons D et al (2007) Lymphocyte-mediated regulation of beta-endorphin in the myenteric plexus. Am J Physiol Gastrointest Liver Physiol 292:G344–G348
Ward JH (1963) Hierarchical grouping to optimize an objective function. J Am Stat Assoc 58:236–244
Weaver W (1948) Science and complexity. Am Scientist 36:536–549
Wills QF, Livak KJ, Tipping AJ et al (2013) Single-cell gene expression analysis reveals genetic associations masked in whole-tissue experiments. Nat Biotechnol 31:748–752
Zheng GX, Terry JM, Belgrader P et al (2017) Massively parallel digital transcriptional profiling of single cells. Nat Commun 8:14049
Zhu J, Yamane H, Paul WE (2010) Differentiation of effector CD4 T cell populations. Annu Rev Immunol 28:445–489

Enhancing Metabolic Models with Genome-Scale Experimental Data

Kristian Jensen, Steinn Gudmundsson, and Markus J. Herrgård

Contents

Abstract Genome-scale metabolic reconstructions have found widespread use in scientific research as structured representations of knowledge about an organism's metabolism and as starting points for metabolic simulations. With few simplifying assumptions, genome-scale models of metabolism can be used to estimate intracellular reaction rates in any organism for which a well-curated metabolic reconstruction is available. However, with the rapid increase in the availability of genome-scale data, there is ample opportunity to refine the predictions made by metabolic models by integrating experimental data. In this chapter, we review different methods for combining genome-scale metabolic models with genome-scale experimental data, such as transcriptomics, proteomics, and metabolomics. Integrating experimental data into the models generally results in more precise and accurate simulations of cellular metabolism.

Keywords Genome-scale modeling · Constraint-based metabolic modeling · Flux balance analysis · Genome-scale data · Transcriptomics · Proteomics · Metabolomics · Shadow prices · Machine learning

---

K. Jensen · M. J. Herrgård (✉)
The Novo Nordisk Foundation Center for Biosustainability, Technical University of Denmark, Kongens Lyngby, Denmark
e-mail: herrgard@biosustain.dtu.dk

S. Gudmundsson
Center for Systems Biology, University of Iceland, Reykjavik, Iceland

N. Rajewsky et al. (eds.), *Systems Biology*, RNA Technologies,
https://doi.org/10.1007/978-3-319-92967-5_17

1 Reconstruction and Analysis of Metabolic Networks

It is essential to study metabolism in order to describe and understand the functioning of living cells. The chemical conversion of nutrients into energy, biomass and secondary products is one of the main components of the cellular phenotype, and a defining characteristic of life. Since the metabolic capabilities of an organism are ultimately determined by its genotype, advances in genome sequencing technologies during the last two decades have had a substantial impact on our knowledge about metabolism. With a fully annotated whole genome sequence of an organism, it is feasible to compile a database of all the biochemical reactions that can be catalyzed inside the cell. Besides a list of reactions and their stoichiometries, such a database, called a *genome-scale metabolic reconstruction*, often includes information that links each reaction to the genes encoding the enzymes that catalyze it (Price et al. 2004). The earliest published genome-scale reconstructions were for organisms with small genomes such as *Haemophilus influenzae* (Schilling and Palsson 2000) and *Escherichia coli* (Edwards and Palsson 2000), but reconstructions for more complex organisms including *Saccharomyces cerevisiae* (Förster et al. 2003), *Arabidopsis thaliana* (de Oliveira Dal'Molin et al. 2010), and *Homo sapiens* (Duarte et al. 2007) have followed. Revised versions of genome-scale metabolic reconstructions are often published when new genes are discovered or annotated functions of known genes are updated.

A genome-scale metabolic reconstruction allows systematic analysis of the metabolic network of an organism, and can even form a starting point for whole-cell simulations (Orth et al. 2010; Karr et al. 2012). In order to perform such analyses, the genome-scale reconstruction must be formulated as a mathematical model, e.g., in the form of a system of differential equations,

$$\frac{\mathrm{d}\boldsymbol{x}}{\mathrm{d}t} = \boldsymbol{S} \cdot \boldsymbol{v}\,(\boldsymbol{x}, \boldsymbol{k}) \tag{1}$$

Here $\boldsymbol{S}$ denotes the stoichiometric matrix, derived from the genome-scale reconstruction with element s_{ij} denoting the stoichiometric coefficient of metabolite i in reaction j, and $\boldsymbol{x}$ is a vector of concentrations of all metabolites in the cell. Reaction rates, $\boldsymbol{v}$, are a function of current metabolite concentrations and kinetic parameters, $\boldsymbol{k}$. Given initial metabolite concentrations, the system of differential equations is readily solved numerically. While the formulation is conceptually simple, its use on the genome-scale is impeded by limited knowledge of the many kinetic parameters (McCloskey et al. 2013).

To avoid the issue of unknown kinetic parameters, *constraint-based metabolic modeling* methods are often used instead. Constraint-based modeling imposes constraints on the system and finds metabolic reaction rates that are consistent with these constraints. The most central constraint is the assumption of steady-state, where the concentrations of internal metabolites are assumed to be constant. This corresponds to setting the left-hand side of Eq. (1) to zero and results in a system of linear equations,

$$\boldsymbol{S} \cdot \boldsymbol{v} = \boldsymbol{0} \tag{2}$$

that can be solved for the reaction rates or metabolic fluxes, $\boldsymbol{v}$ (Orth et al. 2010). The kinetic parameters are not accounted for explicitly in constraint-based models, which only require the stoichiometric matrix to be known. For most genome-scale reconstructions, the system of equations is underdetermined, meaning that an infinite number of flux solutions exist. One way to address this issue is to identify a solution that optimizes a specific objective. This is based on an assumption that the cell has evolved to maximize some biological objective, e.g., production of ATP or production of biomass. Production of biomass is modeled through a bulk-reaction that consumes biomass constituents such as nucleotides and amino acids in empirically determined ratios (Orth et al. 2010). This method is known as *flux balance analysis* (FBA) and has become the foundation of most work in constraint-based metabolic modeling. Performing flux balance analysis requires the solution of a linear optimization problem. The result is a set of reaction rates that satisfy the constraints of the system and is consistent with the defined biological objective.

Despite the simple formulation and strong assumptions, FBA has proven useful in a number of metabolic modeling applications, to predict the rates of metabolic reactions, typically called the *flux distribution* (McCloskey et al. 2013). It can be used for instance to predict essential metabolic genes, i.e., genes that are required for the synthesis of one or more biomass constituents. This is done by simply removing corresponding reactions from the model and performing FBA. If the maximal biomass flux is zero in the knockout model, the gene is expected to be essential. Comparisons with experimental data from single-knockout studies have shown good correspondence with the results of FBA-based essentiality predictions in *E. coli* and other bacteria such as *Pseudomonas aeruginosa* (Edwards and Palsson 2000; Oberhardt et al. 2008). In other organisms, e.g., *S. cerevisiae*, predictions of essentiality are less accurate, and for multiple knockouts in particular there is only a very low correlation between experimental data and FBA predictions (Heavner and Price 2015).

The assumption of maximization of biomass production as a metabolic objective is often reasonable for microorganisms during exponential growth, but it will clearly not hold for most mammalian cells or other multicellular organisms whose evolutionary pressure has selected for far more complex traits than simply growth at the cellular level. As replacement for FBA, Markov chain Monte Carlo (MCMC) methods can be used to uniformly sample the feasible steady-state flux space described by Eq. (2). MCMC methods provide an estimate of the joint probability distribution of fluxes and do not depend on a prespecified biological objective. The applications of random sampling methods include the analysis of red blood cells under storage conditions (Bordbar et al. 2016), aspirin resistance in platelets (Thomas et al. 2015), transcriptional regulation in human adipocytes (Mardinoglu et al. 2014) and in bacterial communities in the human gut (Shoaie et al. 2013), as well as the metabolic rewiring that takes place in epithelial to mesenchymal transition during the development of breast cancer (Halldorsson et al. 2017).

2 Constraining Metabolic Models with Transcriptomics and Proteomics Data

Although mass balance is an essential principle, metabolism is constrained by other factors and physical principles as well. FBA assumes that the cell can use all metabolic reactions at a given time in the combination that gives the highest biomass production. However, this is contradicted by the fact that only a proportion of an organism's genes will be transcriptionally active at the same time. Thus further constraints can be imposed on the model by leveraging information about the transcriptional state of the cell. This can be used to create context-specific models from generic models, such as the generic human reconstruction Recon1 (Duarte et al. 2007), as well as to improve the accuracy of flux predictions. The simplest realization of this idea utilizes the fact that an enzyme cannot catalyze any reaction flux if its encoding gene is not expressed. Reactions catalyzed by genes with transcript levels below a defined threshold can thus be forced to be inactive by removing them from the model. Flux distributions obtained with such a constrained model were found to be more strongly correlated to experimentally measured fluxes in *S. cerevisiae* compared to an unconstrained model (Åkesson et al. 2004). More sophisticated algorithms minimize the difference between the predicted flux distribution and the gene expression data. The gene inactivity moderated by metabolism and expression (GIMME) algorithm (Becker and Palsson 2008) finds flux values which minimize the utilization of reactions with low expression levels, in order to meet prespecified metabolic requirements such as growth. The iMAT method developed by Shlomi and coworkers (2008) alleviates the need for a prespecified cellular objective and is therefore suitable for analyzing mammalian cells and tissues. The method partitions gene expression values into three groups, corresponding to high, moderate and low expression and then maximizes the number of reactions with flux levels in agreement with the expression states. This enabled identification of tissue-specific metabolic activities in different human tissues, and the construction of tissue-specific models of human metabolism. An extension of iMAT was used to construct a model of cancer metabolism from Recon1 and expression data from cancer cell lines in the NCI-60 collection. The cancer model was then used to identify several cytostatic drug targets, and generate a list of potential selective anticancer treatments (Folger et al. 2011).

Since Åkesson and coworkers first used gene-expression data to constrain metabolic models, a large number of methods that integrate expression data and flux predictions have been published. An evaluation of many of these methods, by their ability to predict flux distributions in *E.* coli and *S. cerevisiae*, showed that none of them performed significantly better than parsimonious FBA, an extension of FBA that finds the flux distribution with the smallest sum of fluxes that can support the optimal objective value (Machado and Herrgård 2014). This suggests that gene transcription levels do not correlate strongly with reaction fluxes, at least in microbial cells, which is not surprising considering that translational efficiency, posttranslational modifications, and allosteric regulation all have an effect on fluxes as well.

A step closer to the actual reactions than mRNA abundance is protein concentration. A certain correlation between mRNA and protein concentration is to be expected (Gry et al. 2009), and several methods for integrating gene expression data into metabolic models can indeed use protein abundance data with the same algorithms, simply by replacing gene expression thresholds with protein abundance thresholds (Becker and Palsson 2008; Machado and Herrgård 2014). However, there have also been attempts to more explicitly incorporate proteomics data into the modeling frameworks. A central component of enzyme kinetics is the concept of the catalytic capacity of an enzyme. Each enzyme molecule can only perform a certain number of conversions per second; an increased flux will therefore require a larger number of enzymes at some point. The maximum possible flux, represented by the $V_{\max}$ parameter, can be calculated from the enzyme concentration and catalytic turnover number, k_{cat}

$$V_{\max} = k_{\text{cat}} \cdot [E] \tag{3}$$

If the catalytic turnover parameters are known, this relationship can be used to constrain fluxes using protein concentration data. In the GECKO modeling framework (Sánchez et al. 2017), a constraint is added for each enzyme, representing the enzyme's degree of utilization, where the upper bound is set to the measured enzyme concentration. The utilization of an enzyme is obtained by summing v/k_{cat} for all reactions catalyzed by that enzyme. Using GECKO with a proteomics dataset for *S. cerevisiae*, Sanchez and coworkers showed that the space of possible fluxes was reduced considerably by excluding all flux distributions that were not consistent with the observed enzyme levels. On the other hand, the fluxes predicted for *S. cerevisiae* grown in glucose limited minimal medium did not have a significantly smaller error compared to experimentally measured fluxes than those predicted with FBA. It is possible however, that the advantage of using proteomics data will be larger in cases where the assumption of maximal growth is not valid, e.g., under stress conditions or in genetically perturbed strains. GECKO can also be used in the absence of proteomics data by imposing a single overall constraint on the total enzyme mass. This resulted in more accurate predictions of maximal growth rates on a wide range of different carbon sources, for which FBA tends to overestimate growth rate. Another interesting growth effect that was captured by including an overall protein constraint is the shift from respiration to fermentation at high growth rates. This overflow metabolism, also known as the Crabtree effect in yeast (Crabtree 1929) and the Warburg effect in cancer cells (Warburg et al. 1927), cannot be captured by FBA, where simply the flux distribution with the highest biomass yield is found, independently of growth rate. The overflow effect is most likely caused by respiratory enzymes having a higher proteome cost than fermentative enzymes (Basan et al. 2015), which means that at high growth rates protein allocation becomes limiting and fermentation becomes more efficient even though it has a lower energy/carbon yield. Overflow metabolism has been modeled, e.g., in *E. coli* (Basan et al. 2015), *S. cerevisiae* (Sánchez et al. 2017), and cancer cells (Shlomi et al. 2011), by different models with the common trait of somehow constraining the proteome.

The causes of the Warburg effect in cancer cells were studied using Recon1 by placing a constraint on total enzyme concentration to account for enzyme solvent capacity (Shlomi et al. 2011). To compute the contribution of each enzyme to the total concentration, an estimate of the enzyme turnover number was required. Estimates for 15% of the reactions could be obtained from biochemical databases, while the rest were assigned a fixed value of 25 s^{-1}. Using FBA and random sampling, the Warburg effect was shown to be a consequence of metabolic adaptations to increase biomass productivity. Further analysis revealed the preference of cancer cells to take up glutamine instead of other amino acids.

Resource allocation between cellular processes in *Bacillus subtilis* was recently analyzed using a method that incorporates genome-wide protein quantification data and extracellular nutrient concentrations with a metabolic reconstruction (Goelzer et al. 2015). The method, resource balance analysis (RBA), links flux to enzyme abundance, assuming a relationship similar to Eq. (3), while incorporating information on protein activity and protein localization. The use of RBA is fairly involved compared to the methods described earlier and requires specification of a large number of parameters. The parameters were partly obtained from Uniprot and partly inferred from data. RBA accurately predicted the allocation of resources in *B. subtilis* over a wide range of conditions. In vivo knockouts of enzymes that were expressed but predicted to have zero flux in the model resulted in significantly increased growth (Goelzer et al. 2015). This suggests that the method may be useful for constructing minimal cell factories, e.g., for protein production.

3 Models of Metabolism and Macromolecular Expression

The previously described methods for combining *omics* data and metabolic models are mostly based on heuristically formulated constraints and/or objectives. When the measured quantities—such as mRNA and protein abundances—are not explicitly accounted for in the modeling framework, they cannot be seamlessly integrated into it. To address this problem, an extended modeling framework that explicitly models the expression of macromolecules, such as RNA and protein, has been developed. Construction of such models of metabolism and expression (*ME-models*) began with the reconstruction of the macromolecular expression network of *E. coli*, analogously to the metabolic network (Thiele et al. 2009). Transcription of a given gene to produce mRNA is modeled as a reaction consuming nucleotides in proportions consistent with the specific sequence, and similarly translation is modeled as a reaction consuming charged tRNAs while producing protein and uncharged tRNAs. In order to model how metabolic catalysis is dependent on translation of a specific protein and how translation of a protein is dependent on transcription of its gene to mRNA, these different reactions must be coupled (Thiele et al. 2009; Lerman et al. 2012). A certain quantity of an enzyme can only catalyze a limited reaction flux and Eq. (3) can be rearranged to enable calculation of the minimum amount of enzyme required to catalyze a given flux

$$[E] \geq \frac{v}{k_{\text{cat}}} \tag{4}$$

Equation (4) represents a constraint that can be used to couple metabolic reactions to the enzymes that catalyze them. Identical constraints can be formulated for ribosomes and mRNA in translation reactions and for RNA-polymerase in transcription reactions. A constraint-based modeling framework, however, does not model concentrations of metabolites (or enzymes) and is thus not directly compatible with such constraints. To circumvent this it is necessary to account for growth-related dilution. In a growing cell, metabolite pools are continuously diluted, because of the expanding intracellular volume, by a rate equal to the product of the growth rate and metabolite concentration. This means that in steady-state, catalysis of a reaction requires that the catalyzing enzyme be produced at a rate proportional to the growth rate. Enzymatic conversion of compound A into compound B by enzyme E thus becomes (Lloyd et al. 2017):

$$A + \frac{\mu}{k_{\text{cat}}} E \rightarrow B \tag{5}$$

In FBA the requirement of enzyme production is modeled through the composition of the biomass reaction, but since this reaction is determined a priori, FBA cannot model how biomass composition changes under different growth rates and conditions. With ME-models the empirical biomass reaction is replaced by explicitly modeling the relationship between metabolism and macromolecular expression. ME-models can thus directly predict the expression levels of different proteins, which can be compared with *omics* datasets. A ME-model of the thermophilic bacterium *Thermotoga maritima* (Lerman et al. 2012) found moderate correlations between predicted and experimentally measured mRNA profiles ($r = 0.54$), protein expression profiles ($r = 0.57$), as well as proteome amino acid composition ($r = 0.79$). A ME-model of *E. coli* showed improved prediction of growth rates in different nutrient conditions compared to FBA (Thiele et al. 2012), and could accurately predict several internal fluxes (O'Brien et al. 2013). Additionally, since ME-models explicitly include the cost of producing the enzymes required for various pathways, they implicitly limit the total proteome size and thus also capture metabolic overflow effects, such as the acetate overflow metabolism in *E. coli* (O'Brien et al. 2013).

Whereas traditional constraint-based metabolic models include, and can thus directly predict, growth rate, uptake and secretion rates, and internal fluxes, ME-models can additionally predict expression profiles and proteome composition, and thus they can also be directly constrained by expression and proteomics data. Because of this, ME-models represent an intuitive and theoretically justified method of integrating transcriptomics and proteomics data into metabolic models. They have not yet found broad usage in the metabolic modeling community, presumably because of the time it takes to run simulations (several orders of magnitude higher than with FBA), and the lack of related model and software infrastructure, but these issues are continuously being addressed (Yang et al. 2016; Lloyd et al. 2017).

4 Augmenting Models with Metabolomics Data

In a discussion of data integration in metabolic models, it is impossible not to mention metabolomics. Different analytical methods, e.g., enzymatic assays, chromatography and mass spectrometry, can be used to take snapshots of the cellular metabolism with varying resolution, coverage, precision and throughput. However, they all provide useful information about the concentrations of metabolite pools in the cell. One of the earliest uses of metabolomics data to improve metabolic modeling was *metabolic flux analysis* (MFA), which utilizes time-course metabolite concentration data from cultures fed with isotopically labeled substrates to infer flux values in the metabolic network (Stephanopoulos 1999; Sauer 2006). This is done by monitoring how the isotopes, e.g., ^{13}C or ^{15}N, spread to downstream metabolite pools over time. The advantage of this method is that the resulting fluxes can be used directly to constrain metabolic models or to compare the validity of different simulation methods. However, MFA is labor- and cost-intensive and works best on a smaller subset of the entire metabolic network, typically just the central carbon metabolism (Antoniewicz 2015; Gopalakrishnan and Maranas 2015).

Changes in extracellular metabolite concentrations over time can be used to estimate uptake and secretion rates and constrain the flux space. However, since constraint-based modeling frameworks model fluxes under an assumption of steady-state, internal metabolite concentration data at a single time point without isotopic labeling cannot be directly utilized. Despite this, metabolomics data can still be used to either constrain the models or to provide new insights in combination with the simulation results. In order to model cells that are not in steady-state, such as human blood cells undergoing physiological changes during storage, Bordbar and coworkers devised a method called unsteady-state FBA (Bordbar et al. 2017). Using time-course metabolomics they determined the rate of accumulation or depletion for internal metabolites, which was then modeled by adding source and sink reactions to the metabolic model. These reactions were then constrained to have fluxes corresponding to the experimentally determined rates of concentration changes. Subsequent MFA revealed that the fluxes predicted with this method were more accurate than those obtained by regular FBA.

Aside from enforcing steady-state, a commonly used constraint in constraint-based models is to force certain fluxes to only go in one direction. This is straightforward for some reactions whose thermodynamics make it practically irreversible under biological conditions. Other reactions are closer to equilibrium and can go in both directions depending on specific conditions. The spontaneous direction of a reaction can be calculated by the following formula:

$$\Delta_r G = \Delta_r G^{\circ} + RT \ \log(Q) \quad (6)$$

If the left-hand side (the *reaction Gibbs free energy*) is negative, the reaction will proceed spontaneously in the forward direction, while it will proceed spontaneously in the reverse direction if the reaction Gibbs free energy is positive. $\Delta_r G^{\circ}$ is the reaction Gibbs free energy under standard conditions, RT is the gas constant times the absolute temperature and Q is the reaction quotient, containing the concentrations of the reaction products and substrates. The standard Gibbs free energy must in principle be determined experimentally, but in most cases it can be calculated from the structure of the participating metabolites and already known reaction Gibbs free energies for other reactions (Noor et al. 2013). This means that a dataset of metabolite concentrations can be used to constrain reactions to a specific direction depending on the specific metabolic conditions, reducing the space of feasible fluxes significantly (Soh and Hatzimanikatis 2014). In many simulated growth conditions, it can be sufficient simply to constrain reaction directionalities according to the most common mode of operation without regard to actual metabolite concentrations. Some reactions however, occur in the unconventional direction under extreme conditions, such as very high CO_2 concentrations. In such cases using thermodynamics and metabolite data to inform reaction directionalities will be particularly beneficial and can lead to more accurate simulations (Soh et al. 2012).

Constraint-based simulations can also be combined with metabolomics data in another way. In addition to calculating a flux distribution, simulating a constraint-based model also provides the so-called *shadow prices*. Each shadow price is linked to a metabolite and reflects how much the objective function, e.g., growth, could be improved if the model were allowed to get some of that metabolite "for free." In other words, a shadow price is a measure of how limiting a given metabolite's mass balance is for the objective function. Depending on the algorithm used to solve the FBA problem, shadow prices are either a byproduct of the solution process or can be obtained with modest computational effort.

Zampieri and coworkers investigated the evolution of antibiotic resistance in *E. coli* using adaptive laboratory evolution (Zampieri et al. 2017). By maximizing and minimizing flux through each reaction in the model and calculating the shadow prices, the authors could identify reactions, which, when maximized or minimized, resulted in shadow prices that were consistent with the observed patterns of metabolite concentration changes. Those reactions were hypothesized as being targets of evolution, whose flux should be increased in order to increase antibiotic resistance.

Besides constraint-based modeling, the most common way to simulate cellular metabolism is with kinetic models. This involves the solution of the system of differential equations shown in Eq. (1) from given initial metabolite concentrations. As previously described, one of the challenges with this approach is the requirement of knowing the values of all the kinetic parameters of the system. For small biochemical systems, the kinetic parameters can sometimes be determined individually through in vitro experiments, but for genome-scale models this is not feasible. Additionally there is no guarantee that the in vitro kinetic parameters are representative of how an enzyme functions in vivo (Teusink et al. 2000). Instead of the bottom-up approach of experimentally determining each parameter, a top-down approach may be used, where the model parameters might initially be estimated from prior information, such as in vitro data, but are predominantly selected by fitting simulation results to genome-scale experimental data. This has long been done for small-scale networks, using metabolomics and MFA data (Jamshidi and Palsson 2008; Srinivasan et al. 2015); however, with continual increases in dataset sizes and computing power, it has also become feasible to do this for genome-scale networks. Recently, a genome-scale kinetic model of *E. coli* was published along with estimated values for all kinetic parameters (Khodayari and Maranas 2016). The model parameters were fitted using experimental flux data and model predictions were validated against metabolomics data. In addition, the model could quantitatively predict product yields of 24 different compounds in 320 mutant strains, which was considerably better than the constraint-based simulation methods it was tested against. In another study, kinetic models of human red blood cells were used to investigate individual variations in susceptibility to side effects of the hepatitis B drug Ribavirin (Bordbar et al. 2015). By measuring intracellular metabolite levels in red blood cells of 24 patients, they could determine individual kinetic parameter values for each of the patients, and show that those parameters were predictive of whether the patient was sensitive to side effects. Furthermore, the identified relationships between kinetic parameters and sensitivity to drug side-effect were consistent with known mechanisms of Ribavirin side effects. These results show that kinetic modeling frameworks have the potential to significantly outperform constraint-based simulations, and that with modern *omics* technologies and computer power, it is feasible to parametrize them sufficiently to predict metabolic behavior (Saa and Nielsen 2017).

5 Combining Metabolic Models and Machine Learning Methods

The term *machine learning* covers a broad range of methods where large datasets are used to infer relationships between variables or to predict various outcomes from given input data. Often this is done without much consideration of specific mechanisms of the studied phenomena. Such data-driven methods can of course be applied to metabolic data, but with limited connection to biological mechanisms, the results are often difficult to interpret. Instead, machine learning methods can be combined with domain-specific biological knowledge, such as the information encoded within a genome-scale reconstruction, to create hybrid methods that also take advantage of the metabolic network structure.

Plaimas and coworkers predicted gene essentiality in *E. coli* using a hybrid method (Plaimas et al. 2008). Instead of using FBA to predict essentiality as described previously, they defined a set of features for each reaction, including metrics of network topology, gene expression data and predicted FBA fluxes. These features were fed into a support vector machine classifier together with labels from experimental essentiality data (Baba et al. 2006). The predictive accuracy of gene essentiality was 92%, compared to 85% for FBA. Furthermore, the genes where essentiality was not correctly predicted were retested experimentally, and in several cases the authors identified errors in the original experimental dataset. By removing single features from the input data one at a time, the authors could also identify which features were most important for accurately predicting essentiality. Prediction with FBA suffers mainly from two problems, namely that the metabolic network might be incomplete, and that the assumption of growth optimality does not always hold (O'Brien et al. 2015). A hybrid method can instead learn from data, utilizing the biological context, e.g., in the form of a metabolic network, only when it improves prediction performance. A similar method was recently used to predict drug side effects (Shaked et al. 2016). A list of drugs known to inactivate one or more enzymatic reactions was used as training data, with features corresponding to the minimum and maximum possible FBA flux for each reaction after deactivating the drug's target reaction(s) in the Recon1 model. Support vector machine classifiers were then trained to predict which (if any) side effects the drug would have. Using a feature selection method it was also possible to find the features that were most strongly associated with a given side effect. Many of the results were found to be consistent with the published literature of these drug side effects.

A third example of a combination of machine learning with metabolic network data was used to predict novel drug–reaction interactions for cancer therapy (Li et al. 2010). The method requires the construction of a reaction flux similarity matrix. This matrix was obtained using the GIMME algorithm to predict reaction fluxes from gene expression data in 59 cancer cell lines. Reactions with the same flux profile across the cell lines were said to have a high similarity, while reactions with different flux profiles had a low similarity. The reaction flux similarity matrix was combined with knowledge of existing drug–reaction interactions, using a K-nearest neighbors algorithm, to predict new interactions.

Where purely model-based algorithms may suffer from lack of biological knowledge such as kinetic parameters, the use of machine learning methods in biomedical research is often hampered by difficulties in interpreting the results. The examples above show that the two methodologies can be combined to achieve results that are informed by experimental data, while maintaining biologically relevant relationships between variables. Such hybrid methods can be used to build accurate predictive models, while also providing new biological insights and will without doubt find widespread use in the future.

6 Conclusions

Genome-scale models of metabolism have found applications ranging from industrial biotechnology to human health. These models can now be readily built for any organism to predict metabolic phenotypes such as the effect of a gene knock-out on cell growth. Advanced formulations of genome-scale models allow integrating diverse *omics* data types including transcriptomics, proteomics and metabolomics data into the models. Advanced genome-scale models make more accurate condition-dependent model predictions, and expand the range of predicted intracellular variables from metabolic fluxes to concentrations of metabolites and proteins. Genome-scale mechanistic models can also be combined with purely data-driven machine learning methods to obtain hybrid mechanistic/statistical models with the potential for improving predictive performance. With increasing amounts of different *omics* data types becoming available for all organisms, the modeling approaches described in this chapter can be further improved and extended to obtain highly predictive models of cellular processes.

References

Åkesson M, Förster J, Nielsen J (2004) Integration of gene expression data into genome-scale metabolic models. Metab Eng 6:285–293

Antoniewicz MR (2015) Methods and advances in metabolic flux analysis: a mini-review. J Ind Microbiol Biotechnol 42:317–325

Baba T, Ara T, Hasegawa M, Takai Y, Okumura Y et al (2006) Construction of *Escherichia coli* K-12 in-frame, single-gene knockout mutants: the Keio collection. Mol Syst Biol 2:2006.0008

Basan M, Hui S, Okano H, Zhang Z, Shen Y et al (2015) Overflow metabolism in *Escherichia coli* results from efficient proteome allocation. Nature 528:99–104

Becker SA, Palsson BO (2008) Context-specific metabolic networks are consistent with experiments. PLoS Comput Biol 4:e1000082

Bordbar A, McCloskey D, Zielinski DC, Sonnenschein N, Jamshidi N et al (2015) Personalized whole-cell kinetic models of metabolism for discovery in genomics and pharmacodynamics. Cell Syst 1:283–292

Bordbar A, Johansson PI, Paglia G, Harrison SJ, Wichuk K et al (2016) Identified metabolic signature for assessing red blood cell unit quality is associated with endothelial damage markers and clinical outcomes. Transfusion 56:852–862

Bordbar A, Yurkovich JT, Paglia G, Rolfsson O, Sigurjónsson ÓE et al (2017) Elucidating dynamic metabolic physiology through network integration of quantitative time-course metabolomics. Sci Rep 7:46249

Crabtree HG (1929) Observations on the carbohydrate metabolism of tumours. Biochem J 23:536–545 de Oliveira Dal'Molin CG, Quek L-E, Palfreyman RW, Brumbley SM, Nielsen LK (2010) AraGEM, a genome-scale reconstruction of the primary metabolic network in *Arabidopsis*. Plant Physiol 152:579–589

Duarte NC, Becker SA, Jamshidi N, Thiele I, Mo ML et al (2007) Global reconstruction of the human metabolic network based on genomic and bibliomic data. Proc Natl Acad Sci 104:1777–1782

Edwards JS, Palsson BO (2000) The *Escherichia coli* MG1655 in silico metabolic genotype: its definition, characteristics, and capabilities. Proc Natl Acad Sci 97:5528–5533

Folger O, Jerby L, Frezza C, Gottlieb E, Ruppin E et al (2011) Predicting selective drug targets in cancer through metabolic networks. Mol Syst Biol 7:527–527

Förster J, Famili I, Palsson BO, Nielsen J (2003) Large-scale evaluation of in silico gene deletions in *Saccharomyces cerevisiae*. OMICS J Integr Biol 7:193–202

Goelzer A, Muntel J, Chubukov V, Jules M, Prestel E et al (2015) Quantitative prediction of genome-wide resource allocation in bacteria. Metab Eng 32:232–243

Gopalakrishnan S, Maranas CD (2015) 13C metabolic flux analysis at a genome-scale. Metab Eng 32:12–22

Gry M, Rimini R, Strömberg S, Asplund A, Pontén F et al (2009) Correlations between RNA and protein expression profiles in 23 human cell lines. BMC Genomics 10:365

Halldorsson S, Rohatgi N, Magnusdottir M, Choudhary KS, Gudjonsson T et al (2017) Metabolic re-wiring of isogenic breast epithelial cell lines following epithelial to mesenchymal transition. Cancer Lett 396:117–129

Heavner BD, Price ND (2015) Comparative analysis of yeast metabolic network models highlights progress, opportunities for metabolic reconstruction. PLoS Comput Biol 11:1–26

Jamshidi N, Palsson BØ (2008) Formulating genome-scale kinetic models in the post-genome era. Mol Syst Biol 4:171

Karr JR, Sanghvi JC, MacKlin DN, Gutschow M, Jacobs JM et al (2012) A whole-cell computational model predicts phenotype from genotype. Cell 150:389–401

Khodayari A, Maranas CD (2016) A genome-scale *Escherichia coli* kinetic metabolic model k-ecoli457 satisfying flux data for multiple mutant strains. Nat Commun 7:13806

Lerman JA, Hyduke DR, Latif H, Portnoy VA, Lewis NE et al (2012) In silico method for modelling metabolism and gene product expression at genome scale. Nat Commun 3:929

Li L, Zhou X, Ching W-K, Wang P (2010) Predicting enzyme targets for cancer drugs by profiling human metabolic reactions in NCI-60 cell lines. BMC Bioinformatics 11:501

Lloyd CJ, Ebrahim A, Yang L, King ZA, Catoiu E et al (2017) COBRAme: a computational framework for building and manipulating models of metabolism and gene expression. bioRxiv 106559. https://doi.org/10.1101/106559

Machado D, Herrgård M (2014) Systematic evaluation of methods for integration of transcriptomic data into constraint-based models of metabolism. PLoS Comput Biol 10:e1003580

Mardinoglu A, Agren R, Kampf C, Asplund A, Nookaew I et al (2014) Integration of clinical data with a genome-scale metabolic model of the human adipocyte. Mol Syst Biol 9:649–649

McCloskey D, Palsson BØ, Feist AM (2013) Basic and applied uses of genome-scale metabolic network reconstructions of *Escherichia coli*. Mol Syst Biol 9:661

Noor E, Haraldsdóttir HS, Milo R, Fleming RMT (2013) Consistent estimation of Gibbs energy using component contributions. PLoS Comput Biol 9:e1003098

O'Brien EJ, Lerman JA, Chang RL, Hyduke DR, Palsson BO (2013) Genome-scale models of metabolism and gene expression extend and refine growth phenotype prediction. Mol Syst Biol 9:693–693

O'Brien EJ, Monk JM, Palsson BO (2015) Using genome-scale models to predict biological capabilities. Cell 161:971–987

Oberhardt MA, Puchałka J, Fryer KE, Martins Dos Santos VAP, Papin JA (2008) Genome-scale metabolic network analysis of the opportunistic pathogen *Pseudomonas aeruginosa* PAO1. J Bacteriol 190:2790–2803

Orth JD, Thiele I, Palsson BØ (2010) What is flux balance analysis? Nat Biotechnol 28:245–248

Plaimas K, Mallm J-P, Oswald M, Svara F, Sourjik V et al (2008) Machine learning based analyses on metabolic networks supports high-throughput knockout screens. BMC Syst Biol 2:67
Price ND, Reed JL, Palsson BØ (2004) Genome-scale models of microbial cells: evaluating the consequences of constraints. Nat Rev Microbiol 2:886–897
Saa PA, Nielsen LK (2017) Formulation, construction and analysis of kinetic models of metabolism: a review of modelling frameworks. Biotechnol Adv 35:981–1003
Sánchez BJ, Zhang C, Nilsson A, Lahtvee P, Kerkhoven EJ et al (2017) Improving the phenotype predictions of a yeast genome-scale metabolic model by incorporating enzymatic constraints. Mol Syst Biol 13:935
Sauer U (2006) Metabolic networks in motion: 13C-based flux analysis. Mol Syst Biol 2:1–10
Schilling CH, Palsson BØ (2000) Assessment of the metabolic capabilities of *Haemophilus influenzae* Rd through a genome-scale pathway analysis. J Theor Biol 203:249–283
Shaked I, Oberhardt MA, Atias N, Sharan R, Ruppin E (2016) Metabolic network prediction of drug side effects. Cell Syst 2:209–213
Shlomi T, Cabili MN, Herrgård MJ, Palsson BØ, Ruppin E (2008) Network-based prediction of human tissue-specific metabolism. Nat Biotechnol 26:1003–1010
Shlomi T, Benyamini T, Gottlieb E, Sharan R, Ruppin E (2011) Genome-scale metabolic modeling elucidates the role of proliferative adaptation in causing the warburg effect. PLoS Comput Biol 7:1–8
Shoaie S, Karlsson F, Mardinoglu A, Nookaew I, Bordel S et al (2013) Understanding the interactions between bacteria in the human gut through metabolic modeling. Sci Rep 3:2532
Soh KC, Hatzimanikatis V (2014) Constraining the flux space using thermodynamics and integration of metabolomics data. In: Krömer JO, Nielsen LK, Blank LM (eds) Metabolic flux analysis: methods and protocols. Springer, New York, pp 49–63
Soh KC, Miskovic L, Hatzimanikatis V (2012) From network models to network responses: integration of thermodynamic and kinetic properties of yeast genome-scale metabolic networks. FEMS Yeast Res 12:129–143
Srinivasan S, Cluett WR, Mahadevan R (2015) Constructing kinetic models of metabolism at genome-scales: a review. Biotechnol J 10:1345–1359
Stephanopoulos G (1999) Metabolic fluxes and metabolic engineering. Metab Eng 1:1–11
Teusink B, Passarge J, Reijenga CA, Esgalhado E, Van Der Weijden CC et al (2000) Can yeast glycolysis be understood terms of vitro kinetics of the constituent enzymes? Testing biochemistry. Eur J Biochem 267:5313–5329
Thiele I, Jamshidi N, Fleming RMT, Palsson BO (2009) Genome-scale reconstruction of *Escherichia coli's* transcriptional and translational machinery: a knowledge base, its mathematical formulation, and its functional characterization. PLoS Comput Biol 5:e1000312
Thiele I, Fleming RMT, Que R, Bordbar A, Diep D et al (2012) Multiscale modeling of metabolism and macromolecular synthesis in *E. coli* and its application to the evolution of codon usage. PLoS One 7:e45635
Thomas A, Rahmanian S, Bordbar A, Palsson BØ, Jamshidi N (2015) Network reconstruction of platelet metabolism identifies metabolic signature for aspirin resistance. Sci Rep 4:3925
Warburg O, Wind F, Negelein E (1927) The metabolism of tumors in the body. J Gen Physiol 8:519–530
Yang L, Ma D, Ebrahim A, Lloyd CJ, Saunders MA et al (2016) solveME: fast and reliable solution of nonlinear ME models. BMC Bioinformatics 17:391
Zampieri M, Enke T, Chubukov V, Ricci V, Piddock L et al (2017) Metabolic constraints on the evolution of antibiotic resistance. Mol Syst Biol 13:917

An Integrative MuSiCO Algorithm: From the Patient-Specific Transcriptional Profiles to Novel Checkpoints in Disease Pathobiology

Anastasia Meshcheryakova, Philip Zimmermann, Rupert Ecker, Felicitas Mungenast, Georg Heinze, and Diana Mechtcheriakova

Contents

A. Meshcheryakova · F. Mungenast · D. Mechtcheriakova (✉)
Department of Pathophysiology and Allergy Research, Center for Pathophysiology, Infectiology and Immunology, Medical University of Vienna, Vienna, Austria
e-mail: diana.mechtcheriakova@meduniwien.ac.at

P. Zimmermann
Nebion AG, Zürich, Switzerland

R. Ecker
TissueGnostics, Vienna, Austria

G. Heinze
Section for Clinical Biometrics, Center for Medical Statistics, Informatics, and Intelligent Systems, Medical University of Vienna, Vienna, Austria

N. Rajewsky et al. (eds.), *Systems Biology*, RNA Technologies,
https://doi.org/10.1007/978-3-319-92967-5_18

Abstract Strong efforts are invested in the field of cancer and other multifactorial diseases to evaluate the applicability of gene expression patterns for identification of novel disease-relevant checkpoints and nomination of promising biomarkers for disease and/or targets. Deciphering the disease complexity demands the implementation of a holistic approach, which covers the levels of the biological hierarchy from molecules to functional gene network(s) and biological pathways and further to disease (patho)mechanisms and clinical relevance. In this chapter we describe the systems biology-based integrative algorithm, named by us as *MuSiCO/from **Mu**ltigene **Si**gnature to Patient-Orientated **C**linical **O**utcome*, and discuss its applicability for translational research. This innovative approach is based on the implementation of consecutive analytical modules integrating advanced gene expression profiling of clinical patient specimens, prognostic/predictive modeling, digital pathology, and systems biology. It consolidates in-depth expertise from diverse scientific and medical disciplines and hereby bridges systems biology and systems medicine to maximize the benefit of the patient.

Keywords *MuSiCO* algorithm · Multigene signature · Gene expression profiling · Statistical modeling for survival prediction and therapy response · *AID/APOBEC* gene family · Sphingolipid system · Systems biology · Next generation digital pathology · Personalized medicine

List of Abbreviations

AICDA	Activation-induced cytidine deaminase
CRCLM	Colorectal cancer metastasis in the liver
EMT	Epithelial to mesenchymal transition
MuSiCO	from Multigene signature to patient-orientated clinical outcome
ROIs	Regions of interest

1 Introduction

The implementation of innovative technological solutions and analytical tools for dissecting the pathobiology of complex multifactorial diseases—among those are chronic inflammation, immune disorders, and cancer—and then reconstituting the system networks is a prerequisite for understanding the underlying checkpoints and for developing of novel targeting and clinical decision-making strategies. The complexity of underlying pathomechanisms demands the implementation of integrative, systems biology-based approaches, which cover the levels of the biological hierarchy from molecules to functional gene network(s) and biological pathways and further to the pathobiology of diseases and clinical relevance. Herein we describe the innovative analysis algorithm named by us as *MuSiCO/from* ***Mu****ltigene* ***Si****gnature to Patient-Orientated* ***C****linical* ***O****utcome*, which is based on consecutive analytical modules integrating advanced gene expression profiling of patient specimens, prognostic/predictive modeling, next-generation digital pathology, and systems biology. The cornerstone and the starting point of the integrative *MuSiCO* algorithm is the assembling and application of multigene signature(s) for gene expression profiling of patient material that is well-characterized in terms of clinicopathological parameters. The obtained patient-specific transcriptional profiles are taken as the basis for understanding the relevance of gene perturbations and gene-gene associations in complex, multifactorial disorders. To address the clinical relevance of profiling-derived data sets, we developed a novel strategy for patient stratification and risk assessment/survival prediction based on multivariable modeling. For evaluating the model performance in respect of predictive accuracy and discriminative ability, we propose to use a parameter set enabling the comparison of individual models within and across studies. As outcome, novel survival models, implementing both the patient-specific, profiling-derived variables as well as clinical risk factors, are developed including the nomination of top-ranked candidate/target genes. Within the next module, we explore the advantage of the innovative computerized microscopy-based technology, the TissueFAXS cytometry platform, to assess the magnitude and localization of nominated top candidate molecules in tissue sections in a patient-specific manner. The dissection of complex diseased tissue to individual functional parts allows accounting for tissue anatomy. As a final step, to analyze the data in the context of current knowledge and to get an overview of the underlying pathomechanisms, we apply a systems biology approach using the power of GENEVESTIGATOR, a web-based analysis platform for manually curated transcriptomic data sets, and the Ingenuity Pathway Analysis software, the web-based application for analysis and interpretation of complex "omics" data. We extend the signature-derived knowledge by comparative/similarity analysis of transcriptional profiles with public transcriptomic data sets and by extracting top genes co-expressed with modeling-derived candidate molecules across various physiological/pathophysiological and cell perturbation conditions. Overall, consolidation of the individual modules of *MuSiCO* algorithm provides a unique and advantageous, comprehensive overview, allowing the nomination of novel biomarkers and therapeutic directions as well as patient stratification strategies (Fig. 1). This concept bridges systems biology and systems medicine and is aiming to maximize the benefit of the patient.

Of particular importance is the broad applicability of *MuSiCO*—it can be applied for any biologically relevant gene signature and any disease or perturbation condition.

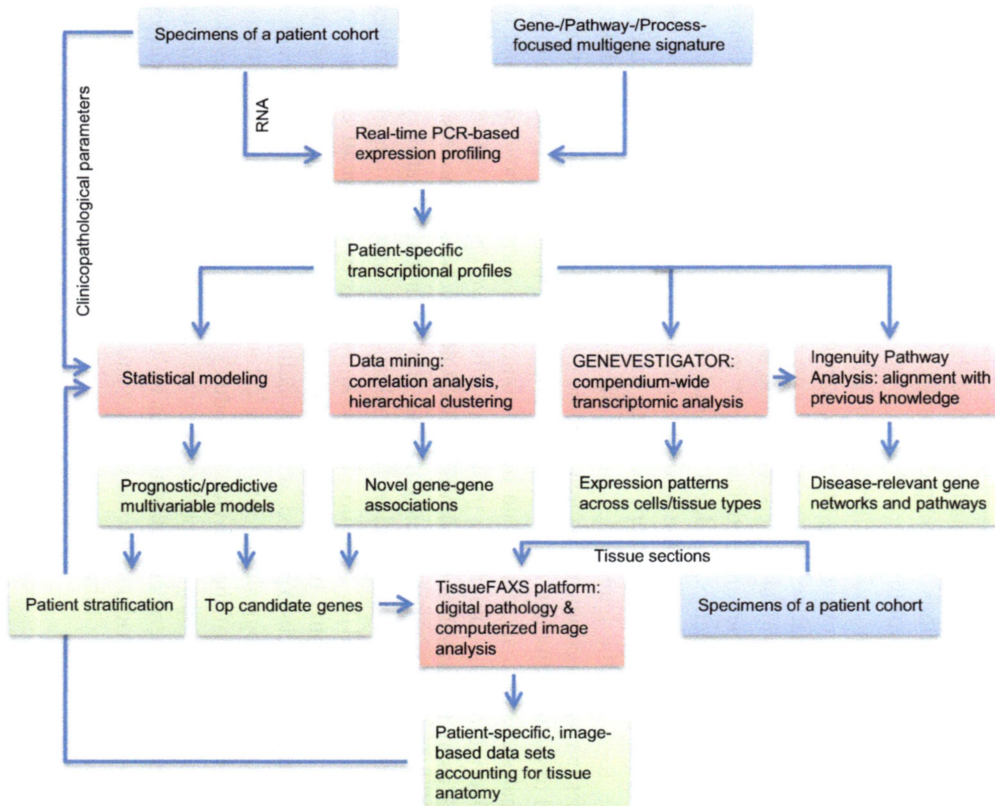

Fig. 1 Overview of *MuSiCO* algorithm. Color code: *blue*, input; *red*, individual *MuSiCO* modules/methodological solutions; *green*, output

2 A Multigene Signature Approach to Assess Patient-Specific Transcriptional Profiles

We and others propose that the multigene signature approach represents an advantageous strategy for addressing the relevance of transcriptional profiles/gene alterations under (patho)physiological circumstances (examples in Mechtcheriakova et al. 2011, 2012, Meshcheryakova et al. 2016; Svoboda et al. 2016; Gillet et al. 2012).

A multigene signature is a biologically relevant, meaningful composition of genes. The composition of the expert-designed multigene signature(s) is knowledge- and biology-driven; it may be assembled around one gene of interest designated as the "core" node gene, cover a particular interconnected gene network, and/or being process- and pathway-focused.

The multigene signature can be applied for expression profiling of clinical specimens to dissect patient-specific transcriptional profiles as well as for characterization of disease-relevant transcriptional programs using in vitro cell-based models. Being performed via real-time PCR-based analysis, it has incontestable advantage of high sensitivity and reproducibility and does not need further methodological validation. It is the most suitable solution for genes showing low expression on the mRNA level and thus being below the detection limit of microarray technology; this also applies to multicomponential/multicellular complex tissue, where a gene of interest might be specifically expressed in a particular cell subpopulation present at low abundance. Additionally, it gives advantage for genes within a family characterized by strong sequence similarity.

The multigene signature approach is a superior solution for understanding of disease complexity, which has indisputable advantage over the concept of "one gene – one outcome" that might lead to an oversimplification.

To give concrete examples, we next describe multigene signatures developed and successfully applied by us both for profiling of patient material and for assessing perturbations in cell-based models.

2.1 AID-Associated Multigene Signature

The signature was assembled around *AICDA* (encoding activation-induced cytidine deaminase, AID) as a core gene. AID has the distinguishing ability to introduce point mutations or other types of DNA damages. Under physiological circumstances, when expressed in B cells, AID functions as a "good", health-protecting factor, being *the molecule* responsible for the diversity of antibody repertoire directed against infectious agents by targeting immunoglobulin genes and driving somatic hypermutations and class switch recombination events (Muramatsu et al. 2000; Conticello 2012). However, under pathophysiological conditions such as chronic inflammation, an aberrant AID expression might be triggered in non-B cells including epithelial cells. In that case, AID may target cancer-related genes genome-wide and thus act as an extremely "bad" factor (Okazaki et al. 2003). The complexity is multiplied by the existence of ten other functionally related molecules, which together compose the *AID/APOBEC* gene family. The self-designed AID-associated multigene signature includes the entire *AID/APOBEC* family as well as genes involved in their regulation, functional cofactors, and target genes.

A multigene signature can be easily adapted for a particular scientific and/or medical question.

In the study where we assessed the role of AID and APOBECs in ovarian cancer pathobiology, we upgraded the signature with (1) *ESR1* and *ESR2* (encoding the estrogen receptors) based on two reasons, the hormone-dependent nature of ovarian cancer and the contribution of estrogen to the transcriptional regulation of AID, and (2) B-cell identity markers enabling the differentiation between the B-cell- and non-B-cell-attributed events in the follow-up analyses (Svoboda et al. 2016).

When we analyzed the impact of AID in the etiology of nasal polyposis (Mechtcheriakova et al. 2011)—a disease with inflammatory/allergic background—the core signature was complemented with genes encoding classical immune cell markers and thus allowing to detect various immune populations within the complex tissue, IgG and IgE mature transcripts to investigate the local AID activity, and genes encoding the low- and high-affinity IgE receptors, which mediate numerous responses of the AID/IgE axis. Thus, in this particular case, we were able to assess not only the expression of the molecule of interest but also prove its functional activity.

2.2 *Sphingolipid-/EMT-Associated Multigene Signature*

The designed sphingolipid-/EMT-associated 41-gene/23-gene signature covers the cellular sphingolipid system and the key players of the epithelial to mesenchymal transition (EMT) program. This is an example of a gene network- and process-derived multigene signature. It was applied to denote gene alterations in an in vitro cell-based model of a cancer-related, pathological EMT (Meshcheryakova et al. 2016). The signature includes genes encoding (1) the key players of the sphingolipid machinery, the enzymes synthesizing/modifying/degrading various sphingolipid mediators as well as the sphingolipid-specific receptors and transporters, and (2) the key players of the EMT program including cell adhesion molecules, cytoskeleton components, different families of transcriptional regulators, and pluripotency markers. This two-component signature allowed us to explore the multidimensional contribution of the sphingolipid machinery to pathological EMT in lung cancer and to nominate novel sphingolipid-related, disease-relevant checkpoints (Meshcheryakova et al. 2016).

3 Patient-Specific Transcriptional Data Sets as Basis for Prognostic/Predictive Models

3.1 Prerequisites

We assume the situation where a number of features (signature-derived gene expression values) have been obtained by the profiling steps explained above. In this subchapter we explain how to relate this gene set to survival outcomes in order to assess its prognostic and/or predictive value.

> The methods can be applied to any field of medicine where survival outcomes and/or treatment responses are of interest.

We assume that a sample of patients, randomly selected from a well-defined population, is available, from which gene expression values of the genes of interest have been measured, and clinicopathological parameters, outcome values (survival times), and treatment response parameters have been assessed. It should be emphasized that gene expression values and clinicopathological parameters have to be taken at baseline, i.e., at the time point at which predictions should be made for a patient (e.g., at cancer diagnosis), and not later during follow-up, which could incur immortal time bias (Gleiss et al. 2017). A particular proportion of survival times in a cohort are usually censored, as a study can terminate before all patients have died or some patients may have been lost to follow-up. We assume that among *n* patients, *m* events (deaths) were observed during follow-up. Cox regression is the method of choice to estimate statistical models that relate explanatory variables to survival outcomes. These associations are expressed as hazard ratios for the explanatory variables, expressing the ratio of mortality hazards between two hypothetical patients differing in an explanatory variable by one unit with all other explanatory variables being equal.

3.2 *Challenges and Solutions*

3.2.1 Challenge 1: Many Variables, Few Events

It has been recognized that when the number of variables approaches the number of events, asymptotic properties of estimators do no longer hold and predictions may be highly unreliable (Peduzzi et al. 1995). As a rule of thumb, the events-per-variable ratio can be computed, and alternatives to the standard maximum likelihood method should be considered if there are less than ten events per explanatory variable (Vittinghoff and McCulloch 2007; Courvoisier et al. 2011; Heinze et al. 2018).

3.2.2 Solution 1: Penalized Likelihood Methods

Such alternatives to estimate statistical models, termed "penalized likelihood methods," subtract a penalty term that is a function of the magnitude of regression coefficients from the log likelihood, i.e., they penalize large regression coefficients. Finally, the penalized log likelihood is maximized with respect to the regression coefficients. Typical choices for such penalties are the ridge penalty, proportional to the sum of squared regression coefficients, or the Lasso penalty proportional to the sum of absolute values of the regression coefficients (Verweij and Van Houwelingen 1994; Tibshirani 1997). The ridge penalty usually leads to regression coefficients being shrunken toward (but not exactly to) zero (i.e., hazard ratios toward 1) which introduces a downward bias but stabilizes the variance. The Lasso penalty is particularly interesting as it shrinks some of the coefficients exactly to 0, thus providing variable selection.

The relative impact of the penalty, the so-called tuning parameter, can be chosen freely but is usually optimized by means of cross-validating the log likelihood (Van Houwelingen and Le Cessie 1990). In cross-validation, data is randomly divided into, say, ten blocks. Models are fit with a prespecified grid of tuning parameters on 9/10 of the data, and predictions are made for the tenth block which is not used for model estimation, and a cross-validated log likelihood is computed for that block. This is done ten times in turn such that each subject finally has been used nine times in the "training set" and once in the "validation set." By averaging the cross-validated log likelihood at each value of the tuning parameter, one can easily determine the optimal tuning parameter value. Finally, the model is reestimated on the full data set fixing the tuning parameter at its optimal value.

3.2.3 Challenge 2: Validation of a Model if No Test Set Is Available

If an independent test cohort is available, the final model could be evaluated in that test cohort in order to assess model performance. For evaluating model performance, measures of discrimination (c-index or area under the ROC), predictive accuracy (the deviation of predictions from observed survival), and explained variation (which proportion of the observed variation of survival times can be explained by the model) can be computed. However, the usual case is that a test set is not available and the data set at hand is too small to save a test set. Moreover, if the data is not very large, a particular training-test split can be quite arbitrary, and different splits may lead to different results. Therefore, we suggest to use bootstrap resampling for validation.

3.2.4 Solution 2: Bootstrap Validation

Bootstrap validation relies on the simple idea that in the same way that our study cohort was sampled from a population, we could (re)sample data from our study cohort. If a model is then trained on such resampled data and validated in the original study cohort, we obtain values for model performance (c-index, predictive accuracy, explained variation, see Table 1) that resemble those that would be obtained if the model trained in our study cohort was validated in the full population. With the use of computer software, we can easily draw a large number of resamples to get rid of random noise in resampling, taking the average of the performance measures. A refined method of bootstrap validation is the so-called .632+ bootstrap for binary outcome data (Efron and Tibshirani 1997). For Cox regression it has been noticed that resampling without replacement (i.e., subsampling) could be more appropriate (De Bin et al. 2016). The resampling validation procedure is schematically depicted in Fig. 2. Other strategies for internal validation to correct optimism in model performance indices are the enhanced bootstrap, tenfold cross-validation, and leave-pair-out or leave-one-out cross-validation (Harrell 2001; Smith et al. 2014).

Table 1 Concepts for evaluating a model's performance

Concept	Description	References
Concordance (c-)index	Expresses the probability that a patient who died earlier has a higher gene expression score than a patient who died later. Also known as area under the ROC. Can be computed for a particular prediction horizon (e.g., for 5 years) or cumulative over time	C-index specific to a particular prediction horizon: Heagerty et al. (2000) Cumulative C-index: Uno et al. (2011)
Predictive accuracy	Describes the discrepancy of observed and predicted survival, averaged over all patients. Can be computed for a particular prediction horizon or cumulative over time	Schemper (2003)
Explained variation	Expresses the relative improvement in predictive accuracy by the gene signature score	Gleiss et al. (2016)

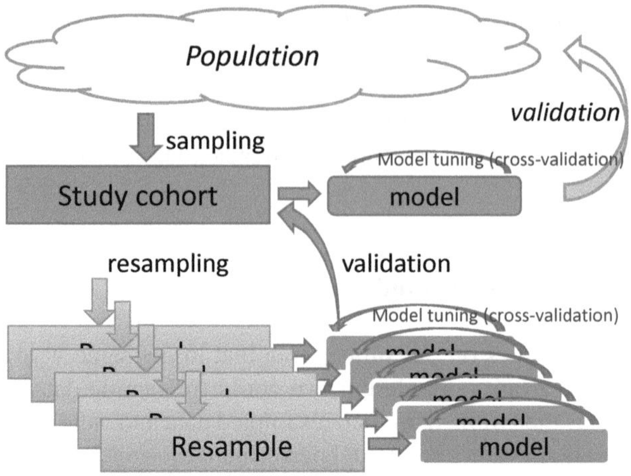

Fig. 2 Schematic model development and validation strategy

3.2.5 Challenge 3: Evaluating the Added Value of a Gene Signature on Top of Established Clinicopathological Predictors

Of particular interest is the ability of gene expression values to add predictive value to routinely assessed clinicopathological parameters. For example, in ovarian cancer variables such as age, grading, histology, and FIGO stage, the presence of residual tumor after operation can already explain a significant part of the outcome variation. Therefore, the costs of evaluating gene expressions only pay off if this improves model performance. Often the added value of gene expression on top of such established variables is low; therefore evaluating added value is of crucial importance (Dunkler et al. 2007).

3.2.6 Solution 3: Model Performance Improvement by Considering the Gene Signature on Top of Established Clinicopathological Predictors

To evaluate the added value of gene signature, one could compare the performance of a model including only clinicopathological parameters (model 1) with that of a model also including genes (model 2). The relevance of the gene signature can then easily be defined as the difference between model performance of model 2 and model 1. These measures should be evaluated in an independent or bootstrap validation and can be expressed in terms of c-index increment, predictive accuracy improvement, and increase in explained variation.

3.2.7 Challenge 4: Assessing Significance of the Gene Signature

Testing is not the ultimate goal in assessing the added value of a gene signature, as *p*-values will always decrease with larger sample sizes (unless the null hypothesis strictly applies). Nevertheless, some evidence is needed to guard against findings entirely caused by random variation. Unfortunately, classical Wald or likelihood ratio tests are not available if penalized likelihood is used for model estimation.

3.2.8 Solution 4: Bootstrap Testing

A bootstrap procedure can also be used to test the added value of the gene signature as follows (De Bin et al. 2014). In a bootstrap resample, we estimate regression coefficients for the clinicopathological parameters and the gene expressions using the penalized likelihood approach, with or without variable selection, as explained above. We now compute a new "gene signature score" variable for each subject in the original data set, which is just the sum of a subject's gene expression values multiplied by their regression coefficients. In the full study cohort, a standard Cox regression model can then be estimated with all clinicopathological parameters and the gene signature score as a single summary variable. The regression coefficient of the gene signature score (adjusted for the clinicopathological parameters) is estimated in the study cohort, and from the distribution of regression coefficients over all bootstrap models, a *p*-value can be derived (see Table 1). Note that the mean of those regression coefficients can be interpreted as calibration slope, which ideally should assume a value of 1. The same set of resamples and the same models estimated in the resamples as in "solution 2" can be reused, with only slightly different subsequent computational steps when applying those models in the original study cohort.

3.2.9 Challenge 5: How to Report a Model?

> Having many explanatory variables to deal with, concise reporting of an estimated model becomes a challenging task.

We propose to present quantities, which describe the overall performance of a model, and quantities, which describe the model itself, making it applicable for future outcome prognostication.

3.2.10 Solution 5: Report Pre-transformations, Effect Sizes, Importance, and Stability

In absence of a test cohort, the prognostic performance of the estimated model itself cannot be assessed. However, bootstrap validation can be used to describe the performance of the model building strategy (i.e., the algorithm that was applied to the data in order to arrive at a final model). One can then assume that applying this strategy to resamples and evaluating the performance of the resulting models in the study cohort are equivalent to applying the strategy to the data of the study cohort and the impossible task of evaluating its performance in the full target population. As model performance descriptors, which allow comparisons of model within and across studies, we propose the discrimination index (c-index), the predictive accuracy, and the explained variation.

To describe the model itself, we propose to report any transformation steps that were performed prior to analysis. For example, a logarithmic transformation to base 2 is often useful to reduce the disproportional impact of high gene expression values and still leads to a straightforward interpretation of regression coefficients as the increase in log hazard per doubling of an original gene expression value. For each variable in the gene signature score, regression coefficient and hazard ratio (both expressed per doubling of gene expression) and standardized regression coefficients (per standard deviation) should be reported. The latter allow to easily rank variables by their importance. It should be kept in mind that penalized likelihood methods are particularly designed for prediction, not for unbiased hazard ratio estimation, and therefore do not supply reliable confidence intervals. In case that the Lasso was used, bootstrap selection frequencies for all variables, including those that were finally not selected for the model, inform about model stability. Further bootstrap-based measures were proposed by Heinze et al. (2018).

The study cohort is a random sample of the population. A model building strategy is chosen, and a model including a gene signature score is estimated using the data from the study cohort. To optimize the model, tenfold cross-validation is applied. The model should be validated in the population, but this is not possible as the population is usually inaccessible. Resamples are taken from the study cohort, and the model building strategy is repeatedly applied to each resample, including tuning by cross-validation. Each of the resampled models is then validated in the original study cohort to evaluate their performance. The averaged model performance is used as estimate of the model performance in the population.

4 Expanding Gene Expression Data to Context-Based Image Cytometry and to a System Approach

To link gene expression data to the visualization of the molecules of interest within diseased tissue, we established a computerized microscopy-based algorithm using the TissueFAXS cytometry platform. We assume that the obtained information is complementary to gene profiling-derived knowledge and gives an additive value to the understanding of pathobiological mechanisms in a patient-specific and patient-orientated manner. Analysis is performed on the basis of whole-slide paraffin-embedded tissue sections from a clinically well-characterized patient cohort. Similarly to the patient-specific transcriptional profiles, the obtained staining- and tissue anatomy-derived data sets can be used as variables for building up the prognostic/predictive models. To maximize the outcome of the digital imaging approach, we consolidate diverse expertizes such as molecular and cellular biology, clinical pathology, programming, and modeling. In this subchapter we describe methodological advantages and application examples.

4.1 Next-Generation Digital Pathology as Novel Platform in Translational Research: Superior Solutions for Unbiased, Standardized, and Quantitative Analysis

With the emergence of automated and increasingly reliable slide scanning technologies, the discipline of digital pathology has slowly spread out into the research market, even less so into clinical routine. The emergence of digital pathology started in the early 1990s, and technologies are still being further developed and improved. Up to today, the main focus of *classical digital pathology* has been to digitize histological sections mounted on glass slides and to generate digital or *virtual slides*, with the vast majority of analyses still being performed visually. In other words, visual analysis can be performed on the monitor and not only through the microscope's oculars.

Current efforts, both in academia and industry, are directed toward *next-generation digital pathology*, which has its focus on digital analysis and extraction of numerical data. Such data shall describe molecular/functional parameters with the aim to provide an alternative to observer-biased image interpretation in form of observer-independent quantification. Instead of verbal descriptions by experienced observers, computerized analyses are performed that as outcome deliver various parameters (e.g., percentage of positive/negative cells, cellular density, staining intensity, size of histological structures) to be used as variables for statistical analysis and complex modeling. The need to complement human experience and the human brain's interpolation power (which is the strength of a human observer) with quality-controlled and standardized tools to quantify multiple optical patterns in parallel and determine statistical interdependencies among and between them (which is the strength of a computer) has already been recognized with the emergence of digital pathology (Baak 1991).

The current ongoing efforts are directed to push digital pathology *from image to data sets* by developing novel technologies for the next-generation digital pathology platforms, with solutions ranging from high-throughput slide scanning, over automated color reproduction on display monitors, to diagnostic apps using contextual tissue cytometry as well as current machine and deep learning approaches.

The TissueFAXS platform-based module within the *MuSiCO* algorithm enables (1) automated scanning of slides in fluorescence and bright-field mode; (2) automated detection of single cells within their native tissue environment by using nuclear markers as master channel; (3) quantification of the stained marker of interest per single cell (cytoplasmic and/or nuclear); (4) automated identification of tissue anatomy-based structures, for instance, epithelial structures including glands or tumor foci, or immune cell-based structures, for instance, immune aggregates or ectopic lymphoid structures; and (5) measurement of 20+ parameters per cell and color channel or marker.

The TissueFAXS platform allows to investigate the relationships between all multicellular meta-structures, single cells, and markers.

The readout is provided in form of numerical data—primary measurements and derived statistical values.

The next two subchapters are intended to illustrate the power of the TissueFAXS-based module to conduct patient-orientated translational research within the emerging field of tumor immunology.

4.2 Tissue Image Cytometry to Determine the Patient-Specific Immunological Imprint in the Context of Tumor Anatomy

Breaking discoveries in the field of tumor immunology highlight the strong impact of the immunological imprint of tumor and tumor microenvironment on development, progression, and clinical outcome in several cancer types. To describe best the cancer type-attributed and patient-orientated immunological imprint, several parameters have to be considered including (1) the immune landscape, characterizing the distribution and organization pattern of immune cells within tumor tissue; (2) the immune contexture, describing the immune cell composition; and (3) the tumor anatomy, representing the morphological complexity of malignant tissue (Becht et al. 2016; Fridman et al. 2012; Galon et al. 2006). In our study (Meshcheryakova et al. 2014), we utilized the power of the computerized image analysis platform to assess the patient-specific B-cell-attributed immunological imprint of patients with colorectal cancer metastasis in the liver (CRCLM). Digitalization of the whole-slide CRCLM sections of the entire cohort of patients allowed us to perform patient-to-patient comparison and by this to define the right strategy for setting of tumor anatomy- and immune landscape-attributed regions of interest (ROIs), thus unifying and standardizing the quantitative analysis algorithm (Fig. 3). We detected the accumulation of B cells, in the form of immune aggregates/ectopic follicular structures, at the tumor-liver interface and quantified their magnitude for each specimen (ranging from 0.3% to 13% with a median value of 2%). The magnitude of B cells, used as variable, showed strong prognostic effect with highly significant patient stratification into low-risk (above median) and high-risk (below median) groups. Detection of such fine inter-patient differences in B-cell numbers would not be possible by an observer-based image interpretation and demands the power of next-generation digital pathology.

4.3 Biomarker and/or Target Proposal and Validation: Immune-Based and Beyond

Comprehensive analyses of gene profiling-derived data such as hierarchical clustering and/or multivariable modeling contribute to the nomination of novel disease-relevant checkpoints and thus propose potential biomarkers and/or targets. To consider, transcriptional profiling of patient material is often done on the basis of whole tissue homogenates of multicomponential diseased tissue.

> How can one define the cell type which contributes the most to the measured gene expression value?

This question is particularly relevant in the field of tumor biology, where the tumor microenvironment, including various infiltrating immune cell populations, plays a decisive role in disease progression and influences treatment regiments. Correlation analysis- and clustering-based identification of close gene-gene associations between the gene of interest and cell identity marker(s) gives a strong indication for expression in a particular cell type within heterogeneous biological sample. Alternatively, cell type deconvolution offers an attractive approach and aims at extracting the cell subpopulation information directly from heterogeneous samples (Shen-Orr and Gaujoux 2013). A prerequisite is to have knowledge about the expression characteristics of individual cell types or their proportions in a given mixed sample. For purpose of verification, the tissue-level expression of a candidate gene can be easily extracted using the GENEVESTIGATOR tool (see Sect. 5).

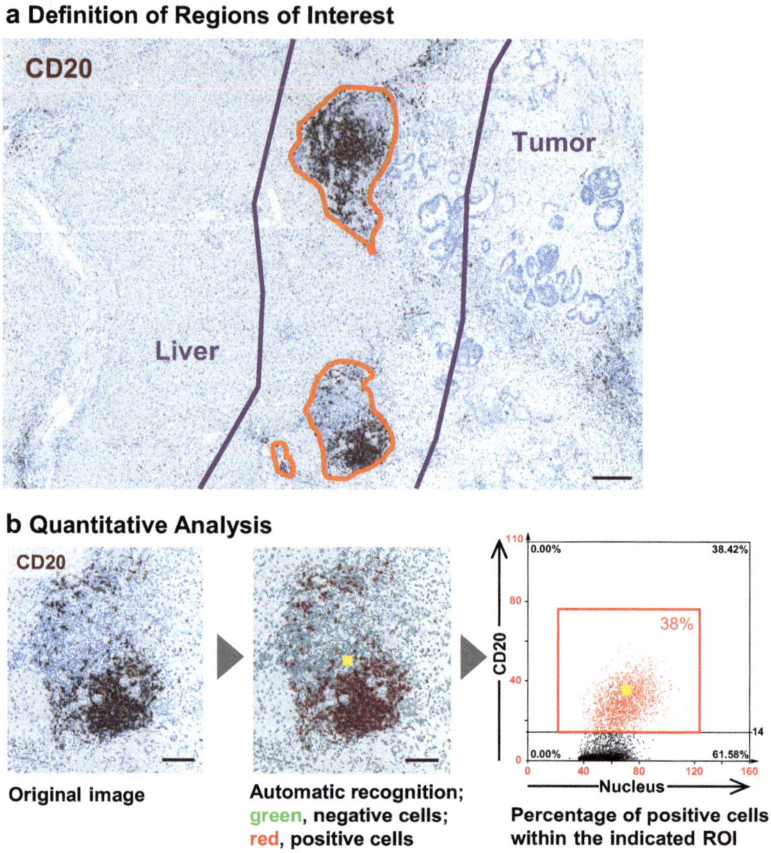

Fig. 3 Software-based qualitative and quantitative analyses of the whole-slide image in respect of resident and/or infiltrating immune cells at the CRCLM site. (**a**) First step of analysis is the definition of regions of interest (ROIs) accounting for tumor anatomy and immune landscape, which are defined and drawn manually or automatically recognized. Shown is the staining of CD20, the classical B-cell marker. Defined ROIs are the tumor-liver border (indicated by a purple line) and immune aggregates/ectopic follicular structures (indicated by an orange line); brown color, CD20 staining; blue color, nuclear counterstaining with hematoxylin. Scale bar: 200 μm. (**b**) For the software-based quantitative analysis, marker-specific profiles are designed; quantification algorithm is based on single-cell recognition strategy and allows simultaneous recognition of various cell types within their complex native tissue environment. The setting of the sample/patient-specific cutoff ensures the proper identification of positively stained cells. Shown is the staining of CD20 within CRCLM tissue (*original image*) and the software-based recognition of each cell as an individual object, based on the nuclear staining, and the recognition of the specific staining (*automatic recognition*); setting of the cutoff/gating allows the discrimination between positive cells (indicated in red) and negative cells (indicated in green). Result of quantitative analysis is displayed by scattergram, which shows the percentage of CD20-positive cells detected within the ROI. The *forward/backward connection algorithm* allows to link each individual cell within the acquired image with the corresponding dot in the scattergram (yellow dot). Scale bar: 100 μm

The tissue image cytometry provides a superior solution to explore the expression in the context of tissue (e.g., tumor) anatomy with the possibility for quantitative assessment.

Thus, the *digital pathology and computerized image analysis* module within the *MuSiCO* algorithm allows to dissect the cell type- and tissue anatomy-attributed localization and expression pattern of any nominated candidate molecule in any type of diseased tissue under investigation.

5 Dissecting Novel Breakpoints in Disease Pathobiology Using Compendium-Wide Analysis

Organisms grow and survive in their environment by activating specific transcriptional programs. The tissues of any complex eukaryotic organism are generated during its life cycle via temporally and spatially regulated processes. Concurrently, transcriptional programs are launched in response to perturbations and diseases to ensure stability, defense, fitness, and ultimately survival. The study of the regulation of these programs is important to further our understanding about the genetic regulatory mechanisms that direct growth and survival but also response to diseases, eventually leading to novel applications in medicine.

5.1 GENEVESTIGATOR: Integrating and Analyzing Public and Proprietary Expression Data

Research on a given disease is most often carried out on data generated specifically to study the disease. However, complementing an analysis with data from other diseases is of increasing interest because it helps interpreting one's own results and enhances our understanding of a particular disease under investigation. To achieve this, data from a broad range of diseases must be readily available in a format, scaling, and transformation that makes them comparable. Although significant efforts have been carried out to make genomic and gene expression experiments publicly available, the heterogeneity of formats, the different levels of quality, and the lack of use of standards for describing them make it very difficult to explore and exploit the data. The purpose of GENEVESTIGATOR is to curate and integrate gene expression data from a wide set of diseases, genotypes, and experimental conditions and to offer analysis tools for single-study, cross-study, and cross-disease analysis. To achieve this, public gene expression studies are thoroughly curated, quality controlled, and annotated by experts using controlled vocabularies and standard operating procedures. Additionally, the measurement data is processed in a way that results can be compared (to a certain degree) between various studies. There are hundreds of different technology platforms for transcriptomic measurement (mainly based on microarray or sequencing technology), and some of them were used for only a few studies. Setting up an entire curation pipeline for a platform is tedious; therefore, a choice must be made on which platforms to perform curation. Figure 4a shows the diversity of deeply curated content in GENEVESTIGATOR (as of October 2017) for mammalian species. The presence of data for over 500 diseases from 19 therapeutic areas allows powerful data interpretation and generation of new hypotheses.

Once the data is curated and integrated, they can be aggregated and displayed in different ways depending on the particular research question (see Fig. 4b). For example, one can visualize expression by sample, or alternatively, by tissue type, in which case all samples from a given tissue are grouped together to deliver a representative expression level (meta-profile). The same aggregation approach can be applied for multiple dimensions of biology, such as organs/tissues/cell types, cell lines, cancers, diseases, genotypes, or other parameters, coming from either in vivo or in vitro studies.

Understanding the spatial ("Anatomy" tool), developmental ("Development" tool), and response ("Perturbations" tool) characteristics of a target gene is crucial for exploring its role and involvement in the disease under study. Moreover, identifying other genes co-regulated with the gene of interest helps identifying pathways affected in the disease.

> Using simple but powerful visualizations, GENEVESTIGATOR allows to easily understand the regulatory context of genes in various biological dimensions, and via enrichment analysis, to correlate them to canonical pathways or biological processes.

An extension of this approach is to compare a multigene disease-relevant signature with an entire compendium of studies to identify diseases, drugs, or other conditions causing similar or opposite results. Such approaches can be used to validate a hypothesis (i.e., a result is confirmed by other studies) or to find connections between the disease of interest and other diseases, drugs, or disease models based on, e.g., mouse knockouts, patient-derived xenografts, or cell lines.

In the next two subchapters, we will illustrate the power and applicability of the GENEVESTIGATOR tool in translational research in two of our studies.

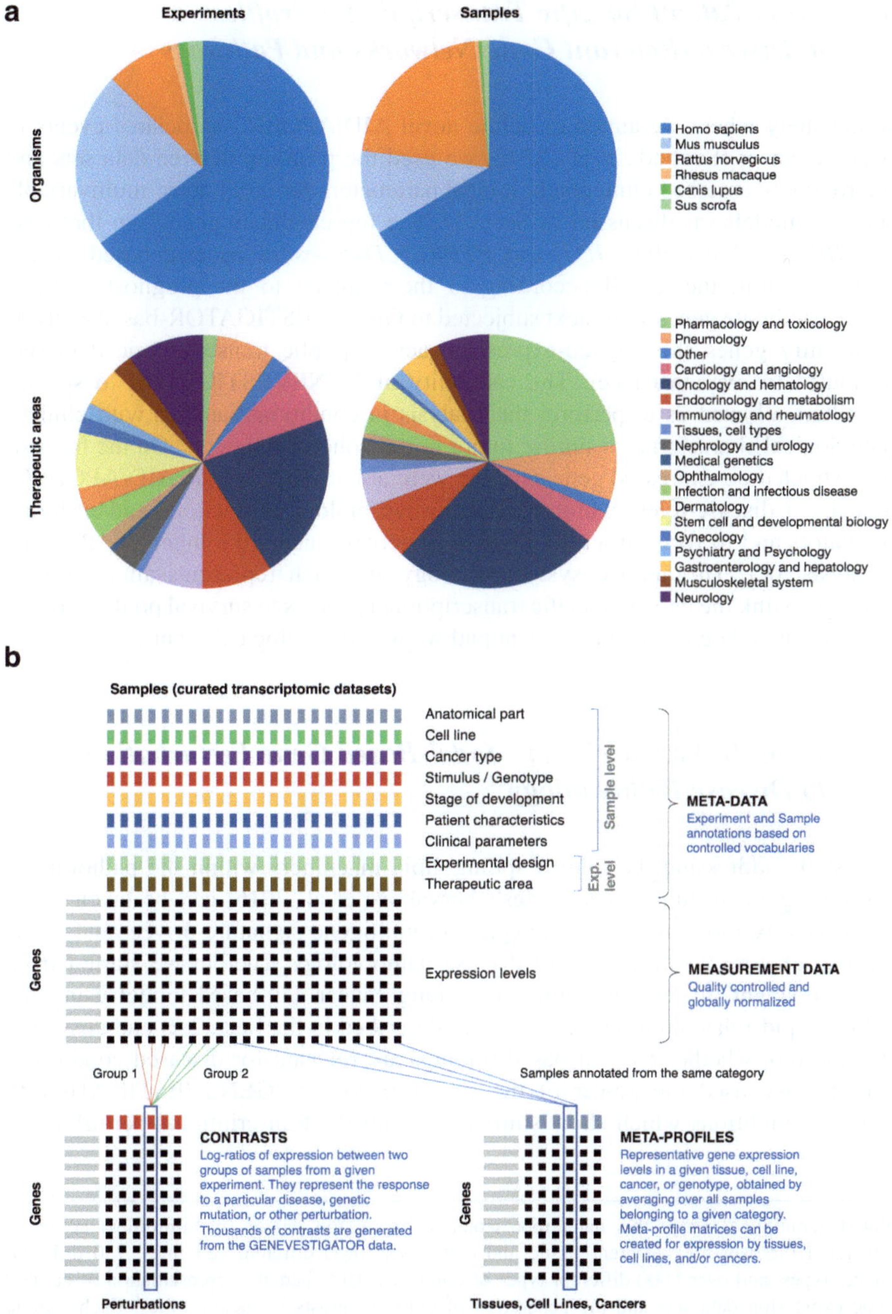

Fig. 4 (**a**) Composition of the GENEVESTIGATOR database (October 2017), without counting the LINCS data sets. The total number of deeply curated studies (i.e., quality controlled, normalized, and described in detail) was 2530 comprising a total of 150,900 expression data sets.

5.2 *From Patient-Specific Transcriptional Profiles to Disease-Relevant Gene Networks and Pathways*

In the study where we aimed to define novel AID/APOBEC-associated events in ovarian cancer (Svoboda et al. 2016), we used the profiling-derived data sets and the routinely assessed clinicopathological parameters for building of multivariable survival models (as discussed in Sect. 3). The top candidate genes—in that case *APOBEC3G*, *ESR1*, *ID2*, *ID3*, and *PTPRC/CD45*—were selected based on the ranking within the model according to their impact to the prognostic effect. These candidate genes were next subjected to GENEVESTIGATOR-based analysis to identify genes showing co-expression across public transcriptomic data sets attributed to ovarian cancer. The capability of GENEVESTIGATOR to specify the filters allowed us to perform the analysis in conditions/samples with clinicopathological characteristics similar to our initial cohort of patients. On the basis of the extracted co-expressed gene sets, we were able to identify the AID/APOBEC-associated, disease-relevant biological pathways, biological functions, and upstream regulators and to reconstruct the regulatory network using the Ingenuity Pathway Analysis tool. The applied systems biology approach represents an innovative strategy to link the patient-specific transcriptional profiles to survival prediction/risk assessment and delineate the relevant pathways/pathobiological events.

5.3 *From In Vitro Cellular Model-Based Gene Perturbations to Disease Pathobiology*

The study addressing the role of sphingolipid machinery within the pathological EMT program in lung cancer (Meshcheryakova et al. 2016) can be taken as an additional example of how the integrative, multigene signature-driven analysis is used to explore novel aspects of disease pathobiology. Therein we started from the assessment of gene alterations in a lung cancer cell-based model using the sphingolipid-related multigene signature (described in the Sect. 2.2) and answered the question whether the cell-based findings are relevant for diseased conditions. For that we used the power of the *signature tool* of GENEVESTIGATOR to identify conditions which show similarities with the transcriptional perturbations

Fig. 4 (continued) From this data, representative expression values (meta-profiles) could be generated for over 560 different tissue and cell types, 1800 different cell lines, 730 different cancer types, and over 5000 different types of contrasts. (**b**) Schematic representation of curated gene expression data sets and the generation of either contrasts or meta-profiles. Each sample is described in detail on each of the levels (anatomy, cell line, cancer or disease type, genotype, stimulus, stage of development, patient characteristics, and clinical parameters). In addition, data is aggregated to form either *contrasts* or *meta-profile*, which are used to characterize existing targets or to identify novel, highly specific targets and biomarkers by global search

defined in the in vitro model. The analysis was performed across various neoplasms (over 600 different neoplasm categories; from more than 24,000 arrays) and demonstrated for the first time that the sphingolipid-associated events do occur in lung adenocarcinoma tissue of patients with non-small cell lung cancer. Thus, the applied strategy allowed us to extend our cell model-based findings to novel disease-relevant sphingolipid-associated checkpoints. Given the druggability of the sphingolipid machinery, this may yield new biomarkers and therapeutic targets in lung cancer.

6 Conclusions

Each individual module of the herein described *MuSiCO* algorithm deepens our understanding of the disease relevance of transcriptional profiles; their consolidation in turn provides a unique and advantageous, comprehensive overview allowing the nomination of novel biomarkers and therapeutic directions as well as patient stratification strategies. We would like to emphasize that the presented algorithm is universal in the sense that it can be applied for any biologically relevant signature and any type of complex multifactorial disorder. Given the multidisciplinary nature of the multimodular analysis strategy, we hope that the chapter is of interest for specialists of diverse disciplines – scientists at research, oncologists and pathologists, biostatisticians, and systems biologists.

References

Baak JPA (1991) Manual of quantitative pathology in cancer diagnosis and prognosis. Springer, Berlin

Becht E, Giraldo NA, Dieu-Nosjean MC, Sautes-Fridman C, Fridman WH (2016) Cancer immune contexture and immunotherapy. Curr Opin Immunol 39:7–13

Conticello SG (2012) Creative deaminases, self-inflicted damage, and genome evolution. Ann N Y Acad Sci 1267:79–85

Courvoisier DS, Combescure C, Agoritsas T, Gayet-Ageron A, Perneger TV (2011) Performance of logistic regression modeling: beyond the number of events per variable, the role of data structure. J Clin Epidemiol 64:993–1000

De Bin R, Herold T, Boulesteix AL (2014) Added predictive value of omics data: specific issues related to validation illustrated by two case studies. BMC Med Res Methodol 14:117

De Bin R, Janitza S, Sauerbrei W, Boulesteix AL (2016) Subsampling versus bootstrapping in resampling-based model selection for multivariable regression. Biometrics 72:272–280

Dunkler D, Michiels S, Schemper M (2007) Gene expression profiling: does it add predictive accuracy to clinical characteristics in cancer prognosis? Eur J Cancer 43(4):745–751

Efron B, Tibshirani R (1997) Improvements on cross-validation: the .632+ bootstrap method. J Am Stat Assoc 92:548–560

Fridman WH, Pages F, Sautes-Fridman C, Galon J (2012) The immune contexture in human tumours: impact on clinical outcome. Nat Rev Cancer 12:298–306

Galon J, Costes A, Sanchez-Cabo F, Kirilovsky A, Mlecnik B et al (2006) Type, density, and location of immune cells within human colorectal tumors predict clinical outcome. Science 313(5795):1960–1964

Gillet JP, Calcagno AM, Varma S, Davidson B, Bunkholt Elstrand M et al (2012) Multidrug resistance-linked gene signature predicts overall survival of patients with primary ovarian serous carcinoma. Clin Cancer Res 18(11):3197–3206

Gleiss A, Zeillinger R, Braicu EI, Trillsch F, Vergote I et al (2016) Statistical controversies in clinical research: the importance of importance. Ann Oncol 27(7):1185–1189

Gleiss A, Oberbauer R, Heinze G (2017) An unjustified benefit: immortal time bias in the analysis of time-dependent events. Transpl Int. https://doi.org/10.1111/tri.13081

Harrell FE (2001) Regression modeling strategies: with applications to linear models, logistic regression, and survival analysis. Springer, New York

Heagerty PJ, Lumley T, Pepe MS (2000) Time-dependent ROC curves for censored survival data and a diagnostic marker. Biometrics 56:337–344

Heinze G, Wallisch C, Dunkler D (2018) Variable selection – a review and recommendations for the practicing statistician. Biom J 60(3):431–449

Mechtcheriakova D, Sobanov Y, Holtappels G, Bajna E, Svoboda M et al (2011) Activation-induced cytidine deaminase (AID)-associated multigene signature to assess impact of AID in etiology of diseases with inflammatory component. PLoS One 6(10):e25611

Mechtcheriakova D, Svoboda M, Meshcheryakova A, Jensen-Jarolim E (2012) Activation-induced cytidine deaminase (AID) linking immunity, chronic inflammation, and cancer. Cancer Immunol Immunother 61:1591–1598

Meshcheryakova A, Tamandl D, Bajna E, Stift J, Mittlboeck M et al (2014) B cells and ectopic follicular structures: novel players in anti-tumor programming with prognostic power for patients with metastatic colorectal cancer. PLoS One 9:e99008

Meshcheryakova A, Svoboda M, Tahir A, Kofeler HC, Triebl A et al (2016) Exploring the role of sphingolipid machinery during the epithelial to mesenchymal transition program using an integrative approach. Oncotarget 7(16):22295–22323

Muramatsu M, Kinoshita K, Fagarasan S, Yamada S, Shinkai Y et al (2000) Class switch recombination and hypermutation require activation-induced cytidine deaminase (AID), a potential RNA editing enzyme. Cell 102:553–563

Okazaki IM, Hiai H, Kakazu N, Yamada S, Muramatsu M et al (2003) Constitutive expression of AID leads to tumorigenesis. J Exp Med 197:1173–1181

Peduzzi P, Concato J, Feinstein AR, Holford TR (1995) Importance of events per independent variable in proportional hazards regression analysis. II. Accuracy and precision of regression estimates. J Clin Epidemiol 48:1503–1510

Schemper M (2003) Predictive accuracy and explained variation. Stat Med 22:2299–2308

Shen-Orr SS, Gaujoux R (2013) Computational deconvolution: extracting cell type-specific information from heterogeneous samples. Curr Opin Immunol 25:571–578

Smith GC, Seaman SR, Wood AM, Royston P, White IR (2014) Correcting for optimistic prediction in small data sets. Am J Epidemiol 180:318–324

Svoboda M, Meshcheryakova A, Heinze G, Jaritz M, Pils D et al (2016) AID/APOBEC-network reconstruction identifies pathways associated with survival in ovarian cancer. BMC Genomics 17:643

Tibshirani R (1997) The lasso method for variable selection in the Cox model. Stat Med 16:385–395

Uno H, Cai T, Pencina MJ, D'Agostino RB, Wei LJ (2011) On the C-statistics for evaluating overall adequacy of risk prediction procedures with censored survival data. Stat Med 30:1105–1117

Van Houwelingen JC, Le Cessie S (1990) Predictive value of statistical models. Stat Med 9:1303–1325

Verweij PJ, Van Houwelingen HC (1994) Penalized likelihood in Cox regression. Stat Med 13:2427–2436

Vittinghoff E, McCulloch CE (2007) Relaxing the rule of ten events per variable in logistic and Cox regression. Am J Epidemiol 165:710–718

Nanocellulose: A New Multifunctional Tool for RNA Systems Biology Research

Elena Bencurova, Meik Kunz, and Thomas Dandekar

Contents

Abstract Bioinformatics techniques allow the monitoring of large-scale interaction data such as gene expression changes. We explain suitable algorithms and databases for the analysis of direct regulatory interactions of RNA, such as micro (mi)RNA–mRNA, long non-coding (lnc)RNA, and RNA–protein complexes. Network analysis and dynamic simulations of RNA interaction networks are described next. RNA interactions probed by experiments are then described. For these interactions, nanocellulose provides a strong scaffolding platform; we evaluate different application modes regarding such uses of nanocellulose. Nanocellulose also provides options with which to probe biomedical RNA interactions. Future perspectives of nanocellulose use in various fields are discussed.

E. Bencurova · M. Kunz · T. Dandekar (✉)
Functional Genomics and Systems Biology Group, Department of Bioinformatics, Biocenter, University of Würzburg, Würzburg, Germany
e-mail: dandekar@biozentrum.uni-wuerzburg.de

N. Rajewsky et al. (eds.), *Systems Biology*, RNA Technologies,
https://doi.org/10.1007/978-3-319-92967-5_19

Keywords Nanocellulose · RNA · Bioinformatics · RNA-interaction networks · Biomedicine · Food industry · Electronics

1 Introduction

This chapter introduces a new and versatile tool for RNA systems biology research—nanocellulose (NC). NC is the molecular form of cellulose, which is particularly easy to modify, as an NC composite, and hence, in particular, it can easily be brought into contact with RNA.

First we provide a basic outline of the useful classical tools we use in RNA research for our special focus, network analysis. In networks, RNA interacts with partner molecules and, hence, these interactions have to rely on the sequence, structure, and stability of the RNA—all of which features can be recognized by specific software and useful tools. However, a systems biological look at regulatory networks also involves a number of tools that are specifically geared to analyze network connections, molecular classifications (e.g., gene ontology terms), and the network's inherent dynamics. The novelty here is the connection, the emergent systems behavior of an RNA-mediated network response. As such interactions can have important biomedical implications and translational features, we also give examples of these interactions. Throughout, we stress eukaryotic examples and, as innovative RNA types are becoming increasingly important as more of their functions are elucidated, we focus on long non-coding (lnc)RNAs and micro (mi)RNAs.

We are aware, of course, that there are other exciting RNAs mediating metabolic effects; for example, prokaryotic riboswitches. A tool to recognize this type of RNA interaction with a metabolic network is the riboswitch finder (http://riboswitch.bioapps.biozentrum.uni-wuerzburg.de/server.html). It best identifies high-affinity guanosine riboswitches (Bengert and Dandekar 2004). Moreover, RNA-network interactions are also evident from bacterial small (s)RNAs of regulatory types and new types of protein–RNA interactions (Smirnov et al. 2017), but this is not the focus of this chapter.

The question we ask is: how can you find an experimental platform for such RNA-network interactions? We supply some answers to the question under the headings: "2. Systems Biology of Networks and Key Tools for RNA Research in this Field" and "3. Introducing Nanocellulose as a New Tool for Studying RNA Interactions".

The text under heading 2 explains why there is a convincing case for establishing this link by using NC as a new platform and tool for RNA interaction studies and similar investigations.

The text under heading 3 embeds these considerations in an up-to-date overview of the methodological capacities of NC and the multitude of applications currently available for NC.

2 Systems Biology of Networks and Key Tools for RNA Research in This Field

With advances in high-throughput technologies, several classes of RNAs that show a fundamental role in biology have been identified. However, although several thousand RNAs have been annotated, most of them are not well understood, as experimental functional characterization is challenging and RNAs often show complex regulatory effects on transcriptional and translational regulation. In this regard, systems biology analysis can help to obtain a comprehensive functional understanding from experimental datasets. How can we perform a systems biology analysis to understand the complex interplay of RNAs with the genome? Such an analysis should follow three steps: (*1*) data collection, (*2*) data integration and analysis, and (*3*) data modeling (Fig. 19.1). Table 19.1 summarizes important resources for systems biology analysis. For more details, see our recent RNA analysis reviews (Kunz et al. 2014, 2016a).

1. RNA datasets can be used from own experiments or downloaded from databases such as Gene Expression Omnibus (GEO; https://www.ncbi.nlm.nih.gov/geo/) and GENEVESTIGATOR (https://genevestigator.com/gv/). Information regarding the genomic position and sequences can be obtained from genome browsers such as Ensembl (https://www.ensembl.org/index.html). More specific information about RNAs can be obtained from the LNCipedia (Volders et al. 2013) and miRBase databases (Kozomara and Griffiths-Jones 2014). Publication and further information such as sequences can be derived from the Medline (https://www.ncbi.nlm.nih.gov/) database. All this information builds the basis for the analysis.
2. Programs such as R (https://www.r-project.org/), Perl (https://www.perl.org/), and mySQL (https://www.mysql.com/de/) help in data integration and analysis. For instance, the program language R allows graphical representations of items

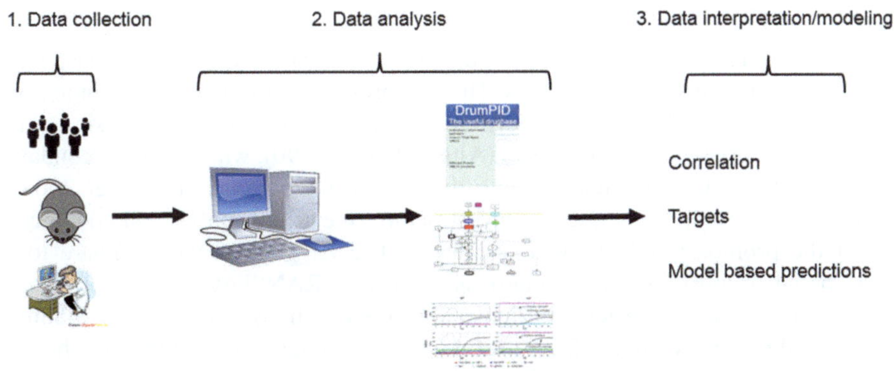

Fig. 1 Overview of systems biology analysis to understand the complex interplay of RNAs with the genome

Table 1 Overview of important databases and software tools for systems biology analysis

Resource	Usage	References
Alggen PROMO	Promoter analysis	Messeguer et al. (2002)
BLAST	Sequence analysis	Altschul et al. (1990)
catRAPID	RNA-interaction analysis	Agostini et al. (2013)
ClueGO	Functional analysis	Bindea et al. (2009)
CPC coding potential calculator	Coding analysis	Kong et al. (2007)
Cytoscape	Network visualization/analysis	http://www.cytoscape.org/
DrumPID	Interaction analysis	Kunz et al. (2016b)
Ensembl	Genome browser	https://www.ensembl.org/index.html
Gene Expression Omnibus (GEO)	Experimental datasets	https://www.ncbi.nlm.nih.gov/geo/
GENEVESTIGATOR	Experimental datasets	https://genevestigator.com/gv/
Jimena	Network simulations	Karl and Dandekar (2013)
KEGG	Interaction analysis	Kanehisa et al. (2010)
LNCipedia	lncRNA database	Volders et al. (2013)
LocARNA package	Sequence structure analysis	Hofacker (2003)
Medline (NCBI)	Database	https://www.ncbi.nlm.nih.gov/
microRNA.org/ miRanda	miRNA database	Enright et al. (2003)
miRBase	miRNA database	Kozomara and Griffiths-Jones (2014)
MySQL	Programming/data warehouse	https://www.mysql.com/de/
Npinter	RNA-interaction analysis	Wu et al. (2006)
Panther	Functional analysis	Mi et al. (2016)
Perl	Programming/analysis	https://www.perl.org/
R	Programming/analysis	https://www.r-project.org/
Reactome	Interaction analysis	Croft et al. (2011)
STRING	Interaction analysis	Szklarczyk et al. (2015)
SQUAD	Network simulations	Di Cara et al. (2007)
Wikipathway	Interaction analysis	Kutmon et al. (2016)

such as a heatmap, but also allows data clustering and statistical analysis, e.g., for differentially expressed RNAs. The Perl program helps in genome annotation and comparisons. A specific algorithm for sequence analysis is the Basic Local Alignment Search Tool (BLAST) (Altschul et al. 1990), whereas the secondary structure can be investigated using the LocARNA package (Hofacker 2003). Besides focusing on sequence and structure, the RNA analysis should focus on the promoter and interaction context. The AlggenPROMO software tool is based on position weight matrices from the TRANSFAC database (release version 8.3) and analyzes a promoter for potential transcription factor binding sites (Messeguer et al. 2002). A useful systems biology tool is Cytoscape (http://www.cytoscape.org/index.html), which contains several plugins for visualization and analysis (Saito et al. 2012). RNA interaction partners can be derived from

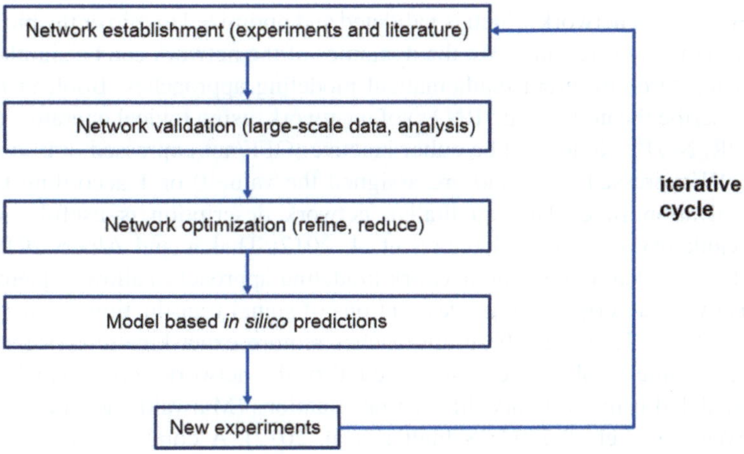

Fig. 2 Overview of in silico modeling. Different steps are sketched. They build on an iterative cycle of theory and experiments

databases such as Wikipathway (Kutmon et al. 2016), KEGG (Kanehisa et al. 2010), Reactome (Croft et al. 2011), and STRING (Szklarczyk et al. 2015). A more specific database that focuses on the drug-target interaction context is the DrumPID database (Kunz et al. 2016b), whereas databases such as Npinter (Wu et al. 2006) and catRAPID (Agostini et al. 2013) focus on non-coding (nc)RNA interactions. The analysis (*1*) should form the basis for the data modeling (*3*). In silico modeling is an important step in systems biology research. This can help to model the data and get new insights; for instance by finding correlations and new targets and making model-based predictions. In this context, the in silico modeling should build on an iterative cycle of theory and experiments (see Fig. 19.2) and should include: (a) network establishment, (b) network validation and optimization, and (c) simulation. For more details see our previously developed systems biology analysis approaches (Gottlich et al. 2016; Kunz et al. 2016a, 2017).

(a) Based on data from experiments and the literature (e.g., GEO; https://www.ncbi.nlm.nih.gov/geo/), differentially regulated genes can be identified and used as a basis to set up an in silico network. The required protein-protein interactions can be derived from KEGG (Kanehisa et al. 2010) and DrumPID (Kunz et al. 2016b). An in silico network should be a simplified view of the cell, e.g., a cancer cell or a cardiac cell, and should reflect the underlying mechanism of the interacting components, e.g., the RNA signaling cascade.
(b) The network should next be iterative, validated by large-scale data such as transcriptome and interactome data (e.g., RNA-Seq, mRNA targets) to optimize the network connectivity and to reduce the complexity (refine and reduce).

(c) Based on a network which is validated and optimized to mirror the biological system (steps (a) and (b)), the dynamics of the network can be simulated by using three different mathematical modeling approaches. Boolean models describe the nodes (e.g., RNAs) of a network, using logical operators (AND, OR, NOT). Nodes can be either inactive (OFF/not expressed = 0) or active (ON/expressed = 1) and are assigned the value 0 or 1 according to their activation state. This qualitative network description is useful for gene regulatory networks (Schlatter et al. 2012; Thakar and Albert 2010). On the other hand, dynamic network modeling approaches allow a quantitative network description, e.g., descriptions of signal transductions or enzymatic reactions. However, these approaches require exact kinetic concentrations (e.g., time-resolved western blot data) of the network nodes, which can be modeled using ordinary differential equations (Maiwald and Timmer 2008; Wangorsch et al. 2011; Schlatter et al. 2012). A combination of both the above modeling approaches is the semi-quantitative modeling approach. This approach allows an intuitive description of the network without accurate kinetic data (Schlatter et al. 2012; Philippi et al. 2009; Di Cara et al. 2007). Example modeling software such as SQUAD (Di Cara et al. 2007) and Jimena (Karl and Dandekar 2013) combine Boolean and dynamic modeling using a mathematical transformation for interpolation between full active and passive nodes (SQUAD: exponential function; Jimena: exponential function, steep Bool cube interpolation or sigmoidal hill cube interpolation between on and off state as well as other possibilities). Interestingly, such software calculates the steady states of the network that describe the stable states of the network to which it can return after stimulation (Di Cara et al. 2007). These steady states are of importance, as they may show how a network can therapeutic return from a disease state to a normal cellular state. The modeling approach, similar to experimental and clinical data, allows us to model the network to get functional insights, e.g., insights into how tumors grow or how cardiac hypertrophy occurs. Furthermore, the network modeling approach also allows us to explore the in silico network changes, e.g., those that occur after drug administration, in order to develop the best therapeutic strategy or to perform in silico knockouts, e.g., by investigating the effect of a mutation (also an iterative process validated by transcriptome and interactome literature data). This modeling approach allows us to design experiments and to pre-evaluate, in silico, the potential effects of a therapy, for example. Subsequently, the model-based predictions can be tested experimentally and used for additional model analysis. This approach not only reduces the number of unnecessary experiments but also reduces costs and saves time.

Nevertheless, experiments are required to validate the predictions and to minimize failures. Such prediction failures are possible, as the network does not consider all components and interactions of the cell. For instance, the cell system contains different biochemical entities, such as metabolites (e.g., RNAs such as iron response elements, riboswitches, and sRNAs) and lipids (Czakai et al. 2017).

However, for already known RNAs and their network components, such modeling approaches are very helpful. Nevertheless, what can we do for newly annotated RNAs, e.g., new RNA players such as non-coding RNAs (ncRNAs), without any knowledge about the functional interaction context? In the following section we demonstrate an example of systems biological modeling of lncRNAs.

Non-coding RNAs such as miRNAs and lncRNAs are important RNAs that regulate biological processes and pathways associated with several diseases, such as cancers and heart disease (Kunz et al. 2014, 2016a). However, although miRNAs for cardiac disease are well characterized, lncRNAs are not well studied (Viereck et al. 2016). lncRNAs (multi-exonic, >200 nt) show complex genomic regulation (e.g., transcriptional and translational regulation, chromatin modifier) (Kunz et al. 2016a; Fiedler et al. 2015; Viereck et al. 2016). Therefore, the systems biology analysis should cover genomic localization, phylogenetic sequence–structure conservation, and functional interaction partners (mRNA, miRNA, protein) (Kunz et al. 2016a) (Fig. 19.3). For example, using bioinformatics, we characterized the lncRNA Chast (cardiac hypertrophy-associated transcript) that has been shown to promote cardiac remodeling (Viereck et al. 2016). As experimental lncRNA profiling shows thousands of deregulated transcripts, a combined bioinformatics screening and experimental validation strategy for lncRNA selection is required. Experimental lncRNA profiling (19,427 of 31,423 transcripts) from pressure-overloaded mice revealed 2860 (1237 up-regulated, 1623 down-regulated) differentially expressed

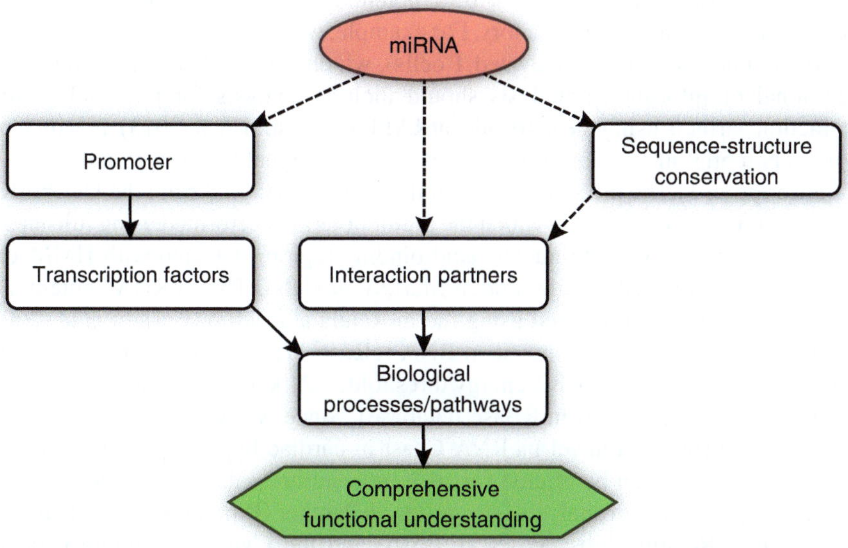

Fig. 3 Overview of systems biology analysis of long non-coding (*lnc*)RNAs (Fig. from M. Kunz et al. 2016a, with permission from the Creative Commons Attribution (CC-BY) license (http://creativecommons.org/licenses/by/4.0/).)

lncRNAs (logFC), in which four lncRNAs (one up-regulated and three down-regulated) showed significant deregulation in the heart by further quantitative polymerase chain reaction (qPCR) analyses. Additional analysis regarding the protein-coding potential, using the coding potential calculator (CPC; Kong et al. 2007), showed no potential for the up-regulated lncRNA (its identifier is ENSMUST00000130556). The lncRNA received the new name Chast. Chast is positioned in the antisense region with the two protein-coding genes Arhgap27 and Plekhm1. Interaction analysis using the IntaRNA program (Busch et al. 2008; locally installed) shows a potential interaction of Chast with Plekhm1. Interestingly, experimental validation shows inverse expression of Plekhm1 and Chast.

Next, the analysis flow should investigate sequence–structure conservation. Owing to their sequence length, lncRNAs are often not conserved, which limits their experimental characterization in model organisms, as well as their potential clinical applications. Sequence analysis using the BLAST algorithm (http://blast.ncbi.nlm.nih.gov/Blast.cgi; BLASTn, nucleotide blast) shows several homologous mammalian sequences (e.g., human, pig, and rat) with a threshold E-value *of* ≤ 0.05 and identity $\geq$80%. Moreover, the LocARNA alignment and folding tool implemented in the ViennaRNA package (Hofacker 2003; parameter: alignment type, global; alignment mode, standard) shows sequence–structure conservation for the homologous sequences in mammals. More interestingly, the human sequence–structure homolog was experimentally validated (66% homology, Chast chromosome 17 NC_000017.11|64783199–64783552; Homo sapiens chr17, GRCh38.p2 primary assembly), highlighting its potential clinical relevance. Promoter analysis of Chast using the Allgen PROMO tool shows potential binding sites for cardiac and prohypertrophic transcription factors. For example, we found a potential binding site for NFAT (nuclear factor of activated T cells), which was experimentally validated. Additional bioinformatics analysis should include analysis for potential protein interaction partners using the software catRAPID (Agostini et al. 2013), in which we found a potential interaction of Chast with STAT1 (signal transducer and activator of transcription 1).The proteomes used in the tool are gathered from the UniProtKB database (release 2012_11), in which the predictions are performed using full-length proteins, or are restricted to nucleic acid binding regions detected with HMMscan (probabilistic statistical profile hidden Markov models). The miRNAs potentially bound by Chast were predicted using the miRanda algorithm (Enright et al. 2003; locally installed; parameter: mouse miRNA release 17; gap open penalty: -2, gap extend: -8, score threshold: 80, energy threshold: -21 kcal/mol, scaling parameter: 2; miRNA sequences from miRBase database). Since we want to illuminate here the functional role of a novel lncRNA Chast in cardiac hypertrophy, the potential interaction partners can be filtered for such associations using a biological process and pathway enrichment analysis. This relies on Panther (**P**rotein **A**nalysis **TH**rough **E**volutionary **R**elationships; Mi et al. 2016) and the Cytoscape plugin ClueGO (Bindea et al. 2009). Alternatively, the interaction partners can be manually mapped against cardiac and hypertrophy pathway data from the databases Reactome (Croft et al. 2011), WikiPathway (Kutmon et al. 2016), and KEGG (Kanehisa et al. 2010). All these results indicate that Chast is predicted to function in transcriptional

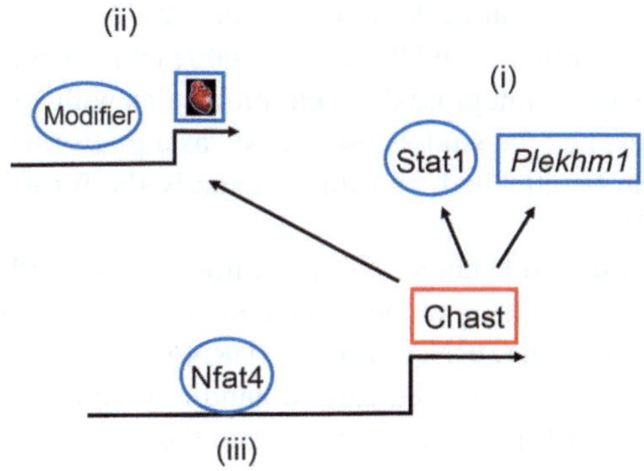

Fig. 4 Systems biology analysis of the lncRNA *Chast* (cardiac hypertrophy-associated transcript). Sketched is the transcription of Chast lncRNA (*red box*), its regulation by *NFAT4* (nuclear factor of activated T cells 4), and key interaction partners *STAT1* (signal transducer and activator of transcription 1) and Plekhm1 (*blue boxes*), as well as further heart- and stress-associated modifiers

regulation through (i) the mRNA targeting of Plekhm1, and/or (ii) Chast can directly bind transcription factors and guide them to the promoter to regulate transcription, e.g., for heart and stress-induced genes, and/or (iii) it can regulate its own transcription in a feedback loop with NFAT (Fig. 19.4). Finally, experimental in vivo inhibition of Chast reverses experimental cardiac hypertrophy (for details see Viereck et al. 2016).

In conclusion, these examples highlight the importance of combined bioinformatic analysis and experiments and how they synergistically help in the characterization of new RNAs and RNA interactions; in particular, this is exemplified by looking at new lncRNAs with novel biological functions.

3 Introducing Nanocellulose as a New Tool for Studying RNA Interactions

3.1 Overview

Nanocellulose (NC) has attracted increasing attention during the past few years. To use it for RNA interactions we first of all need the raw material. Potential sources include refined cellulose from bulk production (e.g., from Innventia, Stockholm, Sweden), as well as bacterial NC produced by sophisticated continuous fermentation, supplied from companies such as JeNaCell (Jena, Germany). Cellulose itself is easily furnished, being the most abundant bioorganic molecule on the planet. However, for molecular interactions and use with RNA, the NC form is required. In particular, to achieve good interaction with RNA, modified forms, such as hairy cellulose nanocrystalloids (van de Ven and Sheikhi 2016), as well as strong, self-standing oxygen barrier films from NCs are available (Sirvio et al. 2014).

Most interaction studies integrate different molecules with NC, thus forming so-called NC composites; i.e., the studies use the NC as a platform or chassis to which further molecules can be attached, and this is exactly the form advocated here for RNA interaction studies.

Moreover, as a natural substance, NC does not damage cells or interfere with biological processes, and this includes RNA in all its different forms [mRNA, miRNA, lncRNA, ribosomal (r)RNA, transfer (t)RNA, sRNA]. This is definitely an advantage and as long as the required care in handling the RNA alone is observed, NC does not destroy RNA molecules and leaves them intact. NC forms that can be used to study RNA interactions include laser-structured bacterial NC (BNC) hydrogels (Ahrem et al. 2014) and three-dimensional (3D) porous BNC scaffolds (Krontiras et al. 2015). These NC platforms can also be used in screening, replacing cytokine profiling (Bhattacharya et al. 2017) by a suitable RNA screen.

As screening can, of course, be done also regarding RNA and RNA-RNA interactions, NC is a very precise and innovative tool with which to study RNA molecules and the processes and interactions these molecules are involved in. Often the direct focus is on the resulting mRNA response and gene expression profiling; for example, regarding the differentiation of liver progenitor cell lines and the establishing of organotypic functionality in nanofibrillar cellulose hydrogels (Malinen et al. 2014) or other cartilage regeneration studies (Pretzel et al. 2013).

However, it is our aim in this book on modern RNA technologies to advocate for the employment of the full potential of the NC compound platform. RNA and DNA can be readily anchored at the NC surface or incorporated in its porous interior (Razaq et al. 2011; Xu et al. 2016). Furthermore, there is a powerful new technique that systematically generates high variation of different RNA molecules—digital PCR technology. Hatch et al. (2011) introduced one-million droplet arrays, together with wide-field fluorescence imaging, for massive analysis of digital PCR products. As in their study, we advocate using the NC chassis as a microfluidic droplet platform. In particular, high-throughput droplet processing arrays are available "on the chip", and these RNA arrays can subsequently be screened, or, using fluorescent labels, be scanned by a low-cost 21-megapixel digital camera and macro lens with an 8- to 12-cm^2 field of view at 1× to 0.85× magnification.

Furthermore, multiplexing is also possible (Zhong et al. 2011). Instead of using standard qPCR together with a first reverse transcriptase (RT) step regarding the RNA, this new multiplex digital PCR method can be applied after the RT step as a new, powerful, and highly sensitive method for the analysis of RNA. Digital PCR starts from mass-synthesized specific primers and the amplification of single-target DNA molecules in thousands of separate reactions. For this purpose, a microstructured surface (Zhong et al. 2011), as available with NC, is highly advantageous and provides a modern method to synthesize, screen, and evaluate thousands, and even millions of RNA molecules. Multiplexing starts from picoliter droplets within emulsions and continues by varying the concentrations of different fluorogenic probes of the same color. This process opens up the attractive possibility of identifying different PCR products on the basis of their fluorescence intensity. Different colors further increase the number of PCR reaction products that can be discerned.

A further attractive option of this combination, RNA and an NC scaffold, is the use of optogenetics to control molecules in transcription or translation. For example, the Rippe group recently reported real-time observation of light-controlled transcription in living cells (Rademacher et al. 2017). They used an optogenetic tool termed BLInCR (Blue Light-Induced Chromatin Recruitment) and, with it, controlled the activity of a reporter gene cluster. This tool can also be applied to in vitro translation systems, and an NC composite carrying different mRNA species can be embedded in such in vitro translation systems.

A number of suitable in vitro translation techniques have been summarized by Steinle et al. (2017) and they also discuss exciting medical applications for such in vitro translated mRNAs, providing helpful mRNAs for diseased cells (Steinle et al. 2017). Even better for our purpose, i.e., studying RNA in a controlled and directed way, is the new tool reported by Baumschlager et al. (2017). It features an artificial blue light-responsive gene construct, T7 RNA polymerase ("Opto-T7RNAPs"). These constructs are not leaky, but are engineered to be almost shutoff in the dark state and they show high RNA expression strength after induction by blue light. Their range of induction is up to more than 300-fold that of baseline intensity. Moreover, screening allows one to obtain a variant that returns to the inactive dark state within minutes once the blue light is turned off. Again, the NC chassis here is highly advantageous, as control of the induction of the RNAs of interest can be studied completely in vitro without any bacterial or other cellular background transcription.

Similarly, specific RNA molecules such as lncRNAs, miRNAs, and sRNAs can be studied advantageously using NC as a scaffold. Suitable treatment options and modifications of NC for this purpose include the magnetic functionalization of bacterial nanocellulose (BNC) (Echeverry-Rendon et al. 2017), the development of NC scaffolds with tunable structures (Liu et al. 2016), the direction of biomimetic composite scaffolds using a combination of mineralization and electrospinning (Si et al. 2016), and even real cellular networks, such as neuronal networks, anchored on the NC (Jonsson et al. 2015). Table 19.1 provides a nice overview of suitable databases from which you can pick your favorite RNA, e.g., miRNA or lncRNA, to be anchored and studied on the NC scaffold.

Next, RNA interactions can be favorably studied on the NC chassis, and novel powerful techniques are available for this purpose. In particular, microfluidic proximity ligation assays allow the profiling of signaling networks with single-cell resolution (Blazek et al. 2015). Moreover, modern printing techniques allow us to rapidly not only print but also to well separate different RNA species (or, applying an RT step, DNA) on a compatible surface or scaffold (Stumpf et al. 2015). Of note, a protocol for self-biotinylation of DNA/RNA quadruplexes has recently been published (Einarson and Sen 2017). The biotinylated DNA/RNA can then be used in a PCR for amplification, and can be fixed to NC, but it can also be used for identifying, labeling, and pulling down further cellular RNA and DNA.

Finally, surface plasmon resonance can be applied to detect specific RNAs, using, for example, the protocol reported by Li et al. (2016).

This part of our chapter, hence, conveys the vision of bringing NC and RNA-network interactions together. As the cited references testify, a number of already available protocols and discoveries will allow us to profit from the combination. These protocols are also given here so that they can be readily applied to our NC composite as a platform to study RNA. We have to stress that there are, of course, many more "lab on a chip" concepts, including, in particular, perhaps the main class of "lab on chip" instruments employing microfluidics, such as acoustic droplet splitting and steering, using a disposable microfluidic chip and doing this on demand (Park et al. 2018).

We are convinced that NC composites have, apart from the examples given here, high potential for even more studies on RNA. Many of the alternative "lab on chip" concepts have the drawback that they do not consist of biocompatible material. Also in many other aspects (price, versatility, long-term durability) NC is an almost ideal component for an "RNA lab on a chip" and is described here in detail. The whole section "3. Introducing Nanocellulose as a New Tool for Studying RNA Interactions" provides an overall reference to support the advantages of NC in all these application aspects.

3.2 Nanocellulose: Structure and Properties

NC is one of the most investigated materials of the past decade. Cellulose is commonly distributed across the world; it can be found in both terrestrial and aquatic environments, and has been detected in bacteria, fungi, and plants. NC can occur in the form of cellulose nanocrystals and cellulose nanofibrils, both of which are derived from plant cellulose, such as wood, bark, cotton, and wheat straw; algae; and bacteria. In nature, BNC plays a key role in the protection of bacteria in their natural environment. Bacteria, mainly Gram-negative types, secrete extracellular pure cellulose, which acts as a self-immobilization tool that is essential for protection from desiccation, ultraviolet (UV) radiation, and also for protection against other microorganisms. BNC also facilitates the effective transport of nutrients and oxygen, and several studies mention its role in virulence modulation (Castiblanco and Sundin 2016; El Haga et al. 2017). NC is non-toxic and hypoallergenic; it is fully biodegradable, biocompatible, cheap to produce, and easy to modify to obtain the desired structure and properties that are crucial for many medical and biological applications. The high stability of BNC allows extreme manipulation, such as treatment with hot acid and alkaline substances, boiling for up to 120 min (depending on the thickness of the cellulose product), and pressing. NC fibers have extreme water-holding capacity, more than 99 wt%, by which they form a hydrogel, and they have a high degree of polymerization, with up to 10,000 repeating cellulose units. Thanks to its unique properties, BNC can be used in very creative ways—as a gel-like material (hydrogel, aerogel), hot-pressed sheet, dry sheet (paper-like form), or freeze-dried sheet, or even in combination with another material. Moreover, BNC can be molded into forms of any shape and size, such as fleeces, foils, tubes, aggregates, or irregularly formed shapes. In its dry form, NC sheets form a thermostable and slightly hydrophilic material with a decomposition temperature of 360°C (Gea et al. 2011).

3.2.1 Structure of NC

NC is a natural polymer containing β-(1,4)-glucan chains of different lengths, which can vary among species, production process, and post-production treatment. Although the molecular formula of cellulose from bacteria is identical to that of cellulose from plants, BNC is often preferred, owing to its ease of production and, mainly, purification. The production of wood-originated NC requires additional purification steps to remove lignin, pectin, and hemicellulose, while bacteria produce NC as a pure product without any contaminants. The typical length of a pure BNC fibril is around 100–700 μm, with a diameter of 100 nm. After biosynthesis, fibrils aggregate into sub-fibrils with a width of about 1.5 nm. The typical distance between the junction points of nanofibrils in the dry form is 523 $\pm$ 0.273 μm, with the orientation of the nanofibrils being 85.64° $\pm$ 0.56° (Grande et al. 2008). However, the morphology and geometry of the network can differ dramatically according to the source of BNC, as well as the life cycle of the bacteria and their growth rate.

The mechanical properties of NC are strongly determined by the source, form, and treatment of the BNC. A study by Grande et al. reported an ultimate tensile strength (UTS) value of 241.42 $\pm$ 21.86 MPa, a maximum elongation of 8.21 $\pm$ 3.01%, and Young's modulus of 6.86 $\pm$ 0.32 GPa for hot-pressed BNC in combination with starch (Grande et al. 2008). However, recently, the tensile strength of BNC was determined as 200–300 MPa, its Young's modulus was 15–35 GPa, and its maximum elongation was up to 2%. Compared with common cellulose materials such as cellophane, BNC in the dry state shows more than three times better tensile strength, but 20 times lower elasticity, which can be beneficial for various purposes, such as biological applications or membranes for loudspeakers and headphones (Gatenholm and Klemm 2010), or even as transparent displays and computer chips (Dandekar 2016). A combination of BNC with polyester resulted in a maximum UTS of 26.7 MPa for three sheets of BNC–polyester material; the modulus of elasticity of the BNC–polyester was significantly increased compared with that of pure polyester. Higher numbers of sheets resulted in lower elasticity, owing to defects such as porosities and micro cracks in the material (Abral and Mahardika 2016). Pure BNC, even in small amounts, can enhance the properties of other cellulose-based materials. A mixture of 10% BNC used as an additive for a birch chemithermomechanical pulp paper sheet resulted in an increase of more than 1.5 MNm/kg in tensile stiffness, and paper sheets containing the additive also showed a 140 % increase in the tear index, with almost no effect on the weight or thickness of the paper sheets (G. Q. Chen et al. 2017). Of note,

BNC hydrogels have significantly lower tensile strength than dry BNC films; the maximum tensile strength was observed in the work of Scionti, with a UTS of 0.9 MPa and an elongation of 64% (Scionti 2010). The cultivation medium also plays a role in the properties of BNC. A BNC hydrogel obtained by the fermentation of *Gluconacetobacter xylinus* from casein hydrolysate medium showed a higher UTS (0.04 MPa) than that of a glucose-based medium [0.03 MPa (Cavka et al. 2013)].

The optical properties of BNC make it a novel prospect for the research of optically functional materials. The first study of reinforcement using nanofibers of electrospun nylon-4,6 showed that a fiber content of only 3.9% was sufficient to build a transparent film (Bergshoef and Vancso 1999); however, for many applications, a higher nanofiber content is necessary. Nowadays, it is possible to construct optically transparent and flexible composites from BNC with a fiber content as high as 70%, with five times the mechanical strength of engineered plastics, and with a low thermal expansion coefficient (Yano et al. 2005). Transparency of BNC can also be obtained using several post-production processes. It was demonstrated that the treatment of dry BNC with acetic anhydride resulted in a decreased cellulose refractive index and lower hygroscopicity compared with findings in the non-treated substrate. Acetylation of BNC was shown to reduce water absorption to less than 0.5%, and the surface became highly transparent without any collapsing of the crystal structure—this occurred with a fiber content of 68% (Ifuku et al. 2007). Also, the treatment of BNC with 2,2,6,6-tetramethylpiperidine-1-oxyl radical-mediated oxidation resulted in the production of a transparent film with low thermal expansion (Wu and Cheng 2017). The development of such treated BNC in the near future is a promising technology for emerging flexible devices, such as displays for smart watches and solar cell substrates.

3.3 *Biosynthesis of Nanocellulose*

Microbial NC can be derived from different species as extracellular secreted fibers (Table 19.2). With respect to species, the cultivation conditions, yield, and purification processes make the species *Komagataeibacter* (syn. *Gluconacetobacter*, *Acetobacter*) preferable for the commercial production of BNC. However, many other species have been studied and optimized as alternative sources of BNC. The synthesis of BNC has been described in detail for several species; however, the best studied is the production of BNC by *Komagataeibacter*. The production of BNC is a multistep process that is regulated and synthesized by the *bcs* operon, involving several enzymes, catalytic complexes, and regulatory units (Fig. 19.5). Exhaustive research on the *bcs* operon has identified two fundamental genes in almost all of the cellulose-producing bacteria: cellulose synthase encoded by *bcsA* (the nomenclature differs according to species; also named *acsA*, *yhjO*, and *celA*), and the bis-(3′–5′)-cyclic dimeric guanosine monophosphate (c-di-GMP) protein encoded by *bcsB* (also named *acsB* and *celB*).

Table 2 Typical sources of nanocellulose and technologies for its extraction

Nanocellulose source	Manufacturing process	References
Wood	Mechanical/chemical pulping, steam explosion, acid hydrolysis, ultrasonication	Abraham et al. (2011), Brinchi et al. (2013), Chakraborty et al. (2005), Jiang and Hsieh (2013), Li et al. (2011)
Cotton	Acidic hydrolysis	Morais et al. (2013)
Fungi (e.g., *Trichoderma* sp., *Aspergillus* sp.)	Hydrolysis	Vigneshwaran and Satyamurthy (2016)
Green algae (e.g., *Cladophora* sp.)	Microfluidization	Xiang et al. (2016)
Red algae (*Gelidium elegans*)	Alkali treatment, bleaching, and acid hydrolysis	Chen et al. (2016)
Kombucha	Microfluidization/atomization	Dima et al. (2017)
Komagataeibacter xylinus (syn. *Gluconacetobacter*)	Biosynthesis via fermentation	Vazquez et al. (2013)
Komagataeibacter rhaeticus	Cultivation	Machadoa et al. (2016)
Komagataeibacter medellinensis	Biosynthesis via fermentation	Molina-Ramirez et al. (2017)
Acetobacter xylinum	Biosynthesis via fermentation	Chao et al. (2000)
Gluconacetobacter hansenii	Cultivation	Costa et al. (2017)

The ***bcsA*** gene is highly conserved among the cellulose-producing species and it is essential for the production of cellulose in vitro in a complex with the BcsB subunit (Omadjela et al. 2013). Crystallographic studies have revealed that BcsA consists of eight transmembrane helices, four of each N- and C-terminal, and the PilZ domain and GT domain, which are located between the fourth and fifth transmembrane helixes (Morgan et al. 2013). The role of the PilZ domain (Pfam domain PF07238) is essential for the functionality of the protein, while it serves as a binding site for the c-di-GMP. PilZ occurs in the C-terminal, consists of six-stranded β-barrels, and recognizes c-di-GMP. The PilZ: c-di-GMP interaction activates the synthesis of bacterial cellulose (Amikam and Galperin 2006; Morgan et al. 2014). Sequence analysis of several bacterial species has also suggested the role of the PilZ domain in regulation and signaling (Amikam and Galperin 2006).

The **BcsB** protein is an essential catalytic subunit of cellulose synthase, with a membrane-anchored periplasmic domain and membrane-associated transmembrane anchor (Fig. 19.2). This protein is associated with indirect interaction with c-di-GMP (Mayer et al. 1991), and with the transport of the β-glucan chains. Mutation of *bcsB* resulted in the disordered organization of glucan fibrils (Omadjela et al. 2013). Despite the fact that subunits BcsA and BcsB are sufficient for the in vitro cellulose expression of *K. xylinus*, BNC production in live bacteria requires the activation of four genes from the operon bcsABCD. Subunits BcsC and BcsD are

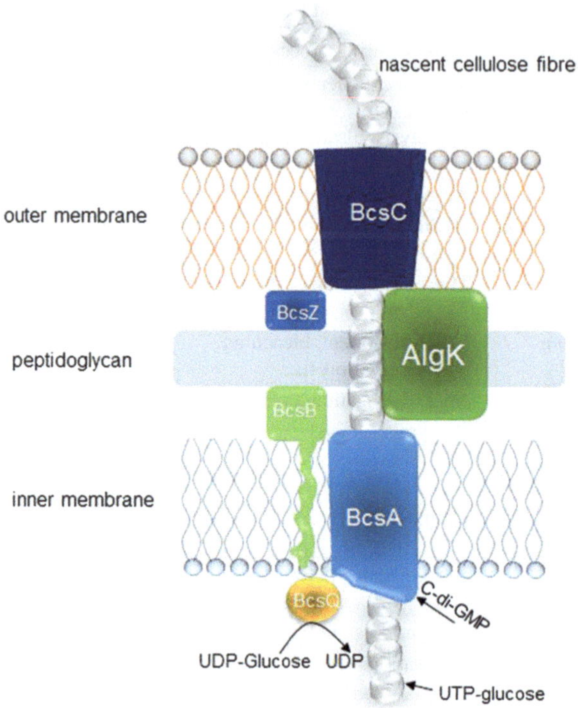

Fig. 5 Simplified schema of multicomponent subunits involved in bacterial nanocellulose (*BNC*) synthesis

necessary for the export of the glucan molecules and for loading them at the cell surface. A *bcsC* mutant exhibited dysfunction of cellulose production, whereas a *bcsD* mutant produced about 40% less cellulose than the wild-type cells (Wong et al. 1990). The next necessary component to be elaborated in BNC production in vivo is periplasmic AlgK protein, which interacts with the peptidoglycan layers. AlgK is involved in secretion processes and mediates protein–protein interactions (Keiski et al. 2010). The *bcsZ* gene (also named *acsC*, *celC*) encodes an endoglucanase that is indirectly involved in the c-di-GMP regulation of cellulose biosynthesis; however, a *bcsZ* mutation showed no significant changes in cellulose production (Castiblanco and Sundin 2016). Finally, the *bcsQ* gene (syn. *yhjQ*, *ccp*) encodes cellulose-complementing factor (Ccp) in *K. xylius*; however, its exact function remains unclear. The Ccp protein is very small, with a molecular weight of 8 kDa, and it seems to be conserved among the genus *Acetobacter* (Sunagawa et al. 2013).

The biosynthesis of cellulose is divided into four steps: (1) phosphorylation of glucose to glucose-6-phosphate catalyzed by glucokinase, (2) isomerization of glucose-6-phosphate to glucose-1-phosphate, a reaction catalyzed by phosphoglucomutase, (3) synthesis of uridine diphosphate (UDP)-glucose from glucose-1-

phosphate (catalyzed by UGPase (UDP-Glucose-Pyrophosphorylase)), and, finally (4) the conversion of UDP-glucose to cellulose in the presence of c-di-GMP. The cellulose fibers are next formed in two intermediary phases; in the first phase, 1,4-β-glucan chains are formed, and in the second phase, nascent chains are assembled and crystallized (reviewed in Lee et al. 2014).

3.4 Production of Nanocellulose

BNC can be produced commercially by several techniques—as a static (stationary) culture (cultivation in plastic plates); by agitated cultivation in a fermentation jar or horizontal fermenter; or by cultivation in internal loop airlift reactors (Chao et al. 2000; Machadoa et al. 2016; Vazquez et al. 2013), with the typical yield being 2–15 g/l in 50 h of cultivation. The choice of cultivation method can be critical for the final properties of BNC. Static culture leads to the production of a gelatinous layer with a 3D network on the surface of the culture medium. BNC obtained from a static culture has a high water content and high crystallinity level, as well as significant mechanical strength; however, the yield is lower than that from agitated cultivation, which can be a restrictive point for some commercial applications. Agitated cultivation is the most commonly used approach for bulk BNC production. During the agitation, bacteria benefit from the direct contact with oxygen, which directly influences their growth and metabolism. With agitation methods, the BNC is produced in the form of pellets/granules; the BNC thus produced has poorer mechanical properties and crystallinity than the BNC produced by stationary cultivation (Czaja et al. 2004). Nevertheless, the efficiency of BNC expression can be increased by optimization of the cultivation conditions. The composition of the medium has a critical impact on the growth of bacteria and their cellulose production. Glucose is typically used as a source of carbon, but alternative sources such as fructose, xylose, arabitol, galactose, sucrose, mannitol, and some other reducing sugars can be used (Cavka et al. 2013). It was shown that mannitol, glucose, and fructose were most effectively metabolized by *G. xylinus*. On the other hand, sucrose and galactose seem to be very unprofitable carbon sources, owing to inefficient uptake from the medium through the bacterial cell membrane. Sucrose and galactose need to first be hydrolyzed to glucose or fructose; otherwise, bacteria are not able to benefit from their use in cellulose biosynthesis (Mikkelsen et al. 2009; Velasco-Bedran and Lopez-Isunza 2007). However, the preference for a certain carbon source is related to the exact species. As an example, sucrose cannot be utilized by *Gluconacetobacter*; however, sucrose was shown to be the preferred substrate for BNC production by *Acetobacter* sp. (Son et al. 2001). The second essential component of the cultivation medium is the nitrogen source. The most commonly recommended nitrogen source for agitated cultivation is corn steep liquor, at a concentration 0.15–2 (v/v)%, while yeast extract, peptone soybean meal, glycine, casein hydrolysate, and glutamic acid have been described as the most effective for stationary cultures (Coban and Biyik 2011; Ramana et al. 2000).

Bacteria can also utilize amino acids, vitamins, and mineral salts (Matsuoka et al. 1996; Ramana et al. 2000). The yield of BNC can also be increased by the addition of glycerol, which positively influences cell growth (Mikkelsen et al. 2009); ethanol (Park et al. 2003); sodium alginate; lignosulfonate; and other substances (Ramana et al. 2000). The crucial factor for the proper growth of bacteria is the pH of the medium. Depending on the strain, pH 4–7 is recommended. The pH of the medium changes during the cultivation time because of the accumulation of secondary metabolites; ergo, continuous control of pH is necessary during the cultivation (Zeng et al. 2011). The production of BNC is also influenced by the speed of agitation (Jeon et al. 2014), temperature (Czaja et al. 2004), and salt concentration.

3.5 Nanocellulose in Biomedical Applications

In the past few years, several NC-based materials with huge potential for medical applications have been created. As early as1986, BNC was proposed as an ideal wound-healing material, in the form of a liquid-loaded pad (Ring et al. 1986), and until now, most NC-based biomedical applications have been related to the development of **wound plasters** and intelligent skins. Bacterial contamination is a serious problem in wound healing. More and more pathogens are becoming resistant to traditional antibiotics, which are being used for prolonged treatment of skin injuries, but are inappropriate for this purpose. In order to increase the effectiveness of such treatment, it is necessary to develop innovative therapeutic methods. BNC has been used in several studies to deliver antimicrobial compounds that promote the acceleration of healing. Silver is a traditional and very effective antibacterial agent. In a study by Berndt et al., BNC was prepared as a porous 3D network with immobilized silver nanoparticleson the top and bottom of the BNC surface. This hybrid exhibited strong antibacterial activity against *Escherichia coli*, and, further, the special design of the wound plaster avoided the release of the BNC into the wound (Berndt et al. 2013). In a recent study, wound dressing material comprising BNC and zinc oxide was tested in vivo on a skin burn mouse model. The BNC-zinc oxide nanocomposite showed 66% higher healing activity than BNC only, and histological analysis showed the regeneration of hair follicles and the development of new blood vessels. Moreover, strong antibacterial activity was observed against *E. coli*, *Pseudomonas aeruginosa*, *Staphylococcus aureus*, and *Citrobacter freundii* (Khalid et al. 2017). BNC has also been tested for possibile use in long-term wound treatment. Poloxamers together with an octenidine were loaded onto BNC and gradual release of the octenidine was observed; the material showed high compression stability and water-binding capacity, as well as strong antimicrobial activity against *S. aureus* and *P. aeruginosa* (Alkhatib et al. 2017). Similarly, material consisting of BNC and chitin nanocrystals presented excellent mechanical properties and bacteriostatic activity (Butchosa et al. 2013).

As BNC promotes chondrocyte adhesion and proliferation, several groups have tested BNC in cartilage **tissue engineering** and in the production of **bioactive**

implants. Nimeskern et al. (2013) showed that an implant material for ear cartilage replacement based only on native BCN did not have efficient mechanical properties; however, when they implemented chemical modification before or after the implantation, relaxation kinetics and fluid flow resistance were improved. In another study, Svensson et al. (2005) found that chemically modified BNC exhibited no effect on chondrocyte proliferation, while native BNC significantly promoted cell growth. Bodin et al. (2007) assessed BNC-based meniscus implants and found that the material had excellent mechanical properties, better than those of collagen-based implants. Material consisting of BNC was also used as an effective approach for the design of artificial blood vessels (Andrade et al. 2010), for bone regeneration (Grande et al. 2009), and even as a substrate for artificial tears (Mencucci et al. 2015). In this context, owing to its low cost, flexibility, and excellent biological properties, such as its promotion of cell migration and proliferation, BNC can be considered as an effective material for regenerative medicine and implants.

Bioprinting is a method of 3D-printing of biological materials, living cells, and functional components with high accuracy, resulting in surprising properties: biomimicry, autonomous self-assembly units, and mini-tissue building blocks. To date, flat tissues such as skin, as well as complex organs such as the liver, have all been bioprinted (reviewed in Dodziuk 2016). One of the challenges in obtaining functional bioprinted material is to fabricate the appropriate biological ink ('bioink'). A mitogenic hydrogel system based on alginate sulfate was shown to be an appropriate candidate for promoting chondrocyte proliferation (Ozturk et al. 2016); however, owing to its rheological properties, the hydrogel could not be printed. Conversion of the alginate sulfate into a bioink was accomplished by the addition of NC. After this, the produced bioink exhibited very good printability. As a non-printed matrix, this material also promoted cell proliferation and the synthesis of collagen II by encapsulated cells (Muller et al. 2017).

NC in the form of nanocrystals was tested as a potential **drug delivery vector**. Ideal drug delivery vectors must have several properties, such as good absorption of the functional molecule, controlled release of the drug, and penetration through the membranes to reach the target. Ethyl cellulose-methyl cellulose (EC–MC) was shown to be a material with excellent absorption properties. In the work of Pan-In et al. (2015), an EC–MC polymer was loaded with α-mangosin, a pharmaceutical molecule used in wound healing, eczema, and skin infections. The polymer was then tested on 20 human volunteers in a patch test against *Propionibacterium acnes*, the causative agent of acne vulgaris. The 4-week treatment revealed that the patch had a therapeutic effect, with minimal skin irritation, suggesting a new method for the treatment and prevention of acne (Pan-In et al. 2015). The capacity of BNC for use as a transdermal drug delivery medium was tested by a Portuguese research group. They used ibuprofen and lidocaine hydrochloride bound to native BNC and tested delivery of the drug in vitro. The lidocaine-BNC membrane was permeated at a lower rate than the classical forms; however, the ibuprofen-BNC membrane showed three times higher delivery compared to the native gel or polyethylene glycol (PEG) 400 usage (Trovatti et al. 2012).

3.6 Nanocellulose and RNA

Small interfering RNAs (siRNAs) are short (20–25 nt) double-stranded RNA molecules that can induce the degradation of specifically targeted RNA. Specifically designed siRNAs represent new drug types for use against viral diseases, cancer, and neurodegenerative diseases, as well as for stem cell research. Several methods have been developed as delivery systems for siRNAs; for example, using viruses, cyclodextrins, and nanoparticles. However, finding the ideal candidate that has high transfection efficiency and minimal toxicity remains challenging. NC, thanks to its biocompatibility, low toxicity, and biodegradability can act as a promising delivery substrate. However, the negatively charged surface of NC can limit the binding of siRNA. To increase the interaction capacity of NC, cellulose nanocrystals (CNC) originating from cotton were mixed with polyethylene imines (PEIs), resulting in positively charged particles. Thanks to electrostatic interactions, siRNA killer was bound to the PEI–CNC matrix. These complexes showed efficient uptake of siRNA in C2C12 murine myoblasts and an in-vitro study demonstrated the inhibition of tumor cell proliferation (Ndong Ntoutoume et al. 2017).

3.7 Nanocellulose in the Food Industry

NC can be used as an effective additive to enhance the properties of various food products. The major reason for using NC as a direct food ingredient is its high viscosity at low concentrations; also, it is low in calories and is heat stable up to 180°C. At a low concentration, the addition of NC does not influence taste, aroma, or texture and it stabilizes water-in-oil emulsions and foams. Bread dough containing native BNC showed increased volume and moisture retention compared with bread dough that did not contain BNC, although the surface of the BNC-containing bread had a reduced browning index and showed reduced firmness of the breadcrumbs (Corral et al. 2017). BNC has also been added to a meat emulsion (sausage) to produce a low-lipid, low-sodium meat product with appropriate sensory characteristics and storage stability (Marchetti et al. 2017). BNC can also be found in other foods, such as the popular desserts Nata de Coco, whipped creams, and waffles.

NC can be combined with various polymers and substances to produce bioactive packaging material, which can potentially eliminate pathogens in ready-to-eat food packaging. Methylcellulose, a chemically derived cellulose-ether, forms strong and clear films with excellent stability in cold conditions. In a study by Piccirillo et al. (2013), methylcellulose was combined with a natural extract from the stems of Ginja cherry (*Prunus cerasus* L.), which contains various polyphenols and terpenes possessing antibacterial activity. The resulting NC film was tested for different Gram-negative and Gram-positive bacteria, and showed positive inhibition of the growth of both groups of bacteria (Campos et al. 2014). A food packaging membrane fabricated of NC, chitosan, and *S*-nitroso-*N*-acetyl-D-penicillamine also showed effective inhibition of the growth of *E. faecalis*, *S. aureus*, and *Listeria monocytogenes*, as well as excellent water-barrier properties; however, the membrane showed a decreased Young's modulus (Sundaram et al. 2016).

3.8 Nanocellulose as a Support Material for Transient Electronics

Printed electronics (PE) is an emerging area of research that has exhibited strong development in the past 15 years, combining printing processes and ink chemistry for the manufacturing of microelectronic devices. The substrate for PE must comply with specific requirements, such as flexibility, transparency, a smooth and non-porous surface, and a low coefficient of thermal expansion (CTE). The sintering temperature for the production of PE is usually very high, up to 250°C, and thus the substrate must withstand this thermal burden without any deformation. Classical synthetic polymers and plastics have a high CTE and thus, high temperature can lead to damage of the plastic substrate and even the destruction of the functional electronic parts. For this reason, researchers are on the hunt for an ideal material with a low CTE that can replace the classical plastic substrates. Sheets of densely packed cellulose nanofibrils, with a width of 15 nm and CTE of 8.5 ppm/K (the CTE of plastic is approximately 40–120 ppm/K), were tested as a prospective material; the tensile strength was as high as 223 MPa and adequate optical transparency was obtained (Nogi et al. 2009). This novel flexible material can be used not only as an electrically conductive material, but also as a gas barrier film or a hard coating.

The role of NC in PE dominates in two areas: the use of NC as a **flexible substrate** for PE, and the direct incorporation of NC in **functional ink** containing conductive polymers, metal particles and flakes, and carbon particles. Ummartyotin et al. (2012) studied the use of BNC in organic light-emitting diode (OLED) technology. A nanocomposite film with BNC at 10- to 50-wt%, combined with a polyurethane-based resin, was manufactured, resulting in a thermally stable substrate for flexible OLED display, with a minimal CTE of less than 20 ppm/K, yet with a visible light transmittance of 80% (Ummartyotin et al. 2012). Similarly, light transmittance of above 75% was obtained by a combination of cellulose nanofibers with indium tin oxide. However, together with the excellent conductivity of indium tin oxide comes its very high cost, a disadvantage for the standard use of this material. Alternatively, vacuum filtration of Ag-nanowire mixed with bamboo cellulose resulted in the production of a transparent nanopaper with even higher light transmittance than that obtained with indium tin oxide, demonstrating the potential role of NC in optoelectronic devices as a low-cost material with remarkably strong adhesion and mechanical flexibility (Song et al. 2015). Moreover, as NC is eco-friendly, complete biological degradation and further recovery of the electronic components is possible. Jung et al. (2015) have demonstrated full fungal degradation of a substrate of cellulose nanofibers coated with epoxy after 84 days, suggesting a promising way to replace plastic substrate. In other works, an NC substrate was successfully tested for energy storage devices, as a binding agent for the production of flexible self-standing graphite anodes, and for lithium-ion battery applications based on NC-graphite nanocomposites (Gerbaldi et al. 2010; Jabbour et al. 2010), as well as for flexible thin transistors (Fujisaki et al. 2014). Several authors have also suggested the use of NC in solar cells; however, the performance of these cells was lower than that of conventional solar cells, with up to 3% power conversion, while the low-cost conventional polymer solar cells achieved 10–15% efficiency (Nogi et al. 2015; Zhou et al. 2013, 2014).

BNC can also be used as a substrate for the immobilization of DNA and proteins, as described by Uth et al. (2014) and Xu et al. (2016). Sensor and actuator molecules, as well as cells and other components, can be integrated with the BNC matrix to produce a novel computer chip or smart card for the active storage of information in DNA or RNA. Alternatively, this information can be processed in an "intelligent plaster" for monitoring wound healing (Dandekar 2016).

Another field where NC can be an alternative to classical materials is in piezoelectric measurements. Piezoelectric materials are typically ceramic elements used in PE. However, these novel applications require a highly flexible, thermally stable, and biocompatible material with optical transparency and low manufacturing cost. Mangayil et al. (2017) tested BNC films expressed by *K. xylinus* to investigate piezoelectric sensitivity. They genetically engineered bacteria with high production of BNC, resulting in a thick film. The film constructed by genetically engineered bacteria was thicker than from the wild type bacteria, with a high crystallinity index of up to 97.5%, and most importantly, with a significant piezoelectric response of up to 20 pC/N. Interestingly, the response depended on the orientation of the bacterial cellulose crystal region. Besides the astonishing mechanical, optical, and sensing properties of BNC, its production cost is far lower than that of classical piezoelectric materials, and it has better flexibility than wood-based cellulose (Mangayil et al. 2017).

4 Perspectives of Nanocellulose-RNA-composites

In many cells RNA-interaction networks are sophisticated regulatory circuits. To elucidate RNA interactions in a systematic way, a combination of high-power bioinformatics and modern experimental approaches is required. Next-generation sequencing techniques provide a number of such approaches, including RNA immunoprecipitation combined with microarray analysis (RIP–ChIP) and UV cross-linking and immunoprecipitation (CLIP), leading to high-throughput sequencing CLIP (HITS-CLIP); alternatively, another approach is photoactivatable ribonucleoside-enhanced CLIP (PAR-CLIP) (see X. Li et al. 2014 for details). For the detailed investigation of RNA-interaction networks, we have identified here the strong scaffolding capacities of NC as an attractive anchor technique, powerfully combined with modern large-scale analysis techniques from bioinformatics. We have also shown the dynamic modeling of such identified interactions.

5 Conclusions

Investigating RNA in systems biology has a clear focus on RNA-mediated interactions. For such investigations, we have presented a number of bioinformatics techniques that can be used to monitor large-scale interaction data, such as gene expression changes; as well, we have presented algorithms and databases for the analysis of the direct regulatory interactions of RNA, such as miRNA–mRNA, lncRNAs, and RNA–protein complexes. In this complex network analysis, dynamic simulations of RNA-interaction networks have been established, and these have been described here. To validate and probe RNA interactions, NC provides a strong scaffolding platform. After evaluating the different application modes for which NC can be employed for investigating RNA interactions, we have placed these against the background of other biomedical uses and applications of NC. Future work will try to broaden such approaches and techniques to improve the methodology used to study the systems biology of RNA-interaction networks.

References

Abraham E, Deepa B, Pothan LA et al (2011) Extraction of nanocellulose fibrils from lignocellulosic fibres: a novel approach. Carbohydr Polym 86:1468–1475

Abral H, Mahardika M (2016) Tensile properties of bacterial cellulose nanofibers-polyester composites. In: IOP conference series: materials science and engineering, vol 137, No 1. IOP Publishing, Bristol, p 012019

Agostini F, Zanzoni A, Klus P et al (2013) catRAPID omics: a web server for large-scale prediction of protein-RNA interactions. Bioinformatics 29:2928–2930

Ahrem H, Pretzel D, Endres M et al (2014) Laser-structured bacterial nanocellulose hydrogels support ingrowth and differentiation of chondrocytes and show potential as cartilage implants. Acta Biomater 10:1341–1353

Alkhatib Y, Dewaldt M, Moritz S et al (2017) Controlled extended octenidine release from a bacterial nanocellulose/Poloxamer hybrid system. Eur J Pharm Biopharm 112:164–176

Altschul SF, Gish W, Miller W et al (1990) Basic local alignment search tool. J Mol Biol 215:403–410

Amikam D, Galperin MY (2006) PilZ domain is part of the bacterial c-di-GMP binding protein. Bioinformatics 22:3–6

Andrade FK, Costa R, Domingues L et al (2010) Improving bacterial cellulose for blood vessel replacement: functionalization with a chimeric protein containing a cellulose-binding module and an adhesion peptide. Acta Biomater 6:4034–4041

Baumschlager A, Aoki SK, Khammash M (2017) Dynamic blue light-inducible T7 RNA polymerases (Opto-T7RNAPs) for precise spatiotemporal gene expression control. ACS Synth Biol 6:2157–2167

Bengert P, Dandekar T (2004) Riboswitch finder – a tool for identification of riboswitch RNAs. Nucleic Acids Res 32(Web Server issue):W154–W159

Bergshoef MM, Vancso GJ (1999) Transparent nanocomposites with ultrathin, electrospun nylon-4,6 fiber reinforcement. Adv Mater 11:1362–1365

Berndt S, Wesarg F, Wiegand C et al (2013) Antimicrobial porous hybrids consisting of bacterial nanocellulose and silver nanoparticles. Cellulose 20:771–783

Bhattacharya K, Kilic G, Costa PM et al (2017) Cytotoxicity screening and cytokine profiling of nineteen nanomaterials enables hazard ranking and grouping based on inflammogenic potential. Nanotoxicology 11:809–826

Bindea G, Mlecnik B, Hackl H et al (2009) ClueGO: a Cytoscape plug-in to decipher functionally grouped gene ontology and pathway annotation networks. Bioinformatics 25:1091–1093

Blazek M, Roth G, Zengerle R et al (2015) Microfluidic proximity ligation assay for profiling signaling networks with single-cell resolution. Methods Mol Biol 1346:169–184

Bodin A, Concaro S, Brittberg M et al (2007) Bacterial cellulose as a potential meniscus implant. J Tissue Eng Regen Med 1:406–408

Brinchi L, Cotana F, Fortunati E et al (2013) Production of nanocrystalline cellulose from lignocellulosic biomass: technology and applications. Carbohydr Polym 94:154–169

Busch A, Richter AS, Backofen R (2008) IntaRNA: efficient prediction of bacterial sRNA targets incorporating target site accessibility and seed regions. Bioinformatics 24:2849–2856

Butchosa N, Brown C, Larsson PT et al (2013) Nanocomposites of bacterial cellulose nanofibers and chitin nanocrystals: fabrication, characterization and bactericidal activity. Green Chem 15:3404–3413

Campos D, Piccirillo C, Pullar RC et al (2014) Characterization and antimicrobial properties of food packaging methylcellulose films containing stem extract of Ginja cherry. J Sci Food Agric 94:2097–2103

Castiblanco F, Sundin GW (2016) Cellulose production, activated by cyclic di-GMP through BcsA and BcsZ, is a virulence factor and an essential determinant of the three-dimensional architectures of biofilms formed by *Erwinia amylovora* Ea1189. Mol Plant Pathol 19:90–103

Cavka A, Guo X, Tang SJ et al (2013) Production of bacterial cellulose and enzyme from waste fiber sludge. Biotechnol Biofuels 6:25

Chakraborty A, Sain M, Kortschot M (2005) Cellulose microfibrils: a novel method of preparation using high shear refining and cryocrushing. Holzforschung 59:102–107

Chao YP, Ishida T, Sugano Y et al (2000) Bacterial cellulose production by *Acetobacter xylinum* in a 50-L internal-loop airlift reactor. Biotechnol Bioeng 68:345–352

Chen YW, Lee HV, Juan JC et al (2016) Production of new cellulose nanomaterial from red algae marine biomass Gelidium elegans. Carbohydr Polym 151:1210–1219

Chen GQ, Wu GC, Alriksson B et al (2017) Bioconversion of waste fiber sludge to bacterial nanocellulose and use for reinforcement of CTMP paper sheets. Polymers 9:458

Coban EP, Biyik H (2011) Effect of various carbon and nitrogen sources on cellulose synthesis by *Acetobacter lovaniensis* HBB5. Afr J Biotechnol 10:5346–5354

Corral ML, Cerrutti P, Vazquez A et al (2017) Bacterial nanocellulose as a potential additive for wheat bread. Food Hydrocoll 67:189–196

Costa AFS, Almeida FCG, Vinhas GM et al (2017) Production of bacterial cellulose by *Gluconacetobacter hansenii* using corn steep liquor as nutrient sources. Front Microbiol 8:2027

Croft D, O'Kelly G, Wu G et al (2011) Reactome: a database of reactions, pathways and biological processes. Nucleic Acids Res 39:D691–D697

Czaja W, Romanovicz D, Brown RM (2004) Structural investigations of microbial cellulose produced in stationary and agitated culture. Cellulose 11:403–411

Czakai K, Dittrich M, Kaltdorf M et al (2017) Influence of platelet-rich plasma on the immune response of human monocyte-derived dendritic cells and macrophages stimulated with *Aspergillus fumigatus*. Int J Med Microbiol 307:95–107

Dandekar T (2016) Modified bacterial nanocellulose and its uses in chip cards and medicine: Google Patents WO 2016174104 A1

Di Cara A, Garg A, De Micheli G et al (2007) Dynamic simulation of regulatory networks using SQUAD. BMC Bioinformatics 8:1471–2105

Dima S, Panaitescu D, Orban C et al (2017) Bacterial nanocellulose from side-streams of kombucha beverages production: preparation and physical-chemical properties. Polymers 9:374

Dodziuk H (2016) Applications of 3D printing in healthcare. Kardiochir Torakochirurgia Pol 13:283–293

Echeverry-Rendon M, Reece LM, Pastrana F et al (2017) Bacterial nanocellulose magnetically functionalized for neuro-endovascular treatment. Macromol Biosci 17:24

Einarson OJ, Sen D (2017) Self-biotinylation of DNA G-quadruplexes via intrinsic peroxidase activity. Nucleic Acids Res 45:9813–9822

El Haga M, Feng Z, Su YY et al (2017) Contribution of the csgA and bcsA genes to *Salmonella enterica* serovar *Pullorum* biofilm formation and virulence. Avian Pathol 46:541–547

Enright AJ, John B, Gaul U et al (2003) MicroRNA targets in *Drosophila*. Genome Biol 5:R1

Fiedler J, Breckwoldt K, Remmele CW et al (2015) Development of long noncoding RNA-based strategies to modulate tissue vascularization. J Am Coll Cardiol 66:2005–2015

Fujisaki Y, Koga H, Nakajima Y et al (2014) Transparent nanopaper-based flexible organic thin-film transistor array. Adv Funct Mater 24:1657–1663

Gatenholm P, Klemm D (2010) Bacterial nanocellulose as a renewable material for biomedical applications. MRS Bull 35:208–213

Gea S, Reynolds CT, Roohpour N et al (2011) Investigation into the structural, morphological, mechanical and thermal behaviour of bacterial cellulose after a two-step purification process. Bioresour Technol 102:9105–9110

Gerbaldi C, Nair JR, Ahmad S et al (2010) UV-cured polymer electrolytes encompassing hydrophobic room temperature ionic liquid for lithium batteries. J Power Sources 195:1706–1713

Gottlich C, Muller LC, Kunz M et al (2016) A combined 3D tissue engineered in vitro/in silico lung tumor model for predicting drug effectiveness in specific mutational backgrounds. J Vis Exp 6:53885

Grande CJ, Torres FG, Gomez CM et al (2008) Morphological characterisation of bacterial cellulose-starch nanocomposites. Compos Sci Technol 16:181–185

Grande CJ, Torres FG, Gomez CM et al (2009) Nanocomposites of bacterial cellulose/hydroxyapatite for biomedical applications. Acta Biomater 5:1605–1615

Hatch AC, Fisher JS, Tovar AR et al (2011) 1-Million droplet array with wide-field fluorescence imaging for digital PCR. Lab Chip 11:3838–3845

Hofacker IL (2003) Vienna RNA secondary structure server. Nucleic Acids Res 31:3429–3431

Ifuku S, Nogi M, Abe K et al (2007) Surface modification of bacterial cellulose nanofibers for property enhancement of optically transparent composites: dependence on acetyl-group DS. Biomacromolecules 8:1973–1978

Jabbour L, Gerbaldi C, Chaussy D et al (2010) Microfibrillated cellulose-graphite nanocomposites for highly flexible paper-like Li-ion battery electrodes. J Mater Chem 20:7344–7347

Jeon S, Yoo YM, Park JW et al (2014) Electrical conductivity and optical transparency of bacterial cellulose based composite by static and agitated methods. Curr Appl Phys 14:1621–1624

Jiang F, Hsieh YL (2013) Chemically and mechanically isolated nanocellulose and their self-assembled structures. Carbohydr Polym 95:32–40

Jonsson M, Brackmann C, Puchades M et al (2015) Neuronal networks on nanocellulose scaffolds. Tissue Eng Part C Methods 21:1162–1170

Jung YH, Chang TH, Zhang HL et al (2015) High-performance green flexible electronics based on biodegradable cellulose nanofibril paper. Nat Commun 6:7170

Kanehisa M, Goto S, Furumichi M et al (2010) KEGG for representation and analysis of molecular networks involving diseases and drugs. Nucleic Acids Res 38:D355–D360

Karl S, Dandekar T (2013) Jimena: efficient computing and system state identification for genetic regulatory networks. BMC Bioinformatics 14:1471–2105

Keiski CL, Harwich M, Jain S et al (2010) AlgK is a TPR-containing protein and the periplasmic component of a novel exopolysaccharide secretin. Structure 18:265–273

Khalid A, Khan R, Ul-Islam M et al (2017) Bacterial cellulose-zinc oxide nanocomposites as a novel dressing system for burn wounds. Carbohydr Polym 164:214–221

Kong L, Zhang Y, Ye ZQ et al (2007) CPC: assess the protein-coding potential of transcripts using sequence features and support vector machine. Nucleic Acids Res 35:W345–W349

Kozomara A, Griffiths-Jones S (2014) miRBase: annotating high confidence microRNAs using deep sequencing data. Nucleic Acids Res 42:D68–D73

Krontiras P, Gatenholm P, Hagg DA (2015) Adipogenic differentiation of stem cells in three-dimensional porous bacterial nanocellulose scaffolds. J Biomed Mater Res B Appl Biomater 103:195–203

Kunz M, Xiao K, Liang C et al (2014) Bioinformatics of cardiovascular miRNA biology. J Mol Cell Cardiol 89:3–10

Kunz M, Wolf B, Schulze H et al (2016a) Non-coding RNAs in lung cancer: contribution of bioinformatics analysis to the development of non-invasive diagnostic tools. Genes (Basel) 8:pii: E8

Kunz M, Liang C, Nilla S et al (2016b) The drug-minded protein interaction database (DrumPID) for efficient target analysis and drug development. Database (Oxford) 2016. https://doi.org/10.1093/database/baw041

Kunz M, Pittroff A, Dandekar T (2017) Systems biology analysis to understand regulatory miRNA networks in lung cancer. Methods Mol Biol

Kutmon M, Riutta A, Nunes N et al (2016) WikiPathways: capturing the full diversity of pathway knowledge. Nucleic Acids Res 44:D488–D494

Lee KY, Buldum G, Mantalaris A et al (2014) More than meets the eye in bacterial cellulose: biosynthesis, bioprocessing, and applications in advanced fiber composites. Macromol Biosci 14:10–32

Li W, Wang R, Liu SX (2011) Nanocrystalline cellulose prepared from softwood kraft pulp via ultrasonic-assisted acid hydrolysis. Bioresources 6:4271–4281

Li X, Song J, Yi C (2014) Genome-wide mapping of cellular protein-RNA interactions enabled by chemical crosslinking. Genomics Proteomics Bioinformatics 12:72–78

Li J, Lei P, Ding S et al (2016) An enzyme-free surface plasmon resonance biosensor for real-time detecting microRNA based on allosteric effect of mismatched catalytic hairpin assembly. Biosens Bioelectron 77:435–441

Liu J, Cheng F, Grenman H et al (2016) Development of nanocellulose scaffolds with tunable structures to support 3D cell culture. Carbohydr Polym 148:259–271

Machadoa RTA, Gutierrez J, Tercjak A et al (2016) *Komagataeibacter rhaeticus* as an alternative bacteria for cellulose production. Carbohydr Polym 152:841–849

Maiwald T, Timmer J (2008) Dynamical modeling and multi-experiment fitting with PottersWheel. Bioinformatics 24:2037–2043

Malinen MM, Kanninen L, Corlu A et al (2014) Differentiation of liver progenitor cell line to functional organotypic cultures in 3D nanofibrillar cellulose and hyaluronan-gelatin hydrogels. Biomaterials 35:5110–5121

Mangayil R, Rajala S, Pammo A et al (2017) Engineering and characterization of bacterial nanocellulose films as low cost and flexible sensor material. ACS Appl Mater Interfaces 9:19048–19056

Marchetti L, Muzzio B, Cerrutti P et al (2017) Bacterial nanocellulose as novel additive in low-lipid low-sodium meat sausages. Effect on quality and stability. Food Struct 14:52–59

Matsuoka M, Tsuchida T, Matsushita K et al (1996) A synthetic medium for bacterial cellulose production by *Acetobacter xylinum* subsp sucrofermentans. Biosci Biotechnol Biochem 60:575–579

Mayer R, Ross P, Weinhouse H et al (1991) Polypeptide composition of bacterial cyclic diguanylic acid-dependent cellulose synthase and the occurrence of immunologically cross-reacting proteins in higher-plants. Proc Natl Acad Sci U S A 88:5472–5476

Mencucci R, Boccalini C, Caputo R et al (2015) Effect of a hyaluronic acid and carboxymethylcellulose ophthalmic solution on ocular comfort and tear-film instability after cataract surgery. J Cataract Refract Surg 41:1699–1704

Messeguer X, Escudero R, Farre D et al (2002) PROMO: detection of known transcription regulatory elements using species-tailored searches. Bioinformatics 18:333–334

Mi H, Poudel S, Muruganujan A et al (2016) PANTHER version 10: expanded protein families and functions, and analysis tools. Nucleic Acids Res 44:D336–D342

Mikkelsen D, Flanagan BM, Dykes GA et al (2009) Influence of different carbon sources on bacterial cellulose production by *Gluconacetobacter xylinus* strain ATCC 53524. J Appl Microbiol 107:576–583

Molina-Ramirez C, Castro M, Osorio M et al (2017) Effect of different carbon sources on bacterial nanocellulose production and structure using the low pH resistant strain Komagataeibacter medellinensis. Materials 10:pii: E639

Morais JPS, Rosa MD, de Souza MDM et al (2013) Extraction and characterization of nanocellulose structures from raw cotton linter. Carbohydr Polym 91:229–235

Morgan JLW, Strumillo J, Zimmer J (2013) Crystallographic snapshot of cellulose synthesis and membrane translocation. Nature 493:181–192

Morgan JLW, McNamara JT, Zimmer J (2014) Mechanism of activation of bacterial cellulose synthase by cyclic di-GMP. Nat Struct Mol Biol 21:489–496

Muller M, Ozturk E, Arlov O et al (2017) Alginate sulfate-nanocellulose bioinks for cartilage bioprinting applications. Ann Biomed Eng 45:210–223

Nimeskern L, Avila HM, Sundberg J et al (2013) Mechanical evaluation of bacterial nanocellulose as an implant material for ear cartilage replacement. J Mech Behav Biomed Mater 22:12–21

Nogi M, Iwamoto S, Nakagaito AN et al (2009) Optically transparent nanofiber paper. Adv Mater 21:1595–1598

Nogi M, Karakawa M, Komoda N et al (2015) Transparent conductive nanofiber paper for foldable solar cells. Sci Rep 5:17254

Ndong Ntoutoume GM, Grassot V, Bregier F et al (2017) PEI-cellulose nanocrystal hybrids as efficient siRNA delivery agents–synthesis, physicochemical characterization and in vitro evaluation. Carbohydr Polym 164:258–267

Omadjela O, Narahari A, Strumillo J et al (2013) BcsA and BcsB form the catalytically active core of bacterial cellulose synthase sufficient for in vitro cellulose synthesis. Proc Natl Acad Sci U S A 110:17856–17861

Ozturk E, Arlov O, Aksel S et al (2016) Sulfated hydrogel matrices direct mitogenicity and maintenance of chondrocyte phenotype through activation of FGF signaling. Adv Funct Mater 26:3649–3662

Pan-In P, Wongsomboon A, Kokpol C et al (2015) Depositing alpha-mangostin nanoparticles to sebaceous gland area for acne treatment. J Pharmacol Sci 129:226–232

Park JK, Jung JY, Park YH (2003) Cellulose production by *Gluconacetobacter hansenii* in a medium containing ethanol. Biotechnol Lett 25:2055–2059

Park J, Jung JH, Park K et al (2018) On-demand acoustic droplet splitting and steering in a disposable microfluidic chip. Lab Chip 18:422–432

Philippi N, Walter D, Schlatter R et al (2009) Modeling system states in liver cells: survival, apoptosis and their modifications in response to viral infection. BMC Syst Biol 3:1752–0509

Piccirillo C, Demiray S, Ferreira ACS et al (2013) Chemical composition and antibacterial properties of stem and leaf extracts from Ginja cherry plant. Ind Crop Prod 43:562–569

Pretzel D, Linss S, Ahrem H et al (2013) A novel in vitro bovine cartilage punch model for assessing the regeneration of focal cartilage defects with biocompatible bacterial nanocellulose. Arthritis Res Ther 15:R59

Rademacher A, Erdel F, Trojanowski J et al (2017) Real-time observation of light-controlled transcription in living cells. J Cell Sci 130:4213–4224

Ramana KV, Tomar A, Singh L (2000) Effect of various carbon and nitrogen sources on cellulose synthesis by *Acetobacter xylinum*. World J Microbiol Biotechnol 16:245–248

Razaq A, Nystrom G, Stromme M et al (2011) High-capacity conductive nanocellulose paper sheets for electrochemically controlled extraction of DNA oligomers. PLoS One 6:15

Ring DF, Nashed W, Dow T (1986) Liquid loaded pad for medical applications, US Patent 4,588,400

Saito R, Smoot ME, Ono K et al (2012) A travel guide to Cytoscape plugins. Nat Methods 9:1069–1076

Schlatter R, Philippi N, Wangorsch G et al (2012) Integration of Boolean models exemplified on hepatocyte signal transduction. Brief Bioinform 13:365–376

Scionti G (2010) Mechanical properties of bacterial cellulose implants. Chalmers University of Technology, Göteborg

Si J, Cui Z, Wang Q et al (2016) Biomimetic composite scaffolds based on mineralization of hydroxyapatite on electrospun poly(varepsilon-caprolactone)/nanocellulose fibers. Carbohydr Polym 143:270–278

Sirvio JA, Kolehmainen A, Visanko M et al (2014) Strong, self-standing oxygen barrier films from nanocelluloses modified with regioselective oxidative treatments. ACS Appl Mater Interfaces 6:14384–14390

Smirnov A, Schneider C, Hor J et al (2017) Discovery of new RNA classes and global RNA-binding proteins. Curr Opin Microbiol 39:152–160

Son HJ, Heo MS, Kim YG et al (2001) Optimization of fermentation conditions for the production of bacterial cellulose by a newly isolated *Acetobacter* sp A9 in shaking cultures. Biotechnol Appl Biochem 33:1–5

Song YY, Jiang YQ, Shi LY et al (2015) Solution-processed assembly of ultrathin transparent conductive cellulose nanopaper embedding AgNWs. Nanoscale 7:13694–13701

Steinle H, Behring A, Schlensak C et al (2017) Concise review: application of in vitro transcribed messenger RNA for cellular engineering and reprogramming: progress and challenges. Stem Cells 35:68–79

Stumpf F, Schoendube J, Gross A et al (2015) Single-cell PCR of genomic DNA enabled by automated single-cell printing for cell isolation. Biosens Bioelectron 69:301–306

Sunagawa N, Fujiwara T, Yoda T et al (2013) Cellulose complementing factor (Ccp) is a new member of the cellulose synthase complex (terminal complex) in *Acetobacter xylinum*. J Biosci Bioengineer 115:607–612

Sundaram J, Pant J, Goudie MJ et al (2016) Antimicrobial and physicochemical characterization of biodegradable, nitric oxide-releasing nanocellulose-chitosan packaging membranes. J Agric Food Chem 64:5260–5266

Svensson A, Nicklasson E, Harrah T et al (2005) Bacterial cellulose as a potential scaffold for tissue engineering of cartilage. Biomaterials 26:419–431

Szklarczyk D, Franceschini A, Wyder S et al (2015) STRING v10: protein-protein interaction networks, integrated over the tree of life. Nucleic Acids Res 43:D447–D452

Thakar J, Albert R (2010) Boolean models of within-host immune interactions. Curr Opin Microbiol 13:377–381

Trovatti E, Freire CS, Pinto PC et al (2012) Bacterial cellulose membranes applied in topical and transdermal delivery of lidocaine hydrochloride and ibuprofen: in vitro diffusion studies. Int J Pharm 435:83–87

Ummartyotin S, Juntaro J, Sain M et al (2012) Development of transparent bacterial cellulose nanocomposite film as substrate for flexible organic light emitting diode (OLED) display. Ind Crop Prod 35:92–97

Uth C, Zielonka S, Horner S et al (2014) A chemoenzymatic approach to protein immobilization onto crystalline cellulose nanoscaffolds. Angew Chem Int Ed Engl 53:12618–12623

van de Ven TG, Sheikhi A (2016) Hairy cellulose nanocrystalloids: a novel class of nanocellulose. Nanoscale 8:15101–15114

Vazquez A, Foresti ML, Cerrutti P et al (2013) Bacterial cellulose from simple and low cost production media by *Gluconacetobacter xylinus*. J Polymers Environ 21:545–554

Velasco-Bedran H, Lopez-Isunza F (2007) The unified metabolism of *Gluconacetobacter entanii* in continuous and batch processes. Process Biochem 42:1180–1190

Viereck J, Kumarswamy R, Foinquinos A et al (2016) Long noncoding RNA chast promotes cardiac remodeling. Sci Transl Med 8:326ra22

Vigneshwaran N, Satyamurthy P (2016) Nanocellulose production using cellulose degrading fungi. Springer, Basel

Volders P-J, Helsens K, Wang X et al (2013) LNCipedia: a database for annotated human lncRNA transcript sequences and structures. Nucleic Acids Res 41:D246–D251
Wangorsch G, Butt E, Mark R et al (2011) Time-resolved in silico modeling of fine-tuned cAMP signaling in platelets: feedback loops, titrated phosphorylations and pharmacological modulation. BMC Syst Biol 5:1752–0509
Wong HC, Fear AL, Calhoon RD et al (1990) Genetic organization of the cellulose synthase operon in *Acetobacter xylinum*. Proc Natl Acad Sci U S A 87:8130–8134
Wu CN, Cheng KC (2017) Strong, thermal-stable, flexible, and transparent films by self-assembled TEMPO-oxidized bacterial cellulose nanofibers. Cellulose 24:269–283
Wu T, Wang J, Liu C et al (2006) NPInter: the noncoding RNAs and protein related biomacromolecules interaction database. Nucleic Acids Res 34:D150–D152
Xiang ZY, Gao WH, Chen LH et al (2016) A comparison of cellulose nanofibrils produced from *Cladophora glomerata* algae and bleached eucalyptus pulp. Cellulose 23:493–503
Xu CG, Carlsson DO, Mihranyan A (2016) Feasibility of of using DNA-immobilized nanocellulose-based immunoadsorbent for systemic lupus erythematosus plasmapheresis. Colloids Surf B Biointerfaces 143:1–6
Yano H, Sugiyama J, Nakagaito AN et al (2005) Optically transparent composites reinforced with networks of bacterial nanofibers. Adv Mater 17:153–155
Zeng XB, Liu J, Chen J et al (2011) Screening of the common culture conditions affecting crystallinity of bacterial cellulose. J Ind Microbiol Biotechnol 38:1993–1999
Zhong Q, Bhattacharya S, Kotsopoulos S et al (2011) Multiplex digital PCR: breaking the one target per color barrier of quantitative PCR. Lab Chip 11:2167–2174
Zhou YH, Fuentes-Hernandez C, Khan TM et al (2013) Recyclable organic solar cells on cellulose nanocrystal substrates. Sci Rep 3:1536
Zhou YH, Khan TM, Liu JC et al (2014) Efficient recyclable organic solar cells on cellulose nanocrystal substrates with a conducting polymer top electrode deposited by film-transfer lamination. Org Electron 15:661–666

MIX
Papier aus verantwortungsvollen Quellen
Paper from responsible sources
FSC® C105338

If you have any concerns about our products,
you can contact us on
ProductSafety@springernature.com

In case Publisher is established outside the EU,
the EU authorized representative is:
Springer Nature Customer Service Center GmbH
Europaplatz 3, 69115 Heidelberg, Germany

Printed by Libri Plureos GmbH
in Hamburg, Germany